METHODS IN MOLECULAR BIOLOGY™

Series Editor
John M. Walker
School of Life Sciences
University of Hertfordshire
Hatfield, Hertfordshire, AL10 9AB, UK

For further volumes:
http://www.springer.com/series/7651

Integrin and Cell Adhesion Molecules

Methods and Protocols

Edited by

Motomu Shimaoka

Immune Disease Institute, Program in Cellular and Molecular Medicine at Children's Hospital Boston, Department of Anesthesia, Harvard Medical School, Boston, MA, USA;
Department of Molecular Pathobiology and Cell Adhesion Biology, Mie University Graduate School of Medicine, Tsu-City, Mie, JAPAN

Humana Press

Editor
Motomu Shimaoka
Immune Disease Institute
Program in Cellular and Molecular Medicine
at Children's Hospital Boston
Department of Anesthesia
Harvard Medical School
Boston, MA, USA
and
Department of Molecular Pathobiology
and Cell Adhesion Biology
Mie University Graduate School of Medicine
Tsu-City, Mie, JAPAN
shimaoka@idi.harvard.edu
shimaoka@doc.medic.mie-u.ac.jp

ISSN 1064-3745 e-ISSN 1940-6029
ISBN 978-1-4939-5808-5 ISBN 978-1-61779-166-6 (eBook)
DOI 10.1007/978-1-61779-166-6
Springer New York Dordrecht Heidelberg London

Printed on acid-free paper

Humana Press is part of Springer Science+Business Media (www.springer.com)

Preface

Integrins are the foremost and largest family of cell adhesion molecules with noncovalently associated α and β subunits that mediate cell–cell and cell–extracellular matrix interactions. To date, 19 different integrin α subunits and 8 different integrin β subunits have been reported in vertebrates, forming at least 24 α/β heterodimers and representing the most structurally and functionally diverse cell adhesion molecules. As their name implies, integrins create an integrated connection between the cytoskeleton and attachment points in the extracellular microenvironment, where they mediate force-resistant adhesion, polarization, and cell migration. Integrins play pivotal roles across not only a wide range of physiological processes including tissue morphogenesis, immune responses, wound healing, and regulation of cell growth and differentiation, but also in numerous pathological phenomena such as autoimmunity, thrombosis, and cancer metastasis/progression. Therefore, investigations on integrins often demand multidisciplinary approaches, making researchers long for a handy collection of comprehensive and practical protocols that detail experimental methods for studying integrin and related cell-adhesion molecule functionality. *Integrins and Cell Adhesion Molecules: Methods and Protocols* is, hence, of great interest to a broad readership, from cell biologists and immunologists to cancer researchers, as well as from molecular and structural biologists to biochemists.

The aim of the second edition of *Integrins and Cell Adhesion Molecules: Methods and Protocols* (f/k/a Integrin Protocols) is to provide readers not only with basic protocols in studying integrin functions, but also with summaries on those state-of-the-art technologies that have been utilized for understanding integrin functionality at the cellular, molecular, structural, and organismal levels.

Part I of this book (Chapters 1–6) contains basic protocols for the study of integrin and related cell-adhesion molecule functionality in vitro. To open **Part I**, *Chigaev and Sklar* provide an overview of several experimental procedures used for investigating integrin-dependent cell adhesion (Chapter 1). The cell-adhesion assay is a standard and important experimental procedure to examine the adhesiveness of cells to substrates. *Weitz-Schmidt and Chreng* detail a protocol of a convenient and highly reproducible cell-adhesion assay using a V-bottom-shaped plate (Chapter 2). Although integrins are the major receptors that, in many aspects, regulate cell migration, the migration of certain cell types in the interstitial space requires either a lesser degree of integrin involvement or none at all. *Shulman and Alon* describe real-time assays for the study of integrin-dependent and independent cell migration (Chapter 3). Efficient transduction methods are required for facilitating the close examination of integrin functionality in primary lymphocytes. *Banerjee and Shimaoka* introduce a simple protocol for lentivirus-mediated gene transduction in primary T cells (Chapter 4). To better understand integrin-ligand interactions, biochemical assays using purified integrin proteins or integrin domains are essential. *Vorup-Jensen* discusses an application of surface plasmon resonance biosensing to study the complex ligand-binding kinetics of the integrin αX I domain (Chapter 5). In addition, *Yuki* presents plate- and bead-based assays to investigate the ligand-binding abilities of purified integrin LFA-1 protein (Chapter 6).

Part II (Chapters 7–11) illustrates structural biology approaches for studying integrins and related cell-adhesion molecules. At the beginning of **Part II**, *Fu, Wang, and Luo* provide a comprehensive review of integrin domains and conformational regulation (Chapter 7). Protein expression remains a formidably difficult obstacle in determining the crystal structures of many integrin–ligand complexes. *Zhang and Wang* describe a protocol to express and purify integrin I domains and IgSF ligands for crystallography (Chapter 8). Electron microscopy has been successfully used to understand how integrin conformations are globally changed. *Iwasaki* discusses an application of electron microscopic imaging that tackles the conformational flexibility of integrins (Chapter 9). NMR is a powerful technique to study protein–protein interactions. *Nishida and Shimada* describe a novel NMR method, termed the cross-saturation (CS) method, and its application in studying the ligand-binding activities of cell-adhesion molecules (Chapter 10). Elucidating the biophysical properties of individual adhesion molecules demands the use of single-molecule techniques. *Seog* utilizes two important single-molecule techniques, atomic force microscopy and optical tweezing, to examine cell-adhesion molecules (Chapter 11).

Part III (Chapters 12–16) focuses on emerging imaging technologies for investigating cell migration. **Part III** begins with *Carman's* comprehensive overview of imaging in the study of integrins (Chapter 12). Analysis of cell motility and migration is one of the most important fields to which imaging technologies have greatly contributed. *Wiemer, Wernimont, and Huttenlocher* describe methods for live time-lapse imaging of T-cell migration on ICAM-1 substrates (Chapter 13). Fluorescence resonance energy transfer (FRET) has emerged as a powerful research tool for investigating integrin conformational changes in living cells. *Lefort, Hyun, and Kim* utilize FRET to monitor structural alterations during integrin activation in leukocytes (Chapter 14). Recent technological advancements in optics and fluorescent dyes have enabled high-resolution imaging of the contact interface between adherent leukocytes and endothelial cells. *Carman* details a protocol for high-resolution fluorescence microscopy in the study of transendothelial migration (Chapter 15). Two-photon intravital imaging has revolutionized our understanding of how immune cells behave and move in living animals. *Murooka and Mempel* discuss an application of multiphoton intravital microscopy to study lymphocyte motility in the lymph nodes of living mice (Chapter 16).

Part IV (Chapters 17–21) presents strategies to elucidate signaling through cell-adhesion molecules. To open **Part IV**, *Kinashi* comprehensively reviews integrin signaling (Chapter 17). Rap1 GTPase is a key signaling molecule in integrin activation. *Katagiri and Kinashi* discuss the roles of Rap1 in integrin inside-out signaling and cell polarity, as well as experimental procedures to study Rap1 functionality (Chapter 18). Focal adhesions are the specialized supramolecular assemblies that contain integrins and various integrin-associated signaling molecules and cytoskeletal proteins, thereby linking intracellular proteins to the extracellular matrices. *Kuo, Han, Yates III, and Waterman* present effective protocols for isolating focal adhesion proteins and performing biochemical and proteomic analyses (Chapter 19). Talin constitutes the crucial intracellular protein that directly binds to the integrin cytoplasmic domain, thereby triggering integrin inside-out signaling. *Bouaouina, Harburger, and Calderwood* describe various methods used to investigate the roles of talin in the regulation of integrin activation (Chapter 20). At the leading edge of those cells migrating on the substrates, a cytoskeletal rearrangement occurs that is regulated by cascades of intracellular signaling events, all of which culminates in the formation of the characteristic membrane protrusion known as pseudopodium (or lamellipodium).

Wang and Klemke detail a proteomics-based method for investigating signaling events, specifically phosphotyrosine proteins at the pseudopodium (Chapter 21).

Part V (Chapters 22–26) covers experimental techniques to investigate integrin functions at organismal levels in a physiological context. *Lowell and Mayadas* begin **Part V** with an overview of integrin functions in vivo, which includes a comprehensive discussion of integrin knockout mice phenotypes observed during the developmental process, as well as under pathophysiological conditions (Chapter 22). Conditional gene targeting is a powerful technology that enables researchers to modify genes of interest in vivo only in specific tissue/cell-types or at a specific time-point during development. *Yamamoto and Takeda* describe a method for generating conditionally gene-targeted mice (Chapter 23). Analyzing how cells migrate to specific tissues is essential to the study of in vivo cell-adhesion molecule and chemo-attractant receptor functionality in these cells. *De Calisto, Villablanca, Wang, Bono, Rosemblatt, and Mora* first provide an overview of the mechanisms by which tissue-specific lymphocyte homing is regulated, and then present a protocol to examine T-cell homing to the gut (Chapter 24). Proteomic analysis utilizing stable isotope labeling with amino acid in cell culture (SILAC) was originally utilized for cell biology studies in vitro. *Zanivan, Krueger, and Mann* discuss SILAC mice technology and its successful in vivo application for quantitative proteomic analysis of β1 integrin-deficient mice (Chapter 25). Dictyostelium discoideum amebae represent an excellent model organism for studying the chemotactic responses of migrating cells. *Cai, Huang, Devreotes, and Iijima* detail the use of this model organism to elucidate the signaling machinery of chemotaxis (Chapter 26).

Part VI (Chapters 27–30) showcases the most promising methods and technologies for the development of novel therapeutics and diagnostics. *Foubert and Varner* begin **Part VI** by providing a concise summary of the role integrins play in tumor angiogenesis and lymphangiogenesis, and then describe the methods used to study integrin functionality in these pathologies (Chapter 27). Radiopharmaceutics targeting integrins have been considered for tumor imaging. *Dearling and Packard* discuss a novel application, the use of β7 integrin-targeted radiopharmaceutics to image gut inflammation in a mouse model of inflammatory bowel disease (Chapter 28). Drug delivery is the major problem preventing clinical realization of RNAi-based medicine. *Ben-Arie, Kedmi, and Peer* describe the application of integrin-targeted nanoparticles for leukocyte-directed siRNA delivery (Chapter 29). Drug candidates targeting integrins in patients often exhibit little or no cross-reactivity with rodent counter-parts, thereby making it difficult to perform preclinical studies. *Kim, Kumar, and Shankar* detail the generation of humanized mice harboring human hematopoietic cells for the study of HIV infection, an approach with potential preclinical applications for validating other human integrin-targeted drug candidates in vivo (Chapter 30).

As can be surmised here, the 30 chapters in this book cover many of the most important topics in the field of integrins and cell-adhesion molecules. I hope that this book will serve as a useful and valuable reference for both experts and nonexperts in the scientific community who wish to study cell-adhesion molecules. Finally, I would like to thank Dr. John Walker for the opportunity to edit this volume. I would also like to acknowledge all of the authors for their outstanding contributions to *Integrins and Cell Adhesion Molecules: Methods and Protocols.*

Contents

Part VI Methods and Technologies Towards Novel Therapeutics and Diagnostics Targeting Integrins

Part VI Methods and Approaches Towards Novel Therapeutics and Diagnostics Targeting Integrins

Contributors

RONEN ALON • *Department of Immunology, The Weizmann Institute of Science, Rehovot, Israel*

DALIYA BANERJEE • *Immune Disease Institute, Program in Cellular and Molecular Medicine at Children's Hospital Boston, Boston, MA, USA; Department of Anesthesia, Harvard Medical School, Boston, MA, USA*

NOA BEN-ARIE • *Laboratory of Nanomedicine, Department of Cell Research and Immunology, George S. Wise Faculty of Life Sciences, Tel Aviv University, Tel Aviv, Israel; Center for Nanoscience and Nanotechnology, Tel Aviv University, Tel Aviv, Israel*

MARIA R. BONO • *Facultad de Ciencias, Universidad de Chile and Fundacion Ciencia para la Vida, Santiago, Chile*

MOHAMED BOUAOUINA • *Department of Pharmacology and Interdepartmental Program in Vascular Biology and Transplantation, Yale University School of Medicine, New Haven, CT, USA*

HUAQING CAI • *Department of Cell Biology, Johns Hopkins University, School of Medicine, Baltimore, MD, USA*

DAVID A. CALDERWOOD • *Department of Pharmacology and Interdepartmental Program in Vascular Biology and Transplantation, Yale University School of Medicine, New Haven, CT, USA*

CHRISTOPHER V. CARMAN • *Center for Vascular Biology Research, Division of Molecular and Vascular Medicine, Department of Medicine, Beth Israel Deaconess Medical Center, Harvard Medical School, Boston, MA, USA*

ALEXANDRE CHIGAEV • *Department of Pathology and Cancer Center, University of New Mexico Health Sciences Center, Albuquerque, NM, USA*

STÉPHANIE CHRENG • *Novartis Institutes for BioMedical Research, Basel, Switzerland*

JAIME DE CALISTO • *Gastrointestinal Unit, Massachusetts General Hospital, Harvard Medical School, Boston, MA, USA*

JASON L. J. DEARLING • *Division of Nuclear Medicine, Department of Radiology, Children's Hospital Boston, Boston, MA, USA; Harvard Medical School, Boston, MA, USA*

PETER N. DEVREOTES • *Department of Cell Biology, Johns Hopkins University, School of Medicine, Baltimore, MD, USA*

PHILLIPE FOUBERT • *Moores UCSD Cancer Center, University of California, San Diego, La Jolla, CA, USA*

GUANYUAN FU • *Department of Biological Sciences, Louisiana State University, Baton Rouge, LA, USA*

XUEMEI HAN • *Cell Biology, Scripps Research Institute, La Jolla, CA, USA*

David S. Harburger • *Department of Pharmacology and Interdepartmental Program in Vascular Biology and Transplantation, Yale University School of Medicine, New Haven, CT, USA*
Chuan-Hsiang Huang • *Department of Cell Biology, Johns Hopkins University, School of Medicine, Baltimore, MD, USA*
Anna Huttenlocher • *Department of Medical Microbiology and Immunology, University of Wisconsin, Madison, WI, USA; Department of Pediatrics, University of Wisconsin, Madison, WI, USA*
Young-min Hyun • *Department of Microbiology and Immunology, David H. Smith Center for Vaccine Biology and Immunology, University of Rochester, Rochester, NY, USA*
Miho Iijima • *Department of Cell Biology, Johns Hopkins University, School of Medicine, Baltimore, MD, USA*
Kenji Iwasaki • *Research Center for Structural and Functional Proteomics, Institute for Protein Research, Osaka University, Suita, Osaka, Japan*
Koko Katagiri • *Department of Life Science, School of Science and Technology, Kansei Gakuin University, Hyogo, Japan*
Ranit Kedmi • *Laboratory of Nanomedicine, Department of Cell Research and Immunology, George S. Wise Faculty of Life Sciences, Tel Aviv University, Tel Aviv, Israel; Center for Nanoscience and Nanotechnology, Tel Aviv University, Tel Aviv, Israel*
Minsoo Kim • *Department of Microbiology and Immunology, David H. Smith Center for Vaccine Biology and Immunology, University of Rochester, Rochester, NY, USA*
Sang-Soo Kim • *Department of Oncology, Lombardi Comprehensive Cancer Center, Georgetown University Medical Center, Reservoir Road NW, Washington, DC, USA*
Tatsuo Kinashi • *Department of Molecular Genetics, Institute of Biomedical Science, Kansai Medical University, Osaka, Japan*
Richard L. Klemke • *Department of Pathology, University of California, San Diego, Basic Science, La Jolla, CA, USA*
Marcus Krueger • *Department of Cardiac Development and Remodeling, Max-Planck-Institute of Heart and Lung Research, Bad Nauheim, Germany*
Priti Kumar • *Department of Internal Medicine, Section of Infectious Diseases, Yale School of Medicine, New Haven, CT, USA*
Jean-Cheung Kuo • *Cell Biology and Physiology Center, National Heart, Lung, and Blood Institute, Bethesda, MD, USA*
Craig T. Lefort • *Department of Microbiology and Immunology, David H. Smith Center for Vaccine Biology and Immunology, University of Rochester, Rochester, NY, USA*
Clifford A. Lowell • *Department of Laboratory Medicine, University of California, San Francisco, San Francisco, CA, USA*
Bing-Hao Luo • *Department of Biological Sciences, Louisiana State University, Baton Rouge, LA, USA*
Matthias Mann • *Department of Proteomics and Signal Transduction, Max-Planck-Institute of Heart and Lung Research, Bad Nauheim, Germany*

Tanya N. Mayadas • *Department of Pathology, Center of Excellence in Vascular Biology, Brigham and Women's Hospital & Harvard Medical School, Boston, MA, USA*

Thorsten R. Mempel • *Center for Immunology and Inflammatory Diseases and Center for Systems Biology, Massachusetts General Hospital and Harvard Medical School, Charlestown, MA, USA*

J. Rodrigo Mora • *Gastrointestinal Unit, Massachusetts General Hospital,* GRJ-815, *Boston, MA, USA*

Thomas T. Murooka • *Center for Immunology and Inflammatory Diseases and Center for Systems Biology, Massachusetts General Hospital and Harvard Medical School, Boston, MA, USA*

Noritaka Nishida • *Graduate School of Pharmaceutical Sciences, The University of Tokyo, Hongo, Bunkyo-ku, Tokyo, Japan*

Alan B. Packard • *Division of Nuclear Medicine, Department of Radiology, Children's Hospital Boston, Boston, MA, USA; Harvard Medical School, Boston, MA, USA*

Dan Peer • *Laboratory of Nanomedicine, Department of Cell Research and Immunology, George S. Wise Faculty of Life Sciences, Tel Aviv University, Tel Aviv, Israel; Center for Nanoscience and Nanotechnology, Tel Aviv University, Tel Aviv, Israel*

Mario Rosemblatt • *Facultad de Ciencias, Universidad de Chile; Facultad de Ciencias Biologicas, Universidad Andres Bello and Foundation Ciencia para la Vida, Santiago, Chile*

Joonil Seog • *Department of Materials Science and Engineering, Fischell Bioengineering Department, University of Maryland, College Park, MD, USA*

Premlata Shankar • *Department of Biomedical Sciences, Center of Excellence for Infectious Diseases, Paul L. Foster School of Medicine, Texas Tech University Health Sciences Center, El Paso, TX, USA*

Ichio Shimada • *Graduate School of Pharmaceutical Sciences, The University of Tokyo, Hongo, Bunkyo-ku, Tokyo, Japan; Japan Biomedicinal Information Research Center (BIRC), National Institute of Advanced Industrial Science and Technology (AIST), Aomi, Koto-ku, Tokyo, Japan*

Motomu Shimaoka • *Immune Disease Institute, Program in Cellular and Molecular Medicine at Children's Hospital Boston, Department of Anesthesia, Harvard Medical School, Boston, MA, USA; Department of Molecular Pathobiology and Cell Adhesion Biology, Mie University Graduate School of Medicine, Tsu-City, Mie, JAPAN*

Ziv Shulman • *Department of Immunology, The Weizmann Institute of Science, Rehovot, Israel*

Larry A. Sklar • *Department of Pathology and Cancer Center, University of New Mexico Health Sciences Center, Albuquerque, NM, USA*

Kiyoshi Takeda • *Department of Microbiology and Immunology, Graduate School of Medicine, Osaka University, Suita, Osaka, Japan; Laboratory of Mucosal Immunology, WPI Immunology Frontier Research Center, Osaka University, Suita, Osaka, Japan*

Judith A. Varner • *Moores UCSD Cancer Center, University of California, San Diego, La Jolla, CA, USA*
Eduardo J. Villablanca • *Gastrointestinal Unit, Massachusetts General Hospital, Harvard Medical School, Boston, MA, USA*
Thomas Vorup-Jensen • *Biophysical Immunology Laboratory, Department of Medical Microbiology and Immunology, University of Aarhus, Aarhus C, Denmark*
Jia-huai Wang • *Dana-Farber Cancer Institute, Harvard Medical School, Boston, MA, USA*
Sen Wang • *Gastrointestinal Unit, Massachusetts General Hospital, Harvard Medical School, Boston, MA, USA*
Wei Wang • *Department of Biological Sciences, Louisiana State University, Baton Rouge, LA, USA*
Yingchun Wang • *Key Laboratory of Molecular Development Biology, Insitute of Genetics and Development Bilogy, Chinese Academy of Sciences, Beijing, China*
Clare M. Waterman • *Cell Biology and Physiology Center, National Heart, Lung, and Blood Institute, Bethesda, MD, USA*
Gabriele Weitz-Schmidt • *Novartis Institutes for BioMedical Research, Basel, Switzerland; University Basel, PharmaCenter, Basel, Switzerland*
Sarah Wernimont • *Program in Cellular and Molecular Biology, University of Wisconsin, Madison, WI, USA*
Andrew J. Wiemer • *Department of Medical Microbiology and Immunology, University of Wisconsin, Madison, WI, USA*
Masahiro Yamamoto • *Department of Microbiology and Immunology, Graduate School of Medicine, Osaka University, Osaka, Japan; Laboratory of Mucosal Immunology, WPI Immunology Frontier Research Center, Osaka University, Osaka, Japan*
John R. Yates III • *Cell Biology, Scripps Research Institute, La Jolla, CA, USA*
Chunting Ye • *Department of Biomedical Sciences, Center of Excellence for Infectious Diseases, Paul L. Foster School of Medicine, Texas Tech University Health Sciences Center, El Paso, Tx, USA*
Koichi Yuki • *Department of Anesthesiology, Perioperative and Pain Medicine, Children's Hospital Boston, Boston, MA, USA; Immune Disease Institute, Boston, MA, USA; Department of Anaesthesia, Harvard Medical School, Boston, MA, USA*
Sara Zanivan • *Vascular Proteomics Group,Beatson Institute for Cancer Research, Glasgow, United Kingdom*
Hongmin Zhang • *Department of Physiology, The University of Hong Kong, Hong Kong, SAR, China*

Part I

Basic Protocols for the Study of Integrin and Related Cell Adhesion Molecule Functionality In Vitro

Chapter 1

Overview: Assays for Studying Integrin-Dependent Cell Adhesion

Alexandre Chigaev and Larry A. Sklar

Abstract

Interaction of the integrin receptors with ligands determines the molecular basis of integrin-dependent cell adhesion. Integrin ligands are typically large proteins with relatively low binding affinities. This makes direct ligand-binding kinetic measurements somewhat difficult. Here we examine several real-time methods, aimed to overcome these experimental limitations and to distinguish the regulation of integrin conformation and affinity. This chapter includes: the use of a small ligand-mimetic probe for studies of inside-out regulation of integrin affinity and unbending, real-time cell aggregation and disaggregation kinetics to probe integrin conformational states and the number of integrin–ligand bonds, as well as the real-time monitoring of ligand-induced epitopes under signaling through G-protein-coupled receptors, and others. Experimental data obtained using these novel methods are summarized in terms of the current model of integrin activation.

Key words: Ligand–receptor interaction, Ligand mimetic, Real-time kinetics, Cells adhesion, Inside-out signal, Monoclonal antibodies, Quantitative approaches

1. Introduction

Understanding how cell adhesion and migration is regulated is essential for describing embryonic development, tissue repair, hemostasis, inflammation, cell mobilization, and metastasis. The ability to rapidly and reversibly modulate cellular adhesive properties serves as the basis for multiple biological functions of multicellular organisms. Several adhesion molecules regulate cell adhesion through de novo expression, rapid upregulation by the means of exocytosis, downregulation through proteolysis, shedding, and other mechanisms that can alter the number of molecules on the cell surface. Methods for studying these molecules

Motomu Shimaoka (ed.), *Integrin and Cell Adhesion Molecules: Methods and Protocols*,
Methods in Molecular Biology, vol. 757, DOI 10.1007/978-1-61779-166-6_1, © Springer Science+Business Media, LLC 2011

are beyond the scope of this chapter. We focus here on integrins, a unique class of adhesion molecules that can rapidly change cell adhesion through a conformational change and/or clustering, without altering molecule expression.

Our current understanding of integrin conformational regulation implies the potential existence of multiple conformational states, with different binding affinities for their ligands, different degrees of unbending (extension), and different positioning of integrin domains (hybrid domain in particular). These states are expected to contribute to the lifetime of the ligand–receptor bond, and the efficiency of the bond formation. Such a model allows us to describe how an integrin such as VLA-4 can be responsible for very diverse cellular behaviors, such as a nonadhesive state, as well as rolling, cell arrest, and firm adhesion (1). The recent discovery that G-protein-coupled receptors can provide a negative (deactivating) signal, which results in cell deadhesion, adds to the number of possible conformational states and highlights the complexity of integrin conformational regulation (2).

In this chapter, we review basic methods that led to the current model of integrin activation and focus on basic techniques that are currently used in our and other laboratories to study integrin-dependent cell adhesion. Because of the limited space we will primarily focus on unique assays specifically developed for integrin studies in our laboratory. We apologize to the others whose studies contributed to the current understanding of integrin regulation and were not cited because of the lack of space.

2. Small Molecules as Tools for Integrin Studies

Interaction of the integrin receptors with ligands determines the molecular basis of integrin-dependent cell adhesion. Methods that allow monitoring of these ligand–receptor interactions in real-time on living cells under physiologically relevant signaling conditions would represent a desirable "gold standard" for these types of studies. In the best case scenario a scientist should be able to purify cells of interest, add labeled ligand, and monitor binding of the probe in real time after activation/deactivation through other types of receptors ("inside-out" or "outside-in" signal). Unfortunately, soluble integrin ligands are large proteins that have relatively low binding affinities. Therefore, direct kinetic measurements of natural integrin ligand binding are technically difficult.

One of the solutions to this problem is the development of small molecule probes that exhibit higher binding affinities and, at the same time, reflect the binding of the natural ligand.

These probes can be used as reporters of the affinity state of the integrin-binding pocket, as well as in other applications (see below). Fluorescently labeled molecules of this type can be used in a conventional flow cytometer to make homogeneous real-time measurements of ligand–receptor interactions (3, 4). Drug-like small molecules also appear to be good candidates for these assays.

Integrins represent an attractive target for treatment of several diseases. Therefore, a number of drug-like small molecules (direct and allosteric integrin antagonists) have been developed by several pharmaceutical companies (5). Fluorescent antagonists for GPIIb/IIIa (RGD peptidomimetics) were described and used in a flow cytometer by Dr. Bednar et al. from Merck Research Labs (4). The binding of fluorescent LFA-1 antagonists has been described by Dr. Keating et al. from Genentech, Inc. (6). We took advantage of the published structure of LDV-based competitive antagonists developed by Biogen Idec Inc. (BIO1211) (7, 8), and created a fluorescent probe that mimics binding of a natural VLA-4 ($\alpha4\beta1$-integrin) ligand (9). This probe has been used for determination of rapid affinity changes of the integrin ligand-binding pocket in real time in our laboratory and others (9, 10). The assay is performed directly in a tube attached to a flow cytometer and cells are continuously sampled for periods up to several tens of minutes. For a short period of time the tube is removed from the cytometer and a signaling molecule of interest is added. Because the fluorescent probe is added at a concentration sufficient to occupy only high-affinity VLA-4 sites, additional binding of the probe is observed in response to an affinity change. The presence of the affinity change can be verified using dissociation rate analysis, where a large excess of the unlabelled competitor is added to prevent rebinding of the fluorescent probe. A strong correlation between dissociation rates for the probe and natural ligand, as well as cellular dissociation rates has been observed for the case of multiple affinity states (11, 12).

The same fluorescent probe can be used to assess integrin unbending (Fig. 1). The ability to independently measure the affinity state of the ligand-binding pocket and molecular unbending permitted us to study the regulation of these two processes through "inside-out" signaling. Surprisingly, this resulted in the observation that affinity and unbending are regulated by two independent signaling pathways (1). According to these types of measurements "inside-out" signaling through different G-protein-coupled receptors results in a plethora of conformational states, at a minimum the four combinations of high and low affinity with independently regulated bent and unbent states (2). Thus, the idea that a single integrin molecule can adopt states suitable for rolling (extended and low affinity of the binding pocket), arrest (high affinity), and nonadhesive (low affinity bent with hidden

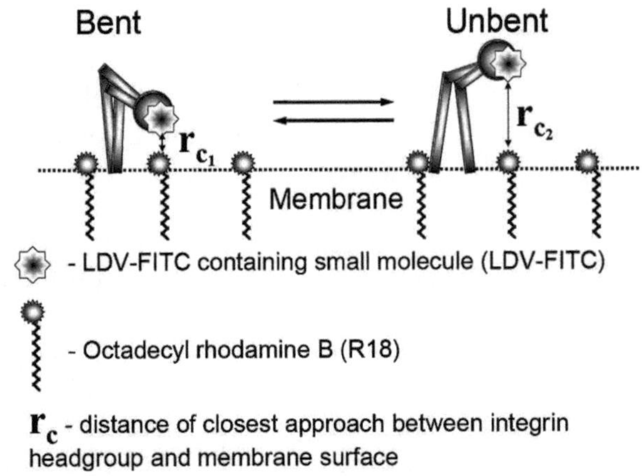

Fig. 1. Schematic depicting the FRET assay for assessing VLA-4 conformational unbending (modified from (1)). Energy transfer between VLA-4 head groups and lipid probes incorporated into the plasma membrane provides a way of studying integrin conformational unbending. The LDV-FITC probe that specifically binds to the head group of VLA-4 is used as a fluorescent donor at a high enough concentration to saturate all low-affinity resting binding sites. A change in VLA-4 affinity would not affect probe binding. Octadecyl rhodamine B (R18), a lipophilic probe, inserts into the membrane as an acceptor. Upon activation, VLA-4 assumes an unbent (upright) conformation. r_{C1} and r_{C2} are the distances of closest approach before and after molecular unbending. Changes in the fluorescence of the donor were measured on live cells in real time at 37°C by flow cytometry.

binding pocket) may be realistic for non I-domain-containing integrins (such as VLA-4) (13). For integrins with an inserted domain (such as LFA-1), the situation is more complicated.

The development of similar fluorescent ligand-mimicking probes for other integrins appears to be very beneficial. Small molecule probes with appropriate affinity (in the nM range) can be used for detecting affinity changes and unbending in real-time on live cells after activation and/or deactivation through signaling receptors. However, only competitive antagonists, which mimic the binding of a natural ligand, can be used for the detection of the affinity change of the ligand-binding pocket. We have also used a fluorescent allosteric antagonist of LFA-1 (fluorescent derivative of BIRT-377) to probe vertical extension upon activation in a FRET-based assay analogous to Fig. 1 (14). Only the reducing agent DTT caused a large FRET signal change, in a manner analogous to DTT-induced extension of VLA-4 (15). The absence of a large conformational change was explained by the fact that BIRT was shown to stabilize the inactive (bent) conformation of LFA-1 (14). Nevertheless, the question remains open why β1-, and β3-integrin-specific small molecules are predominantly competitive antagonists, while the majority of β2-integrin antagonists are allosteric (at least for LFA-1) (5).

3. Single Bond Life-Times

Rapid kinetic measurements of natural integrin ligands binding and other protein–protein interaction are possible with the use of a rapid-mix flow cytometer (16–18). In a conventional flow cytometer several seconds are required for the delivery of a sample from a test tube to the flow chamber. Modern automated rapid-mix devices allow mixing and delivery under a second using microliter volume of samples (55–600 ms, 35–45 μl aliquots) (16, 18). We used a rapid-mix flow cytometer to determine the dissociation rate of soluble fluorescently labeled recombinant human VCAM from a rapidly dissociating intermediate affinity state of VLA-4 integrin. However, a direct measurement of the VCAM dissociation rate for resting VLA-4 (without activation and with physiological concentrations of divalent cations) using this technique is still elusive (12). Nonetheless, the single molecule dissociation rates appear to provide insight into the duration of cell adhesion as described below.

Single bond life-times have also been evaluated with the bioforce probe (19). When these measurements are extrapolated to 0 force, the bioforce probe and flow cytometry measurements give comparable results (Evan Evans, unpublished data).

4. Real-Time Aggregation and Disaggregation Kinetics

Another powerful method for studying real-time integrin activation and cell adhesion is the cell-suspension adhesion assay. Two types of cells, one population expressing the integrin of interest along with activating or inhibiting pathway receptors (G-protein-coupled receptor) and the other cell population expressing an integrin ligand, can be stained with two fluorescent dyes (e.g., green and red). For the case of homotypic aggregation, such as neutrophil aggregation, a single color stain is sufficient (20). After cells are mixed in a tube maintained at 37°C with constant stirring, they are continuously sampled over several tens of minutes. Aggregates, which are formed over a period of time, are detected as double-positive (green and red co-fluorescent) events. Because flow cytometers also detect single cells (only green or red events), it is possible to follow cell aggregation in real time by evaluating the aggregates or depletion of "singlets." This allows eliminating the effect of multicellular aggregates that present in the double-positive gate (11).

Using this methodology it is possible to observe GPCR-dependent activation of integrin-dependent cell adhesion ("inside-out" activation), as well as rapid deactivation and cell disaggregation (1, 2, 20) (Fig. 2). Moreover, it was possible to

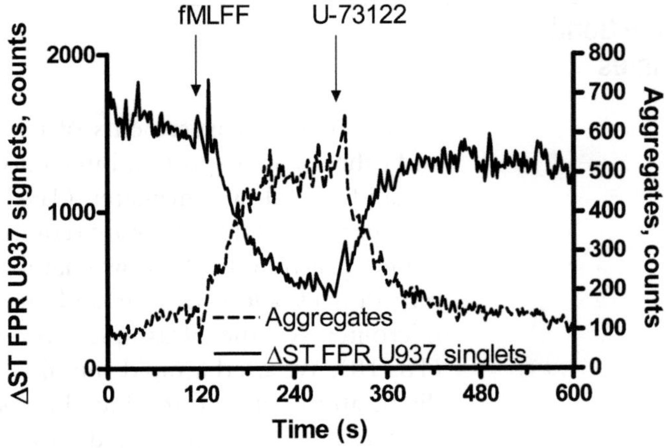

Fig. 2. Changes in cell adhesion between formyl peptide transfected U937 cell and VCAM-1-transfected B78H1 cells at resting state and in response to receptor stimulation (modified from (1)). Addition of fMLFF (formyl peptide) induces cell aggregation. This results in U937 singlets depletion. PLC inhibitor U-73122 has the opposite effect.

establish a relationship between cellular disaggregation rates and ligand dissociation rates for different affinity states. Quantitative analysis of molecular and cellular dissociation rates revealed that only a small number of VLA-4-VCAM-1 bonds (~1.5 on average) was sufficient to hold together cellular aggregates (11).

This method can be also adopted to study cell aggregation and disaggregation under force. We and others have used devices which create defined shear in cone and plate as well as parallel-plate conditions (i.e., Ravenfield model EM Shear Generator (Ravenfield Designs Ltd., Heywood, UK) (12, 21). As expected, shear stress had a significant effect on cellular disaggregation rates (12). The work of Simon et al. showed how the contributions of L-selectin with PSGL and β2-integrin with ICAM-1 worked together under shear in neutrophil aggregation (20, 22).

The method can be also used to determine cellular association rates (analogous to the "forward kinetics" for the ligand binding). Based on our measurements of integrin molecule extension (using a FRET-based assay, see below), we postulated that molecular extension could facilitate integrin ligand recruitment because of the better exposure of the integrin ligand-binding pocket. We established experimental conditions to enhance integrin extension (determined using a FRET-based assay) while maintaining the affinity state of the ligand-binding pocket (determined in a ligand dissociation assay). We found that the initial rate of cell aggregation was dramatically elevated for the case of "extended" integrins (see Fig. 9 in (1)).

5. Parallel-Plate Flow Chamber

The parallel-plate flow chamber with immobilized integrin ligands or cells bearing integrin ligands has been actively used by many groups to study adhesive interactions under shear (23, 24). Several important findings include the role of shear in the induction of transmigration, and the role of force exerted on the bond in the formation of shear resistant adhesion (22, 25–27). In the case of high ligand density, where multiple bonds between the cell and the ligand can be formed, it is difficult to determine the molecular mechanisms that are responsible for the changes in cell adhesion avidity. Rapid clustering of integrin molecules and formation of multivalent contacts could be indistinguishable from the changes in the properties of individual integrin–ligand contacts. Thus, the results of Alon et al. (28) are also consistent with our model in which extension regulates captures frequency while affinity regulates tether duration (1).

Using ligand at low density creates conditions where predominantly only one integrin–ligand contact is formed. This type of experiment allows determination of individual bond kinetics based upon tether frequencies and bond life-times. The increase in the life-time of the bond can be interpreted as a decrease in bond dissociation rate and elevated affinity state of the ligand-binding pocket. It is not surprising that values of bond life-times determined in these experiments are comparable to the life-time determined in a soluble system (compare dissociation rates for different activation states from (28) and (12)). This will be true only if a relatively small force applied is applied to the integrin bond. According to theory, bond life-time would exponentially decrease with the force. However, a recent report from Cheng Zhu group showed that force (in the 10–30 pN range) applied to the integrin actually prolonged bond life-times (29). This so-called catch bond behavior, could potentially be tested in a parallel-plate chamber as well. In unpublished experiments we observed cell–cell adhesion with a very long life-times, where we did not distinguish catch bonds from altered avidity due to multiple bonds formation.

Another interesting phenomenon observed in parallel-plate experiments is the rapid change in tether frequency following "inside-out" activation (e.g., see (28)). Currently, two different interpretations are considered: (1) rapid clustering of integrins and (2) rapid conformational change of individual integrin molecules. Clustering is usually determined using staining with mAbs on fixed cells. To our knowledge no widely used method for rapid real-time determination of integrin clustering exists. Fluorescence resonance energy transfer (FRET)-based methods are the most likely option right now (30). FRET has been used to study integrin subunit separation upon activation, molecule unbending (extension), and microclustering (31). Rapid unbending (extension) of the integrin molecules upon activation, which leads to the rapid exposure of previously hidden ligand-binding site, can also be considered as possible mechanism for a rapid change in the tether frequency. Combining a parallel-plate flow chamber with FRET-based measurements of integrin activation would be a logical next step for studying intimate mechanisms of integrin activation under shear.

6. mAbs and Integrin Conformation

Historically, monoclonal antibodies were the first critical tool for identifying cell adhesion molecules (32). Blocking antibodies that inhibit cell–cell, or cell-soluble or immobilized ligand interactions remain invaluable for determining the specificity of molecular contributions to cell adhesion. For multiple anti-integrin antibodies, the binding epitopes are finely mapped and overlaid on 3D models of integrins (33). From a practical point of view it is important to note that certain blocking mAbs with epitopes mapped close to the ligand-binding site (but not exactly covering it) can be successfully used to block the binding of a large protein ligand, and yet can fail to block binding of a small ligand-mimicking probe.

A conformational change that enhances integrin–receptor interaction upon binding of "activating" antibodies can directly lead to the affinity change of the ligand-binding pocket. However, in our experience, the affinity state generated by this treatment can be several orders of magnitude higher than the "physiological" high-affinity state generated upon G-protein-coupled receptor signaling ("inside-out" activation) (9). Thus, the physiological role of this "artificially" generated state is not certain. Other "activating" antibodies can mimic "inside-out" activation by inducing a separation of integrin subunits, or by inducing an extended integrin conformation (34, 35). This type of mAb is particularly useful in the absence of a proper "physiological" activating pathway, such as G-protein-coupled or other receptors.

Antibodies that recognize the "activated" integrin conformation, also termed "activation reporter" mAbs, often report the ligand bound conformation of the integrin (13, 32). Therefore, they are also termed "ligand-induced binding sites" or LIBS mAbs. Fine mapping of LIBS epitopes, together with mutagenesis studies has provided valuable information about the nature of conformational changes resulting from integrin activation and ligand binding (13, 33, 36). Generation and screening of mAbs, (LIBS-type and others) is described in (32).

A major drawback for using mAbs to detect rapid conformational changes is the long incubation time, which is required to

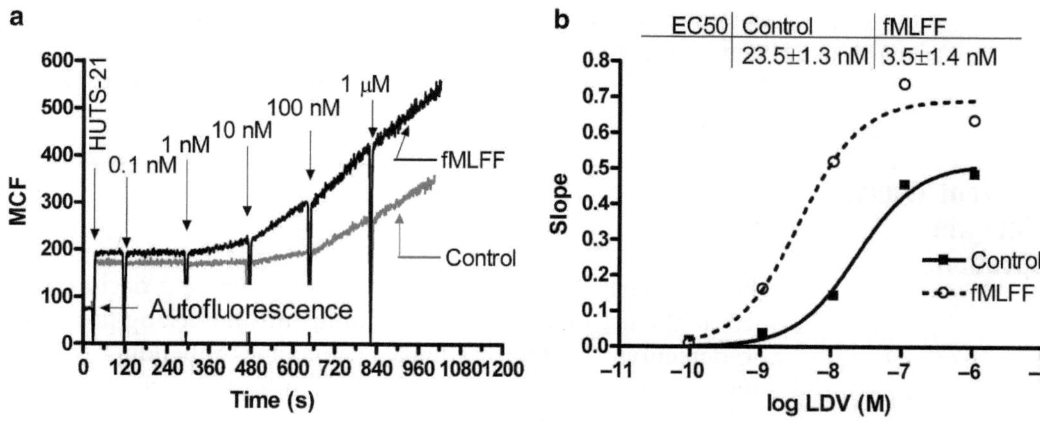

Fig. 3. Kinetics of real-time binding of LIBS mAbs (HUTS-21) demonstrates a difference in VLA-4 ligand (LDV) binding affinity before (control) and after "inside-out" activation through formyl peptide receptor (modified from (13)). (**a**) U937 cells expressing the formyl peptide receptor were treated with 100 nM fMLFF (activated) or vehicle before the start of the experiment. The addition of HUTS-21 antibodies (*first arrow*) resulted in the rapid low-level nonspecific binding of the antibody. Next, increasing amounts of LDV ligand were added. This induced binding of mAbs resulted in different rates of antibody binding (compare slopes after LDV additions). (**b**) absolute rates of HUTS-21 binding (slopes of lines between sequential LDV additions calculated from panel a plotted vs. concentration of LDV in solution). The data were fitted using the sigmoidal dose–response equation with variable slope. Differences in EC50 values for resting and activated cells indicated the affinity change for LDV binding.

reach binding equilibrium. This is caused by a slow mAb dissociation rate, as well as the size of the protein that largely determines its slow diffusion. One possible solution to this problem is to perform binding experiments under conditions that are far from equilibrium (Fig. 3). For these types of studies, real-time detection of mAb binding can be done in a homogeneous assay in a conventional flow cytometer. Since the exposure of the LIBS antibody epitope is proportional to the occupancy of the integrin-binding pocket, the rate of LIBS mAb binding at a given concentration of the ligand will be proportional to the concentration of the ligand–receptor complex. By plotting mAbs binding rate vs. ligand concentrations it is possible to determine ligand-binding affinities at rest and after "inside-out" activation through a GPCR (13). This approach has been verified for multiple VLA-4 ligands. The EC50s for β1-integrin LIBS mAbs (HUTS-21) binding determined in the mAb-binding experiments were identical to K_is, determined in the competition assay with fluorescent VLA-4 ligand (37). This novel approach can be used for determination of ligand-binding affinities for unlabeled ligands, as well as for previously unknown integrin ligands. It was also adapted to a high-throughput screening format for identification of integrin antagonists (PubChem, AID: 2617, Summary of HTS for Identification of VLA-4 Allosteric Modulators). Moreover, it is worthwhile considering, that under conditions that resolves the four states of affinity and extension regulation, the presence of ligand under these conditions creates a unique LIBS state, allowing the potential of eight distinct physiological VLA-4 states (13).

7. Current Models of Integrin Activation

The combination of several novel real-time approaches that include the use of a small fluorescent ligand-mimicking probe for the detection of integrin affinity change and unbending (extension), real-time analysis of LIBS mAb binding, together with the analysis of real-time aggregation and disaggregation kinetics resulted in a model of integrin activation with several novel features (see Fig. 6 in (13)). Here without going into details we highlight specific methods that provided evidence for each specific conclusion (Table 1).

Taken together, a combination of real-time assays performed on live cells at 37°C under physiologically relevant "inside-out" signaling through multiple GPCRs have provided a unique opportunity to study integrin conformational regulation and its role in the modulation of integrin-dependent cell adhesion and deadhesion. This approach can be extended to other integrins, and will provide valuable information about mechanisms of integrin conformational regulation.

Table 1
General approaches for studying integrin-dependent cell adhesion

Experimental methods	Conclusions	References
Ligand-mimetic binding and dissociation, FRET-based extension method, real-time cellular aggregation, binding of a soluble integrin ligand (VCAM-1), rapid-mix flow cytometry	Independent regulation of integrin affinity and unbending (extension). Affinity regulates the life-time of the interaction, unbending regulates initial rate of aggregate formation	(1, 2, 17) and unpublished data
Real-time binding of LIBS antibodies, ligand mimetic under activation through Galphai-coupled GPCRs	Affinity state of the ligand-binding pocket is independent of hybrid domain movement, which is solely determined by ligand occupancy	(13, 37)
Ligand mimetic binding and dissociation, FRET-based extension method under activation through Galphai-coupled and Galphas-coupled GPCRs, real-time cellular aggregation/disaggregation, parallel-plate flow chamber	Galphai-coupled GPCRs provide "pro-adhesive signal," Galphas-coupled GPCRs provide "anti-adhesive signal"	(2) and unpublished data

Acknowledgments

This work was supported by R01HL081062 and U54MH084690.

References

1. Chigaev, A., Waller, A., Zwartz, G. J., Buranda, T., and Sklar, L. A. (2007) Regulation of cell adhesion by affinity and conformational unbending of alpha4beta1 integrin. *J. Immunol.* **178**, 6828–6839.
2. Chigaev, A., Waller, A., Amit, O., and Sklar, L. A. (2008) Galphas-coupled receptor signaling actively down-regulates alpha4beta1-integrin affinity: a possible mechanism for cell de-adhesion. *BMC. Immunol.* **9**, 26.
3. Sklar, L. A., Edwards, B. S., Graves, S. W., Nolan, J. P., and Prossnitz, E. R. (2002) Flow cytometric analysis of ligand-receptor interactions and molecular assemblies. *Annu. Rev. Biophys. Biomol. Struct.* **31**, 97–119.
4. Bednar, B., Cunningham, M. E., McQueney, P. A., Egbertson, M. S., Askew, B. C., Bednar, R. A., Hartman, G. D., and Gould, R. J. (1997) Flow cytometric measurement of kinetic and equilibrium binding parameters of arginine-glycine-aspartic acid ligands in binding to glycoprotein IIb/IIIa on platelets. *Cytometry* **28**, 58–65.
5. Shimaoka, M. and Springer, T. A. (2003) Therapeutic antagonists and conformational regulation of integrin function. *Nat. Rev. Drug Discov.* **2**, 703–716.
6. Keating, S. M., Clark, K. R., Stefanich, L. D., Arellano, F., Edwards, C. P., Bodary, S. C., Spencer, S. A., Gadek, T. R., Marsters, J. C., Jr., and Beresini, M. H. (2006) Competition between intercellular adhesion molecule-1 and a small-molecule antagonist for a common binding site on the alpha1 subunit of lymphocyte function-associated antigen-1. *Protein Sci.* **15**, 290–303.
7. Lin, K., Ateeq, H. S., Hsiung, S. H., Chong, L. T., Zimmerman, C. N., Castro, A., Lee, W. C., Hammond, C. E., Kalkunte, S., Chen, L. L., Pepinsky, R. B., Leone, D. R., Sprague, A. G., Abraham, W. M., Gill, A., Lobb, R. R., and Adams, S. P. (1999) Selective, tight-binding inhibitors of integrin alpha4beta1 that inhibit allergic airway responses. *J. Med. Chem.* **42**, 920–934.
8. Chen, L. L., Whitty, A., Lobb, R. R., Adams, S. P., and Pepinsky, R. B. (1999) Multiple activation states of integrin alpha4beta1 detected through their different affinities for a small molecule ligand. *J. Biol. Chem.* **274**, 13167–13175.
9. Chigaev, A., Blenc, A. M., Braaten, J. V., Kumaraswamy, N., Kepley, C. L., Andrews, R. P., Oliver, J. M., Edwards, B. S., Prossnitz, E. R., Larson, R. S., and Sklar, L. A. (2001) *J. Biol. Chem.* **276**, 48670–48678.
10. Chan, J. R., Hyduk, S. J., and Cybulsky, M. I. (2003) Detecting rapid and transient upregulation of leukocyte integrin affinity induced by chemokines and chemoattractants. *J. Immunol. Methods* **273**, 43–52.
11. Zwartz, G., Chigaev, A., Foutz, T., Larson, R. S., Posner, R., and Sklar, L. A. (2004) Relationship between molecular and cellular dissociation rates for VLA-4/VCAM-1 interaction in the absence of shear stress. *Biophys. J.* **86**, 1243–1252.
12. Chigaev, A., Zwartz, G., Graves, S. W., Dwyer, D. C., Tsuji, H., Foutz, T. D., Edwards, B. S., Prossnitz, E. R., Larson, R. S., and Sklar, L. A. (2003) Alpha4beta1 integrin affinity changes govern cell adhesion. *J. Biol. Chem.* **278**, 38174–38182.
13. Chigaev, A., Waller, A., Amit, O., Halip, L., Bologa, C. G., and Sklar, L. A. (2009) Real-time analysis of conformation-sensitive antibody binding provides new insights into integrin conformational regulation. *J. Biol. Chem.* **284**, 14337–14346.
14. Larson, R. S., Davis, T., Bologa, C., Semenuk, G., Vijayan, S., Li, Y., Oprea, T., Chigaev, A., Buranda, T., Wagner, C. R., and Sklar, L. A. (2005) Dissociation of I domain and global. conformational changes in LFA-1: refinement of small molecule-I domain structure-activity relationships. *Biochemistry* **44**, 4322–4331.
15. Chigaev, A., Zwartz, G. J., Buranda, T., Edwards, B. S., Prossnitz, E. R., and Sklar, L. A. (2004) Conformational regulation of alpha 4 beta 1-integrin affinity by reducing agents. "Inside-out" signaling is independent of and additive to reduction-regulated integrin activation. *J. Biol. Chem.* **279**, 32435–32443.
16. Wu, Y., Zwartz, G., Lopez, G. P., Sklar, L. A., and Buranda, T. (2005) Small-volume rapid-mix device for subsecond kinetic analysis in flow cytometry. *Cytometry A* **67**, 37–44.
17. Nolan, J. P., Posner, R. G., Martin, J. C., Habbersett, R., and Sklar, L. A. (1995) A

rapid mix flow cytometer with subsecond kinetic resolution. *Cytometry* **21**, 223–229.

18. Graves, S. W., Nolan, J. P., Jett, J. H., Martin, J. C., and Sklar, L. A. (2002) Nozzle design parameters and their effects on rapid sample delivery in flow cytometry. *Cytometry* **47**, 127–137.
19. Evans, E. (2001) Probing the relation between force--lifetime--and chemistry in single molecular bonds. *Annu. Rev. Biophys. Biomol. Struct.* **30**, 105–128.
20. Simon, S. I., Chambers, J. D., Butcher, E., and Sklar, L. A. (1992) Neutrophil aggregation is beta 2-integrin- and L-selectin-dependent in blood and isolated cells. *J. Immunol.* **149**, 2765–2771.
21. Zwartz, G. J., Chigaev, A., Dwyer, D. C., Foutz, T. D., Edwards, B. S., and Sklar, L. A. (2004) Real-time analysis of very late antigen-4 affinity modulation by shear. *J. Biol. Chem.* **279**, 38277–38286.
22. Simon, S. I. and Goldsmith, H. L. (2002) Leukocyte adhesion dynamics in shear flow. *Ann. Biomed. Eng* **30**, 315–332.
23. Brown, D. C. and Larson, R. S. (2001) Improvements to parallel plate flow chambers to reduce reagent and cellular requirements. *BMC. Immunol.* **2**, 9.
24. DiVietro, J. A., Brown, D. C., Sklar, L. A., Larson, R. S., and Lawrence, M. B. (2007) Immobilized stromal cell-derived factor-1alpha triggers rapid VLA-4 affinity increases to stabilize lymphocyte tethers on VCAM-1 and subsequently initiate firm adhesion. *J. Immunol.* **178**, 3903–3911.
25. Alon, R. and Feigelson, S. W. (2009) Chemokine signaling to lymphocyte integrins under shear flow. *Microcirculation.* **16**, 3–16.
26. Alon, R. and Ley, K. (2008) Cells on the run: shear-regulated integrin activation in leukocyte rolling and arrest on endothelial cells. *Curr. Opin. Cell Biol.* **20**, 525–532.
27. Ley, K. (2009) Cell adhesion under flow. *Microcirculation.* **16**, 1–2
28. Grabovsky, V., Feigelson, S., Chen, C., Bleijs, D. A., Peled, A., Cinamon, G., Baleux, F., Arenzana-Seisdedos, F., Lapidot, T., van Kooyk, Y., Lobb, R. R., and Alon, R. (2000) Subsecond induction of alpha4 integrin clustering by immobilized chemokines stimulates leukocyte tethering and rolling on endothelial vascular cell adhesion molecule 1 under flow conditions. *J. Exp. Med.* **192**, 495–506.
29. Kong, F., Garcia, A. J., Mould, A. P., Humphries, M. J., and Zhu, C. (2009) Demonstration of catch bonds between an integrin and its ligand. *J. Cell Biol.* **185**, 1275–1284.
30. Kim, M., Carman, C. V., Yang, W., Salas, A., and Springer, T. A. (2004) The primacy of affinity over clustering in regulation of adhesiveness of the integrin {alpha}L{beta}2. *J. Cell Biol.* **167**, 1241–1253.
31. Lefort, C. T. and Kim, M. (2009) Fluorescence resonance energy transfer in the studies of integrin activation. *Curr. Top. Membranes* **64**, 359–388.
32. Cabanas, C. and Sanchez-Madrid, F. (1999) Monoclonal antibodies specific for leukocyte adhesion molecules. Selective protocols of immunization and screening assays for generation of blocking, activating and activation reporter antibodies. *Methods Mol. Biol.* **96**, 1–9.
33. Humphries, M. J., Symonds, E. J., and Mould, A. P. (2003) Mapping functional residues onto integrin crystal structures. *Curr. Opin. Struct. Biol.* **13**, 236–243.
34. Kim, M., Carman, C. V., and Springer, T. A. (2003) Bidirectional transmembrane signaling by cytoplasmic domain separation in integrins. *Science* **301**, 1720–1725.
35. Mould, A. P., Travis, M. A., Barton, S. J., Hamilton, J. A., Askari, J. A., Craig, S. E., Macdonald, P. R., Kammerer, R. A., Buckley, P. A., and Humphries, M. J. (2005) Evidence that monoclonal antibodies directed against the integrin beta subunit plexin/semaphorin/integrin domain stimulate function by inducing receptor extension. *J. Biol. Chem.* **280**, 4238–4246.
36. Mould, A. P., Barton, S. J., Askari, J. A., McEwan, P. A., Buckley, P. A., Craig, S. E., and Humphries, M. J. (2003) Structure of an integrin-ligand complex deduced from solution X-ray scattering and site-directed mutagenesis. *J. Biol. Chem.* **278**, 17028–17035.
37. Njus, B. H., Chigaev, A., Waller, A., Wlodek, D., Ostopovici-Halip, L., Ursu, O., Wang, W., Oprea, T. I., Bologa, C. G., and Sklar, L. A. (2009) Conformational mAb as a tool for integrin ligand discovery. *Assay. Drug Dev. Technol.* **7**, 507–515.

Chapter 2

Cell Adhesion Assays

Gabriele Weitz-Schmidt and Stéphanie Chreng

Abstract

Standard adhesion assays measure cell binding either to immobilized ligands or to cell monolayers in flat-well microtiter plates under static conditions. Typically, these test systems require several washing steps to separate adherent from nonadherent cells. Here, we describe an adhesion assay which avoids these washing steps by employing V-bottom 96-well plates. In this assay, fluorescently labeled leukocytes are allowed to adhere to V-well plates coated with soluble ligand for a fixed time. Thereafter, centrifugal force is applied to separate adherent cells from nonadherent cells. Nonadherent cells accumulate in the nadir of the V-shaped wells and are quantified using a fluorometer with a narrow aperture. This simple and reproducible method has been validated with different classes of adhesion molecule families (selectins and integrins) and is adaptable to several other adhesive interactions. The assay format is suitable for screening applications and may also be used for diagnostic testing. The receptor/ligand interaction chosen as an example to describe the assay methodology is the interaction between the integrin lymphocyte function-associated molecule-1 (LFA-1, $\alpha_L\beta_2$) and intercellular adhesion molecule-1 (ICAM-1).

Key words: Cell adhesion, V-bottom plates, Screening, Integrins, Selectins, LFA-1, ICAM-1

1. Introduction

During an inflammatory response, a multistep process (also referred to as "adhesion cascade") recruits leukocytes from the bloodstream to sites of inflammation. The major steps of the adhesion cascade are mediated by receptors of the selectin, integrin, and immunoglobulin families. According to a widely accepted model, selectins attract circulating leukocytes and mediate their rolling along the vascular endothelium. Once captured, the leukocytes are activated by chemokines presented on endothelial cells. This activation triggers intracellular signals that convert leukocyte integrins from a low- to a high-affinity state. Subsequently,

Motomu Shimaoka (ed.), *Integrin and Cell Adhesion Molecules: Methods and Protocols*,
Methods in Molecular Biology, vol. 757, DOI 10.1007/978-1-61779-166-6_2, © Springer Science+Business Media, LLC 2011

the activated integrins interact with endothelial counter receptors of the immunoglobulin superfamily to mediate firm cell adhesion and extravasation (1).

Various in vitro test systems have been established to investigate the role of selectins and integrins in inflammation (2–4). These assays can be divided into two major classes: the static assays in which cells are allowed to settle on the adhesive substrate or the flow-based assays in which cells are perfused over the substrates. The traditional static assays are easy to set up; however, they are time consuming and exhibit substantial well-to-well variability due to several washing steps required to separate adherent from nonadherent cells. Furthermore, the static assays lack the controlled shear force that leukocytes encounter during their adhesion in the bloodstream. In contrast, the flow-based test systems can simulate this shear force and thus are thought to generate more "physiological" data (2). Moreover, the flow-based test systems involve microscopic observation that allows to analyze the different steps of the adhesion process individually. However, the flow-based assays are difficult to set up and are not suitable for high-throughput applications.

In this chapter, we describe a semi-homogenous cell adhesion assay that measures the ability of cells to bind to ligand immobilized on V-bottom 96-well plates (5). In brief, fluorescently labeled cells are added to ligand-coated plates and allowed to settle randomly on the V-bottom. Thereafter, centrifugal force is applied to separate adherent from nonadherent cells. The force produced by the centrifugation step results in the accumulation of free or loosely attached cells in the tip of the V-shaped wells. The nonadherent cells are quantified using a fluorescent reader (Fig. 1).

The V-well assay provides several advantages over the conventional adhesion assays. The assay involves precisely controlled centrifugal force to separate adherent from nonadherent cells, instead of turbulent, uncontrolled shear (washing steps) as applied in the flat-well assay. In consequence, well-to-well and plate-to-plate variations are reduced, making the V-assay more reliable and reproducible than the flat-well assay. In addition, elimination of the washing steps reduces the time necessary to perform the assay. Moreover, the V-well method can be readily utilized for high throughput applications. This property clearly distinguishes the V-well from the flat-well or flow-based test systems, which are less suitable for automation. The assay is adaptable to a number of integrin- and selectin-mediated interactions. Table 1 provides a list of adhesion molecules that can be analyzed using the test system (5). In summary, the V-well format overcomes major limitations of the flat-well format (e.g., high variability, time consumption, and operator-dependent results) and adopts major advantages of the flow-based assay (e.g., application of controlled force to study attachment and detachment of cells).

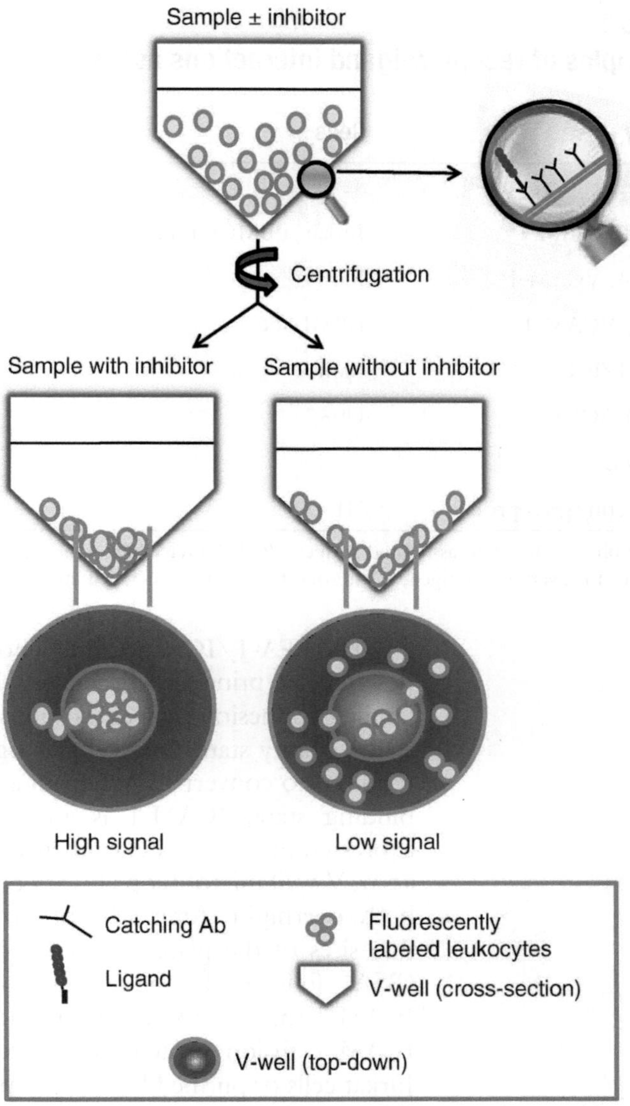

Fig. 1. Schematic presentation of the V-well adhesion assay. Leukocytes labeled with a fluorescent dye are added to ligand-coated V-well microtiter plates in the presence or absence of an inhibitor (pre-coating of the V-wells with anti-mouse c-κ antibody (catching Ab) is recommended for ligands that are expressed as mouse c-κ fusion proteins). Centrifugal force is applied to separate bound from free cells. Nonadherent cells accumulate in the tip of the well and are measured using a fluorescence reader. In the presence of an inhibitor, the cells pellet into the nadir of the V-well, resulting in an increased fluorescence signal. In the absence of an inhibitor, most of the cells adhere to the wall of the V-well and do not pellet, thus the signal is low.

The V-well assay is suitable for the identification and analysis of integrin and selectin inhibitors (5). It can also be used to assess the regulation of cell adhesiveness (6, 7). Moreover, it may be applied to diagnose disease associated with cell adhesion defects such as leukocyte adhesion deficiencies (LAD) (8).

Table 1
Examples of receptor/ligand interactions assessed in the V-well adhesion assay

Assay	Cells	Ligand
Integrins		
LFA-1/ICAM-1	Jurkat, PBMCs, transfectants	ICAM-1 m c-κ fusion protein
VLA-4/VCAM-1	Ramos	VCAM-1 m c-κ fusion protein
α4β7/VCAM-1	RPMI 866	VCAM-1 m c-κ fusion protein
VLA-4/FN	Ramos	Purified plasma FN
Mac-1/iC3b	U937	Purified iC3b
Selectins		
E-selectin/glycoprotein	HL-60	E-selectin m c-κ fusion protein

LFA lymphocyte function-associated antigen, *ICAM* intercellular adhesion molecule, *VCAM* vascular cell adhesion molecule, *VLA* very late antigen, *FN* fibronectin. For more detailed information, see ref. 5

The LFA-1/ICAM-1 interaction is used here as an example to describe the principle of the assay. LFA-1 belongs to the integrin family of adhesion molecules and is expressed on all leukocytes in a low-affinity state. Intracellular signaling or divalent cations are required to convert LFA-1 from a low- to a high-affinity, ligand-binding state. ICAM-1 is an inducible cell surface molecule expressed by activated endothelial cells and leukocytes (9). In this assay, V-well microtiter plates are coated with anti-mouse c-κ antibody overnight. After adsorption of the catching antibody, the free sites of the plates are blocked with bovine serum albumin (BSA), followed by the addition of recombinant soluble human ICAM-1/mouse c-κ fusion protein. After immobilization of the ICAM-1 fusion protein, fluorescently labeled, cation-activated Jurkat cells or purified human peripheral blood mononuclear cells (PBMCs) are transferred to the V-wells. The cells are allowed to adhere for 10 min before the plates are centrifuged, and nonadherent cells are quantified by fluorescence in the tip of the wells. Moreover, modifications to the basic protocol are described in order to analyze cell adhesion in the presence of low molecular weight LFA-1 inhibitors and blocking monoclonal antibodies.

2. Materials

2.1. Jurkat Cell Culture

1. Cell line: Jurkat cells clone E6-1 from the American Type Culture Collection (ATCC).
2. Culture flasks: T-75-cm^2 flasks and T-175-cm^2 flasks (tissue culture-treated polystyrene, with vented screw cap).

3. Stericup™-GP filter unit, 0.22 μm, 500-ml bottle with cap.
4. Culture medium: RPMI-1640 supplemented with 10% heat-inactivated fetal bovine serum (FBS), 1% glutamax, 1% MEM nonessential amino acids, 10 mM HEPES, and 1% penicillin/streptomycin. Sterile filter using Stericup™ filter unit (0.22 μm) and store at 4°C.

2.2. Isolation and Cryopreservation of PBMCs

1. Source of PBMCs: Human buffy coat or human blood samples from healthy donors taken into natrium-heparin tubes.
2. Separation medium: Ficoll-Paque™ Plus, density 1.077 g/ml, store at room temperature (RT).
3. Buffer: Dulbecco's divalent cation-free phosphate-buffered saline with or without 0.5% (w/v) BSA, store at 4°C.
4. Freezing solution: Mix 10% dimethyl sulfoxide (DMSO) and 90% heat-inactivated, sterile filtered fetal calf serum (FCS). The freezing medium is stable for up to 3 months when stored at 4°C.
5. Polypropylene tubes Leucosep™, 30 ml, sterile, for 15–30 ml sample volume.
6. Trypan blue 0.4% solution.
7. Cryovials, sterile, 1.8 ml.
8. BD 50-ml Falcon™ centrifuge tubes, sterile.
9. BD Falcon™ cell strainer, 70-μm pores.
10. NALGENE Cryo 1°C freezing container.
11. Thawing medium: RPMI medium-1640, supplemented with 2.5% FCS, 10 mM HEPES, and 1% penicillin/streptomycin. Filter using a Stericup™ filter unit and store at 4°C.

2.3. V-Well Adhesion Assay

General remark: Filter all buffers below using a Stericup™ filter unit (0.22 μm) and store at 4°C.

1. Tris-buffered saline (TBS)/BSA buffer: 50 mM Tris base and 150 mM NaCl, pH 7, supplemented with 1.5% BSA (w/v).
2. Carbonate buffer: 10 mM Na_2CO_3 and 35 mM $NaHCO_3$, pH 8.5.
3. Blocking buffer: TBS supplemented with 1.5% BSA (w/v) and 0.5% Tween 20 (v/v), pH 7.2–7.4.
4. Binding buffer: TBS/BSA buffer containing 2 mM $MgCl_2$, 2 mM $MnCl_2$, and 5 mM D-glucose monohydrate, pH 7.2–7.4.
5. Catching antibody (Ab): Goat anti-mouse c-κ Ab, 1 mg/ml, store at 4°C.
6. Ligand: Human ICAM-1 mouse c-κ fusion protein (ICAM-1 m c-κ) (Novartis). A stock solution of 0.3 mg/ml in TBS/BSA buffer is prepared and aliquots are stored at –20°C (see Note 1).

7. Cell labeling agent: Prepare a stock solution of 2′,7′-bis-(2-carboxyethyl)-5-(and-6)-carboxyfluorescein, acetoxymethyl ester (BCECF-AM, Invitrogen) in anhydrous DMSO at 10 mg/ml and store in aliquots at −20°C. The DMSO used for the stock solution is stored at RT with desiccant beads (Molecular sieves, 4 Å, Roth) (see Note 2).
8. Micronic U-tubes (1.4 ml), non-coded, in polypropylene.
9. Tube sealing.
10. 96-Well plates, V-bottom, transparent in polystyrene (V-well plates, Corning).
11. Adhesive Clear Polyester Seal film Uniseal™, 0.05-mm thickness.
12. Fluorescence plate reader: Victor 2™ (Perkin Elmer).

2.4. Inhibitors

1. Test compounds: Prepare stock solutions of test compounds (e.g., LFA878, Novartis) at 10 mM in DMSO and store at 4°C (10).
2. Blocking Ab: mouse antihuman ICAM-1 monoclonal antibody (mAb), clone LB-2, IgG2b κ (Becton Dickinson), store at 4°C.

3. Methods

3.1. Jurkat Cell Culture

General remark: All solutions and equipment coming in contact with the cells must be sterile, and proper sterile techniques must be used accordingly. Culture conditions of Jurkat cells are critical for the quality of the assay (see Note 3).

1. Maintain Jurkat cells in 75-cm^2 tissue culture flasks in a humidified incubator at 37°C in an atmosphere of 5% CO_2/95% air.
2. Split the cells three times a week and ensure that the cell density is maintained below 1×10^6 cells/ml. Use either 75-cm^2 or 175-cm^2 culture flasks depending on the number of cells needed for the experiment (see Note 3).
3. One day prior to the experiment, split the cells in fresh medium such that the cell density at the day of the experiment does not exceed 8×10^5 cells/ml.
4. Every second or third month, replace the cell cultures by a new batch of cells with a low passage number.

3.2. Isolation and Cryopreservation of PBMCs

General remarks: *Biosafety practices (BL-2) must be followed when working with primary human material.* The method below outlines the isolation of PBMCs from one buffy coat. The isolation of PBMCs from fresh human blood samples is described in Note 4. Avoid mechanical stress (e.g., resuspend and pipette gently, minimize pellet time and avoid bubbles). Steps 1–6 are performed at RT followed by washing steps at 4°C.

1. Fill a 50-ml Leucosep™ tube with 15 ml Ficoll separation medium. Six Leucosep™ tubes are needed for one buffy coat.
2. Close the tubes containing the separation medium with the screw cap and centrifuge for 1 min at 1,000 × *g* (RT). The separation medium is now located below the porous barrier.
3. Dilute the buffy coat 1:3 with DPBS and add 25 ml of sample material directly on the porous barrier of the Leucocep™ tube.
4. Switch off the brake of the centrifuge and centrifuge tubes for 10 min at 1,000 × *g* at RT in a swinging bucket rotor. After centrifugation, the sequence of layers occurs as follows (seen from top to bottom): plasma (yellow) – enriched cell fraction (opaque interphase consisting of lymphocytes/PBMCs) – Ficoll (colorless) – porous barrier – Ficoll – pellet (containing erythrocytes and granulocytes).
5. Harvest the enriched cell fraction (PBMCs) including most of the plasma fraction using a pipette (results in ~15 ml suspension per Leucosep™ tube) and distribute the total of ~90 ml cell suspension in four sterile 50-ml centrifuge tubes.
6. Add DPBS to each tube up to the 50-ml graduation, centrifuge for 10 min at 400 × *g* at RT (brake switched on), and remove supernatant (containing platelets) by a total flip of the tube.
7. Resuspend the cells by tapping until no larger clumps can be seen and add 5 ml cold DPBS containing 0.5% BSA to each tube.
8. Combine cell suspensions in one 50-ml tube, add cold DPBS/BSA buffer (4°C) up to the 50-ml graduation, and centrifuge for 10 min at 300 × *g* at 4°C (with brake).
9. Repeat washing step with another 50 ml cold DPBS/BSA (see Note 5).
10. Use an inverted microscope to inspect the morphology of the cells and examine their viability by trypan blue exclusion. If the cells pass both inspections, proceed with the freezing step below or start the adhesion experiment (see Note 6).
11. Resuspend the cells in cold freezing solution (3×10^7 cells/ml) and transfer 0.8–1 ml of the cell suspension to cryovials. Cap the vials and immediately place them in the Nalgene Cryo 1°C freezing container. Store at −80°C freezer. One buffy coat results in approximately 30 cryovials filled with the PBMC preparation. For short-term storage (up to 2 months),

leave the cells at –80°C. For long-term storage, transfer the cells to a liquid nitrogen freezer. Separated PBMCs can be cryopreserved without significant loss of function (see Note 7).

3.3. V-Well Adhesion Assay

3.3.1. Immobilization of ICAM-1 m c-κ

1. Dilute the goat anti-mouse c-κAb in carbonate buffer to a concentration of 1 μg/ml. Use a multichannel pipette to dispense 100 μl of catching Ab to each well of a 96-well V-bottom microtiter plate. Cover the plate with adhesive seal film to control humidity and to reduce evaporation of samples during the incubation step below (see Note 8).
2. Incubate the plates for 2 h at 37°C.
3. Remove the antibody solution and add 200 μl of blocking buffer (TBS/BSA/Tween) to each well using a multichannel pipette. Cover the plates with adhesive seal film.
4. Incubate at 4°C overnight or at 37°C for 90 min to block nonspecific binding sites (see Note 9).
5. Wash the wells once with 150 μl TBS/BSA buffer.
6. Dilute ICAM-1 m c-κ in TBS/BSA buffer to a concentration of 100 ng/ml and add 100 μl to the appropriate wells (10 ng/well). Add to some wells 100 μl TBS/BSA buffer without ICAM-1 m c-κ for use as negative controls. Cover the plates with seal film (see Note 10) (Fig. 2).

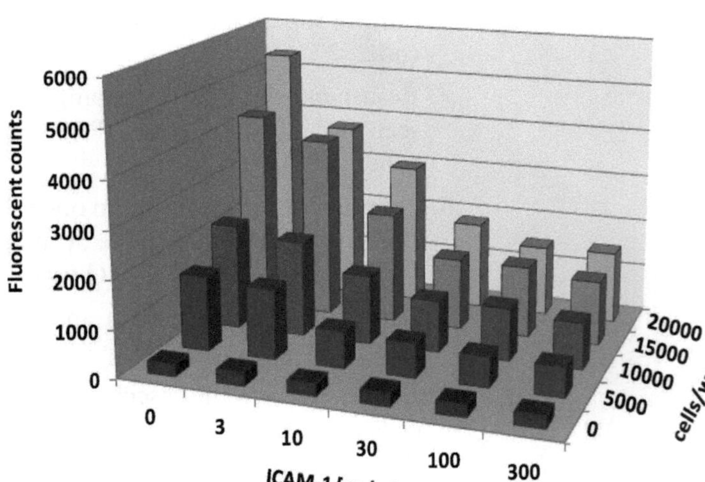

Fig. 2. Optimization of the V-well adhesion assay. ICAM-1 m c-κ was immobilized on V-well microtiter plates (via an anti-mouse c-κ Ab) at indicated concentrations, followed by the addition of fluorescently labeled PBMCs at indicated numbers. Adhesion of the cells to ICAM-1 m c-κ was quantified as described under Subheading 3. Each bar represents the mean value of triplicates. The best S/N ratios were obtained by using 100–300 ng/ml ICAM-1 m c-κ (10–30 ng/well) combined with 20,000 cells/well.

7. Incubate the plates for 90 min at 37°C.
8. Remove unbound ligand by three washing steps using binding buffer (150 μl/well).

3.3.2. Labeling and Activation of Jurkat Cells or PBMCs

Jurkat cells (see Note 11)

1. Centrifuge Jurkat cells at 300×*g* for 5 min at RT, remove supernatant, and resuspend at 1×10^6 cells/ml in DPBS containing 5 μg/ml BCECF-AM (see Note 2).
2. Incubate at 37°C for 30 min in the dark and gently rotate the tube every 10 min (cell labeling step).
3. Add an equal volume of cold DPBS and centrifuge at 300×*g* (1,200 rpm, Megafuge 2.0R) for 5 min at RT (see Note 12).
4. Resuspend the cells in binding buffer (containing divalent cations) at a density of 2×10^5 cells/ml and incubate at RT (not 4°C or 37°C) in the dark for 30–45 min (activation step) (see Note 13).
5. Immediately before addition to the plate, resuspend the cells by pipetting ~6 times up and down to reduce the number of cell aggregates and to make sure that the cells are uniformly distributed in the liquid.

PBMCs

1. Retrieve cryovial containing the PBMCs from storage (−80°C or liquid nitrogen freezer) and place on dry ice to minimize thawing.
2. Thaw the cells quickly by placing the vial in a 37°C water bath for 30 s to 1 min. Swirl the vial gently until cells are thawed.
3. Wipe the vial with 70% (v/v) ethanol and take it to the tissue culture hood.
4. Add 0.8–1.0 ml pre-warmed thawing medium and carefully pipette up and down, avoid foaming.
5. Transfer the cell suspension quickly into a 50-ml Falcon tube containing 4 ml pre-warmed thawing medium. Rinse the vial with 1 ml medium to recover all cells. Dilute the cells further by adding another 4 ml of pre-warmed medium (see Note 7).
6. Add another 20 ml of pre-warmed medium and centrifuge the Falcon tube at 300×*g*, for 10 min at RT (with brake) and discard supernatant.
7. Gently tap the tube to break up the pellet before resuspending the cells in 10 ml of medium (pre-warmed) and count the cells.

8. Centrifuge $0.6–1 \times 10^7$ cells at $300 \times g$ for 10 min at RT and resuspend in 10 ml of TBS containing 1.5% BSA and 10 μg/ml BCECF (see Note 2).
9. Incubate the cells in a water bath, at 37°C, for 30 min in the dark.
10. Centrifuge the cells at $300 \times g$ for 10 min at RT, wash with 10 ml TBS/1.5% BSA, and centrifuge at $300 \times g$, for 5 min at RT. Repeat the washing step twice.
11. For activation, resuspend the cells in binding buffer (containing Mn^{2+} and Mg^{2+}) at a density of 2×10^5 cells/ml and incubate in a water bath for 30 min at 37°C in the dark. After the incubation step, the cells are ready to use.

3.3.3. Measurement of Cell Adhesion

1. Add 100 μl of cell suspensions (Jurkat cells or PBMCs in binding buffer) to each well of the microtiter plates (20,000 cells/well) (Fig. 2).
2. Incubate the plates at 37°C for 10 min.
3. Centrifuge the plate at $209 \times g$ for 10 min at RT (see Note 14).
4. Quantify nonadherent cells in the tip of the V-bottom well by using a fluorescence reader with filter sets allowing excitation at 485 nm and quantification of emission at 535 nm. An example of device settings (Victor 2™) is as follows:

• Name of the label: fluorescein (0.1 s)
• Label technology: prompt fluorometry
• CW-lamp filter name: F485 (fluorescein)
• CW-lamp filter slot: A5
• Emission filter name: F535 (fluorescein)
• Emission filter slot: A5
• Measurement time: 0.1 s
• Measurement height: 8 mm
• Emission aperture: small
• CW-lamp energy: 25,288
• Second measurement CW-lamp energy: .0
• Emission side: below
• CW-lamp control: stabilized energy

3.4. Analyzing Cell Adhesion in the Presence of Inhibitors

3.4.1. Dilution of Test Agents

1. Low molecular weight compounds (e.g., LFA878): Prepare compound dilutions on the day of the experiment at 2× concentration. Serially pre-dilute the compounds in DMSO (100%) to avoid precipitation, before performing final dilution steps in assay buffer. Keep the final DMSO concentration in the assay constant in all samples (see Note 15).
2. Dilute protein – test candidates directly in binding buffer.

3.4.2. Test Effect of Compounds and Abs on Cell Adhesion

Prepare ligand-coated plates as described in Subheading 3.3.1 and label cells as described in Subheading 3.3.2. Then follow the steps below.

Compounds

1. Dilute labeled PBMCs or Jurkat cells at 4×10^5 cells/ml in binding buffer (instead of 2×10^5 cells/ml), incubate the cells for 30 min at 37°C (activation step), and transfer the cell suspension to micronic tubes.
2. Dilute the cells 1:2 with binding buffer containing the solvent control or test compound at 2× concentration.
3. Preincubate the cells with the solvent control or the compound for the appropriate time in the dark at 37°C (e.g., 10 min in case of LFA878).
4. After the pre-incubation step, add 100 μl of the cell suspension to each well of the ligand-coated plates (see Subheading 3.3.1).
5. Incubate the plates at 37°C for 10 min.
6. Centrifuge the plates at 209 × *g* for 10 min at RT.
7. Quantify nonadherent cells as described above.
8. Determine the half maximal inhibitory concentration (IC_{50}) of the compound using an appropriate software such as the ORIGIN 7.5 software program (Fig. 3 and Table 2).

Antibody

1. Transfer 50 μl of the anti-ICAM-1 mAb diluted in binding buffer to the desired concentration to the ICAM-1 coated V-bottom microtiter plates and cover the plates with seal film.
2. Incubate the plates for 30 min at 37°C (do not wash).
3. Dilute labeled and activated PBMCs or Jurkat cells with binding buffer to a concentration of 4×10^5 cells/ml and transfer 50 μL of the cell suspension to the microtiter plates containing the antibody.
4. Incubate the plates at 37°C for 10 min.
5. Centrifuge the plates at 209 × *g* for 10 min at RT.
6. Quantify nonadherent cells as described above (Fig. 3).

The V-well adhesion assay described can be easily adapted to the flat-well format (see Note 16).

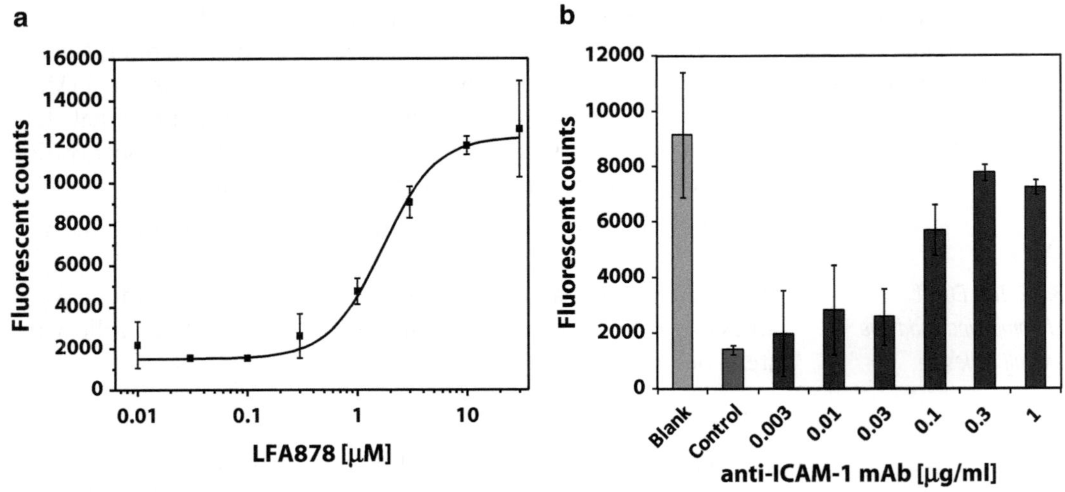

Fig. 3. Effect of inhibitors on PBMC binding to ICAM-1 in the V-well adhesion assay. The activity of the small molecule LFA-1 inhibitor LFA878 (**a**) and the anti-ICAM-1 mAb (**b**) was tested at indicated concentrations in the V-well assay, as described under Subheading 3. Each point or bar represents the mean value ±SD of triplicates. Blank, no ICAM-1; control: plus ICAM-1, no inhibitor. Both inhibitors blocked cell adhesion to immobilized ligand in a concentration-dependent manner.

Table 2
Properties of the PBMC/ICAM-1 V-well adhesion assay

Donor[a]	S/N	LFA878, IC_{50} (μM)
1	3.0; 3.6; 4.8; 4.6	1.7, 1.1
2	4.4; 3.5; 3.7; 4.1; 6.6	2.0, 0.8, 0.7
3	4.5	1.9
4	4	n.d.

Values from independent experiments are shown. The donor-to-donor variation of the S/N ratios or the IC_{50} values is low.
S/N signal-to-noise ratio of the assay calculated as described in Note 3, *LFA878* low molecular weight LFA-1 inhibitor, *n.d.* not determined;
[a]1–3: PBMCs isolated from buffy coats of different donors, 4: PBMCs isolated from a fresh blood sample

4. Notes

1. Human ICAM-1 m c-κ was cloned and expressed in the insect cell line Sf9 using a baculovirus expression system. The protein was purified using immobilized rat anti-mouse c-κ Ab. The protein is composed of the extracellular domain of ICAM-1 and the c-κ light chain of murine IgG. ICAM-1 m c-κ

can be replaced by commercially available human ICAM-1 (e.g., recombinant human sICAM-1 from R&D systems, ADP4-200) or purified tonsil ICAM-1 (6) (see Note 10).

2. BCECF-AM is intrinsically nonfluorescent until it is cleaved by intracellular esterases to BCECF, the fluorescent-free acid form of BCECF-AM. Due to negative charges, BCECF retains in cells for several hours. At −20°C, the stock solution of BCECF-AM is stable for at least 6 months. If the solution shows strong fluorescence and coloration, it should be discarded. BCECF-AM stock diluted in aqueous buffer or medium should be used immediately and not be stored. Avoid exposure of BCECF-AM to FCS, because endogenous esterases present in FCS may cleave BCECF-AM. Alternatively, similar dyes such as calcein-AM (Invitrogen), or membrane intercalating dyes such as carboxyfluorescein succinimidyl ester (CFSE, Sigma) can be used to label the cells.

3. The quality of the assay is influenced by the density of the Jurkat cell culture. If Jurkat cells were grown at high density ($>8 \times 10^5$ cells/ml) before use, a significant decrease in the signal-to-noise (S/N) ratio of the assay was observed. The S/N ratio of the assay was calculated according to the following formula:

$$S/N = \frac{\text{Fluorescence in the absence of ICAM - 1 (highest reading)}}{\text{Fluorescence in the presence of ICAM - 1 (lowest reading)}}$$

Thus, it is very important to ensure careful and consistent cell passage procedures.

4. PBMCs can also be isolated from fresh human blood samples. In brief, blood (10–50 ml) is obtained by simple venipuncture and collected into Na-Heparin tubes (B. Braun, 46613, 5,000 I.E./50 ml blood). The heparinized blood is diluted 1:2 in DPBS. The samples (25 ml) are loaded on the porous barrier of the Leucosep™ tube. After centrifugation, PBMCs are isolated as described. Normal human blood contains $\sim 2.5 \times 10^6$ PBMCs/ml.

5. The washing steps are required to remove most of the platelets from the PBMC fraction.

6. One buffy coat contains approximately 10^9 PBMCs. Approximately 60–70% of the mononuclear cells are lymphocytes. If the preparation contains an unusually high number of aggregated monocytes, it is filtered through a cell strainer to remove the aggregates.

7. DMSO can cause osmotic damage. Therefore, uninterrupted working is important during freezing and thawing.

8. The goat anti-mouse c-κ Ab is used as a catching Ab to ensure proper orientation and presentation of the ICAM-1 m c-κ. In the absence of the catching Ab (direct coating), the concentration of ICAM-1 m c-κ needs to be increased to 1–3 μg/ml (100–300 ng/well) to support optimal cell adhesion (see Note 10).
9. Blocked plates covered with a lid can be stored at 4°C up to 1 week.
10. If ICAM-1 m c-κ is replaced by other ICAM-1 preparations, it is recommended to optimize assay conditions experimentally by varying the concentrations of ICAM-1 as shown in Fig. 2. For example, recombinant human sICAM-1 (R&D systems, ADP4-050) directly coated onto the plates has been successfully used in our laboratory at a concentration of 10 μg/ml (1 μg/well). Commercially available human ICAM-1 Fc chimera preparations may be captured via anti-Fc antibody to reduce the amount of protein necessary to support adhesion.
11. Several LFA-1-expressing cell lines were tested for binding to immobilized ICAM-1 m c-κ in the V-well format (including HuT-78, HL-60, Molt-3). Jurkat cells bound to ICAM-1 m c-κ with the best S/N ratio and were selected for the assay.
12. Fluorescently labeled Jurkat cells can be cryopreserved until use: after the labeling step and addition of DPBS, the cells are centrifuged at 300 × *g* and resuspended at 1×10^7 cells/ml in RPMI-1640 containing 10% DMSO and 20% FCS. The cells are frozen in aliquots of 0.5 ml at −80°C overnight and then stored in a liquid nitrogen freezer for not longer than 2 months. Prior use, the cells are rapidly thawed and then resuspended in 1 ml pre-warmed RPMI-1640 (37°C), followed by the addition of another 1 ml medium (see Note 7). This cell suspension is transferred to 8 ml RPMI-1640 medium and centrifuged. The cells are resuspended in binding buffer and activated as described. This cryopreservation step enables the rapid preparation of the cells on the day of the experiment.
13. Mn^{2+}/Mg^{2+} are used in the present protocol to induce LFA-1-dependent cell adhesion. In the absence of divalent cations, leukocytes do not bind to ICAM-1. Note that Ca^{2+} can be a negative regulator of the cation-induced adhesion (11). Leukocyte adhesion via LFA-1 can also be induced by other stimuli, for example, PMA (6, 7). However, if using a different stimulus, the assay needs to be optimized regarding the cell type, cell number, centrifugal force, temperature, and timing of the centrifugation step.

14. The centrifugal force applied in the assay is very critical for the S/N ratio. At 209×*g*, the S/N ratio of the PBMC/ICAM-1 m c-κ assay is 4.25 (mean value of 11 independent experiments) (Table 2). At higher centrifugal force, the S/N ratio starts to decrease. Dependent on the adhesion molecule pair studied, the optimal centrifugal force varies from 75×*g* to 1,820×*g* (5).
15. The assay as described (using divalent cations to stimulate adhesion) tolerates DMSO up to 1%, e.g., the addition of solvent up to 1% causes no detectable inhibition of Jurkat cell or PBMC binding to immobilized ICAM-1 m c-κ. Ethanol and methanol are also tolerated up to 1%. Higher concentrations of the solvents result in a declining fluorescence signal in the presence and absence of immobilized ICAM-1. The reduced signal is probably due to cell lysis leading to the release and dilution of BCECF (5).
16. Alternate protocol using flat-well microtiter plates: Coat flat-well microtiter plates with goat anti-mouse c-κ mAb, block with BSA, and add ICAM-1 m c-k as described. Wash the plates and add labeled and cation-activated Jurkat cells (100,000 cells/well). Incubate the cells for 30 min at 37°C to allow the cells to adhere. Separate non-adherent from adherent cells by four to eight washing steps (use binding buffer for these steps) and quantify fluorescence as described (12).

Acknowledgments

The authors would like to thank Dr. Karl Welzenbach, Novartis, for critical reading of the manuscript.

References

1. Ley, K., Laudanna, C., Cybulsky, M.I., and Nourshargh, S. (2007) Getting to the site of inflammation: the leukocyte adhesion cascade updated. *Nat. Rev. Immunol.* 7, 678–689.
2. Butler, L.M., McGettrick H.M., and Nash G.B. (2009) Static and dynamic assays of cell adhesion relevant to the vasculature. *Methods Mol. Biol.* **467**, 211–228.
3. Mobley, J.L., and Shimizu Y. (2001) Measurement of cellular adhesion under static conditions. *Curr. Protoc. Immunol.* Chapter 7: Unit 7.28
4. Humphries, M.J. (2009) Cell adhesion assays. *Methods Mol. Biol.* **522**, 203–210.
5. Weetall, M., Hugo, R., Friedman, C., Maida, S., West, S., Wattanasin, S., Bouhel, R., Weitz-Schmidt, G., and Lake, P. (2001) A homogeneous fluorometric assay for measuring cell adhesion to immobilized ligand using V-well microtiter plates. *Anal. Biochem.* **293**, 277–287.
6. Kim, M., Carman, C.V., and Springer, T.A. (2003) Bidirectional transmembrane signaling by cytoplasmic domain separation in integrins. *Science* **301**, 1720–5.
7. Kim, M, Carman, C.V., Yan, W., Salas, A., and Springer, T.A. (2004) The primacy of affinity

over clustering in regulation of adhesiveness of the integrin αLβ2. *J. Cell. Biol.* **167,** 1241–1253.

8. Etzioni, A. (2010) Defects in the leukocyte adhesion cascade. *Clinic. Rev. Allergy. Immunol.* **38,** 54–60.
9. Evans, R., Patzak, I., Svensson, L., De Filippo, K., Jones, K., McDowall, A., and Hogg, N. (2009) Integrins in immunity. *J. Cell Sci.* **122,** 215–225.
10. Weitz-Schmidt, G., Welzenbach, K., Dawson, J., and Kallen, J. (2004) Improved lymphocyte function-associated antigen-1 (LFA-1) inhibition by statin derivatives: molecular basis determined by x-ray analysis and monitoring of LFA-1 conformational changes in vitro and ex vivo. *J. Biol. Chem.* **279,** 46764–46771.
11. Dransfield, I., Cabañas, C., Craig, A., Hogg, N. (1992) Divalent cation regulation of the function of the leukocyte integrin LFA-1. *J. Cell Biol.* **116,** 219–226.
12. Kallen, J., Welzenbach, K., Ramage, P., Geyl, D., Kriwacki, R., Legge, G., Cottens, S., Weitz-Schmidt, G., and Hommel, U. (1999) Structural basis for LFA-1 inhibition upon lovastatin binding to the CD11a I-domain. *J. Mol. Biol.* **292,** 1–9.

Chapter 3

Real-Time Analysis of Integrin-Dependent Transendothelial Migration and Integrin-Independent Interstitial Motility of Leukocytes

Ziv Shulman and Ronen Alon

Abstract

The role of integrins in leukocyte migration across endothelial barriers is widely accepted. In contrast, the contribution of integrins to interstitial motility of leukocytes is still elusive. Chemokine binding to G-protein-coupled receptors expressed on the surface of leukocytes plays key roles in both of these processes by directly activating integrin conformations favorable for ligand binding and integrin micro-clustering. Chemokines can also serve as weak adhesive ligands and potent inducers of actin cytoskeleton remodeling. Real-time assays utilizing live imaging microscopy have been implemented to dissect these versatile roles of chemokines in different leukocyte migration processes. Here, we review several in vitro assays useful for exploring the contribution of chemokine signals and shear forces to integrin activation and function during various stages of leukocyte transendothelial migration. In addition, we describe a new assay that assesses the contribution of chemokines to integrin-independent interstitial leukocyte motility. These assays can also follow the outcome of specific genetic or biochemical manipulations of either the leukocyte or the endothelial barrier on distinct migratory steps. Following fixation, subcellular changes in the distribution of integrin subsets and of specific integrin-associated adaptors can be further dissected by immunofluorescence tools and by ultrastructural electron microscopic analysis.

Key words: Chemokines, G-protein-coupled receptors, Migration, Endothelium, Inflammation, Trafficking

1. Introduction

The ability of specific leukocyte integrins to generate and maintain resistance to detachment by disruptive shear forces is instrumental for leukocyte crawling on and crossing through endothelial barriers (1–4). As these integrins are largely nonadhesive, leukocytes must respond to signals from endothelial chemotactic cytokines, termed chemokines, which transduce sequential and highly dynamic changes

Motomu Shimaoka (ed.), *Integrin and Cell Adhesion Molecules: Methods and Protocols*, Methods in Molecular Biology, vol. 757, DOI 10.1007/978-1-61779-166-6_3, © Springer Science+Business Media, LLC 2011 in integrin conformation and clustering critical for firm leukocyte adhesion and crawling on endothelial cells (5). These chemokines also activate a variety of actin remodeling Rho family GTPases in leukocytes (6–8), and thereby link actin remodeling events to dynamic integrin adhesion (8, 9). Recent findings also suggest that leukocytes navigating on stromal networks and collagen fibers use similar chemokine-driven actin remodeling programs to locomote rapidly along these tissue components, but do so in an integrin-independent manner (10, 11).

Despite the increasing evidence highlighting the role of integrin adhesiveness in transendothelial leukocyte migration rather than in interstitial motility, this dichotomy in integrin usage has been difficult to dissect by conventional macroscopic assays (12). Over the last few years, real-time *in vitro* imaging assays were optimized for transendothelial migration (TEM) studies and for monitoring interstitial leukocyte motility (13–15). These assays have made use of both phase contrast and fluorescence microscopy to track, at a single cell level, the ability of leukocytes to migrate over and through activated endothelial monolayers under physiological conditions of shear flow (Fig. 1).

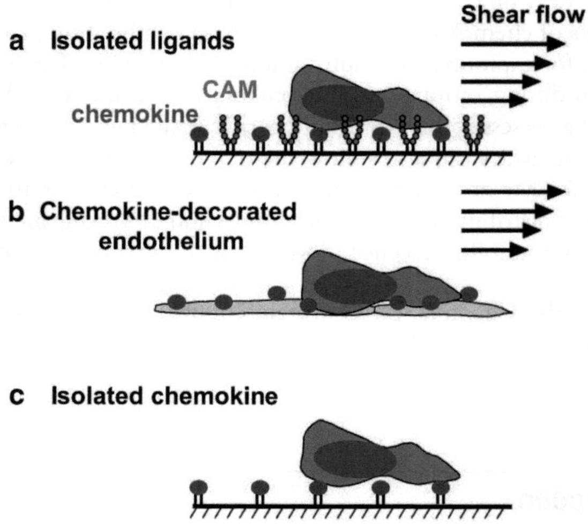

Fig. 1. Adhesive and chemotactic model surfaces used for single cell analysis in the different experimental configurations described in Subheading 3. (**a**) A leukocyte spreading and crawling under shear flow over isolated endothelial integrin ligands (CAM: ICAM-1 or VCAM-1) and chemokines. The ligands contain an Fc tag and are immobilized in a uniform configuration on the substrate via a protein A anchor (not shown). The integrin ligands are co-immobilized with integrin stimulatory chemokines of interest that activate the leukocyte integrins and facilitate actin remodeling in the adherent leukocyte, promoting leukocyte crawling and resistance to detachment by shear forces. (**b**) A leukocyte spreading and crawling under shear flow over cytokine-activated endothelium (expressing endogenous ICAM-1 and VCAM-1) decorated with integrin stimulatory chemokines of interest. (**c**) A leukocyte locomoting over surface-presented chemokine. Note that in this configuration, unlike under shear flow, the chemokine functions both as a weak adhesive ligand and as a signaling molecule. The substrate can be co-coated with additional extracellular matrix components and chemokine-presenting glycosaminoglycan scaffolds.

Shear forces were recognized as a major element in rapid integrin activation by endothelial chemokines (16). New methodologies to assess integrin activation and function in leukocytes interacting with specific isolated integrin ligands reconstituted with chemokines of interest were, therefore, introduced in order to monitor integrin function independently of the contribution of active endothelial machineries (Fig. 1) (1, 17, 18). These cell-free assays allow the researcher to dissect the role of surface-presented chemokines versus that of soluble chemokines in the dynamic triggering of integrin adhesiveness and microclustering during leukocyte adhesion and crawling under disruptive shear forces. These assays are also useful in addressing how surface immobilized chemokines on their own and together with extracellular matrix integrin ligands can promote different modalities of integrin-independent interstitial (i.e., extravascular) motility of specific subsets of leukocytes (Fig. 1c) (16, 17, 19, 20).

2. Materials

2.1. Integrin-Dependent Lymphocyte Crawling on Purified Endothelial Ligands Under Shear Flow

1. Glass-bottom dishes: Polystyrene Petri dishes, 60 × 15 mm (Falcon, Franklin Lakes, NJ), are prepared with a 37-mm diameter hole (custom made). A microscope cover glass (#1), 45-mm in diameter (Marienfeld GMBH & Co., Germany), is glued to the bottom of the dish with silicone glue. To improve protein adsorption to the plate surface, the glass dishes are treated with hexamethyldisilazane for approximately 10 min, until it is fully evaporated.
2. Protein A: 1 mg/ml in PBS and stored at −20°C.
3. Chemokines CXCL12 and CCL21 (R&D Systems Inc., Minneapolis, MN): 100 μg/ml in PBS containing 0.1% (w/v) human serum albumin (HSA, Sigma–Aldrich) and stored at −20°C.
4. ICAM-1-Fc or VCAM-1-Fc fusion proteins (R&D systems): 1 mg/ml in PBS containing 0.1% HSA and stored at −20°C.
5. Binding medium: cation-free Hank's balanced solution supplemented with 10 mM HEPES, pH 7.4, 2% (w/v) bovine serum albumin (BSA), and 1 mM each of Ca^{2+} and Mg^{2+}.
6. An inverted microscope: An instrument (DeltaVision System, Delta Vision Spectris RT, Applied Precisions, Issaquah, WA) equipped with 20× PlanApo differential interference contrast (DIC) or phase contrast objectives, and a motorized stage that allows recordings of multiple microscopic fields of view with high spatial precision.

7. A flow chamber: a parallel plate flow chamber kit (GlycoTech, Rockville, MD, gasket thickness 0.01 in., channel width 2.5 mm with 1/16 in. intake and output tubes) connected to an automated syringe pump (Harvard Apparatus, Natick, MA) and a vacuum pump (MasterFlex, Cole-Palmer Instrument Co., Niles, IL). For details on this apparatus, please refer to Fig. 1 in ref. 21. Alternative flow settings can also be used (e.g., that of Ibidi, Munchen, Germany).

2.2. Integrin-Dependent Lymphocyte Crawling on and Transendothelial Migration Through Activated Endothelial Cells

1. Glass-bottom dishes: as described in Subheading 2.1.
2. Human umbilical vein endothelial cells (HUVECs) or microvascular dermal endothelial cells (HMVDEC) (Heidelberg, Germany).
3. HUVEC culture medium: M-199 supplemented with 10% LPS-free fetal calf serum, penicillin (100 ng/ml), streptomycin (100 U/ml), endothelial mitogen (Biomedical Technologies Inc, Stoughton, MA), and porcine heparin (Sigma–Aldrich, 5 U/ml).
4. Fibronectin (FN, Sigma–Aldrich): 1 mg/ml in PBS and stored at –20°C.
5. Trypsin-EDTA solution (Sigma–Aldrich): stored at –20°C.
6. TNF-α and chemokines (CXCL12 and CCL19, R&D Systems): prepared as stock solutions of 1 mg/ml and 0.1 mg/ml, respectively, in PBS containing 0.1% HSA, and kept at –20°C.

2.3. Integrin-Independent Lymphocyte Motility on Surface Chemokines

1. μ-Slides VI (Munchen, Germany).
2. Chemokines CCL21 and CXCL12: prepared as described in Subheading 2.2. The working solution is prepared in PBS at a final concentration of 0.5–2 μg/ml.
3. Cell staining solution: BCECF-AM (2′,7′-bis(carboxyethyl)-4(or 5)-carboxyfluorescein diacetoxymethyl ester, Invitrogen, Carlsbad, CA (16)), in PBS at a concentration of 1 μM.

2.4. Analysis of Integrin Redistribution in Actively Migrating Lymphocytes

1. Paraformaldehyde (PFA): 4% (w/v) in PBS, supplemented with 2% (w/v) sucrose.
2. Blocking solution: serum (specific to the species the mAb was obtained from) in PBS to 20% (v/v).
3. Saponin (Sigma–Aldrich): 0.1% (w/v) in PBS.
4. Integrin antibodies: For staining of LFA-1, the non-blocking anti-a_L mAb, clone TS2.4, should be used. For staining of VLA-4, the non-blocking anti-α_4 mAb, clone G5B10, is optimal (1, 22).
5. Mounting medium (Dako, Carpinteria, CA, USA).
6. An inverted fluorescence microscope equipped with 60× DIC objective (e.g., DeltaVision system).

3. Methods

3.1. Integrin-Dependent Lymphocyte Crawling on Purified Endothelial Ligands Under Shear Flow

This methodology (Fig. 1a) allows assessing integrin function in leukocytes interacting with specific isolated integrin ligands and chemokines independently of active endothelial machineries. It also allows the introduction of reversible blockers of leukocyte machineries which cannot be used in the presence of endothelial monolayers due to their toxic effects on endothelial cell integrity. Some inhibitors exert dual effects on both the endothelial cell and the leukocyte studied, and this approach allows dissecting the net effects of a given inhibitor on integrin activation and distribution on the leukocyte side.

1. Prepare 20 μg/ml Protein A in PBS containing 20 μM sodium bicarbonate (pH 8.5) and add a 5-ml drop in the middle of the glass dish. Incubate the dishes for 2 h at 37°C under humidified conditions to avoid evaporation. Mark the location of the drop on the outer side of the dish with a dot or a circle.
2. Gently remove the drop of Protein A solution and wash by adding a 5-μl drop of PBS. Repeat this washing procedure three times. Avoid dehydration of the drop.
3. Prepare chemokine solution in PBS (2 μg/ml). Remove the final PBS drop from the dish and add 7 ml of chemokine solution. Incubate the dishes at 4°C for at least 3 h.
4. Prepare the desired integrin ligand in PBS containing 2% HSA. For resting T cells derived from peripheral blood, optimal ligand concentrations vary between 1 and 5 μg/ml (ICAM-1-Fc) and 0.05–2 μg/ml (VCAM-1-Fc).
5. Remove the drop containing chemokine from the dishes, and immediately add 30 ml of integrin ligand-Fc solution. Incubate the dishes for 16–24 h at 4°C.
6. Place the lower part of a test dish on the stage of an inverted microscope equipped with a 10× or 20× phase contrast or DIC objectives. The microscope should be equipped with a controlled temperature chamber or a heated stage set at 37°C. Place the flow chamber in the dish.
7. To assemble the flow chamber apparatus, attach the inlet tube to the inlet hole of the chamber and immerse in a 50-ml reservoir tube filled with the binding medium. Attach the outlet tube to a 3-way Luer-lock that is connected to the automated syringe pump. Connect a 10-ml disposable syringe to the third outlet of the lock. The test dish (uncoated) must be assembled as the lower plate of the flow chamber. Connect the vacuum pump to the vacuum outlet of the chamber and pump medium throughout the system. Ensure that the vacuum tightly seals the chamber to the dish (for details on this setting, please refer

to Fig. 1 in ref. 21). Use the disposable syringe to pump the binding medium manually through the system. Disconnect the vacuum and remove the test dish. The system is now ready for use.

8. Add 3 ml binding medium to the ligand-coated dish and place it on the microscope stage. Locate the dot containing the coated integrin ligand and chemokine on the dish. Place the flow chamber directly over this dot and connect the chamber to the vacuum pump. Make sure that the chamber is tightly sealed to the dish.
9. Choose three to five fields of view within the coated spot.
10. Place 0.5×10^6 fresh or cultured lymphocytes in a 1.5-ml tube and mix with an equal volume of EDTA solution to remove cell-bound integrin ligands. Wash the cells by centrifugation ($200 \times g$, 4 min) and resuspend pellet in 50 ml binding medium.
11. Perfuse the lymphocyte suspension until the entire volume enters the inlet tube. Return the inlet to the reservoir containing 50 ml binding medium. Ensure that no air bubbles enter the inlet tube during this manipulation. Set the automated pump to provide a constant shear force of 5 dyn/cm^2 until the cells enter the chamber.
12. Once the cell flux enters the field of view, stop the pump and immediately start the flow program of the pump (see Note 1). Record lymphocyte spreading and crawling at 1 frame per 10 s.
13. Analysis of lymphocyte crawling can be preformed either manually or by using a computerized tracking software (e.g., SoftWoRx (Applied Precisions) or Volocity (Improvision, Waltham, MA)). We define a crawling cell as a polarized cell that has moved at least 30 μm during the 8-min period of the assay. Round adherent cells which passively move with the flow direction are not counted as crawling cells.

3.2. Integrin-Dependent Lymphocyte Crawling on and Transendothelial Migration Through Activated Endothelial Cells

1. Wipe glass-bottom dishes with ethanol and air dry under sterile conditions.
2. Add a 15-μl drop of FN to the center of the dish and incubate at 37°C for 1 h. Wash three times with 15 μl PBS (sterile). Avoid any dehydration of the FN drop.
3. Harvest HUVECs from a long-term tissue culture plate by washing with 3 ml Ca^{2+}- and Mg^{2+}-free PBS. Remove the PBS and add warm trypsin-EDTA solution. Incubate for 2 min. Add 5 ml HUVEC medium and collect cells by gentle pipetting. Wash the cells in HUVEC medium ($200 \times g$, 4 min).
4. Resuspend the pellet in HUVEC medium, and adjust volume to reach a final concentration of 3×10^6 cells/ml. Add a 15-ml

drop of HUVECs to the FN-spotted glass dishes and incubate for 1 h in a humidified incubator. This procedure is designed to minimize the number of HUVECs required to form a monolayer. Furthermore, the flow chamber attaches more tightly to dish areas free of HUVECs and maintains a tighter vacuum seal.

5. Add 3 ml of HUVEC medium containing 2 ng/ml TNF-α (alone or in combination with other inflammatory cytokines such as IFNγ) and incubate at 37°C in a humidified 5% CO_2 atmosphere for 18–26 h.
6. Use the same flow settings described in Subheading 3.1. Place the HUVEC-coated dish on the inverted microscope (see Note 2) and locate three to five fields of view on the monolayer. The fields should comprise a confluent layer of HUVECs and be positioned near the upstream edge of the monolayer to minimize leukocyte rolling or crawling into the field of view from non-recorded upstream fields. Place the flow chamber directly over the monolayer. Connect the chamber to the vacuum pump. Make sure that the chamber is tightly sealed to the dish and no air bubbles are introduced through the chamber, since any air perfused over the HUVEC monolayer will irreversibly damage its viability or functionality.
7. Wash the monolayer with binding medium. If required, adsorb a chemokine of interest onto the HUVEC monolayer (see Note 3). Briefly, perfuse 100 ml of chemokine-containing binding medium using a disposable syringe. The entire volume of medium should enter the chamber, thereby keeping air out. Allow the chemokine to remain on the HUVEC monolayer for at least 5 min. Collect the unbound chemokine by back-pumping the chemokine-containing medium in the reverse direction with the disposable syringe. Tissue-expanded effector T cells and freshly isolated blood neutrophils do not require addition of exogenous chemokines to the TNF-α-stimulated HUVECs (see Note 3). Return the inlet tube to the binding medium reservoir and wash the system extensively (with at least 1 ml) in order to remove any traces of unbound chemokine.
8. Wash 1×10^6 cultured lymphocytes by mixing with an equal volume of EDTA solution to remove cell-bound integrin ligands. Centrifuge the cells ($200 \times g$, 4 min) and resuspend pellet in 50 μl binding medium. Perfuse the lymphocyte suspension until the entire volume enters the inlet tube. Return the inlet to the reservoir containing 50 ml of binding medium. Ensure that no air bubbles enter the inlet tube during this manipulation.
9. Once the lymphocyte flux enters the field of view, stop the pump and immediately activate the flow program (see Note 4).

Set the automated pump to provide a constant shear flow of 2–5 dyn/cm^2 for the required time period. Record at 1 frame per 15 s (migration phase).

10. Analysis of lymphocyte crawling and TEM: motion analysis should be performed manually on all cells interacting with the endothelial monolayer in at least three fields of view. Lymphocytes are individually tracked from their site of interaction with the endothelial surface at the end of the accumulation phase and throughout their migration phase. Only leukocytes that have accumulated in the field of view during the accumulation phase are analyzed. Lymphocytes rolling or crawling into the field of view from upstream fields or out of the field to downstream fields, as well as lymphocytes captured onto the endothelial cells (ECs) during the migration phase, should not be included in the analysis. Four distinct categories of accumulating lymphocytes are generally defined in this type of analysis: (1) Lymphocytes that roll away or detach from the EC monolayer during the migration phase are considered "detached." (2) Lymphocytes that remain stationary throughout the migration phase, or crawl a distance that is less than their diameter, are considered as "resisting detachment." (3) Lymphocytes that spread and migrate over the EC surface throughout the assay period without crossing the EC barrier are considered "crawling." (4) Lymphocytes that migrate for variable distances on the EC and eventually transmigrate (cross) through the monolayer are considered "crawling and transmigrating." Since lymphocytes (and other leukocytes) may turn dark and can be falsely counted as transmigrating cells, only lymphocytes that undergo stepwise darkening of their leading edge and retain their dark appearance while crawling underneath the EC should be considered transmigrating cells. The different categories are either presented as a percentage of the originally accumulated lymphocytes or normalized to the number of cells interacting with the endothelial cells during the first accumulation phase.

3.3. Integrin-Independent Lymphocyte Motility on Surface Chemokines

This method allows assessing how specific surface-presented chemokines can promote random (chemokinetic) leukocyte motility in the interstitium on their own and in the context of specific extracellular and/or stromal ligands.

1. Inject into a μ-slide slot 30 μl PBS solution containing 0.5–2 mg/ml chemokine and incubate for 1 h at 37°C or overnight at 4°C (see Notes 5 and 6).
2. Apply a clean absorbent paper to the edge of the μ-slide and absorb the chemokine solution. To avoid dehydration, continue immediately to the next step.
3. Wash the μ-slide by injecting 150 μl PBS and absorb the fluid with an absorbent paper wipe.

4. Extracellular matrix (ECM) components such as fibronectin (5 mg/ml) and collagen (5 μg/ml) can be sequentially co-immobilized after the primary chemokine coating step by injecting 30 μl of ECM solution into the chemokine coated μ-slide and incubating for 30 min in room temperature. Alternatively, these or other ECM components can be first coated and serve as scaffolds for specific chemokines (e.g., fibronectin for CXCL12; collagen IV for CCL21).
5. Block the μ-slide by injecting 30 μl of 2% HSA solution and incubating for 30 min at room temperature.
6. Wash the μ-slide by injecting 150 μl PBS. Do not remove the PBS fluid from the μ-slide until use. The slides can be kept at 4°C for up to 24 h.
7. Wash the cells taken from culture with 1:1 volume of EDTA solution.
8. Resuspend the cells in 100 ml of Ca^{2+}- and Mg^{2+}-free PBS containing 1 μM BCECF-AM. Incubate at room temperature in the dark for 2 min.
9. Wash the cells with 200 μl binding medium and resuspend with binding medium at the desired volume (30 μl per run).
10. Remove PBS from the μ-slide with an absorbent paper. Inject 30 μl cell solution in into the μ-slide and immediately mount it on the stage of an inverted microscope equipped with 20× or 40× DIC objective and maintained at 37°C (see Note 7). Typically, 0.3×10^6 cells per slot are sufficient for imaging.
11. Allow the cells to settle at the bottom of the slide by gravitation and to thereby come in contact with the substrate. Meanwhile, mark three to five fields of view using the motorized stage of the microscope. Track the cells by recording both DIC and fluorescence images at four frames per min for 10–30 min (see Note 8). The slide configuration minimizes any fluid convection or vortexing, and thereby maintains a shear-free environment.
12. Analysis of lymphocyte motility over the substrate is performed in multiple fields of view similar to that in the analysis of lymphocyte crawling on integrin ligand-coated surfaces under shear flow (Subheading 3.1).

3.4. Analysis of Integrin Redistribution in Actively Migrating Leukocytes

1. Actively migrating cells must be fixed in situ in the flow chamber before staining. Remove the inlet tube of the flow chamber from the binding medium reservoir to a tube containing 4% PFA and 2% sucrose in PBS kept at 37°C. Once the perfused fixative enters the chamber, the fixative must be injected for an additional 5-min period. Disconnect the chamber from the vacuum source and carefully lift it from the glass dish attached at its bottom. If a μ-slide setting is used for shear-free motility

experiments, gently inject into it 30 μl PFA and incubate the slide for 5 min.

2. Wash the dish three times with 3 ml PBS. The fixed sample on the glass slide is now ready for further processing.
3. Wash the chamber extensively to remove all traces of fixative. The fixed dishes should be kept at 4°C until further use.
4. Remove PBS from the dish and mark a circle around the coated dot with a liquid blocker Pap pen (a water-repellent marking pen, Daido Sangyo Co. (Tokyo, Japan)). This procedure greatly reduces the volumes of the solutions subsequently required to stain the fixed sample.
5. Block the samples with 50 μl of PBS supplemented with appropriate serum (20%, see Note 9) for 20 min at 37°C.
6. Remove the serum without washing and incubate the samples for 45 min at 37°C with 30–50 μl of PBS containing either unlabeled or directly labeled primary antibodies (see Note 10).
7. Wash the cells with 3 ml PBS for 5 min. This must be repeated at least three times.
8. If relevant, incubate the cells with a secondary antibody for an additional 30 min at room temperature.
9. Wash the cells in PBS five times for 5 min.
10. For co-staining of intracellular molecules together with cell surface integrins, permeabilize the fixed cells with 0.1% saponin for 5 min. All of the solutions subsequently used must contain 0.1% saponin (see Note 11). After permeabilization, block the cells with 10% serum in PBS/0.1% saponin and incubate the cells with primary and secondary antibodies, as described above.
11. Wash the cells five times with 3 ml PBS/saponin for 5 min, followed by two washes with saponin-free PBS. Add 20% serum in PBS for 10 min in order to quench both saponin activity and the nonspecific binding of secondary antibodies.
12. Image fixed cells with either confocal or wide-field fluorescence microscopy. In order to detect microclustering of integrins, use an oil 60×/1.4 PlanApo (DIC) objective (see Notes 12 and 13).

4. Notes

1. In order to avoid a biased accumulation of highly adhesive superactivated leukocyte subsets on the chemokine–integrin ligand-coated substrate, the automated pump program is set

to provide a low shear stress of 0.5 dyn/cm^2 for 1 min to allow efficient attachment of cells even in the absence of selectins or selectin ligands on the substrate followed by a mid-range physiological shear stress (typically 5 dyn/cm^2) for an additional 8–10-min period. The inclusion of these primary adhesion molecules would allow the introduction of the studied leukocytes in higher flow rates due to their higher efficiencies of leukocyte capture under flow.

2. We recommend using a 20× phase contrast objective for tracking leukocyte crawling and TEM in real time. This magnification allows accurate monitoring of both the motion and morphological changes of transmigrating leukocytes (30–50 cells in a typical field of view). DIC objectives can also be used, but TEM analysis is more difficult. It is, therefore, recommended to pre-label the leukocytes with a fluorescent dye (such as in Subheading 3.3) when DIC microscopy is used.
3. Lymphocyte TEM not only depends on the type of chemokine added, but also on the chemokine dose. We found out that low levels (less than 0.1 mg/ml) of CXCL12, sufficient to trigger rapid integrin-mediated adhesion strengthening and crawling, are insufficient to trigger invasive filopodia and subsequent TEM (1, 22). Different chemokines may induce different degrees of invasive filopodia and TEM due to differences in the efficacy of their presentation on the endothelial surface. Interestingly, effector T cells and neutrophils do not require exogenous chemokines, as their activated integrins and endogenous endothelial chemokines are sufficient to trigger transmigration through TNF-α and other cytokine-stimulated HUVECs.
4. The automated pump program used to study lymphocyte TEM should be set to provide a low shear stress of 0.75 dyn/cm^2 for 1 min (accumulation phase) and then a mid-range physiological shear stress of 2–5 dyn/cm^2 for an additional 15-min period. The geometry of the flow chamber does not support lymphocyte accumulation on adhesive endothelial surfaces at this high shear stress; however, once accumulated, lymphocytes remain adhesive and shear resistant for prolonged periods, at a very broad range of shear stresses (5–20 dyn/cm^2). The short accumulation phase is also required to synchronize the leukocyte arrest, such that the majority of the postarrest events and the subsequent crawling and TEM take place within a narrow time window (typically within 4–7 min following initial accumulation of resting lymphocytes, and 2–5 min following accumulation of effector lymphocytes and neutrophils). In addition, using low shear flow, one avoids selection of a small fraction of highly adhesive leukocyte subsets within heterogeneous populations.

5. Each slot in the μ-slide coated with chemokine is used for a single run. Coat one slot in each μ-slide and keep the unused μ-slides at 4°C. It is not recommended to coat unused slots in the same μ-slide ahead of experiments, since their exposure to 37°C before use may inactivate any coated chemokines or integrin ligands.
6. The chemokine and the working concentration should be chosen according to the cell type investigated. Typically, human peripheral blood lymphocytes (PBL) are highly responsive to CXCL12 (up to 90% polarization and motility (19)). An example of the effect of surface-presented CXCL12 on lymphocyte polarization and motility is shown in Fig. 2. PBL are also highly responsive to CCL21 (up to 70% polarization and motility (16)). Murine T splenocytes are moderately responsive to CXCL12, but are highly responsive to CCL21 (up to 90% 17). Neutrophils derived from peripheral blood are highly responsive to surface bound IL-8 (Feigelson and Alon, unpublished). Whenever the Mac-1 integrin of these isolated cells gets highly activated and spontaneously binds the albumin-coated substrate, a medium containing Ca^{2+} but no Mg^{2+} should be used in order to reduce the contribution of this activated Mac-1 to IL-8-promoted neutrophil motility.
7. For imaging a large number of motile cells, it is recommended to use either phase contrast or 20× DIC objectives. For imaging of cell shape and small structures, a 40× DIC objective should be used.
8. If BCECF-AM or another fluorescent marker is used, set the fluorescence signal to the minimal level that enables detection (i.e., 1.5-fold above background). Exposure to strong excitation light is partially phototoxic and perturbs the degree and rate of motility.

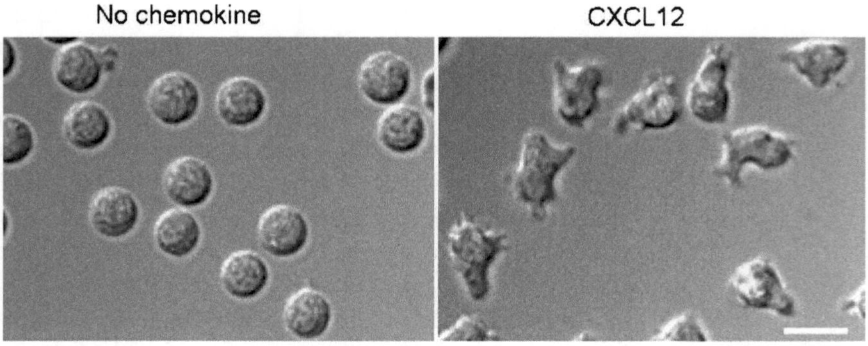

Fig. 2. Surface-presented CXCL12 supports lymphocyte adhesion, spreading, and motility. Human PBL were injected into μ-slide pre-coated with CXCL12 (2 μg/ml), and both their shape and position were monitored in real time. Lymphocytes that came into contact with the CXCL12-coated surface under these shear-free conditions immediately polarized and started to crawl on the isolated chemokine at an average velocity of 10 μm/min (*right*). Lymphocytes settled on an identically non-coated surface (*left*) or coated with a non-signaling CXCL12 mutant remained round (not shown). Scale bar = 10 μm.

9. Cells fixed on glass coated with Protein A should be treated with human serum or with human IgG in order to block any potentially unoccupied Protein A sites, which might adsorb fluorescent primary and secondary antibodies.
10. To visualize functional LFA-1 or VLA-4 subsets occupied by endothelial ICAM-1 or VCAM-1, standard integrin-blocking mAbs cannot be used for fluorescence imaging since their binding sites are already occupied by their ligand. The anti-α_L, TS2.4, mAb (23) and the anti-α_4, B5G10 (24), mAb are both non-blocking mAbs that efficiently stain both ligand-free and ligand-occupied integrins at high efficiency. Non-blocking mAbs that recognize activation of neoepitopes associated with high-affinity integrin conformational states, and which tolerate mild fixation procedures can also be used (1, 25).
11. Permeabilization of cells with Triton X-100 severely damages the outer cell membrane. Saponin is a reversible cholesterol chelator that disrupts cellular membranes more gently and is preferred over Triton X-100. In order to minimize disruption of membrane proteins and to avoid nonspecific mAb staining of cytosolic components, cells should first be permeabilized, stained for the cytoplasmic molecules of interest, subsequently washed, and immediately incubated in serum to remove the saponin. Membrane integrity is largely restored after saponin removal. Under these conditions, integrins and their membrane or cortical cytoskeletal assemblies are generally retained in a native configuration.
12. DIC microscopy allows high-quality imaging of morphological changes in the leukocyte shape (such as leading edge and trailing edge and the nucleus), and thus, is preferred over regular phase contrast microscopy. Combined with fluorescence microscopy, images can be acquired as Z-stack sections (4–6-μm thick and 0.2 μm apart). In order to reduce background noise and enhance the fluorescence signal, the Z-stack sections should be subjected to digital deconvolution. For example, we use the SoftWoRx software (Applied Precision) for deconvolution processing. An example of integrin distribution detected by fluorescence microcopy using such deconvolution processing is shown in Fig. 3.
13. Fixed samples can be processed for transmission electron microscopy (EM) or scanning EM, as described (23, 26, 27). Whereas scanning EM is suitable for analysis of large fields, transmission EM is a more complex method designed for individual cell analysis. It requires sampling numerous thin sections cut through individual lymphocytes and/or endothelial cells. The major advantage offered by transmission EM is the ability to probe the direct interface between two adherent cells (e.g., the bottom of a lymphocyte crawling on an endothelial cell). Another advantage is the ability to combine

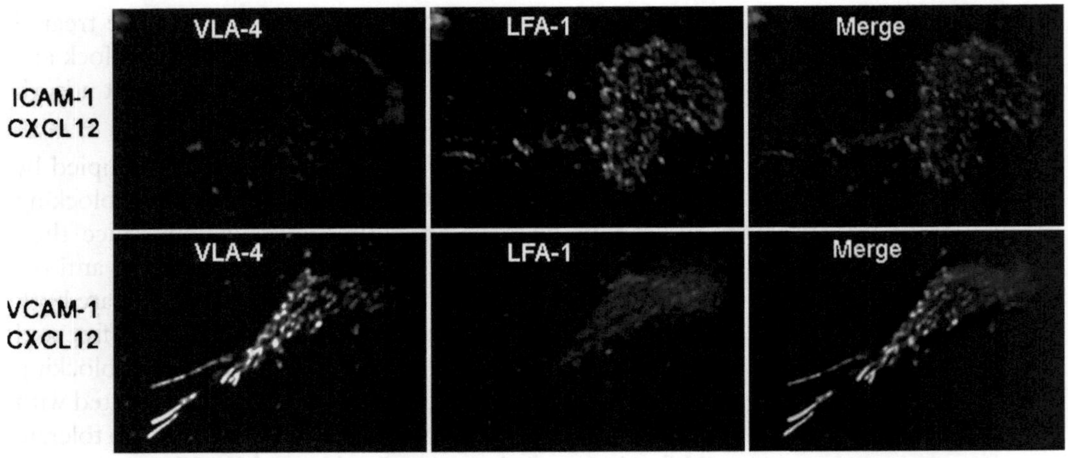

Fig. 3. The distribution of LFA-1 and VLA-4 in lymphocytes crawling on either ICAM-1 or VCAM-1 under shear flow. Lymphocytes (PBL) were accumulated for 1 min under low shear flow on a dish containing either ICAM-1-Fc (5 μg/ml) or VCAM-1-Fc (2 μg/ml), and overlaid on protein A co-immobilized with CXCL12 (2 μg/ml). Adherent T cells were then subjected to high shear flow (5 dyn/cm^2) for 7 min, and the dish was fixed in situ by perfusion of PFA into the flow chamber and then removed for further processing. Samples were stained with both anti-LFA-1 (TS2.4) and anti-α4 (G5B10) mAbs and analyzed by fluorescence microscopy. α4 integrin dots were not detected in lymphocytes crawling on ICAM-1, while T cells crawling on VCAM-1 generated α4 dots and microclusters mainly at their rear. Numerous dots of LFA-1 were detected at the ventral lymphocyte surface during active crawling on ICAM-1. In lymphocytes crawling on VCAM-1, LFA-1 was enriched in macroclusters at the leading edge. Scale bar = 6 μm.

ultrastructural analysis of cellular membranes with immunolabeling of specific molecules of interest (e.g., by immunogold labeling). Scanning EM imaging of large fields of view is highly recommended as a first step of any ultrastructural analysis.

Acknowledgments

We thank S. Schwarzbaum for editorial assistance. R.A. is the Incumbent of The Linda Jacobs Chair in Immune and Stem Cell Research. This research is supported by the Israel Science Foundation, the Minerva Foundation, Germany, and the U.S. Israel binational science foundation (BSF).

References

1. Shulman, Z., Shinder, V., Klein, E., et al. (2009) Lymphocyte crawling and transendothelial migration require chemokine triggering of high-affinity LFA-1 integrin. *Immunity* **30**, 384–496.
2. Ley, K., Laudanna, C., Cybulsky, M.I., Nourshargh, S. (2007) Getting to the site of inflammation: the leukocyte adhesion cascade updated. *Nat. Rev. Immunol.* **7**, 678–689.
3. Schenkel, A.R., Mamdouh, Z., Muller, W.A. (2004) Locomotion of monocytes on endothelium is a critical step during extravasation. *Nat. Immunol.* **5**, 393–400.
4. Phillipson, M., Heit, B., Colarusso, P., Liu, L., Ballantyne, C.M., Kubes, P. (2006) Intraluminal crawling of neutrophils to emigration sites: a molecularly distinct process from adhesion in the recruitment cascade. *J. Exp. Med.* **203**, 2569–2575.

5. Laudanna, C., Alon, R. (2006) Right on the spot. Chemokine triggering of integrin-mediated arrest of rolling leukocytes. *Thromb. Haemost.* **95**, 5–11.
6. Giagulli, C., Scarpini, E., Ottoboni, L., et al. (2004) RhoA and zeta PKC control distinct modalities of LFA-1 activation by chemokines. Critical role of LFA-1 affinity triggering in lymphocyte in vivo homing. *Immunity* **20**, 25–35.
7. Bolomini-Vittori, M., Montresor, A., Giagulli, C., et al. (2009) Regulation of conformer-specific activation of the integrin LFA-1 by a chemokine-triggered Rho signaling module. *Nat. Immunol.* **10**, 185–194.
8. Ridley, A.J., Schwartz, M.A., Burridge, K., et al. (2003) Cell migration: integrating signals from front to back. *Science* **302**, 1704–1709.
9. Vicente-Manzanares, M., Sanchez-Madrid, F. (2004) Role of the cytoskeleton during leukocyte responses. *Nat. Rev. Immunol.* **4**, 1–14.
10. Bajenoff, M., Egen, J.G., Koo, L.Y., et al. (2006) Stromal cell networks regulate lymphocyte entry, migration, and territoriality in lymph nodes. *Immunity* **25**, 989–1001.
11. Lammermann, T., Bader, B.L., Monkley, S.J., et al. (2008) Rapid leukocyte migration by integrin-independent flowing and squeezing. *Nature* **453**, 51–55.
12. Ding, Z., Xiong, K., Issekutz, T.B. (2000) Regulation of chemokine-induced transendothelial migration of T lymphocytes by endothelial activation: differential effects on naive and memory T cells. *J. Leukoc. Biol.* **67**, 825–833.
13. Barreiro, O., Yanez-Mo, M., Serrador, J.M., et al. (2002) Dynamic interaction of VCAM-1 and ICAM-1 with moesin and ezrin in a novel endothelial docking structure for adherent leukocytes. *J. Cell. Biol.* **157**, 1233–1245.
14. Carman, C.V., Jun, C.D., Salas, A., Springer, T.A. (2003) Endothelial cells proactively form microvilli-like membrane projections upon intercellular adhesion molecule 1 engagement of leukocyte LFA-1. *J. Immunol.* **171**, 6135–6144.
15. Millan, J., Hewlett, L., Glyn, M., Toomre, D., Clark, P., Ridley, A.J. (2006) Lymphocyte transcellular migration occurs through recruitment of endothelial ICAM-1 to caveola- and F-actin-rich domains. *Nat. Cell Biol.* **8**, 113–123.
16. Woolf, E., Grigorova, I., Sagiv, A., et al. (2007) Lymph node chemokines promote sustained T lymphocyte motility without triggering stable integrin adhesiveness in the absence of shear forces. *Nat. Immunol.* **8**, 1076–1085.
17. Shulman, Z., Pasvolsky, R., Woolf, E., et al. (2006) DOCK2 regulates chemokine-triggered lateral lymphocyte motility but not transendothelial migration. *Blood* **108**, 2150–2158.
18. Morin, N.A., Oakes, P.W., Hyun, Y.M., et al. (2008) Nonmuscle myosin heavy chain IIA mediates integrin LFA-1 de-adhesion during T lymphocyte migration. *J. Exp. Med.* **205**, 195–205.
19. Hartmann, T.N., Grabovsky, V., Pasvolsky, R., et al. (2008) A crosstalk between intracellular CXCR7 and CXCR4 involved in rapid CXCL12-triggered integrin activation but not in chemokine-triggered motility of human T lymphocytes and CD34+ cells. *J. Leukoc. Biol.* **84**, 1130–1140.
20. Buxboim, A., Geron, E., Alon, R., Bar-Ziv, R. (2009) A biochip model of lymphocyte locomotion on confined chemokine tracks. *Small* **5**, 1723–1726.
21. Shulman, Z., Alon, R. (2009) Chapter 14. Real-time in vitro assays for studying the role of chemokines in lymphocyte transendothelial migration under physiologic flow conditions. *Methods Enzymol.* **461**, 311–332.
22. Cinamon, G., Shinder, V., Alon, R. (2001) Shear forces promote lymphocyte migration across vascular endothelium bearing apical chemokines. *Nat. Immunol.* **2**, 515–522.
23. Shaw, S.K., Ma, S., Kim, M.B., et al. (2004) Coordinated redistribution of leukocyte LFA-1 and endothelial cell ICAM-1 accompany neutrophil transmigration. *J. Exp. Med.* **200**, 1571–1580.
24. Kamata, T., Puzon, W., Takada, Y. (1995) Identification of putative ligand-binding sites of the integrin alpha 4 beta 1 (VLA-4, CD49d/CD29). *Biochem. J.* **305**, 945–951.
25. Smith, A., Carrasco, Y.R., Stanley, P., Kieffer, N., Batista, F.D., Hogg, N. (2005) A talin-dependent LFA-1 focal zone is formed by rapidly migrating T lymphocytes. J. *Cell. Biol.* **170**, 141–151.
26. Carman, C.V., Sage, P.T., Sciuto, T.E., et al. (2007) Transcellular diapedesis is initiated by invasive podosomes. *Immunity* **26**, 784–797.
27. Majstoravich, S., Zhang, J., Nicholson-Dykstra, S., et al. (2004) Lymphocyte microvilli are dynamic, actin-dependent structures that do not require Wiskott-Aldrich syndrome protein (WASp) for their morphology. *Blood* **104**, 1396–1403.

Chapter 4

Lentiviral Gene Transfer Method to Study Integrin Function in T Lymphocytes

Daliya Banerjee and Motomu Shimaoka

Abstract

Integrins play critical roles in adhesion and migration of T cells during an immune response and inflammation. It is of great importance to understand the molecular pathways that regulate integrin function in T cells. Lentiviral vector-based gene transfer method has emerged in the past decade as an efficient means of transferring genes into both resting and activated hard-to-transfect cells, including T cells to knock-down gene expression. Therefore, this technology could be utilized effectively to study different aspects of integrin function or even to perform genome-wide RNAi screens to look globally for regulators of integrin function in T cells. In this chapter, we provide the simplest protocol to infect activated $CD4^{+}$ human T cells with high efficiency.

Key words: Human T lymphocytes, Integrins, Cell adhesion, Migration, Lentiviral, RNA interference

1. Introduction

Integrins constitute the major family of cell adhesion receptors whose functions are conserved from simple invertebrates (e.g., *Drosophila* or *Caenorhabditis elegans*) to highly complex vertebrates (e.g., humans), with respect to their role in development and organogenesis (1). Consequently, genetic screens have primarily been performed in more genetically tractable model organisms – *Drosophila* and *C. elegans* – to understand the cell–matrix interactions and cell migration during development, a more fundamental role mediated by integrins (2–4). However, integrin functions in higher vertebrates are more diverse and complex than those in invertebrate animals (5, 6). This necessitates the study of integrin

Motomu Shimaoka (ed.), *Integrin and Cell Adhesion Molecules: Methods and Protocols*,
Methods in Molecular Biology, vol. 757, DOI 10.1007/978-1-61779-166-6_4, © Springer Science+Business Media, LLC 2011

function in vertebrates such as mice and humans in the context of different cell and tissue types.

In the mammalian immune system, T lymphocytes are key initiators of an effective immune response that requires their efficient migration to the sites of inflammation and a successful interaction with the antigen-presenting cells. These functions are primarily mediated by the integrin adhesion receptors. Lymphocytes function-associated antigen-1 (LFA-1) is a key integrin receptor that is highly expressed on lymphocytes and is critical for leukocyte migration, antigen presentation, and cellular cytotoxicity (7). LFA-1 is nonadhesive in the resting lymphocytes and acquires a dynamic reversible adhesive state upon activation with a variety of stimuli, by altering either its affinity or avidity toward its ligands (6, 8). Therefore, deciphering the signal-transduction pathway initiated by each stimulus that culminates into activation of LFA-1 is critical for our understanding of regulation of integrin-mediated adhesion.

Gene knockout technology in mice has been utilized to study cell adhesion in mammals, but in some cases, the animals die during late embryogenesis or shortly after birth (9, 10). Although this problem has been partially circumvented by the development of tissue-specific knockouts, the approach relies on the appropriate tissue-specific promoters. Also, the generation of such mice is time consuming and expensive. Available T cell-specific promoters that utilize the "lox-cre" system for conditional gene knockout in T cells include CD4-Cre and lck-Cre, both of which are turned on early during T-cell development in the thymus. However, some of the mediators of integrin signaling are also involved in thymocyte development, for example, RAC1 and RAC2 (11), and VAV1 (12). Thus, if the gene of interest to study integrin function is also essential for T-cell development, it may not be an appropriate system to study its function in adhesion and migration in the mature peripheral T cells. Studies in the humans are also limited to the naturally occurring genetic mutations that lead to defects in adhesion and migration and the identification of such conditions in patients. The known human genetic disorders, associated with defective adhesion and migration of leukocytes, are LAD-I, LAD-II, LAD-III, and CDG II syndrome (13). Studies with T cells from LAD-III patients, which are generally defective in integrin activation due to mutation in Kindlin-3 (14), have further revealed Kindlin-3 to act in an inside-out signaling cascade that activates LFA-1 (15). On the contrary, LAD-I is associated with mutation in β2 integrin and LAD-II with defective selectin function as a result of mutation in the gene encoding the fucosyl transporter that affects selectin ligand (13). Although these patients could be used as model systems, they may not yield enough information to understand the mechanism of integrin activation in the humans fully. Therefore, other means of turning on or off of gene expression is needed to achieve this end.

In the past decade, RNA interference (RNAi) has emerged as a novel technology to knockdown gene expression specifically and efficiently in mammalian cells, including hard to transfect T lymphocytes, either by electroporation/transfection of short interfering RNAs (siRNAs) or by expression of short hairpin RNAs (shRNAs) from expression vectors and retroviruses (16, 17). RNAi is, therefore, a useful tool for the functional analyses of genes in humans, thereby developing a potential therapeutic strategy for various diseases.

A number of gene delivery viral vectors have been generated and utilized to date to infect and express shRNAs or other genes efficiently in the T lymphocytes. The adenovirus- and lentivirus-based vectors are most frequently used. Their advantages and disavantages are shown in Table 1. Lentiviruses, which are derived from the HIV-1, are clearly advantageous over adenoviruses or other gene delivery systems because of two reasons – first, stable integration and expression of genes in both nondividing and differentiated cells (18), and second, they can be used to generate transgenic animals by infecting embryonic stem cells or embryos (19).

Lentivirus/retrovirus-based RNAi technology has been effectively utilized recently to study the regulation of integrin function in adhesion and/or migration in both T cells and other cell types (20–25) (see Note 1). One interesting approach to utilize this powerful tool productively would be genome-wide lentiviral shRNA- or siRNA-based screens in T cells, aiming to identify all the players critically involved in regulating LFA-1 activation in

Table 1
Comparison of advantages and disadvantages of lentiviral and adenoviral vectors

Viruses	Packaging capacity	Advantages	Disadvantages
Adenoviral	High (upto ~40 kbs)	• Infects most cell types with very high efficiency including primary dividing and nondividing T cells • High packaging capacity	• Does not integrate with the host genome, so expression is transient • Induces high inflammatory response inside the human body
Lentiviral	Low (upto ~8 kbs)	• Infects both dividing and nondividing cells (including T cells) with moderate efficiency • Integrates with host genome, so good for both transient and stable gene expression • Does not induce inflammatory response in the host	• Low packaging capacity

response to both activating signals, as well as in maintaining basal low affinity status of LFA-1 in resting cells (see Note 2).

2. Materials

2.1. Media and Reagents for 293T Transfection

1. Complete DMEM supplemented with 10% fetal bovine serum (FBS), 100 U/ml penicillin and streptomycin, and 5 mM glutamine.
2. OPTI-MEM reduced serum medium.
3. Genejuice transfection reagent (Novagen).
4. Trypsin EDTA.

2.2. Media and Reagents for T-Cell Lentiviral Infection

1. Complete RPMI medium supplemented with 10% FBS, 100 U/ml penicillin and streptomycin, and 5 mM glutamine.
2. Sterile phosphate-buffered saline (PBS).
3. Histopaque 1077 (Sigma-Aldrich).
4. Phytahemagglutin (PHA) Remel.
5. Interleukin-2 R and D System.
6. Puromycin Sigma-Aldrich.
7. Polybrene (Sigma-Aldrich): Prepare 8 mg/ml stock solution.
8. CD4 T-cell isolation kit II (Miltenyi Biotec).
9. MACs buffer: PBS with 0.5% BSA and 2 mM EDTA.
10. MACs LS columns (Miltenyi Biotec).
11. Lenti-X-Concentrator (Clontech).

2.3. Cells

1. 293T packaging cell line.
2. $CD4^+$ T cells derived from PBMCs isolated from fresh adult human blood.

2.4. DNA

1. Lentiviral packaging vector psPAX2 (addgene).
2. Lentiviral VSV G envelope vector pMD2.G (addgene).
3. Lentiviral vector-expressing hsa-mir-200c (System Biosciences).

3. Methods

3.1. Lentivirus Production

1. *Day 0.* Seed 293T packaging cells at a density of $(4–5)\times10^5$ cells/ml.
2. *Day 1.* Carry out transfection using the Genejuice transfection reagent according to the manufacturer's protocol. Briefly,

the packaging vector psPAX2 (1.8 μg), envelope vector pMD2.G (0.2 μg), and hsa-mir-200c lentiviral vector (2 μg) are mixed together in 300 μl of OPTI-MEM along with 15 μl of Genejuice. The mixture is pipetted up and down several times and incubated at room temperature (RT) for 15 min and added dropwise to the 293T cells (see Note 3).

3. *Day 3.* Change the medium after 12 h of transfection.
4. *Day 4.* Collect the supernatant at 48 h. The supernatant is spun down to remove dead cells, aliquoted into cryovials, and stored at −80°C. For CD4 T cell infection, freshly collected viral supernatants are concentrated using the lenti-X-concentrator (according to manufacturer's protocol). Typically, 8 mls of concentrated viral sup is used for each infection condition.

3.2. CD4⁺ T Cell solation

1. Divide 60 ml of whole blood into four 50-ml falcon tubes (15 ml per tube) and dilute twofold with PBS. Carefully layer 30 ml of the diluted blood on top of 10 ml of Histopaque.
2. Centrifuge the samples at 805 × *g* for 30 min at RT without brake.
3. Discard the upper yellow phase. Carefully remove the underlying white "buffy coat," which contains the peripheral blood mononuclear cells (PBMCs), into a fresh tube.
4. Add ~30 ml of PBS to the extracted buffy coat and mix thoroughly.
5. Centrifuge at 400 × *g* for 10 min at RT with brakes on low.
6. Resuspend the white pellet in 1 ml of ice-cold MACs buffer. If the pellet is reddish, use ACK lysis buffer to remove any contaminating red blood cells.
7. Add antibody cocktail (from CD4 T-cell isolation kit II) and negatively select $CD4^+$ T cells according to the manufacturer's protocol.
8. After antibody cross-linking, add the cells to magnetic columns. Collect the flow-through fraction that contains $CD4^+$ T cell population at ~90–95% purity.
9. Resuspend the $CD4^+$ T cells in complete RPMI medium at a density of $\sim 1 \times 10^6$/ml. Activate the cells with 1 μg/ml PHA for 2–3 days.

3.3. Lentiviral Spin-Infection of Activated T Cells

1. Seed PHA-activated $CD4^+$ T cells (see Note 2) in a 24-well plate, 1 h prior to infection. Add to each well 0.25×10^6 cells in 0.5 ml of fresh RPMI medium containing 10 ng/ml of IL-2.
2. Either thaw out aliquots of viral supernatant on ice or use freshly concentrated viral sups (preferred). Add 0.5 ml of concentrated viral supernatant to each well. Add polybrene at a final concentration of 8 μg/ml (see Note 4).

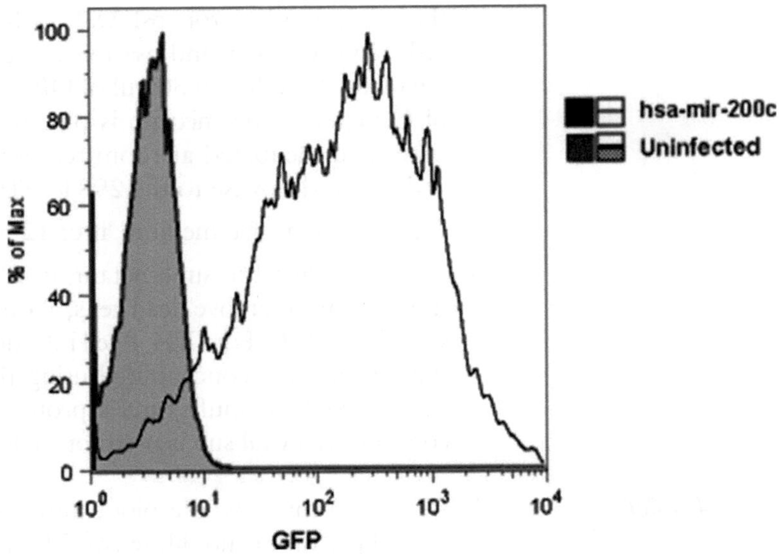

Fig. 1. Lentiviral transduction of PHA-activated human CD4+ T cells. PHA-activated CD4+ T cells were spinoculated with lentiviruses containing hsa-mir-200c microRNA. Percentage of GFP-expressing cells was determined by flow cytometry 48 h after infection.

3. Mix the cells with the viral supernatant gently.
4. Spinoculation: Infect the cells while spinning them at 800 × *g* for 90 min at 30°C (see Note 5).
5. After spinoculation is completed, remove 1.0 ml of the media and replace with 1 ml of fresh RPMI medium and 10 ng/ml of IL-2.
6. Incubate the cells at least for 24 h before analyzing GFP expression using flow cytometry (Fig. 1). For trouble shooting, see Note 6.
7. If the lentiviral vectors contain puromycin resistance gene, the cells can be selected in puromycin at a concentration of 1 μg/ml, added 24–48 h postinfection.
8. Puromycin selection takes 4–5 days to complete and is assessed by monitoring cell death in mock-infected controls.

4. Notes

1. Lentiviral vectors are derived from HIV that is one of known human retroviruses. Compared to other retroviruses such as murine leukemia viruses (MLVs), their ability for stable integration in both dividing and nondividing cells and their long-term expression of the transgene make lentiviral vectors ideal gene transfer vehicles.

2. For a protocol to infect resting T cells, see ref. 26. Other efficient means to transfect resting T cells efficiently includes electroporation using Amaxa technology (17). However, the transgene expression is short lived.
3. Later generation lentiviral vector systems provide for a great margin of personal and public safety, as they use heterologous coat proteins in place of the native HIV-1 envelope protein; they separate vector and packaging functions onto four or more plasmids; and they include additional safety features (e.g., they do not encode Tat, which is essential for replication of wild-type HIV-1).
4. Polybrene enhances hydrophobic interaction between virus and the host, thereby increasing infection efficiency.
5. Spinoculation is an important step for greatly increasing the infection efficiency of suspension cells such as lymphocytes.
6. Preparation of a good viral stock is highly critical to ensure good infection efficiency and the following points should be taken into consideration.
   (a) The ratio of the viral DNA vector to envelope and packaging vectors is important.
   (b) The viral stock should always be on ice upon thawing before infection.
   (c) The spinoculation should be carried out at 30°C.
   (d) Viral stock should be carefully titrated to ensure a high multiplicity of infection (MOI) for infection of T lymphocytes. Typically, MOI = 50 is good to use.

References

1. Brown, N. H. (2000) Cell-cell adhesion via the ECM: Integrin genetics in fly and worm. *Matrix Biol.* **19**, 191–201.
2. Cram, E. J., Shang, H., and Schwarzbauer, J. E. (2006) A systematic RNA interference screen reveals a cell migration gene network in C. elegans. *J. Cell Sci.* **119**, 4811–4818.
3. Wang, X., Bo, J., Bridges, T., Dugan, K. D., Pan, T. C., Chodosh, L. A., and Montell, D. J. (2006) Analysis of cell migration using whole-genome expression profiling of migratory cells in the Drosophila ovary. *Dev. Cell* **10**, 483–495.
4. Cram, E. J., Clark, S. G., and Schwarzbauer, J. E. (2003) Talin loss-of-function uncovers roles in cell contractility and migration in C. elegans. *J. Cell Sci.* **116**, 3871–3878.
5. Hemler, M. E. (1990) VLA proteins in the integrin family: structures, functions, and their role on leukocytes. *Annu. Rev. Immunol.* **8**, 365–400.
6. Luo, B. H., Carman, C. V., and Springer, T. A. (2007) Structural basis of integrin regulation and signaling. *Annu. Rev. Immunol.* **25**, 619–647.
7. Springer, T. A. (1990) Adhesion receptors of the immune system. *Nature* **346**, 425–434.
8. Askari, J. A., Buckley, P. A., Mould, A. P., and Humphries, M. J. (2009) Linking integrin conformation to function. *J. Cell Sci.* **122**, 165–170.
9. Stephens, L. E., Sutherland, A. E., Klimanskaya, I. V., Andrieux, A., Meneses, J., Pedersen, R. A., and Damsky, C. H. (1995) Deletion of beta 1 integrins in mice results in inner cell mass failure and peri-implantation lethality. *Genes. Dev.* **9**, 1883–1895.
10. Yang, J. T., Rayburn, H., and Hynes, R. O. (1995) Cell adhesion events mediated by alpha 4 integrins are essential in placental and cardiac development. *Development* **121**, 549–560.

11. Radtke, F., Wilson, A., Mancini, S. J., and MacDonald, H. R. (2004) Notch regulation of lymphocyte development and function. *Nat. Immunol.* **5**, 247–253.
12. Fischer, K. D., Zmuldzinas, A., Gardner, S., Barbacid, M., Bernstein, A., and Guidos, C. (1995) Defective T-cell receptor signalling and positive selection of Vav-deficient CD4+ CD8+ thymocytes. *Nature* **374**, 474–477.
13. Etzioni, A. (2009) Genetic etiologies of leukocyte adhesion defects. *Curr. Opin. Immunol.* **21**, 481–486.
14. Svensson, L., Howarth, K., McDowall, A., Patzak, I., Evans, R., Ussar, S., Moser, M., Metin, A., Fried, M., Tomlinson, I., and Hogg, N. (2009) Leukocyte adhesion deficiency-III is caused by mutations in KINDLIN3 affecting integrin activation. *Nat. Med.* **15**, 306–312.
15. Manevich-Mendelson, E., Feigelson, S. W., Pasvolsky, R., Aker, M., Grabovsky, V., Shulman, Z., Kilic, S. S., Rosenthal-Allieri, M. A., Ben-Dor, S., Mory, A., Bernard, A., Moser, M., Etzioni, A., and Alon, R. (2009) Loss of Kindlin-3 in LAD-III eliminates LFA-1 but not VLA-4 adhesiveness developed under shear flow conditions. *Blood* **114**, 2344–2353.
16. Rubinson, D. A., Dillon, C. P., Kwiatkowski, A. V., Sievers, C., Yang, L., Kopinja, J., Rooney, D. L., Zhang, M., Ihrig, M. M., McManus, M. T., Gertler, F. B., Scott, M. L., and Van Parijs, L. (2003) A lentivirus-based system to functionally silence genes in primary mammalian cells, stem cells and transgenic mice by RNA interference. *Nat. Genet.* **33**, 401–406.
17. Zhao, Y., Zheng, Z., Cohen, C. J., Gattinoni, L., Palmer, D. C., Restifo, N. P., Rosenberg, S. A., and Morgan, R. A. (2006) High-efficiency transfection of primary human and mouse T lymphocytes using RNA electroporation. *Mol. Ther.* **13**, 151–159.
18. Naldini, L., Blomer, U., Gallay, P., Ory, D., Mulligan, R., Gage, F. H., Verma, I. M., and Trono, D. (1996) In vivo gene delivery and stable transduction of nondividing cells by a lentiviral vector. *Science* **272**, 263–267.
19. Lois, C., Hong, E. J., Pease, S., Brown, E. J., and Baltimore, D. (2002) Germline transmission and tissue-specific expression of transgenes delivered by lentiviral vectors. *Science* **295**, 868–872.
20. Baker, R. G., Hsu, C. J., Lee, D., Jordan, M. S., Maltzman, J. S., Hammer, D. A., Baumgart, T., and Koretzky, G. A. (2009) The adapter protein SLP-76 mediates "outside-in" integrin signaling and function in T cells. *Mol. Cell. Biol.* **29**, 5578–5589.
21. Gong, H., Shen, B., Flevaris, P., Chow, C., Lam, S. C., Voyno-Yasenetskaya, T. A., Kozasa, T., and Du, X. G protein subunit Galpha13 binds to integrin alphaIIbbeta3 and mediates integrin "outside-in" signaling. *Science* **327**, 340–343.
22. Primo, L., di Blasio, L., Roca, C., Droetto, S., Piva, R., Schaffhausen, B., and Bussolino, F. (2007) Essential role of PDK1 in regulating endothelial cell migration. *J. Cell Biol.* **176**, 1035–1047.
23. Jani, K., and Schock, F. (2007) Zasp is required for the assembly of functional integrin adhesion sites. *J. Cell Biol.* **179**, 1583–1597.
24. Lafuente, E. M., van Puijenbroek, A. A., Krause, M., Carman, C. V., Freeman, G. J., Berezovskaya, A., Constantine, E., Springer, T. A., Gertler, F. B., and Boussiotis, V. A. (2004) RIAM, an Ena/VASP and Profilin ligand, interacts with Rap1-GTP and mediates Rap1-induced adhesion. *Dev. Cell* **7**, 585–595.
25. Zhang, Z., Rehmann, H., Price, L. S., Riedl, J., and Bos, J. L. (2005) AF6 negatively regulates Rap1-induced cell adhesion. *J. Biol. Chem.* **280**, 33200–33205.
26. Verhoeyen, E., Costa, C., and Cosset, F. L. (2009) Lentiviral vector gene transfer into human T cells. *Methods Mol. Biol.* **506**, 97–114.

Chapter 5

Surface Plasmon Resonance Biosensing in Studies of the Binding Between β_2 Integrin I Domains and Their Ligands

Thomas Vorup-Jensen

Abstract

Measurements on the kinetic aspects of binding between macromolecular species such as proteins have been greatly advanced by the application of surface plasmon resonance (SPR) biosensors. In studies of ligand binding by integrin I domains, technologies such as the BIAcore instruments have provided important insights into the role of conformational regulation. This chapter describes a protocol for studying the binding between the I domain from integrin $\alpha_X\beta_2$ and its ligand iC3b. Also included are topics on the interpretation of data. Integrin I domains appear to support heterogeneous interactions with ligands, which pose significant challenges in deriving valid information on the binding kinetics from the SPR measurements. Fortunately, new algorithms are available that may resolve even complex ligand-binding reactions; with the application to data on the binding between the α_X I domain, a more consistent and unambiguous result is obtained compared to those obtained by classical approaches for analyzing SPR biosensor data.

Key words: Integrins, Surface plasmon resonance, I domain, Heterogeneous surface binding

1. Introduction

The affinity of interactions between biological macromolecules is a parameter of major importance in their physiological role. Not long ago, accurate measurements on the affinity of even a simple binding reaction involving two proteins would be technically challenging and time consuming. Although, for instance, protocols for equilibrium dialysis with radiolabeled proteins historically has contributed enormously to our understanding of protein–protein interactions, the setup is not well suited for measuring the binding between large protein species and fails to provide detailed

Motomu Shimaoka (ed.), *Integrin and Cell Adhesion Molecules: Methods and Protocols*,
Methods in Molecular Biology, vol. 757, DOI 10.1007/978-1-61779-166-6_5, © Springer Science+Business Media, LLC 2011

information on binding kinetics (1, 2). Furthermore, more modern approaches often call for a high-throughput capability that is not met by classic biochemical assays. The early 1990s saw the advance of so-called label-free detection of interaction between biological macromolecules. In particular, techniques based on excitation of surface plasmons have had a major impact on studies of affinity and binding kinetics with biological macromolecules. The name of the Swedish company BIAcore AB providing some of the first commercial equipment has become almost eponymous with surface plasmon resonance (SPR) biosensors in molecular biology, although several other companies now provide similar instruments. BIAcore has developed several lines of instrumentation, including some suited for drug screening and highly automated operation. In the scientific laboratory, however, the focus is often on flexibility, where, in particular, the tried and tested BIAcore 3000 still remains a very strong candidate for data acquisition by investigators requiring a platform for SPR biosensing.

The purpose of the application of SPR biosensors in molecular biology is to obtain quantitative data on the binding of a macromolecule in solution (the analyte) to immobilized ligands on a surface. The underlying physical principles of the measurements have been described in several publications (3–5). Following laser excitation, the surface plasmon wave propagates along a thin film typically made from gold or silver and its electromagnetic field probes the medium adjacent to the surface. Changes in the refractive index in the proximity of the surface plasmon result in a change in the wave propagation, and when the surface plasmon wave hits, a local irregularity part of the energy can be reemitted as light (3). In the BIAcore methodology, ligands for the analyte molecules are immobilized on a gold surface covered with a layer of dextran, which enhances the capacity for coupling of ligand and limits nonspecific binding to the gold surface (6). When a liquid sample of analyte is applied to the sensor surface, molecules of analyte are captured by the ligand-coupled surface. The sensor consists of a gold-coated glass surface, which is illuminated on the back with a laser while the flow stream passes over the front surface with the gold layer. Binding of analyte to the gold surface gives rise to a change in the refraction of the laser light, which is measured by the optical reader. The BIAcore instruments are equipped with a computer-controlled pump system which permits precise regulation of the injection of analyte into the flow stream over the ligand-coupled surface. The change in the refractive index, or response (usually denoted R measured in arbitrary resonance units, RU), when analyte binds to the surface is a read-out linearly dependent on the amount of analyte bound. Provided that certain criteria are met, notably that the kinetics of binding are not faster than what can be measured by the instrument and that the interaction is sufficiently strong for detection using the concentration of analyte, R together with analyte concentration can then be fitted into models of the binding reaction that will provide information of kinetics and affinity (often expressed by the dissociation constant, K_D). Reliable determination of the kinetic constants usually requires that information on the amount of bound analyte is sampled over a time period of 100–200 s. The data are collected into a so-called sensorgram with R plotted as a function of time. The binding experiment is divided into an injection or association phase with the analyte at a fixed concentration in the flow stream, followed by a dissociation phase, where running buffer without analyte is passed over the ligand surface. While R increases during the injection phase, the signal drops during the dissociation phase; the rate of increase and decrease is related to the specific type of binding reaction. During the injection phase, a simple first-order reaction (Eq. 1) with the dissociation constant K_D.

$$\text{Analyte} + \text{Ligand} \leftrightarrow \text{Analyte} \cdot \text{Ligand} \tag{1}$$

follows the kinetics of binding described by Eq. 2

$$\frac{dR}{dt} = k_{on} \cdot C_{analyte}(R_{Max} - R) - k_{off} \tag{2}$$

where R is the response at time point t following injection initiation, R_{max} is the response at binding saturation, $C_{analyte}$ the concentration of analyte in the flow stream, and k_{on} and k_{off} the binding kinetic constants (on and off rates, respectively). k_{off} can be determined from the exponential decay of R during the dissociation phase; with $C_{analyte} = 0$, Eq. 2 simplifies to

$$k_{off} = \frac{1}{R} \cdot \frac{dR}{dt} \tag{3}$$

In case a steady-state equilibrium is reached during the injection phase, i.e., $dR/dt = 0$, Eq. 2 can be rearranged to

$$R_{Eq} = \frac{R_{Max} \cdot C_{analyte}}{K_D + C_{analyte}}, \tag{4}$$

where R_{Eq} is the response at steady-state equilibrium. While estimation of K_D is frequently the direct aim of SPR studies, R_{Max} is surprisingly often ignored as a result of the analysis. However, R_{Max} provides an interesting information on the stoichiometry (N) of interaction between the analyte and the immobilized ligand as shown by Eq. 5:

$$N = \frac{M_{r,Ligand}}{M_{r,Analyte}} \cdot \frac{R_{Max}}{R_I}, \tag{5}$$

where $M_{r,Ligand}$ is the relative molecular weight of the ligand, $M_{r,Analyte}$ is the weight of the analyte, and R_I is the amount of immobilized ligand (in RU). Surface immobilization will typically destroy some binding sites within the ligand either through introduced chemical modifications or through simple steric aspects of binding the ligand to the surface; for ligand:analyte interaction with an expected 1:1 binding stoichiometry, the observed stoichiometry rarely exceeds 0.6. However, as discussed below, stoichiometric calculations are nevertheless capable of providing valuable information.

The application of SPR for studies of integrin ligand binding is tightly linked with the many recent advances made in understanding the structural biology of these receptors. Integrins are heterodimeric receptors expressed on cell surfaces where they support a diverse range of functions in cellular adhesion, including cell–cell, cell–extra cellular matrix, and cell–pathogen adhesion (7, 8). Briefly, based on several lines of evidence, Springer concluded that all alpha chains of integrins contain a seven-bladed beta-propeller domain (9); in some integrin alpha chains, a separate domain, identified earlier (10) due to its similarity with the von Willebrand factor A domain, is inserted between blades of the beta-propeller domain (9, 11). The domain forms contact with ligands through a Mg^{2+} ion found in the metal ion-dependent adhesion site (MIDAS); in addition to hydroxyl or carboxyl groups in side chains from I domain residues, this Mg^{2+} ion coordinates a carboxyl group in the side chain of a glutamate residue in protein ligands (12, 13). The I domain was suitable for expression in prokaryotic expression systems, and characterization by X-ray crystallography of this domain from five integrin chains has been achieved within the past 15 years (12, 14–17). The integrin I domains are the major ligand interaction site in the receptors, which carry them. Although constructs have been made for expression of soluble β_2 integrin receptors containing all of the ecto domain, quantitative binding studies on the interaction between integrin and ligand have mainly been made with the isolated I domains.

Considerable evidence now exists to suggest that ligand binding by the I domain is conformationally regulated (18), apparently as part of many larger conformational changes in the entire ecto domain of the integrin receptor that accompanies activation. Based on conformational difference between two crystallographic structures of the I domain from integrin $\alpha_M\beta_2$ (also referred to as Mac-1, complement receptor 3, or CD11b/CD18), it was speculated that one of these conformations, named the "open conformation," represented a ligand-binding active form, while, by contrast, the alternate conformation, named "closed," would not support ligand binding (12, 13). The major structural differences between the conformations involve changes in the side chains coordinate, the Mg^{2+} ion in the MIDAS, and a 10-Å movement of the C-terminal helix down the side of the domain (12, 13).

Several strategies for controlling the conformation of the I domain by site-directed mutagenesis have been reported, and many of these studies have used SPR biosensor technology for studying changes in the affinity induced by the conformational regulation. Based on the crystallographic structures (12, 13), Li et al. suggested that mutation of a single residue (Thr-209 to alanine) in the α_M I domain generated an I domain stabilized in the ligand-binding conformation (19). This was investigated by SPR studies of the binding between the wild-type and mutated I domains to iC3b, a proteolytic fragment of complement factor C3 generated through processes of complement deposition on target surface that inactivates C3 proteolysis. However, while the I domain conformers differed somewhat with regard to their binding kinetics, the K_Ds for the interactions were almost identical. A direct demonstration of the influence of I domain conformation on ligand-binding properties through SPR measurements was provided by studies on the I domain from $\alpha_L\beta_2$ (also referred to as LFA-1 or CD11a/CD18). By introducing disulfide bonds engineered to keep the I domain in fixed conformations, Shimaoka et al. showed that the open conformation of the α_L I domain bound its ligand intercellular adhesion molecule (ICAM)-1 with almost 100-fold higher affinity and markedly changed kinetics toward a significantly reduced off-rate compared to a wild-type construct or a construct locked in the closed conformation (20, 21). Crystallographic studies later clearly confirmed that these constructs assumed the expected conformations (21). An important contribution to the design of high-affinity I domain constructs was made by Xiong et al.; in the C-terminal alpha helix of the α_M I domain, the side chain of an isoleucine residue is kept in a socket that restricts the movement of the helix and hence the conformation of the domain. Mutagenesis of this residue to glycine made the construct crystalize in the open conformation and enhanced the affinity for ligand (iC3b) approximately 50-fold compared to a wild-type domain (22). Integrin $\alpha_X\beta_2$ (also named p150, 95, complement receptor 4, or CD11c/CD18) also binds iC3b. The same principle was applied to the α_X I domain with similar results concerning the binding to iC3b and a conformational change in the domain sufficiently stable for detection by gel permeation chromatography (14). In the case of the α_L I domain, single point substitutions of residues was more recently reported to enhance the affinity as much as 100,000-fold (23).

In addition to the structural properties of the I domain, other biochemical factors have been reported to affect the apparent affinity of the domain for ligands. The requirement for divalent cations, i.e., Mg^{2+}, is ubiquitous and addition of EDTA to buffers for the binding experiments usually abolishes ligand binding by I domains. Titration of the Mg^{2+} ion concentration in the range of the concentration in human body fluids (~1 mM) changes the apparent affinity of the α_L I domain as described elsewhere (24). While this raises some interesting questions on the role of Mg^{2+} ions in integrin biology, it points also to potential problems in setting up binding experiments where Mg^{2+} chelating substances are included. These substances include phosphates, sulfates, and free amino acids with negatively charged side chains. Furthermore, some I domains, including the α_M and α_X I domains, but not the α_L I domain, bind acidic compounds such as the γ-COOH group of the side chain of glutamate or even acetic acid with appreciable affinity ($K_D \sim 100–200\ \mu M$) (25, 26). This phenomenon has both interesting scientific implications and practical consequences for the design and interpretation of binding experiments, as discussed below.

In summary, several strategies now exist for making integrin I domains with a high affinity for ligand, and a range of experiments have been reported in the literature with the application of SPR biosensors. It should be noted, however, that all wild-type I domains so far tested show binding detectable with SPR and such studies are consequently not critically dependent on the availability of high-affinity constructs. In Subheadings 2 and 3, the experimental procedure for measuring the affinity of the binding between the high-affinity α_X I domain and iC3b by SPR is presented. The procedure is based on the BIAcore 3000 instrument from GE Health care, but analogous procedures would also be supported by the BIAcore 1000 and BIAcore 2000 instruments. Throughout the text, several references are provided where additional detail on the experimental or analytical procedures can be found.

2. Materials

2.1. Utensils

1. Glass vials, Ø 16 mm/4.0 ml.
2. Rounded polypropylene vials, Ø 7 mm/0.8 ml.
3. Polyethylene caps, 7 mm.

2.2. Preparation of iC3b-Coupled Surfaces

1. 10 mM acetate buffer, pH 4.5.
2. CM-4 chip (GE Health care): Compared to CM-5 chips, the dextran layer of the CM-4 chip contains fewer carboxyl groups.
3. Amine coupling kit with 1-ethyl-3-(3-dimethylaminopropyl) carbodiimide hydrochloride (EDC), *N*-hydroxysuccinimide (NHS), and 1.0 M ethanolamine–HCl, pH 8.5: Following reconstitution with water of the vials containing NHS and EDC, these solutions should be aliquoted into tubes in volumes appropriate for the planned use, e.g., 100 μl. The tubes should be stored at −20°C.

4. iC3b (Calbiochem): iC3b is formed by the cleavage of C3b by Factor I in the presence of Factor H, CRl, or membrane cofactor protein; iC3b fragment is a glycoprotein composed of two C3α′ polypeptides with M_rs of 43,000 and 63,000, respectively, that are disulfide bonded to the intact C3 β-chain (M_r~75,000).

2.3. Binding Between the I314G αX I Domain and iC3b

1. His_5-tagged αX I314G I domain prepared as described (14). Stock solutions of I domain, diluted in HBS/Mg, should be at a protein concentration of 6–10 mg/ml. The protein can be stored at −80°C following aliquoting of samples of ~50 μl into tubes and snap freezing on ethanol-dry ice (see Note 1).
2. Running buffer: HEPES-buffered saline with Mg^{2+} (HBS/Mg) 1 mM $MgCl_2$, 150 mM NaCl, and 10 mM HEPES, adjusted with NaOH to pH 7.4.
3. Regeneration buffer: 50 mM EDTA, 1.5 M NaCl, and 100 mM HEPES, pH 7.4.

2.4. Software Required for Analysis of Sensorgram Data

1. BIAevaluation (GE Healthcare): This is the only software currently available that can read the result files generated by BIAcore instruments.
2. Matlab™ (The MathWorks, Inc. Natick, MA) (see Note 2).

3. Methods

3.1. Preparation of CM-4 Chip Surfaces with iC3b Ligand

1. Running buffer is vacuum-filtered through a 0.2-μm filter and allowed to degas further under air suction for 5–10 min. The pump system is purged and filled with the running buffer.
2. The frozen solutions (described in Subheading 2.3, item 3) of NHS and EDC are thawed and mixed at a 1:1 ratio; the mixture is filled into a 0.8-ml polypropylene vial in a volume in slight excess of the minimum required; e.g., the contact time for the injection of NHS/EDC should be at least 120 s corresponding to an injected volume of 20 μl with a flow rate of 10 μl/min (Fig. 1). In the present experiment, a contact time of 200 s was used corresponding to an injected volume of 60 μl. iC3b was diluted to 10 μg/ml in acetate buffer. All samples required for the coupling of iC3b onto the carboxymethylated surface should be prepared immediately before use (see Note 3).
3. Injection of the iC3b solution was carried out for approximately 400 s. As evident from Fig. 1, the change in R during

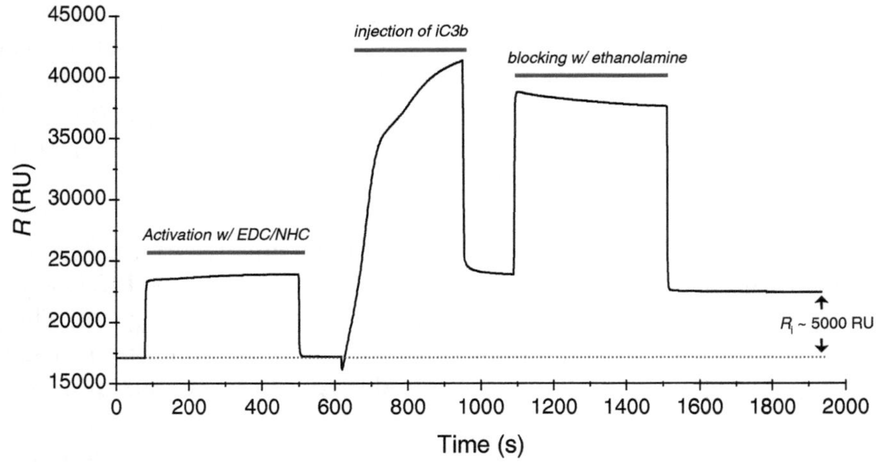

Fig. 1. Immobilization of iC3b by covalent coupling onto carboxymethylated dextran. Activation by the surface was carried out by injection of a mixture of NHS and EDC for 400 s. The ligand (iC3b) diluted in acetate buffer to a concentration of 10 μg/ml was then injected over the surface, leading to the formation of a covalent bond between the activated carboxyl groups and primary amines in iC3b; a sharp rise in the SPR signal was observed, which, however, was considerably reduced when injection ceased. Finally, uncoupled reactive groups on the surface were blocked with an injection of ethanolamine. The amount of immobilized ligand (R_I in units of RU) was determined by comparing the baseline level for the SPR signal with the level obtained after completion of the iC3b immobilization.

the injection phase was not proportional to the final level of immobilization (R_I) due to the influence of the bulk change in the refractive index from differences in the buffer composition of the running buffer and the acetate buffer with iC3b. Following injection of protein ligand, the residual reactive sucinimide esters were inactivated by application of ethanolamine to the flow stream.

4. A reference flow cell, i.e., without ligand protein, was prepared by completing the injection of NHS/EDC immediately followed by injection of ethanolamine (see Note 4). The CM4 chips contain four flow cells, which are linked sequentially in the flow stream; in the present setup, flow cell 2 was coupled with iC3b, while flow cell 1 was prepared as the reference flow cell.

3.2. Binding Between the I314G αX I Domain and iC3b

The design of the binding experiment is critically dependent on the kinetics of interaction between ligand and analyte. A robust estimate of K_D can be obtained from measurements of the response at steady-state equilibrium over a concentration range of analyte determined via Eq. 4. The time required, and hence the amount of analyte used per injection, to reach a steady-state equilibrium can be demonstrated to be proportional to $1/k_{off}$. This implies that interactions with a slow release of bound analyte,

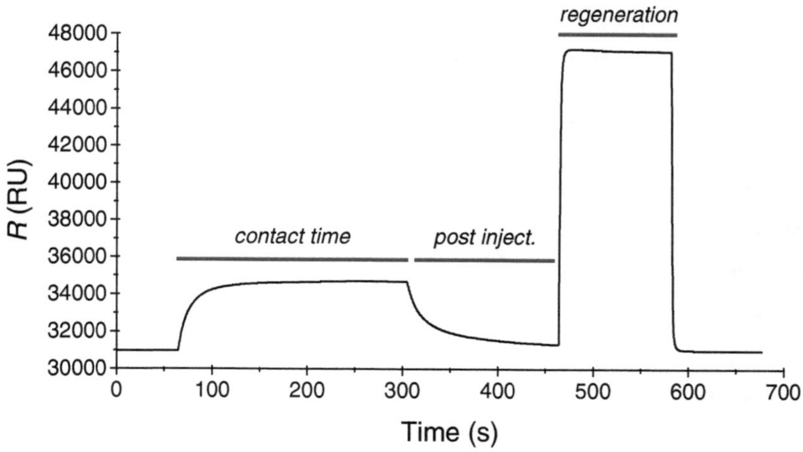

Fig. 2. Measurement cyclin SPR analysis of the high-affinity α_X I domain to iC3b. The domain was diluted in running buffer and applied to the surface with a total contact time of approximately 250 s. Following cessation of injection, the release of the I domain was recorded over a period of 150 s. The surface was regenerated by injection of EDTA-containing buffer with a high-salt concentration.

e.g., antibodies bound to antigen, would require a contact time usually in excess of 300 s. The kinetics of the interaction between β_2 integrin I domains has, however, in general showed sufficiently fast kinetics, permitting determination of K_D by steady-state equilibrium. One example on a study of the interaction between the I314G α_X I domain and iC3b is shown in Fig. 2.

1. The I domain diluted to 10 μM in running buffer was injected over the iC3b-coated surface with a contact time of 250 s, followed by collection of data from postinjection phase, with release of the I domain from surface for another 150 s.
2. Regeneration of the surface, i.e., complete removal of all bound material, was carried out by injection of regeneration buffer.
3. Information on the K_D and kinetics of the interaction is typically made from a set of injections of analyte; in the present experiment, ten injections were made at a concentration range from 278 nM to 10 μM. Obviously, the concentrations of analyte applied to the ligand-coated surfaces should be kept within the titrable range of the response in order to obtain valuable information, usually corresponding to $K_D/10 < C_{\text{analyte}} < 10 \times K_D$. However, in some cases, intrinsic characteristics of the analyte or aspects of its preparation may prevent the application of high concentrations of analyte. In order to automate the injections, the autosample with the BIAcore 3000 instrument is very useful. A program for the present setup is indicated below:

```
MAIN
FLOWCELL 2-1                  ! Data are collected from Flowcells 1 and 2 and a result !file with the signal from Flowcell 1
                              subtracted the signal !from flowcell 2 is generated

        APROG inject r2a1     !specifies the position of the sample to injected, i.e. rack 2, !position a1
        APROG inject r2a2
        APROG inject r2a3
        APROG inject r2a4
        ........................
        APROG inject r2a10
END
DEFINE APROG inject           !Defines the program for injection of samples
PARAM %position
MODE -d0.1                    ! Sets the frequency of data collection
FLOW     10                   ! Sets flow rate to 10μl/min
0:30 RPOINT    -b Baseline
*0:50  INJECT   %position 40  ! injects 40 μl of sample from the specified position
3:50 RPOINT    bound
*6:00 INJECT r2f3 20          ! injection of regeneration buffer
EXTRACLEAN
3:30 RPOINT    dissociated
END
```

Dilutions of the I domain are loaded into the rack in rounded polypropylene vials with polyethylene caps; the caps are necessary to prevent evaporation during the experiment and contains a septum that can be penetrated by the injection needle. The regeneration buffer is placed in a glass vial, which fits into the larger slots of the rack (here at position r2f3).

3.3. Analysis of Data

As mentioned above, the BIAcore 300 software generates sensorgrams where the signal from the reference cell, i.e., the flow cell without coupled ligand, has been subtracted from the signal from the flow cell with coupled ligand. The resulting sensorgrams, shown for 10 injections of I domain in the concentration range from 278 nM to 10 μM in Fig. 3a, consequently reflect the specific binding to iC3b. The simplest analysis for estimation of the K_D is based on the steady-state equilibrium-response, which was clearly reached after a contact time of about 200 s for all concentrations (Fig. 3a). By fitting Eq. 4 to the response levels plotted as a function of the I domain concentration (Fig. 3b), a K_D of 3.1 μM for the interaction between the I314G α_X I domain was

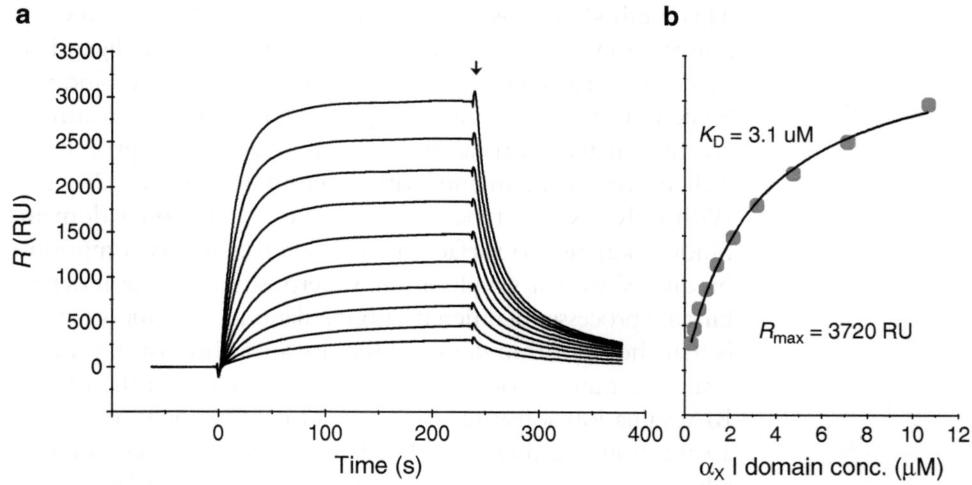

Fig. 3. Measurement of the dissociation constant (K_D) for the binding between the high-affinity α_X I domain and iC3b. (**a**) The α_X I domain was applied to the iC3b-coupled surface and a control surface in ten concentrations ranging from 278 nM to 10.6 μM. Following completion of the cycles as described in Fig. 2, the signals from the iC3b-coupled surface and the control surface were aligned, and the signal from the control surface was subtracted from the signal from the iC3b-coupled surface. The resulting ten sensorgrams are shown in (**a**) with an *arrow head* indicating the end of the injection. (**b**) The steady-state equilibrium responses (R_{Eq}) recorded toward the end of the contact time were plotted as a function of the I domain concentration (c) and fitted to Eq. 4 for the determination of K_D and R_{Max}.

found and a R_{Max} for the surface of 3720 RU (Fig. 3b). All calculations were made with the BIAevaluation™ software according to the manufacturer's instruction. Although the fitting of the data is clearly excellent, the numbers determined must nevertheless give rise to some concerns over the relevance of this procedure. The increment in R_{Eq} with an I domain concentration at 10 μM compared to the R_{Eq} at 8 μM was larger than the increment from 6 μM to 8 μM; this would not be the case if this was a 1:1 interaction with a K_D of 3 μM, since all of these analyte concentrations are higher than K_D. An even more disturbing finding came from calculation of the stoichiometry according to Eq. 5. With the M_r of the α_X I domain at 23,400 and the M_r of iC3b at 180,000, N becomes 5.6, i.e., suggesting that approximately six binding sites for I domains may exist within the iC3b molecule. This observation clearly points that the interaction between the I domain and the iC3b ligand is not a simple 1:1 interaction.

While the BIAevaluation software actually contains a model for the binding between the analyte and a surface with heterogeneous, i.e., multiple types or classes of, binding sites, the number of types covered by this approach is limited to two. Since the stoichiometry of the binding of the I314G α_X I domain to iC3b is of higher complexity, a more general model is required to analyze this interaction fully. Based on earlier observations from the binding between this I domain and fibrinogen, one approach is the application of the algorithms developed by Peter Schuck and colleagues (27, 28).

This methodology is, in fact, a landmark in obtaining meaningful information from binding experiments involving ligand-coated surfaces with a heterogeneous composition of analyte-binding sites. Since both technical and biological parameters may contribute to creating such circumstances, the relevance of this approach extends well beyond experiments with integrin I domains (25, 29–31). With only a single type of analyte, i.e., α_X I314G I domain, the binding kinetics to surfaces with a heterogeneous composition of binding sites is a simple linear superposition of the independent binding processes of different subpopulations. One important feature is that the algorithm does not require an ad hoc assumption about a specific number of discrete sites (e.g., two as in the case of the BIAevaluation software), and instead is based only on the weaker assumption that there could be heterogeneity of sites with rate and affinity constants in an operator-specified range (28). The exact methodology of this algorithm and the implementation of MatLab code for analysis of the binding data are beyond the scope of this chapter, but a brief outline is given below (more information can be found in excellent reviews on the topic (32, 33)). As shown by Svitel et al., R_{Eq} plotted as a function of the analyte concentration contains in principle information on the composition of binding sites on a surface (28). However, a far more rigorous analysis of binding site heterogeneity can be accomplished by considering the kinetics of the interaction (28). The data in the recorded sensorgrams can be exported through the BIAevaluation software through a tab-delimited file format, which is readable in Microsoft Excel or in Matlab.

By providing appropriate information on the concentrations of analyte applied to the surface, contact time, and the postinjection time interval required to be analyzed, the experimental data are characterized. For the analysis, the range for determination of the kinetic parameters, i.e., k_{on} and k_{off}, must also be specified. For instance, if the lower and upper limits of likely values of k_{off} are considered to be 10^{-3}/s and 10^{-1}/s, respectively, these values are supplied together with a value specifying how many k_{off} values (grid points) within the specified interval must be included in the analysis. Although the number of points could be raised to a very high number creating a nearly continuous grid, in practice this value must be kept sufficiently low to meet the computational power of the system used for the calculation. The same information is provided for k_{on}; however, for convenience this information is supplied by stipulating a range for K_D with an associated number of grid points. As will be noted, the information on k_{on} is contained in this information through the well-known relationship $K_D = k_{off}/k_{on}$. Together this information makes a two-dimensional grid, with each grid point corresponding to a type of interaction specified by k_{off} and K_D. Following analysis of the supplied data, the algorithm for each grid point associates a number (in RU) indicating the abundance of interactions at binding saturation with kinetic parameters corresponding to the grid point. Through Eq. 5, the number of such binding sites within a ligand molecule can be calculated; a value similar to R_{Max}, and hence the total number of binding sites with a single ligand molecule, can be calculated as the sum of the abundances of all grid points. The representation of the analysis is typically carried out by constructing plots with k_{off} on the ordinate and K_D on the abscissa, with contours or color representing the abundance of interactions for each grid point.

The above strategy was applied to the data obtained on the binding of α_X I314G I domain to iC3b (Fig. 3a). Information on the concentrations of injected I domain was given for each sensorgram. The grid consisted of points with 20 k_{off} values regularly spaced within the interval 10^{-3} to 10^{-1}/s and 20 K_D values regularly spaced within the interval 10^{-7} to 10^{-3} M. The calculated distribution is shown in Fig. 4b. It is clear that the most abundant type of interaction is characterized by a K_D of ~100 μM, i.e., with a low affinity. As reported earlier (25), this type of interaction is likely to originate from the binding between I314G α_X I domain and acidic side chains from residues in the ligand (in this case, iC3b). A smaller part of the interactions, roughly corresponding to six grid points carrying 360 RU of the I domain binding, was characterized by a K_D in the order of 1 μM, i.e., with almost a 100-fold higher affinity than the most abundant type of interaction. With the R_I and the M_rs for iC3b and the I domain as stated above, application of Eq. 5 suggested approximately two such high-affinity binding sites per iC3b molecule. In the analysis of the I314G α_X I domain binding to iC3b, the types of binding sites identified in this analysis do not appear as distinct with regard to their kinetic properties as have been reported for other similar studies (25, 34). Nevertheless, the finding of binding sites with a K_D ~ 1 μM is consistent with the outcome of the determination of K_D from R_{eq}, although with some important additional observations. Since the top concentration of I domain applied in this experiment was approximately 10 μM, the saturation of the low-affinity sites would be only 10% (mol/mol) at this concentration, while, by contrast, the high-affinity sites would be approximately 90% saturated according to Eq. 5; this condition is even more pronounced at lower concentration of I domain and the analysis in Fig. 3b was consequently dominated by the high-affinity interactions.

In conclusion, heterogeneity in the interaction between surface-exposed binding sites can lead to masking of low-affinity interactions. Although the biological implication of the heterogeneous interactions between the α_X I domain and ligands is still poorly understood, recent evidence suggests that a similar phenomenon is the case for α_M I domain (34), while the α_L I domain binds

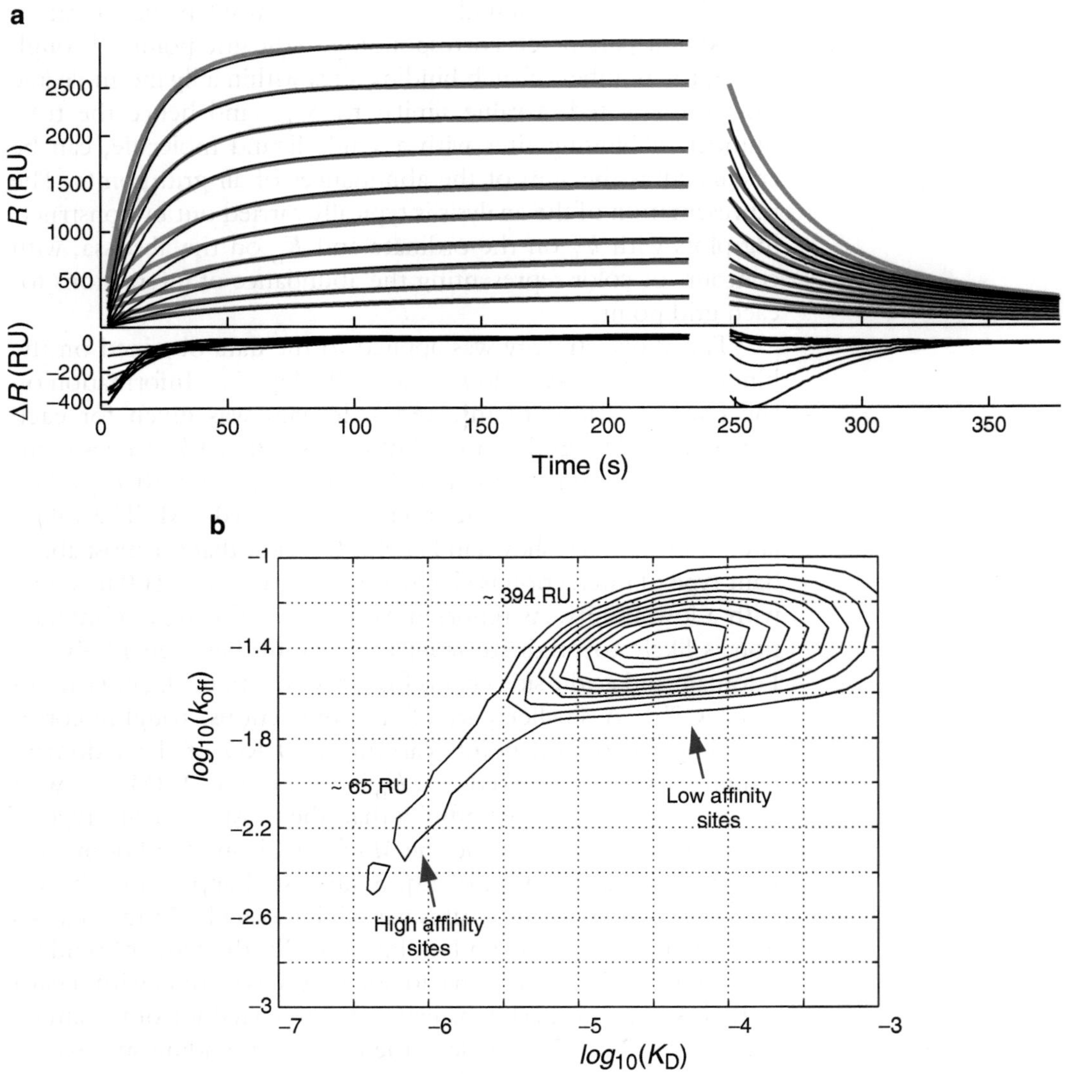

Fig. 4. Binding kinetics for the interaction between the high-affinity α_X I domain and iC3b. (**a**) The sensorgrams shown in Fig. 2a were analyzed with the method described by Svitel et al. (28), which assumes the binding signal to be a superposition of independent parallel binding reactions with a continuous distribution of thermodynamic and kinetic binding constants. The fit of the model is indicated with *solid gray lines* and the experimental data are indicated with *solid black lines*; the quality of the fit is shown in the lower panel as the residuals, i.e., the difference (ΔR) between the model and the experimental data. Although the identification of high-affinity sites agrees well with the analysis based on the response levels at steady-state equilibrium, it is also clear that the model does not account for all kinetic properties of the interaction. This is likely due to the low concentration of analyte compared to the K_D of the low-affinity interactions, which implies that the kinetic properties for these sites are ill-determined by the present analysis. (**b**) The distribution was tested over a grid of 20 values of k_{off} and 20 values of K_D. The grid points (not indicated) comprised values of k_{off} from 10^{-3} to 10^{-1}/s and values of K_D from 10^{-7} to 10^{-3} M. The abundance of each type of interaction is indicated with contours – in order to facilitate the interpretation of the contours, peak values are indicated. The R_{max} for all binding sites, i.e., the total binding capacity of the surface, was estimated to 10,950 RU.

ICAM-1 in a relatively homogeneous way (25). These findings clearly suggest that care must be taken in the analysis of I domain:ligand interactions, with as few as possible assumptions limiting the outcome of the analysis.

4. Notes

1. It is quite important that the analyte preparation is free of aggregates. Although I domain aggregation is unlikely to increase the ligand affinity compared to monomeric protein and thereby confound the analysis as reported for other adhesion molecules (35), aggregates may still negatively affect the outcome of the experiment. Aggregates can be removed by gel permeation chromatography.
2. The program should be supplemented by so-called tool boxes as specified by Svitel et al. (28).
3. Care should be taken to avoid salts in the coupling buffer since this tends to reduce the contact between the active chip surface and the protein; usually, this implies that the stock of ligand proteins (often kept in buffers with a physiologic salt concentration) should be kept at a high protein concentration, e.g., 1 mg/ml or higher, to allow for sufficient dilution of the adventitious salt when diluting the protein to the appropriate concentration for coupling.
4. The choice of reference or control flow cell can be made in several ways. One strategy would be to choose a nonbinding protein. However, for reasons made clear elsewhere (25), such protein substances are probably impossible to find in the case of α_X I domain. A simpler way is carried out here by simply preparing a surface, which differs from the ligand-coupled surface only with regard to the absence of any coupled protein.

Acknowledgments

I am indebted to Drs. Timothy A. Springer and Motomu Shimaoka for introducing me to integrin biology and the details of SPR instrumentation. Likewise, I thank Dr. Junichi Takagi for several helpful suggestions and intellectual support on the experimental investigations of integrin–ligand binding. Drs. Peter Schuck and

Juraj Svitel kindly provided their MatLab code for SPR data analysis, and helped the initial application of this approach to our data. Dr. Malcolm W. Turner kindly helped the writing of this chapter. This work was supported by grants from the Carlsberg Foundation and LEO Pharma Research Foundation.

References

1. Freifelder, D. (1999) *Physical Biochemistry*, 2nd ed., W. H. Freeman & Co., New York.
2. Price, N. C., and Dwek, R. A. (1974) *Principles and Problems in Physical Chemistry for Biochemists*, Clarendon Press, Oxford.
3. Piliarik, M., Vaisocherova, H., and Homola, J. (2009) Surface plasmon resonance biosensing, *Methods Mol. Biol.* **503**, 65–88.
4. Rich, R. L., and Myszka, D. G. (2007) Higher-throughput, label-free, real-time molecular interaction analysis, *Anal. Biochem.* **361**, 1–6.
5. Schuck, P. (1997) Use of surface plasmon resonance to probe the equilibrium and dynamic aspects of interactions between biological macromolecules, *Annu. Rev. Biophys. Biomol. Struct.* **26**, 541–566.
6. Karlsson, R., and Larsson, A. (2004) Affinity measurement using surface plasmon resonance, *Methods Mol. Biol. 248*, 389–415.
7. Hynes, R. O. (2002) Integrins: Bidirectional, Allosteric Signaling Machines, *Cell* **110**, 673–687.
8. Springer, T. A. (1990) Adhesion receptors of the immune system, *Nature* **346**, 425–434.
9. Springer, T. A. (1997) Folding of the N-terminal, ligand-binding region of integrin alpha-subunits into a beta-propeller domain, *Proc. Natl. Acad. Sci. USA.* **94**, 65–72.
10. Pytela, R. (1988) Amino acid sequence of the murine Mac-1 alpha chain reveals homology with the integrin family and an additional domain related to von Willebrand factor, *EMBO J.* 7, 1371–1378.
11. Oxvig, C., and Springer, T. A. (1998) Experimental support for a beta-propeller domain in integrin alpha-subunits and a calcium binding site on its lower surface, *Proc. Natl. Acad. Sci. USA.* **95**, 4870–4875.
12. Lee, J. O., Bankston, L. A., Arnaout, M. A., and Liddington, R. C. (1995) Two conformations of the integrin A-domain (I-domain): a pathway for activation? *Structure* **3**, 1333–1340.
13. Lee, J. O., Rieu, P., Arnaout, M. A., and Liddington, R. (1995) Crystal structure of the A domain from the alpha subunit of integrin CR3 (CD11b/CD18), *Cell* **80**, 631–638.
14. Vorup-Jensen, T., Ostermeier, C., Shimaoka, M., Hommel, U., and Springer, T. A. (2003) Structure and allosteric regulation of the alpha X beta 2 integrin I domain, *Proc. Natl. Acad. Sci. USA.* **100**, 1873–1878.
15. Nolte, M., Pepinsky, R. B., Venyaminov, S., Koteliansky, V., Gotwals, P. J., and Karpusas, M. (1999) Crystal structure of the alpha1beta1 integrin I-domain: insights into integrin I-domain function, *FEBS Lett.* **452**, 379–385.
16. Emsley, J., King, S. L., Bergelson, J. M., and Liddington, R. C. (1997) Crystal structure of the I domain from integrin alpha2beta1, *J. Biol. Chem.* **272**, 28512–28517.
17. Qu, A., and Leahy, D. J. (1995) Crystal structure of the I-domain from the CD11a/CD18 (LFA-1, alpha L beta 2) integrin, *Proc. Natl. Acad. Sci. USA.* **92**, 10277–10281.
18. Shimaoka, M., Takagi, J., and Springer, T. A. (2002) Conformational regulation of integrin structure and function, *Annu. Rev. Biophys. Biomol. Struct.* **31**, 485–516.
19. Li, R., Rieu, P., Griffith, D. L., Scott, D., and Arnaout, M. A. (1998) Two functional states of the CD11b A-domain: correlations with key features of two Mn^{2+}-complexed crystal structures, *J. Cell. Biol.* **143**, 1523–1534.
20. Shimaoka, M., Lu, C., Palframan, R. T., von Andrian, U. H., McCormack, A., Takagi, J., and Springer, T. A. (2001) Reversibly locking a protein fold in an active conformation with a disulfide bond: integrin alphaL I domains with high affinity and antagonist activity in vivo, *Proc. Natl. Acad. Sci. USA.* **98**, 6009–6014.
21. Shimaoka, M., Xiao, T., Liu, J. H., Yang, Y., Dong, Y., Jun, C. D., McCormack, A., Zhang, R., Joachimiak, A., Takagi, J., Wang, J. H., and Springer, T. A. (2003) Structures of the alpha L I domain and its complex with ICAM-1 reveal a shape-shifting pathway for integrin regulation, *Cell* **112**, 99–111.
22. Xiong, J. P., Li, R., Essafi, M., Stehle, T., and Arnaout, M. A. (2000) An isoleucine-based allosteric switch controls affinity and shape shifting in integrin CD11b A-domain, *J. Biol. Chem.* **275**, 38762–38767.

23. Jin, M., Song, G., Carman, C. V., Kim, Y. S., Astrof, N. S., Shimaoka, M., Wittrup, D. K., and Springer, T. A. (2006) Directed evolution to probe protein allostery and integrin I domains of 200,000-fold higher affinity, *Proc. Natl. Acad. Sci. USA.* **103**, 5758–5763.
24. Vorup-Jensen, T., Waldron, T. T., Astrof, N., Shimaoka, M., and Springer, T. A. (2007) The connection between metal ion affinity and ligand affinity in integrin I domains, *Biochim. Biophys. Acta.* **1774**, 1148–1155.
25. Vorup-Jensen, T., Carman, C. V., Shimaoka, M., Schuck, P., Svitel, J., and Springer, T. A. (2005) Exposure of acidic residues as a danger signal for recognition of fibrinogen and other macromolecules by integrin alphaX-beta2, *Proc. Natl. Acad. Sci. USA.* **102**, 1614–1619.
26. Vorup-Jensen, T., Chi, L., Gjelstrup, L. C., Jensen, U. B., Jewett, C. A., Xie, C., Shimaoka, M., Linhardt, R. J., and Springer, T. A. (2007) Binding between the integrin alphaXbeta2 (CD11c/CD18) and heparin, *J. Biol. Chem.* **282**, 30869–30877.
27. Svitel, J., Boukari, H., Van Ryk, D., Willson, R. C., and Schuck, P. (2007) Probing the functional heterogeneity of surface binding sites by analysis of experimental binding traces and the effect of mass transport limitation, *Biophys J.* **92**, 1742–1758.
28. Svitel, J., Balbo, A., Mariuzza, R. A., Gonzales, N. R., and Schuck, P. (2003) Combined affinity and rate constant distributions of ligand populations from experimental surface binding kinetics and equilibria, *Biophys J.* **84**, 4062–4077.
29. O'Shannessy, D. J., and Winzor, D. J. (1996) Interpretation of deviations from pseudo-first-order kinetic behavior in the characterization of ligand binding by biosensor technology, *Anal. Biochem.* **236**, 275–283.
30. Lipschultz, C. A., Li, Y., and Smith-Gill, S. (2000) Experimental design for analysis of complex kinetics using surface plasmon resonance, *Methods* **20**, 310–318.
31. Collins, A. V., Brodie, D. W., Gilbert, R. J., Iaboni, A., Manso-Sancho, R., Walse, B., Stuart, D. I., van der Merwe, P. A., and Davis, S. J. (2002) The interaction properties of costimulatory molecules revisited, *Immunity* **17**, 201–210.
32. Sundberg, E. J., Andersen, P. S., Gorshkova, I. I., and Schuck, P. (2007) Surface Plasmon Resonance Biosensing in the Study of Ternary Systems of Interacting Proteins, in *Protein Interactions* (Schuck, P., Ed.), pp 97–141, Springer US, New York.
33. Schasfoort, R. B. M., and Schuck, P. (2008) Future Trends in SPR Technology, in *Handbook of Surface Plasmon Resonance* (Schasfoort, R. B. M., and Tudos, A. J., Eds.), pp 354–392, RSC Publishing, Cambridge.
34. Stapulionis, R., Oliveira, C. L., Gjelstrup, M. C., Pedersen, J. S., Hokland, M. E., Hoffmann, S. V., Poulsen, K., Jacobsen, C., and Vorup-Jensen, T. (2008) Structural insight into the function of myelin basic protein as a ligand for integrin alpha M beta 2, *J. Immunol.* **180**, 3946–3956.
35. Davis, S. J., Ikemizu, S., Wild, M. K., and van der Merwe, P. A. (1998) CD2 and the nature of protein interactions mediating cell-cell recognition, *Immunol. Rev.* **163**, 217–236.

Chapter 6

Cell-Free Ligand-Binding Assays for Integrin LFA-1

Koichi Yuki

Abstract

Integrin LFA-1 plays an important role in leukocyte trafficking to inflammatory sites as well as the interaction of T cells with antigen-presenting cells. Inhibition of the LFA-1:ICAM-1 interactions is a promising therapeutic approach for alleviating inflammation and autoimmune diseases. This chapter describes cell-free assays to study the LFA-1:ICAM-1 interactions. These assays may be used for screening novel LFA-1 antagonists.

Key words: LFA-1, ICAM-1, Cell-free assay, Inflammation, Autoimmune diseases

1. Introduction

LFA-1 (lymphocyte function-associated antigen-1, integrin $\alpha L\beta 2$) is a member of $\beta 2$ integrin family and expressed on all leukocyte subsets. LFA-1 binds to intercellular adhesion molecules (ICAMs) including its major endothelial ligand ICAM-1. The interaction of LFA-1 with its ligands is important in recruiting leukocytes to sites of inflammation, as well as in stabilizing the immunological synapse formed between T cells and antigen-presenting cells (APCs) (1, 2). Therefore, the inhibition of LFA-1:ICAM-1 binding is a promising therapeutic target to treat inflammation and autoimmune diseases. A humanized antibody to LFA-1 efalizumab (Raptiva™) was approved for the therapy of moderate to severe forms of psoriasis by the FDA in 2003 (3). Although efalizumab is effective in alleviating inflammation in psoriasis, this antibody treatment was associated with a rare but fatal disease of the central nerve system called progressive multifocal leukoencephalopathy, which is caused by reactivaton of latent JC virus infection possibly due to unwanted global immune

Motomu Shimaoka (ed.), *Integrin and Cell Adhesion Molecules: Methods and Protocols*,
Methods in Molecular Biology, vol. 757, DOI 10.1007/978-1-61779-166-6_6, © Springer Science+Business Media, LLC 2011

suppression (4). Therefore, novel LFA-1 antagonists without global immune suppression are warranted.

The overall strength of integrin-mediated cell adhesion is enhanced by the combination of both affinity and valency upregulation. Affinity upregulation refers to an increase in individual integrin receptor affinity and is mediated by the conformational alterations. In contrast, valency upregulation refers to an increase in the number of adhesive bonds formed at cell-to-cell or cell-to-matrix contact sites and is mediated by an increase in either cell surface receptor diffusivity or local density (5). Cell adhesion assays usually study the overall binding strength of integrin-expressing cells to ligand substrates. Thus, cell adhesion assays fail to distinguish the impact of antagonists or chemicals on affinity regulation from that on valency upregulation. For example, a drug candidate that affects the plasma membrane diffusivity can modify integrin-mediated cell adhesion without directly acting on integrin protein. In addition, another drug candidate that affects intracellular signaling cascades leading to integrin activation could also modify integrin-mediated cell adhesion without directly acting on integrin protein. To study direct impact of drug candidates on integrin and/or its affinity regulation, this chapter describes two (ELISA- and bead-based) cell-free LFA-1 ligand-binding assays (see Note 1).

2. Materials

2.1. ELISA Plate-Based ICAM-1 Binding Assay

1. Capturing antibody: Use function non-blocking antibodies such as CBR LFA1/2 (6) or TS2/4 (7).
2. Recombinant human LFA-1 extracellular portion (R&D systems).
3. ICAM-1-Fc fusion protein: Use ICAM-1-Fcα (8) or -Fcγ (R&D systems).
4. Detection antibody: HRP-labeled goat anti-human IgA antibody (for ICAM-1-Fcα) or HRP-labeled goat anti-human IgG antibody (for ICAM-1-Fcγ).
5. Color reagent A and B (R&D systems).
6. Coating buffer: 15 mM Na_2CO_3, 35 mM $NaHCO_3$, pH at 9.6.
7. Tris-buffered saline with Tween (TBS-T): 25 mM Tris–HCl, pH 7.4, 150 mM NaCl, 2 mM KCl, 0.05% (v/v) Tween20.
8. HEPES-buffered saline (HBS): 25 mM HEPES, 150 mM NaCl, pH 7.4.
9. Blocking buffer: HBS with 2% (w/v) fraction V bovine serum albumin (BSA).

10. 2× nonstimulation buffer: HBS, 2 mM $CaCl_2$ and 2 mM $MgCl_2$ (see Note 2).
11. 2× stimulation buffer: HBS, 2 mM $MnCl_2$ (see Note 2).
12. 96-well flat bottom ELISA plate.

2.2. Bead-Based ICAM-1 Binding Assay

1. Recombinant human LFA-1 extracellular portion (R&D systems).
2. ICAM-1-Fc fusion protein: Use ICAM-1-Fcα or -Fcγ (R&D systems).
3. FITC-labeled goat anti-human IgA or IgG antibody.
4. Washing buffer: HBS.
5. Coating buffer: HBS with 0.1% (w/v) BSA.
6. Blocking buffer: HBS with 2% (w/v) BSA.
7. 2× nonstimulation buffer: HBS, 2 mM $CaCl_2$ and 2 mM $MgCl_2$.
8. 2× stimulation buffer: HBS, 2 mM $MnCl_2$.
9. Silica beads amine coated (5 μm, Bang Lab, Inc.) (see Note 3).

3. Methods

3.1. ELISA Type ICAM-1 Binding Assay

1. Add 50-μl of capturing antibody (20 μg/ml in a coating buffer) to each well of ELISA plates. Incubate at 4°C overnight.
2. Wash plates three times with 150 μl/well of TBS-T.
3. Add 150 μl of blocking buffer to each well and incubate at room temperature for 1 h.
4. Wash plates three times with 150 μl/well of TBS-T.
5. Add 50 μl of LFA-1 protein (5 μg/ml in HBS/0.5% BSA) to each well. Incubate at room temperature for 2 h.
6. Wash plates five times with 150 μl/well of TBS-T.
7. Add to each well either 25 μl of 2× nonstimulation or stimulation buffer. In the case of testing antagonists, include 2× concentrations of antagonists.
8. Add to each well 25 μl of HBS containing 10 μg/ml ICAM-1-Fc fusion protein. This makes final concentrations of 5 μg/ml ICAM-1-Fc, 1 mM $CaCl_2$ and $MgCl_2$ (for nonstimulation buffer) or 1 mM $MnCl_2$ (for stimulation buffer); in the presence or absence of 1× concentrations of antagonists. During steps 9–11, maintain the same kind of cations (i.e., 1 mM $CaCl_2$/$MgCl_2$ or 1 mM $MnCl_2$) in TBS-T and detection antibody solution.

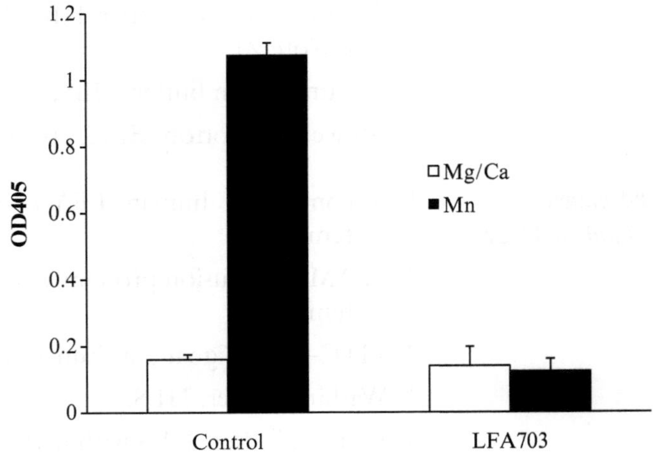

Fig. 1. ELISA plate-based ICAM-1 binding assay testing the LFA-1 small molecule antagonist LFA703 (11). LFA703 (1 μM) inhibited ICAM-1:LFA-1 binding in the presence of 1 mM $MnCl_2$.

9. Wash plates five times with 150 μl/well of TBS-T containing 1 mM $CaCl_2/MgCl_2$ or 1 mM $MnCl_2$.
10. Add to each well 50 μl of detection antibody solution (i.e., HRP-labeled goat anti-human IgA or IgG antibody diluted at 1:1,000–1:10,000 in HBS 1 mM $CaCl_2/MgCl_2$ or 1 mM $MnCl_2$). Incubate at 4°C for 30 min.
11. Wash plates ten times with 150 μl/well of TBS-T containing 1 mM $CaCl_2/MgCl_2$ or 1 mM $MnCl_2$.
12. Mix equal amount of color reagent A and B to make a substrate solution. Add 150 μl of the substrate solution to each well. Let color develop for 15 min.
13. Read absorbance at 405 nm using an ELISA reader (Fig. 1).

3.2. Bead-Based ICAM-1 Binding Assay

1. Suspend silica beads in 1 ml of washing buffer in a 1.5-ml microtube and vortex. Spin down at 2,200 × *g* for 3 min. Discard supernatant.
2. Repeat step 1.
3. Suspend a silica bead pellet at a density of 5×10^6 beads/ml with 500 μl of coating buffer containing 20 μg/ml LFA-1 protein. Incubate at 4°C overnight, while continuously rotating tubes with a laboratory rotator.
4. Preparation of multimeric ICAM-1-Fc solution. Mix 20 μl/ml ICAM-1-Fc fusion protein and 50 μg/ml FITC-labeled anti-human IgA or IgG antibody in 2× nonstimulation or 2× stimulation buffer for 15 min at room temperature. This multimeric ICAM-1-Fc solution is used at step 10.
5. Wash LFA-1-coated beads (from step 3) two times with washing buffer.

6. Incubate beads with blocking buffer at room temperature for 3 h.
7. Wash beads three time with washing buffer, and resuspend bead pellets in washing buffer.
8. Aliquot 5×10^5 of LFA-1-coated beads to each new microtube, spin down, and discard supernatant.
9. Add to each bead pellet 25 μl of 2× nonstimulation or 2× stimulation buffer in the presence or absence of 2× concentrations of antagonists. Add another 25 μl of the multimeric ICAM-1-Fc solution (from step 4). Incubate at room temperature for 30 min.
10. Wash three times with either nonstimulation or stimulation buffer.
11. Resuspend beads with 300 μl of nonstimulation or stimulation buffer.
12. Analyze with flow cytometry.

4. Notes

1. LFA-1 is used as a model integrin. Similar protocols can be applied for cell-free assays of other integrins.
2. LFA-1 is in a default low-affinity conformation in 1 mM $CaCl_2/MgCl_2$, whereas LFA-1 is converted to a high-affinity conformation in 1 mM $MnCl_2$ (9).
3. We used silica beads, as we sought to study the inhibitory effects of volatile anesthetics on LFA-1 (10) and polystyrene beads might not stand for volatile anesthetics. Silica microspheres tend to be hydrophilic, and absorb less proteins. Thus, polystyrene beads may be preferable for testing other antagonists.

Acknowledgments

The author would like to thank Dr. Motomu Shimaoka for supporting the research.

References

1. Van Seventer, G. A., Shimizu, Y., Horgan, K. J., and Shaw, S. (1990) The LFA-1 ligand ICAM-1 provides an important co-simulatory signal for cell receptor-mediated activation of resting T cells. *J. Immunol.* **144**, 4579–4586.
2. Dustin, M. L., and Springer, T. A. (1989) T-cell receptor cross-linking transiently stimulates adhesiveness through LFA-1. *Nature* **341**, 619–624.
3. Marecki, S., and Kirkpatrick, P. (2004) Efalizumab. *Nat. Rev. Drug. Discov.* **3**, 473–474.

4. Molloy, E. S., and Calabrese, L. H. (2009) Therapy: Targeted but not trouble-free: efalizmab and PML. *Nat. Rev. Rheumatol.* **5**, 418–419.
5. Kim, M., Carman, C. V., Yang, W., Salas, A., and Springer, T. A. (2004) The primary of affinity over clustering in regulation of adhesiveness of the integrin αLβ2. *J. Cell Biol.* **167**, 1241–1253.
6. Petruzzeli, L., Maduzia, L., and Springer, T. A. (1995) Activation of lymphocyte function-associated molecule-1 (CD11a/CD18) and Mac-1 (CD11b/CD18) mimicked by an antibody directed against CD18. *J. Immunol.* **155**, 854–866.
7. Sanchez-Madrid, F., Krensky, A. M., Ware, C. F., Robbins, E., Strominger, J. L., Burakoff, S. J., and Springer, T. A. (1982) Three distinct antigens associated with human T-lymphocyte-mediated cytolysis: LFA-1, LFA-2, and LFA-3. *Proc. Natl. Acad. Sci. USA.* **79**, 7489–7493.
8. Martin, S., Casasnovas, J. M., Staunton, D. E., and Springer, T. A. (1993) Efficient neutralization and disruption of rhinovirus by chimeric ICAM-1/immunoglobulin molecules. *J. Virol.* **67**, 3561–3568.
9. Shimaoka, M., and Springer, T. A. (2003) Therapeutic antagonists and conformational regulation of integrin function. *Nat. Rev. Drug Discov.* **2**, 703–716.
10. Yuki, K., Astrof, N.S., Bracken, C., Soriano, S. G., and Shimaoka, M. (2010) Sevoflurane binds and allosterically blocks integrin lymphocyte function-associated antigen-1. *Anesthesiology* **113**, 600–609.
11. Weitz-Schmidt, G., Welzenbach, K., Brinkmann, V., Kamata, T., Kallen, J., Bruns, C., Cottens, S., Takada, Y., and Hommel, U. (2001) Statins selectively inhibit leukocyte function antigen-1 by binding to a novel regulatory integrin site. *Nat. Med.* **7**, 687–692.

Part II

Structural Biology Approaches for Studying Integrins and Related Cell Adhesion Molecules

Chapter 7

Overview: Structural Biology of Integrins

Guanyuan Fu, Wei Wang, and Bing-Hao Luo

Abstract

Integrins are cell adhesion molecules that play important roles in many biological processes including hemostasis, immune responses, development, and cancer. Their adhesiveness is dynamically regulated through a process termed inside-out signaling. In addition, ligand binding transduces outside-in signals from the extracellular domain to the cytoplasm. Advances in the past several years have shed light on structural basis for integrin regulation and signaling, especially how the large-scale reorientations of the ectodomain are related to the inter-domain and intra-domain shape shifting that changes ligand-binding affinity. Experiments have also shown how the conformational changes of the ectodomain are linked to changes in the α- and β-subunit transmembrane and cytoplasmic domains.

Key words: Cell adhesion, Conformational change, Allosteric regulation, Inserted domain, Hybrid domain swing-out, Bidirectional signaling

1. Introduction

Integrin family of adhesion receptors are non-covalently associated α/β heterodimers. They are so named because they serve to integrate the extracellular and intracellular environments by binding to ligands outside the cell and to cytoskeletal components and signaling molecules inside the cell, and to transmit signals bidirectionally across the plasma membrane (1, 2). Integrins mediate cell–cell, cell–extracellular matrix, and cell–pathogen interactions. They, together with their ligands, play diverse and important roles in many biological processes including hemostasis, immune responses, development, and cancer.

Integrins are restricted to metazoans, with different subunit sets in different phyla (3). In vertebrates, 18 α-subunits and 8 β-subunits have been identified, forming at least 24 distinct αβ pairs (Fig. 1). Although all integrins connect to the actin-based

Motomu Shimaoka (ed.), *Integrin and Cell Adhesion Molecules: Methods and Protocols*,
Methods in Molecular Biology, vol. 757, DOI 10.1007/978-1-61779-166-6_7,

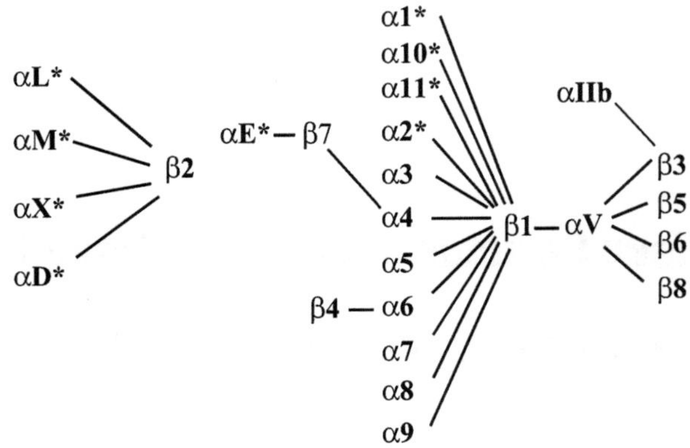

Fig. 1. The mammalian integrin receptor family. 18 α- and 8 β-subunits form 24 heterodimers. Half of the α-subunit-containing I domains are indicated by *asterisks*. The figure is adapted from the review (8).

microfilaments of the cytoskeleton with the exception of α6β4, which makes connection to the intermediate filaments, different integrin families with distinct subunit combinations have diverse functions due to diversity in ligand-binding specificity and submembrane linkers to cytoskeleton.

The uniqueness of integrins as adhesion molecules is that their adhesiveness can be dynamically regulated through a process termed inside-out signaling. Thus, stimuli received by cell surface receptors for chemokines, cytokines, and foreign antigens initiate intracellular signals that impinge on integrin cytoplasmic domains and alter adhesiveness for extracellular ligands. In addition, ligand binding transduces signals from the extracellular environment to the cytoplasm and activates many intracellular signaling pathways, a process known as outside-in signaling. In this chapter, our understandings of integrin structure and the conformational signaling pathway are summarized.

2. An Overview of Integrin Domain Structure

2.1. Integrin Global Topology

Both the α- and β-integrin subunits are type I transmembrane (TM) glycoproteins with large extracellular domains, single spanning TM domains, and, with the exception of β4, short cytoplasmic domains (Fig. 2). From electron microscopy (EM) studies, it has been known for years that the overall topology of integrins includes an extracellular globular N-terminal ligand-binding head domain, representing a critical α- and β-subunit interface, standing on two long and extended C-terminal legs or stalks, which connect to the TM and cytoplasmic domains of each subunit (4). However, X-ray crystal structures of the extracellular domain of the integrin αVβ3 provided the surprising finding that the legs were severely bent at the "genu" or knee, generating a V-shaped

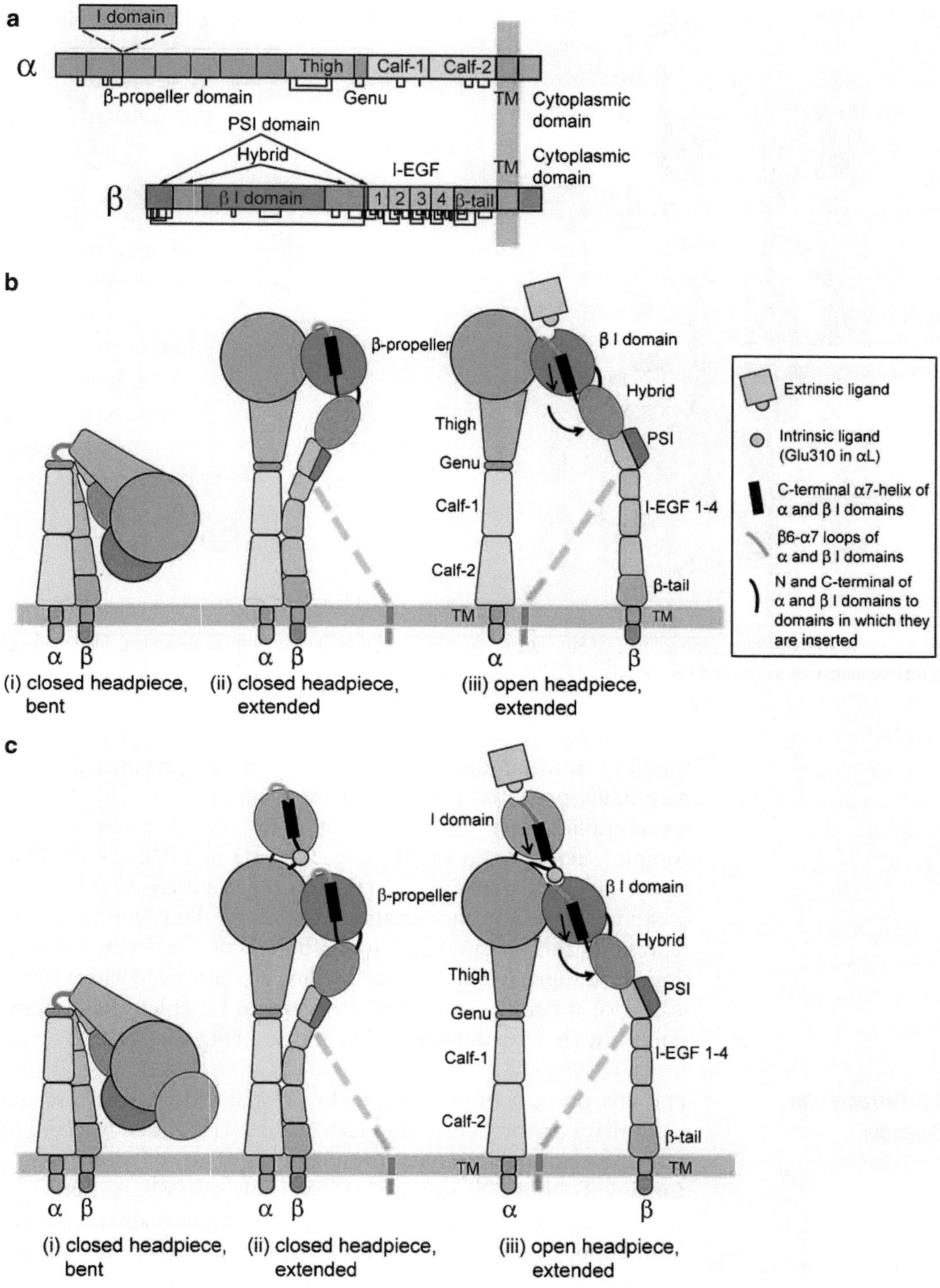

Fig. 2. Integrin architecture. (**a**) Organization of domains within the primary structures. Some α-subunits contain an I domain inserted in the position denoted by the dotted lines. Cysteines and disulfides are shown as lines below the stick figures. (**b** and **c**) Domain rearrangement of integrins lacking (**b**) or containing (**c**) an α I domain during activation. The β-subunit lower legs are flexible and are, therefore, shown in what may be the predominant (*solid representation*) and less predominant (*dashed lines*) orientations. The figures are adapted from the review (8).

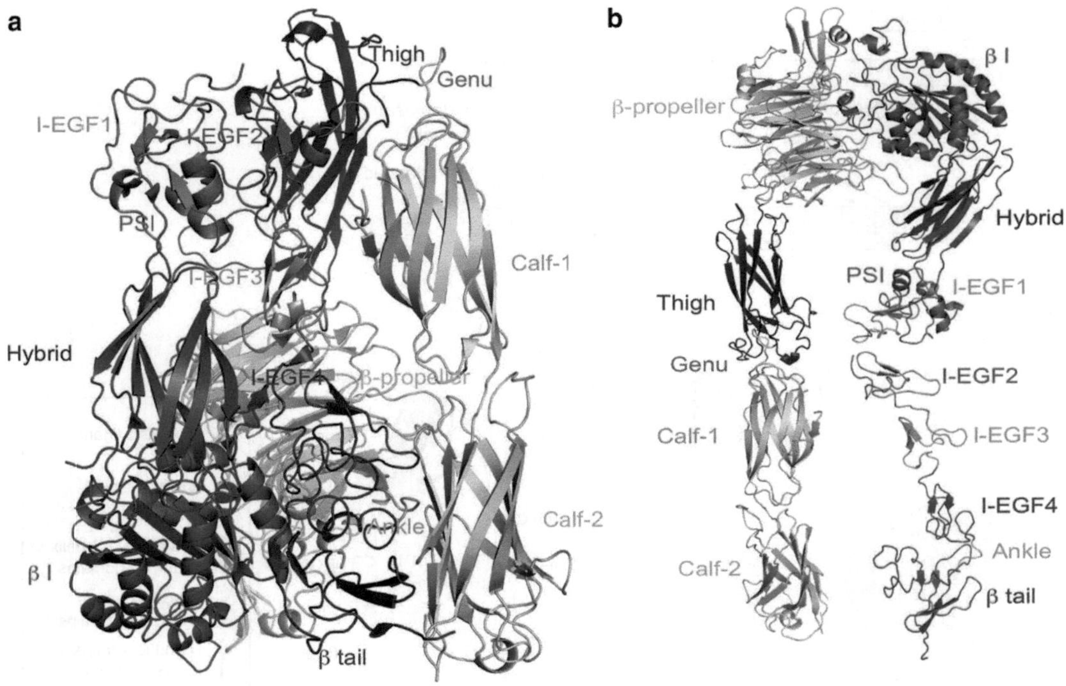

Fig. 3. Crystal structure of integrin αIIbβ3. (**a**) Structure of the αIIbβ3 in the bent conformation. (**b**) An extended model of αIIbβ3 by torsion at the α- and β-knees (From PDB ID code 3FCS).

topology in which the head domain was closely juxtaposed to the membrane-proximal portions of the stalks (5, 6). This finding was consolidated in the more recent X-ray crystal structure of the complete ectodomain and the negatively stained EM of αIIbβ3 in a physiologically resting state (Fig. 3) (7). An increasing number of studies have together established that the bent conformation represents the physiological low affinity state, whereas priming (inside-out signaling) and ligand binding are associated with a large-scale global conformational rearrangement in which the integrin extends with a "switchblade"-like motion (Fig. 2) (7–13).

2.2. Extracellular Domains

2.2.1. The α-Subunit

The first domain of integrins to be crystallized was the inserted (I) domain or von Willebrand factor A domain which is found in half of integrin α-subunits (14) (Fig. 1). The α I domain is a domain of about 200 amino acids and is the major ligand-binding site in integrins in which it is present. The α I domain adopts the dinucleotide-binding or Rossmann fold, with α-helices surrounding a central β-sheet (Fig. 4). There are seven major α-helices, and several short α-helices that differ between I domains in various α-subunits. The β-sheet contains five parallel β-strands and one antiparallel β-strand. β-strands and α-helices tend to alternate in the secondary structure, with the α-helices wrapping around the domain in counterclockwise order when viewed from the top.

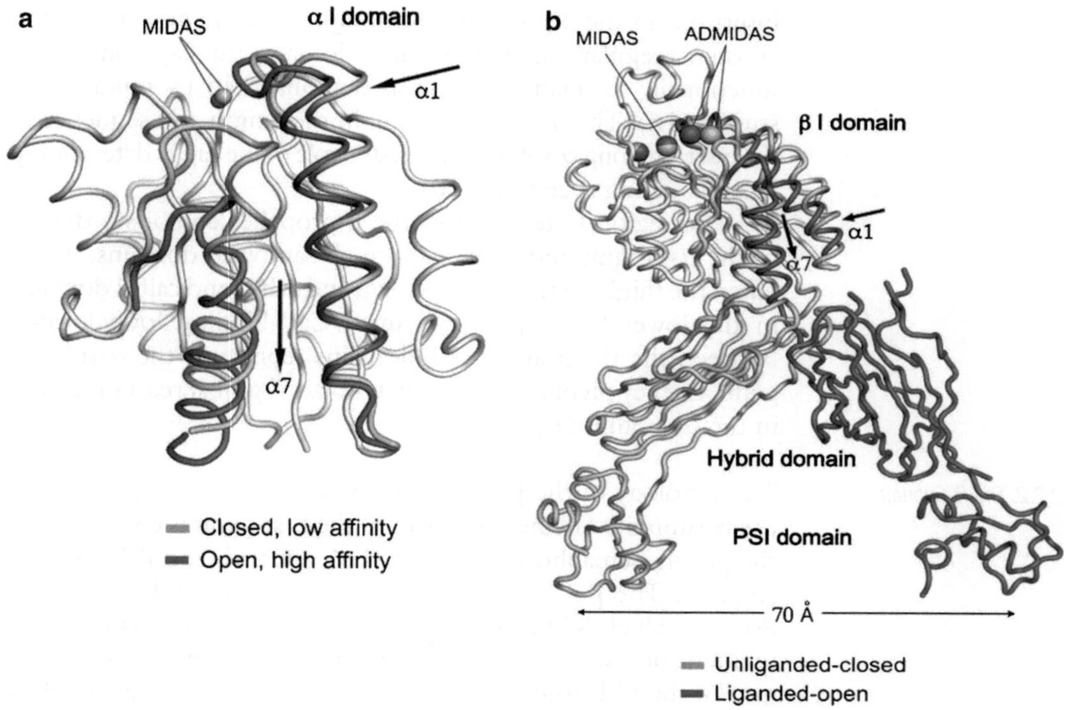

Fig. 4. Conformational change and allosteric transmission by I domains. The moving segments of the backbone and the MIDAS metal ions are labeled. The direction of movement is indicated with *arrows*. (**a**) Superposition of closed and open structures of the α I domains (from PDB ID codes 1JLM and 1IDO). (**b**) Superposition of closed and open structures of the β I domains and their linkages to hybrid and PSI domains (from PDB ID codes 3FCS and 2VDL).

A divalent cation (Mg^{2+}) sits on the top of the domain, ligated by five side chains located in three different loops. The first of these loops, β1-α1 (between β-strand 1 and α-helix 1), contains three coordinating residues in a sequence that is a signature of I domains, DXSXS. The second loop donates a coordinating Thr residue, and the third loop donates an Asp residue. Divalent cations are universally required for ligand binding by integrins. In α I domains, the metal-coordinating residues and the residues surrounding the metal-binding site are important for ligand binding. Therefore, this site has been designated the metal ion-dependent adhesion site (MIDAS).

The N-terminal region of the integrin α-subunit contains seven segments of about 60 amino acids, each with a weak sequence homology to one another. These residues were initially predicted (15) and later confirmed by crystal structures (6, 7, 10) to fold into a seven-bladed β-propeller domain. The β-propeller forms the major α-subunit contribution to the head domain and provides a critical interface with the β-subunit. When present, α I domain is inserted between β-sheets 2 and 3 of the β-propeller (Fig. 2). Interestingly, the crystal structure of αIIbβ3 headpiece, which lacks an α I domain, reveals a cap sub-domain that comprises four insertions in the β-propeller. Although probably not involved in allosteric regulation, as with the I domain, the cap contributes functionally to ligand binding as demonstrated by mutagenesis studies (16). The marked variation in the length and sequence of the inserts among α-subunits suggests a role of the cap in determining ligand-binding specificity (10).

The region C-terminal to the β-propeller comprises the leg of the α-subunit and contains three β-sandwich domains, designated the thigh domain in upper leg, and calf-1 and calf-2 domains in the lower leg (Fig. 3). A small Ca^{2+}-binding loop located between the thigh and calf-1 domains represents the α-subunit genu, the key pivot point for "switchblade" global rearrangement in the α-subunit (Fig. 3).

2.2.2. The β-Subunit

The topology of the β-subunit is more complicated than that of the α-subunit. The head domain of the β-subunit is comprised of the plexin/semaphorin/integrin (PSI), hybrid, and β I domains (Fig. 2). The β I domain is inserted in the β-sandwich hybrid domain, which is in turn inserted in the PSI domain. The second segment of the PSI domain is very short, but can be assigned as part of the PSI domain because it contains β3-Cys435, which is involved in a long-range disulfide bond to β3-Cys11 in the first segment of the PSI domain and this disulfide is structurally conserved in other PSI domains.

The integrin β I domain is a highly conserved domain of about 240 residues and is analogous in structure to the α I domain (Fig. 4). There are two additional segments in the β I domain: one is known as the specificity-determining loop because of its role in ligand binding and the other helps form a critical interface with α-subunit β-propeller. Mutations in the β2 I domain that disrupt this interface has been known to cause leukocyte adhesion deficiency (17, 18). Like the α I domain, the β I domain contains a MIDAS for binding negatively charged residues, which physically binds Mg^{2+}. In addition, two adjacent metal ion-binding sites, which physically bind Ca^{2+}, share some coordinating residues with the MIDAS. The two sites are termed the synergistic metal ion-binding site (SyMBS) and the adjacent to metal ion-dependent adhesion site (ADMIDAS). In integrins lacking α I domains, the β I domain MIDAS appears to bind ligand directly, whereas in α I domain-containing integrins, it acts to regulate the ligand-binding activity of the α I domain. The inserted topology of the β I domain plays a critical role in its allosteric regulation and signaling, as discussed below.

The region C-terminal to the hybrid domain comprises the cysteine-rich β-subunit leg/stalk region, which contains four integrin epidermal growth factor-like (I-EGF) domains, a β-ankle, and a β-tail domain. In the αVβ3 crystal structure, the I-EGF domains 1 and 2 were not resolved (6). However, subsequent

NMR studies resolved the β2 I-EGF domains 2 and 3 (19). Later, a high-resolution crystal structure of the β2 PSI, hybrid, and I-EGF1 domains was solved (20). More recently, the entire extracellular domain of αIIbβ3 in a physiologically resting state has been resolved (7). In this 2.55 Å resolution crystal structure, the conformation of I-EGF domains 1 and 2 at the knee is revealed at the epicenter of the conformational change involved in integrin activation (Fig. 3a). In the bent conformation of integrin αIIbβ3, there is a highly acute bend between the I-EGF domains 1 and 2. This structure confirms previous hypothesis and is consistent with results from other studies. I-EGF domains 2, 3, and 4 in the lower β3 leg extend in an almost straight orientation, with about a 90° left-handed twist between successive domains.

2.3. Transmembrane and Cytoplasmic Domains

In the crystal structures of the full-length αVβ3 and αIIbβ3 ectodomains, the integrins are in the bent conformation with the α- and β-subunit C-termini only a few angstroms apart (6, 7), consistent with association of the α- and β-subunit transmembrane domains. Recently, a newly developed method combining disulfide scanning with Rosetta computational modeling has been used to solve the structure of the integrin TM and cytoplasmic domains (9). Since the structure is obtained based on experimental data using intact integrins with the extracellular and cytoplasmic domains that regulate TM association on the mammalian cell surface, we believe that it most likely represents the physiological structure in the resting state. In this structure, the αIIb GXXXG motif and their β3 counterparts of the TM domains associate with a ridge-in-groove packing (Fig. 5a). The αIIb TM α-helix extends beyond the 23-residue TM hydrophobic segment and then Gly-991 of GFFKR is a turn which changes the TM right-handed α-helix to left-handed one, making Phe-992 and Phe-993 sit in the interface of αIIb and β3 at the membrane/cytoplasm interface (Fig. 5b), and thus, this motif is critical for α/β association. A salt bridge between αIIb Arg-995 and β3 Asp-723 was proposed previously based on mutagenesis data (21). In the structure, Arg-995 is close to both Asp-723 and Glu-726, consistent with this electrostatic interaction. However, there are a variety of different conformations of the side chains of Arg-995 and Asp-723, indicating that this salt bridge is not absolutely necessary for the association.

3. Conformational Regulation of Integrin

3.1. Conformational Regulation of the α I Domains

Structural studies of α I domains in the presence and absence of a ligand, in varied cation conditions, and with mutations that stabilize distinct affinity states have provided a mechanistic understanding of conformational regulation during both priming and ligand binding.

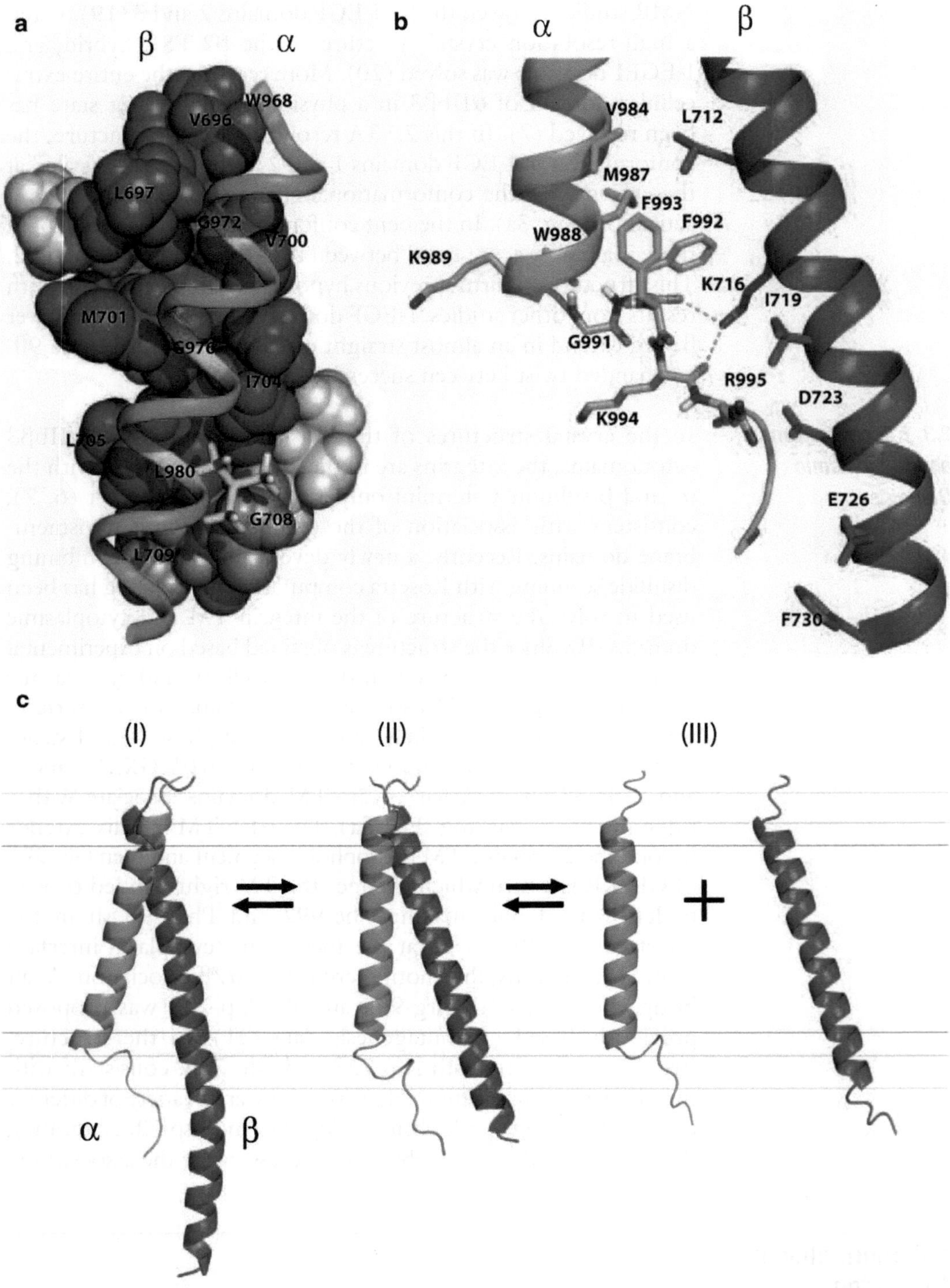

Fig. 5. Structure of the integrin αIIbβ3 transmembrane (TM) and cytoplasmic domains. (**a**) The interface between two associating TM domains on the cell surface. (**b**) Cytoplasmic fragment association of integrin αIIb-and β3-subunits in the Disulfide/Rosetta structure. (**c**) Predicted model of integrin TM activation. (I) The resting state represented as Disulfide/Rosetta structure; (II) the "intermediate" or "transient" state represented as NMR structure of integrin αIIbβ3 TM and cytoplasmic domain peptides (PDB ID 2K9J); (III) the activated state represented as monomeric α_{IIb} and β_3 NMR structures (PDB ID codes 2K1A and 2RMZ). The outer bounds of the hydrophobic, interface, and polar region of the membrane are shown.

The α I domains have been crystallized in three distinct forms, termed closed, intermediate, and open conformations (14, 22, 23). These demonstrate distinct coordination of the metal in the MIDAS, arrangement of the β6-α7 loop, and axial disposition of the C-terminal α7 helix along the side of the α I domain (14, 22, 23) (Fig. 4a). At the α I domain MIDAS, five residues and several water molecules contribute oxygen atoms to the primary and secondary coordination spheres surrounding the metal. In the open conformation of the MIDAS, two serines and one threonine are in the primary coordination sphere, whereas two aspartic acid residues are in the secondary coordinating sphere and fix the positions of coordinating water molecules. Notably, the glutamic acid contributed by the ligand mimetic residue donates the only negatively charged oxygen to the primary coordination sphere in the open conformation. The lack of any charged group in the primary coordination sphere donated by the α I domain is hypothesized to enhance the strength of the metal–ligand bond. In the closed conformation, the threonine moves from the primary to the secondary coordination sphere, and one of the aspartic acid residues moves from the secondary to the primary coordination sphere. This movement is consistent with the idea that an energetically favorable MIDAS requires at least one primary coordination to a negatively charged oxygen, and when this is not provided by a ligand, there is a structural rearrangement within the I domain to provide this from within the MIDAS. The backbone and side chain rearrangements in the I domain are accompanied by a 2.3-Å "sideways" movement of the metal ion away from the threonine and toward the aspartic acid on the opposite side of the coordination shell.

The structural rearrangement of the MIDAS is coupled to backbone movements of the loops that bear the coordinating residues. Linked structural shifts occur in neighboring loops on the top of the α I domain, which are also coupled movements in α-helices on the "side" of the domain. In the largest movement in the transition from the closed to the open structure, the C-terminal α7-helix moves 10 Å down the side of the domain. The axial displacement of the α7-helix represents the critical linkage for transmission of conformational signals both within the α I domain and throughout the integrin, as discussed later (Fig. 4a). Introduction of pairs of cysteines that stabilize the α7-helix in intermediate and open conformations (shifted axially downward by ~5 or 10 Å relative to the closed conformation, respectively) induced rearrangements in the MIDAS and surrounding loops, which were coupled to 500- and 10,000-fold increases in affinity for ICAM-1, respectively (22). Thus, downward pull on the α7-helix is sufficient for priming the α I domain into higher affinity states. Conversely, the act of binding a ligand directly induces MIDAS rearrangements that lead to downward displacement of the α7-helix. Thus, the transmission of inside-out and outside-in signaling within the α I domain occurs along the same pathway, but flows in opposite directions.

3.2. Conformational Activation of the β I Domains

The β I domain directly binds ligands in integrins that lack α I domains, and indirectly regulates ligand binding by integrins that contain α I domains. The structure of the β I domain was first solved in the context of αVβ3 extracellular domains in the absence of ligands (6). Subsequent mutagenesis studies (24–28) and the structure of the αIIbβ3 headpiece co-crystallized with different ligand mimetic drugs (10) revealed conformational change in the open, high-affinity state of the β I domain. With the solution of the αIIbβ3 complete ectodomain crystal structure (7), more detailed conformational change in the β3 I domain can be obtained by superposition of the headpieces from this physiologically resting, unliganded, closed structure with the high-affinity, liganded, open structure (Fig. 4b), revealing the structural similarity of domain allostery between the α I and the β I domain activation. In the high-affinity, liganded β I domain compared with the low-affinity, unliganded β I domain, movements of β1-α1 and β6-α7 loops and of the α1 and α7 helices occur. Coordination of the Met335 backbone carbonyl in the β6-α7 loop to the ADMIDAS metal ion (Ca^{2+} in physiologic condition) in the low-affinity, unliganded conformation is broken in the high-affinity, liganded conformation. The breaking of this coordination in turn enables the movements of β1-α1 loop and the ADMIDAS Ca^{2+} toward the MIDAS metal ion (Mg^{2+} in physiologic condition), which is the major difference between the high- and low-affinity conformations of the β I domain ligand-binding site, agreeing with earlier findings (5, 10). In contrast to that in the α I domains, no lateral movement of the MIDAS metal ion across the ligand-binding site of the β I domain appears to exist. Movements of the α1-helix, β6-α7 loop, and α7-helix are tightly coupled, so that reshaping to the high-affinity, ligand-binding site is allosterically linked to downward movement of the α7-helix. This linkage is critical for propagation of conformational signals from the ligand-binding pocket to the other integrin domains and vice versa.

In the αIIbβ3 complete ectodomain crystal structure, the physiologically important divalent cations Mg^{2+} and Ca^{2+} were found to be loaded in the resting state, prior to ligand binding (7). It has been known for years that cations can regulate integrin ligand-binding affinity. For most integrins, Ca^{2+} has both positive and negative regulatory effects. High concentrations of Ca^{2+} inhibit adhesion, whereas low concentrations of Ca^{2+} augment adhesion at suboptimal Mg^{2+} concentrations, which is called the synergistic effect. Compared with results in the physiologic divalent cations, addition of Mn^{2+} strikingly increases ligand-binding affinity and adhesiveness of almost all integrins. Mutagenesis studies revealed that metal-binding sites in the α-subunit β-propeller domain, α I domain, and α-genu were not responsible for the regulatory effects of Ca^{2+} and Mn^{2+} (29–32). By contrast, the effect of metal ions occurs mainly through the three metal ion sites that form the ligand-binding pocket in the β I domain, namely MIDAS, SyMBS, and ADMIDAS (33–35). The SyMBS functions as a positive regulatory site responsible for Ca^{2+} synergy with Mg^{2+} for ligand binding at MIDAS (35). When Ca^{2+} is absent, Mn^{2+} binding at SyMBS also synergizes with ligand binding at MIDAS, as revealed in the αvβ3 crystals. The basis for the synergy between SyMBS and MIDAS metal ions may lie in the coordination of the Glu-220 side chains to both cations. In the absence of either of these divalent cations, the Glu-220 would probably reorient and lose the proper coordination to the other site.

3.3. Communication Between the α I and β I Domains

Compared to the integrins lacking an α I domain, conformational regulation of integrins containing an α I domain requires the additional step of transmission of allostery from the β I domain to the α I domain (Fig. 2). The α I domain α7-helix has been proposed to transmit allostery between the α I MIDAS and the β I MIDAS; that is, in the active state, downward movement of the α I domain α7-helix enables an invariant Glu residue that is present a few residues after the α7-helix to act as an "intrinsic ligand" and engage the β I MIDAS (36, 37). Mutations of amino acids in the αL linker (e.g., Tyr-307 and Glu-310) to alanine have been shown to abolish ligand binding, suggesting that contacts between the linker and other domains modulate the conformation of the α I domain (38). Yang et al. showed that individual mutation of the αL linker residue Glu-310 or β2 MIDAS residues Ala-210 or Tyr-115 to cysteine abolishes I domain activation, whereas the double mutations of αL-E310C with either β2-A210C or β2-Y115C form disulfide bonds that constitutively activate ligand binding (36). The activation effect of the disulfide mutant is susceptible to small molecule antagonists that bind underneath the I domain α7-helix and certain allosteric antagonistic antibodies. This study provides direct evidence for an activating interaction between αL residue Glu-310 and the β2 MIDAS (36).

3.4. Signal Amplification by the Hybrid Domain Swing-Out

The orientation between the β I and hybrid domains appears to be the critical "translator", converting global conformational change into local intra-domain conformational changes that regulate affinity (Fig. 4b). As a consequence of the inserted topology of the β I domain into the hybrid domain, the piston-like displacement of the α7-helix in the high-affinity, liganded crystal structure results in complete remodeling of the interface between these domains, leading to the swing-out of the hybrid domain (10). As in the α7-helix, the C-terminal connection to the hybrid domain shifts its position, while the N-terminal linkage serves as a pivot point about which the inter-domain angle hinges. Relative to the closed headpiece conformation, the hybrid domain swings out about 60° in the open headpiece conformation, causing the knees of the α- and β-subunits to separate by 70 Å. The two conformations of the integrin headpiece were supported by EM studies of integrins αVβ3 (13) and α5β1 (11, 39). Electron tomography of negatively stained, active detergent-soluble αIIbβ3 purified on an Arg-Gly-Asp peptide affinity column reveals an extended conformation, with >90% of particles showing an open headpiece structure that matches perfectly (40) the open, liganded αIIbβ3 headpiece crystal structure (10). In addition to the structural investigations (10, 11, 13, 40–42), integrin hybrid domain swing-out is supported by a range of other studies (28, 39, 43–45).

Structures of the β3 and β2 integrin PSI domains and β2 I-EGF1 domain (10, 20, 46) demonstrate that during the rearrangement of the headpiece between the closed and open conformations, no change occurs in the hybrid/PSI domain interface. Therefore, this rigid interface, which is reinforced by the two polypeptide chain connections, nearby disulfide bonds, and an Arg deeply buried in the interface, enables the PSI domain to amplify the leg separation triggered by the swing-out of the hybrid domain (10, 20). The PSI and I-EGF1 domains are also shown to be intimately associated so that the hybrid and PSI/I-EGF1 domains move as a rigid unit (20). Some activating antibodies bind to the PSI domain and induce the high-affinity state (47, 48). A group of drug-dependent anti-β3 antibodies associated with quinine-induced immune thrombocytopenia was mapped to the PSI domain, indicating the importance of the PSI domain during integrin activation (49).

3.5. Integrin Extension and Separation of the Legs

The swing-out of the hybrid domain necessitates the existence of the extended conformation because the hybrid domain is central in the interfaces that are buried in the bent conformation; these interfaces are completely disrupted by hybrid domain swing-out (10). The knee of the β-subunit occurs between the PSI/I-EGF1 and I-EGF2 domains; the knee or genu of the α-subunit is a small Ca^{2+}-binding loop between the thigh and calf-1 domains. The mapping of an αL antibody that reports extension to the inner face of the thigh domain and requires the genu and a Ca^{2+}-coordinating residue donated by the calf-1 domain provided evidence that integrin extension occurs by a rearrangement at the thigh–genu interface (29). Integrin extension on the cell surface was confirmed by studies using FRET between fluorescent ligand-bound integrins and lipophilic probes (50).

Crystal (5–7) and EM (13) structures provide direct evidence that in the resting state, the membrane-proximal portions of the extracellular domains of the α- and β-subunits are in close juxtaposition. Enforced association of the two stalks with acid/base coiled coils renders integrin low affinity, whereas release of these constraints promotes high-affinity ligand binding (51). Introducing a 10-residue flexible spacer between the extracellular domain and the TM domain enhanced αMβ2 ligand binding on the cell surface (52). Crystal structure of the open αIIbβ3 headpiece (10) and EM structure of the entire αVβ3 extracellular domains (13) confirmed that the two stalks separate during integrin activation or ligand binding. The lower β-leg in the averaged EM images of the open conformation tended to disappear, suggesting that the lower β-leg is highly flexible and varied in conformations among individual particles (13). Therefore, even though the crystal structure of the open αIIbβ3 headpiece indicated that the swing-out of the hybrid domain results in a 70-Å separation at the knees (10), the distance between the two C-terminal stalks in the open conformations may vary, and this variation will result in a spectrum of different conformations (Fig. 2). The stalk separation is a key step for integrins to transmit signals bidirectionally across the plasma membrane.

3.6. Conformational Change of the Integrin TM and Cytoplasmic Domains

Recently, the basis for integrin activation across the plasma membrane has also been studied. Many studies showed that deletions or mutations in the α- and β-subunit TM and cytoplasmic domains, which are expected to destabilize α/β association, activate integrins (21, 53–56). FRET study shows that in the resting state, the integrin α- and β-subunit cytoplasmic domains are close to one another, but undergo significant spatial separation upon inside-out activation induced by phorbol ester or talin head domain, or outside-in signaling induced by ligand binding (12). NMR studies of the integrin cytoplasmic tails suggest that their association is weak, with significant differences observed between published structures (57–59), or that association is undetectable (60). These studies imply that the cytoplasmic interaction is modest and/or transient. Binding of intracellular proteins such as RAPL (61) and the talin head domain (62–64) to the integrin cytoplasmic tails induces tail separation and activates integrins for ligand binding (59). The structural basis for talin head domain and filamin binding to the integrin β cytoplasmic domain and the resulting integrins' activation has been demonstrated by NMR studies (62, 63, 65, 66).

It has been shown that during integrin activation, the two TM domains separate rather than rearrange after activation of integrins from inside the cell (67). Introduction of disulfide bridges to prevent or reverse separation abolished the activating effect of cytoplasmic mutations (67), whereas mutations that disrupt the TM interface activate integrins (68–70).

The NMR structures of isolated αIIb and β3 TM/cytoplasmic domain fragments were resolved in phospholipid bicelle (71, 72). In contrast to the findings that similar α_{IIb} and β_3 constructs formed homomultimers in detergents, the investigators found that these subunits were monomeric in the more bilayer-like bicelle environment. These structures are believed to represent the physiologically active state in which the TM and cytoplasmic domains are in the dissociated monomeric conformation (Fig. 5c).

The isolated αIIb and β3 TM/cytoplasmic NMR structures are similar to the Disulfide/Rosetta structure. The dissociated β3 TM domain solved by NMR appeared to be a 30-residue linear α-helix extended into the cytoplasm, and instead of the 23 TM residues in the Disulfide/Rosetta complex structure, 29 residues appeared to be embedded in the bicelle core. In the NMR structure, β3 Lys-716 is followed by a 5-residue hydrophobic segment ($L_{717}LITI_{721}$), and the continuous helix spanning the TM and justamembrane segments could undergo a substantial tilt in the membrane, with snorkeling of the Lys-716 side chains into the polar environment (71). By contrast, the β3-helix embedded in lipid membrane in the Disulfide/Rosetta complex structure is significantly shorter, suggesting that after dissociating from the αIIb-helix, the β3-helix is tilted at an angle of 20–30° due to the insertion of five to six additional hydrophobic residues to the hydrophobic lipid environment. This tilting of the β3-helix may be important for integrin activation and signaling.

Interestingly, the structure of integrin αIIbβ3 TM and cytoplasmic domain complex was also solved by NMR in the presence of phospholipid bicelles, and it was found to have an inter-helical interface similar to that of the Disulfide/Rosetta structure (Fig. 5a) (73). However, the NMR structure was solved using an artificial hydrogen-bond constraint between the αIIb(R995) and β3(D723). The presence of the salt bridge was based on the fact that mutations of either residue affected the helix–helix interaction as monitored by NMR. We propose that this electrostatic interaction is important for the priming of helix–helix interaction. After forming a more stable helix–helix interaction, the salt bridge is probably not critical for further stabilization. Therefore, the NMR structure might represent an "intermediate" or "transient" state between the physiological resting state (represented as the Disulfide/Rosetta structure) and the dissociated active state (represented as the NMR structure of the isolated monomers as discussed below) (Fig. 5c). It is interesting that the NMR structures of the complex have structures and angles with membrane almost identical to those of the isolated monomers, and there were substantial amounts of αIIb and β3 monomers present in the solution used for determining the NMR complex structures.These observations confirm our hypothesis that the NMR structure of the complex is an intermediate or "transient" state.

4. An Integrate View of Integrin Bidirectional Conformational Signaling

Numerous studies from different laboratories suggest that integrin bidirectional signaling across the plasma membrane is accomplished by coupling extracellular conformational change to an unclasping and separation of the α- and β-TM and cytoplasmic domains. It is thus envisioned that binding of proteins, such as talin or possibly RAPL, initiates separation of the cytoplasmic and TM domains, which destabilizes the extracellular α/β tail interface, concomitantly perturbing the tail–head interface and facilitating the switchblade-like opening (13). As a consequence, the hybrid domain swings out, which is coupled directly to the downward movement of the β I domain $\alpha7$-helix and thus the MIDAS rearrangement (13, 19). For integrins that lack I domains, this action represents the final step of priming, whereas for I domain-containing integrins, the β I domain next binds to the internal ligand in the linker between the C-terminal helix of the I domain and the β-propeller, thereby exerting a downward pull on the I domain $\alpha7$ leading to affinity modulation of its MIDAS. Evidence exists for similarity in the conformational rearrangements that result from modulation of the cytoplasmic/TM domain association (inside-out signaling) to those from binding of ligand (outside-in signaling) (12, 13, 22, 51). Thus, ligand binding stabilizes integrin in the extended conformation with open headpiece and two separate legs, resulting in the separation of the two cytoplasmic tails. Then, multivalent ligand binding brings several integrins close to each other, leading to integrin clustering, and kinases are recruited and activate each other, leading to intracellular signaling. Due to the space limit, readers interested in details of the many signaling pathways emanating from integrins are referred to several extensive reviews (74–77).

References

1. Hynes, R. O. (1992) Integrins: Versatility, modulation, and signaling in cell adhesion. *Cell* **69**, 11–25.
2. Springer, T. A. (1994) Traffic signals for lymphocyte recirculation and leukocyte emigration: the multi-step paradigm. *Cell* **76**, 301–314.
3. Hynes, R. O. (2002) Integrins: bi-directional, allosteric, signalling machines. *Cell* **110**, 673–687.
4. Nermut, M. V., Green, N. M., Eason, P., Yamada, S. S., and Yamada, K. M. (1988) Electron microscopy and structural model of human fibronectin receptor. *EMBO J.* **7**, 4093–4099.
5. Xiong, J. P., Stehle, T., Zhang, R., Joachimiak, A., Frech, M., Goodman, S. L., and Arnaout, M. A. (2002) Crystal structure of the extracellular segment of integrin alpha Vbeta3 in complex with an Arg-Gly-Asp ligand. *Science* **296**, 151–155.
6. Xiong, J. P., Stehle, T., Diefenbach, B., Zhang, R., Dunker, R., Scott, D. L., Joachimiak, A., Goodman, S. L., and Arnaout, M. A. (2001) Crystal structure of the extracellular segment of integrin alphaVbeta3. *Science* **294**, 339–345.
7. Zhu, J., Luo, B. H., Xiao, T., Zhang, C., Nishida, N., and Springer, T. A. (2008) Structure of a complete integrin ectodomain in a physiologic resting state and activation and deactivation by applied forces. *Mol. Cell* **32**, 849–861.
8. Luo, B.-H., Carman, C. V., and Springer, T. A. (2007) Structural basis of integrin regulation and signaling, *Annu. Rev. Immunol.* **25**, 619–647.
9. Zhu, J., Luo, B. H., Barth, P., Schonbrun, J., Baker, D., and Springer, T. A. (2009) The structure of a receptor with two associating transmembrane domains on the cell surface: integrin alphaIIbbeta3. *Mol. Cell* **34**, 234–249.
10. Xiao, T., Takagi, J., Coller, B. S., Wang, J. H., and Springer, T. A. (2004) Structural basis for allostery in integrins and binding to fibrinogen-mimetic therapeutics. *Nature* **432**, 59–67.

11. Takagi, J., Strokovich, K., Springer, T. A., and Walz, T. (2003) Structure of integrin alpha-5beta1 in complex with fibronectin. *EMBO J.* **22**, 4607–4615.
12. Kim, M., Carman, C. V., and Springer, T. A. (2003) Bidirectional transmembrane signaling by cytoplasmic domain separation in integrins. *Science* **301**, 1720–1725.
13. Takagi, J., Petre, B. M., Walz, T., and Springer, T. A. (2002) Global conformational rearrangements in integrin extracellular domains in outside-in and inside-out signaling. *Cell* **110**, 599–511.
14. Lee, J. O., Rieu, P., Arnaout, M. A., and Liddington, R. (1995) Crystal structure of the A domain from the alpha subunit of integrin CR3 (CD11b/CD18). *Cell* **80**, 631–638.
15. Springer, T. A. (1997) Folding of the N-terminal, ligand-binding region of integrin alpha-subunits into a beta-propeller domain. *Proc. Natl. Acad. Sci. USA.* **94**, 65–72.
16. Kamata, T., Tieu, K. K., Irie, A., Springer, T. A., and Takada, Y. (2001) Amino acid residues in the alpha IIb subunit that are critical for ligand binding to integrin alpha IIbbeta 3 are clustered in the beta-propeller model. *J. Biol. Chem.* **276**, 44275–44283.
17. Huang, C., Zang, Q., Takagi, J., and Springer, T. A. (2000) Structural and functional studies with antibodies to the integrin beta 2 subunit. A model for the I-like domain. *J. Bio. Chem.* **275**, 21514–21524.
18. Bilsland, C. A., Diamond, M. S., and Springer, T. A. (1994) The leukocyte integrin p150,95 (CD11c/CD18) as a receptor for iC3b. Activation by a heterologous beta subunit and localization of a ligand recognition site to the I domain. *J. Immunol.* **152**, 4582–4589.
19. Beglova, N., Blacklow, S. C., Takagi, J., and Springer, T. A. (2002) Cysteine-rich module structure reveals a fulcrum for integrin rearrangement upon activation. *Nat. Struct. Biol.* **9**, 282–287.
20. Shi, M., Sundramurthy, K., Liu, B., Tan, S. M., Law, S. K., and Lescar, J. (2005) The crystal structure of the plexin-semaphorin-integrin domain/hybrid domain/I-EGF1 segment from the human integrin beta2 subunit at 1.8-A resolution. *J. Biol. Chem.* **280**, 30586–30593.
21. Hughes, P. E., Diaz-Gonzalez, F., Leong, L., Wu, C., McDonald, J. A., Shattil, S. J., and Ginsberg, M. H. (1996) Breaking the integrin hinge. A defined structural constraint regulates integrin signaling. *J. Biol. Chem.* **271**, 6571–6574.
22. Shimaoka, M., Xiao, T., Liu, J. H., Yang, Y., Dong, Y., Jun, C. D., McCormack, A., Zhang, R., Joachimiak, A., Takagi, J., Wang, J. H., and Springer, T. A. (2003) Structures of the alpha L I domain and its complex with ICAM-1 reveal a shape-shifting pathway for integrin regulation. *Cell* **112**, 99–111.
23. Emsley, J., Knight, C. G., Farndale, R. W., Barnes, M. J., and Liddington, R. C. (2000) Structural basis of collagen recognition by integrin alpha2beta1. *Cell* **101**, 47–56.
24. Yang, W., Shimaoka, M., Chen, J., and Springer, T. A. (2004) Activation of integrin beta-subunit I-like domains by one-turn C-terminal alpha-helix deletions. *Proc. Natl. Acad. Sci. USA.* **101**, 2333–2338.
25. Luo, B. H., Takagi, J., and Springer, T. A. (2004) Locking the beta3 integrin I-like domain into high and low affinity conformations with disulfides. *J. Biol. Chem.* **279**, 10215–10221.
26. Jannuzi, A. L., Bunch, T. A., West, R. F., and Brower, D. L. (2004) Identification of integrin beta subunit mutations that alter heterodimer function in situ. *Mol. Biol. Cell* **15**, 3829–3840.
27. Barton, S. J., Travis, M. A., Askari, J. A., Buckley, P. A., Craig, S. E., Humphries, M. J., and Mould, A. P. (2004) Novel activating and inactivating mutations in the integrin beta1 subunit A domain. *Biochem. J.* **380**, 401–407.
28. Mould, A. P., Barton, S. J., Askari, J. A., McEwan, P. A., Buckley, P. A., Craig, S. E., and Humphries, M. J. (2003) Conformational changes in the integrin beta A domain provide a mechanism for signal transduction via hybrid domain movement. *J. Biol. Chem.* **278**, 17028–17035.
29. Xie, C., Shimaoka, M., Xiao, T., Schwab, P., Klickstein, L. B., and Springer, T. A. (2004) The integrin alpha-subunit leg extends at a Ca2+-dependent epitope in the thigh/genu interface upon activation. *Proc. Natl. Acad. Sci. USA.* **101**, 15422–15427.
30. Lu, C., Shimaoka, M., Ferzly, M., Oxvig, C., Takagi, J., and Springer, T. A. (2001) An isolated, surface-expressed I domain of the integrin alphaLbeta2 is sufficient for strong adhesive function when locked in the open conformation with a disulfide bond. *Proc. Natl. Acad. Sci. USA.* **98**, 2387–2392.
31. Pujades, C., Alon, R., Yauch, R. L., Masumoto, A., Burkly, L. C., Chen, C., Springer, T. A., Lobb, R. R., and Hemler, M. E. (1997) Defining extracellular integrin alpha-chain sites that affect cell adhesion and adhesion strengthening without altering soluble ligand binding. *Mol. Biol. Cell* **8**, 2647–2657.
32. Knorr, R., and Dustin, M. L. (1997) The lymphocyte function-associated antigen 1 I domain is a transient binding module for inter-

cellular adhesion molecule (ICAM)-1 and ICAM-3 in hydrodynamic flow. *J. Exp. Med.* **186**, 719–730.

33. Chen, J., Takagi, J., Xie, C., Xiao, T., Luo, B. H., and Springer, T. A. (2004) The relative influence of metal ion binding sites in the I-like domain and the interface with the hybrid domain on rolling and firm adhesion by integrin alpha4beta7. *J. Biol. Chem.* **279**, 55556–55561.
34. Mould, A. P., Barton, S. J., Askari, J. A., Craig, S. E., and Humphries, M. J. (2003) Role of ADMIDAS cation-binding site in ligand recognition by integrin alpha 5 beta 1. *J. Biol. Chem.* **278**, 51622–51629.
35. Chen, J., Salas, A., and Springer, T. A. (2003) Bistable regulation of integrin adhesiveness by a bipolar metal ion cluster. *Nat. Struct. Biol.* **10**, 995–1001.
36. Yang, W., Shimaoka, M., Salas, A., Takagi, J., and Springer, T. A. (2004) Intersubunit signal transmission in integrins by a receptor-like interaction with a pull spring. *Proc. Natl. Acad. Sci. USA.* **101**, 2906–2911.
37. Alonso, J. L., Essafi, M., Xiong, J. P., Stehle, T., and Arnaout, M. A. (2002) Does the integrin alphaA domain act as a ligand for its betaA domain? *Curr. Biol.* **12**, R340-342.
38. Huth, J. R., Olejniczak, E. T., Mendoza, R., Liang, H., Harris, E. A., Lupher, M. L., Jr., Wilson, A. E., Fesik, S. W., and Staunton, D. E. (2000) NMR and mutagenesis evidence for an I domain allosteric site that regulates lymphocyte function-associated antigen 1 ligand binding. *Proc. Natl. Acad. Sci. USA.* **97**, 5231–5236.
39. Luo, B. H., Strokovich, K., Walz, T., Springer, T. A., and Takagi, J. (2004) Allosteric beta1 integrin antibodies that stabilize the low affinity state by preventing the swing-out of the hybrid domain. *J. Biol. Chem.* **279**, 27466–27471.
40. Iwasaki, K., Mitsuoka, K., Fujiyoshi, Y., Fujisawa, Y., Kikuchi, M., Sekiguchi, K., and Yamada, T. (2005) Electron tomography reveals diverse conformations of integrin alphaIIbbeta3 in the active state. *J. Struct. Biol.* **150**, 259–267.
41. Nishida, N., Xie, C., Shimaoka, M., Cheng, Y., Walz, T., and Springer, T. A. (2006) Activation of leukocyte beta2 integrins by conversion from bent to extended conformations. *Immunity* **25**, 583–594.
42. Mould, A. P., Symonds, E. J., Buckley, P. A., Grossmann, J. G., McEwan, P. A., Barton, S. J., Askari, J. A., Craig, S. E., Bella, J., and Humphries, M. J. (2003) Structure of an integrin-ligand complex deduced from solution x-ray scattering and site-directed mutagenesis. *J. Biol. Chem.* **278**, 39993–39999.
43. Luo, B. H., Springer, T. A., and Takagi, J. (2003) Stabilizing the open conformation of the integrin headpiece with a glycan wedge increases affinity for ligand. *Proc. Natl. Acad. Sci. USA.* **100**, 2403–2408.
44. Tng, E., Tan, S. M., Ranganathan, S., Cheng, M., and Law, S. K. (2004) The integrin alpha L beta 2 hybrid domain serves as a link for the propagation of activation signal from its stalk regions to the I-like domain. *J. Biol. Chem.* **279**, 54334–54339.
45. Tang, R. H., Tng, E., Law, S. K., and Tan, S. M. (2005) Epitope mapping of monoclonal antibody to integrin alphaL beta2 hybrid domain suggests different requirements of affinity states for intercellular adhesion molecules (ICAM)-1 and ICAM-3 binding. *J. Biol. Chem.* **280**, 29208–29216.
46. Xiong, J. P., Stehle, T., Goodman, S. L., and Arnaout, M. A. (2004) A novel adaptation of the integrin PSI domain revealed from its crystal structure. *J. Biol. Chem.* **279**, 40252–40254.
47. Mould, A. P., Travis, M. A., Barton, S. J., Hamilton, J. A., Askari, J. A., Craig, S. E., Macdonald, P. R., Kammerer, R. A., Buckley, P. A., and Humphries, M. J. (2005) Evidence that monoclonal antibodies directed against the integrin beta subunit plexin/semaphorin/integrin domain stimulate function by inducing receptor extension. *J. Biol. Chem.* **280**, 4238–4246.
48. Honda, S., Tomiyama, Y., Pelletier, A. J., Annis, D., Honda, Y., Orchekowski, R., Ruggeri, Z., and Kunicki, T. J. (1995) Topography of ligand-induced binding sites, including a novel cation-sensitive epitope (AP5) at the amino terminus, of the human integrin beta 3 subunit. *J. Biol. Chem.* **270**, 11947–11954.
49. Peterson, J. A., Nyree, C. E., Newman, P. J., and Aster, R. H. (2003) A site involving the "hybrid" and PSI homology domains of GPIIIa (beta 3-integrin subunit) is a common target for antibodies associated with quinine-induced immune thrombocytopenia. *Blood* **101**, 937–942.
50. Chigaev, A., Buranda, T., Dwyer, D. C., Prossnitz, E. R., and Sklar, L. A. (2003) FRET detection of cellular alpha4-integrin conformational activation. *Biophys J.* **85**, 3951–3962.
51. Takagi, J., Erickson, H. P., and Springer, T. A. (2001) C-terminal opening mimics 'inside-out' activation of integrin alpha5beta1. *Nat. Struct. Biol.* **8**, 412–416.

52. Xiong, Y. M., Chen, J., and Zhang, L. (2003) Modulation of CD11b/CD18 adhesive activity by its extracellular, membrane-proximal regions. *J. Immunol.* **171**, 1042–1050.

53. Lu, C., Takagi, J., and Springer, T. A. (2001) Association of the membrane proximal regions of the alpha and beta subunit cytoplasmic domains constrains an integrin in the inactive state. *J. Biol. Chem.* **276**, 14642–14648.

54. Lu, C. F., and Springer, T. A. (1997) The alpha subunit cytoplasmic domain regulates the assembly and adhesiveness of integrin lymphocyte function-associated antigen-1. *J. Immunol.* **159**, 268–278.

55. O'Toole, T. E., Katagiri, Y., Faull, R. J., Peter, K., Tamura, R., Quaranta, V., Loftus, J. C., Shattil, S. J., and Ginsberg, M. H. (1994) Integrin cytoplasmic domains mediate inside-out signal transduction. *J. Cell Biol.* **124**, 1047–1059.

56. O'Toole, T. E., Mandelman, D., Forsyth, J., Shattil, S. J., Plow, E. F., and Ginsberg, M. H. (1991) Modulation of the affinity of integrin alpha IIb beta 3 (GPIIb-IIIa) by the cytoplasmic domain of alpha IIb. *Science* **254**, 845–847.

57. Vinogradova, O., Vaynberg, J., Kong, X., Haas, T. A., Plow, E. F., and Qin, J. (2004) Membrane-mediated structural transitions at the cytoplasmic face during integrin activation. *Proc. Natl. Acad. Sci. USA.* **101**, 4094–4099.

58. Weljie, A. M., Hwang, P. M., and Vogel, H. J. (2002) Solution structures of the cytoplasmic tail complex from platelet integrin alpha IIb- and beta 3-subunits. *Proc. Natl. Acad. Sci. USA.* **99**, 5878–5883.

59. Vinogradova, O., Velyvis, A., Velyviene, A., Hu, B., Haas, T., Plow, E., and Qin, J. (2002) A structural mechanism of integrin alpha(IIb) beta(3) "inside-out" activation as regulated by its cytoplasmic face. *Cell* **110**, 587–597.

60. Ulmer, T. S., Yaspan, B., Ginsberg, M. H., and Campbell, I. D. (2001) NMR analysis of structure and dynamics of the cytosolic tails of integrin alpha IIb beta 3 in aqueous solution. *Biochemistry* **40**, 7498–7508.

61. Katagiri, K., Maeda, A., Shimonaka, M., and Kinashi, T. (2003) RAPL, a Rap1-binding molecule that mediates Rap1-induced adhesion through spatial regulation of LFA-1. *Nat. Immunol.* **4**, 741–748.

62. Tadokoro, S., Shattil, S. J., Eto, K., Tai, V., Liddington, R. C., de Pereda, J. M., Ginsberg, M. H., and Calderwood, D. A. (2003) Talin binding to integrin beta tails: a final common step in integrin activation. *Science* **302**, 103–106.

63. Garcia-Alvarez, B., de Pereda, J. M., Calderwood, D. A., Ulmer, T. S., Critchley, D., Campbell, I. D., Ginsberg, M. H., and Liddington, R. C. (2003) Structural determinants of integrin recognition by talin. *Mol. Cell* **11**, 49–58.

64. Calderwood, D. A., Zent, R., Grant, R., Rees, D. J., Hynes, R. O., and Ginsberg, M. H. (1999) The Talin head domain binds to integrin beta subunit cytoplasmic tails and regulates integrin activation. *J. Biol. Chem.* **274**, 28071–28074.

65. Kiema, T., Lad, Y., Jiang, P., Oxley, C. L., Baldassarre, M., Wegener, K. L., Campbell, I. D., Ylanne, J., and Calderwood, D. A. (2006) The molecular basis of filamin binding to integrins and competition with talin. *Mol. Cell* **21**, 337–347.

66. Ulmer, T. S., Calderwood, D. A., Ginsberg, M. H., and Campbell, I. D. (2003) Domain-specific interactions of talin with the membrane-proximal region of the integrin beta3 subunit. *Biochemistry* **42**, 8307–8312.

67. Luo, B. H., Springer, T. A., and Takagi, J. (2004) A specific interface between integrin transmembrane helices and affinity for ligand. *PLoS Biol.* **2**, e153.

68. Partridge, A. W., Liu, S., Kim, S., Bowie, J. U., and Ginsberg, M. H. (2005) Transmembrane domain helix packing stabilizes integrin alphaIIbbeta3 in the low affinity state. *J. Biol. Chem.* **280**, 7294–7300.

69. Luo, B. H., Carman, C. V., Takagi, J., and Springer, T. A. (2005) Disrupting integrin transmembrane domain heterodimerization increases ligand binding affinity, not valency or clustering. *Proc. Natl. Acad. Sci. USA.* **102**, 3679–3684.

70. Li, W., Metcalf, D. G., Gorelik, R., Li, R., Mitra, N., Nanda, V., Law, P. B., Lear, J. D., Degrado, W. F., and Bennett, J. S. (2005) A push-pull mechanism for regulating integrin function. *Proc. Natl. Acad. Sci. USA.* **102**, 1424–1429.

71. Lau, T. L., Partridge, A. W., Ginsberg, M. H., and Ulmer, T. S. (2008) Structure of the integrin beta3 transmembrane segment in phospholipid bicelles and detergent micelles. *Biochemistry* **47**, 4008–4016.

72. Lau, T. L., Dua, V., and Ulmer, T. S. (2008) Structure of the integrin alphaIIb transmembrane segment. *J. Biol. Chem.* **283**, 16162–16168.

73. Lau, T. L., Kim, C., Ginsberg, M. H., and Ulmer, T. S. (2009) The structure of the integrin alphaIIbbeta3 transmembrane complex explains

integrin transmembrane signaling. *EMBO J.* **28**, 1351–1361.

74. Shattil, S. J., and Newman, P. J. (2004) Integrins: dynamic scaffolds for adhesion and signaling in platelets. *Blood* **104**, 1606–1615.
75. Guo, W., and Giancotti, F. G. (2004) Integrin signalling during tumour progression. *Nat. Rev. Mol. Cell Biol.* 5, 816–826.
76. Grashoff, C., Thievessen, I., Lorenz, K., Ussar, S., and Fassler, R. (2004) Integrin-linked kinase: integrin's mysterious partner. *Curr. Opin. Cell Biol.* **16**, 565–571.
77. Ridley, A. J., Schwartz, M. A., Burridge, K., Firtel, R. A., Ginsberg, M. H., Borisy, G., Parsons, J. T., and Horwitz, A. R. (2003) Cell migration: integrating signals from front to back. *Science* **302**, 1704–1709.

Chapter 8

Protein Expression and Purification of Integrin I Domains and IgSF Ligands for Crystallography

Hongmin Zhang and Jia-huai Wang

Abstract

Cell adhesion depends on combinational expression and interactions of a large number of adhesion molecules at cell-to-cell or cell-to-matrix contact sites. Integrins and their immunoglobulin superfamily (IgSF) ligands represent foremost classes of cell adhesion molecules in immune system. Structural study is critical for a better understanding of the interactions between integrins and their IgSF ligands. Here we describe protocols for protein expression of integrin αL I domain and its IgSF ligand ICAM-5 D1D2 fragment for crystallography.

Key words: Mammalian cell expression, Integrins, ICAMs, Crystallization, Molecular replacement

1. Introduction

Integrins are major cell adhesion molecules that mediate cell–cell and cell–extracellular matrix interactions, thereby playing a key role in development, immune responses, leukocyte trafficking, homeostasis, and cancer metastasis. Integrins transduce signals across the plasma membrane bi-directionally in an allosteric fashion. Ligand binding to integrins transmits signals to the cytoplasm ("outside-in" signaling). Conversely, integrins are activated in response to intracellular signaling cascades elicited by other receptors ("inside-out" signaling) (1, 2). The structural basis of integrin allostery has been extensively reviewed (3). Structural studies of the binding domain of leukocyte integrins (αI domain) and their ligands have been extensively performed to understand the allosteric regulation of integrins. For example, the complex structures between integrin αL I domain and its ligands ICAM-1, ICAM-3

Motomu Shimaoka (ed.), *Integrin and Cell Adhesion Molecules: Methods and Protocols*,
Methods in Molecular Biology, vol. 757, DOI 10.1007/978-1-61779-166-6_8, © Springer Science+Business Media, LLC 2011

and ICAM-5 (4–6) have revealed a basic binding model between integrins and their ligands. An acidic residue from the ligand coordinates to the metal ion-dependent adhesion site (MIDAS) of the I domain (7), thereby triggering the conformational changes of the MIDAS, which is allosterically linked to an axial movement of the α7-helix at the other end of the I domain. This eventually leads to a large-scale reorientation of the ecto-domains up to 200 Å, and the separation of the integrin α and β subunits by as much as 70 Å (8, 9). In this chapter, we will use the α I domain of αLβ2 integrin and its natural ligand, ICAM-5 as examples to describe the expression and structure determination of integrin I domain in complex with its ligand.

2. Materials

2.1. Expression and Purification of αL I Domain in Escherichia coli

1. Expression vector: pET22b with an inserted fragment encoding the αL I domain (residue N129 to Y307 with a stop codon following Y307) (10, 11).
2. BL21 (DE3) competent cells.
3. LB-amp medium: Dissolve 10 g tryptone, 5 g yeast extract, and 10 g sodium chloride in 1 L of water (see Note 1). Autoclave at 121°C for 20 min, and add 1 ml of 100 mg/ml ampicillin when it is cooled to room temperature.
4. Rich medium: 20 g/L tryptone, 10 g/L yeast extract, 5 g/L NaCl, 20 ml/L glycerol, 50 mM K_2HPO_4, 10 mM $MgCl_2$, 10 g/L glucose, and 100 μg/L ampicillin.
5. 0.1 M Isopropyl β-d-1-thiogalactopyranoside (IPTG).
6. Lysis buffer: 50 mM Tris–HCl pH 8.0, 1 mM $MgCl_2$, 0.4 μg/ml DNase I, 0.4 μg/ml RNase A, 1 μg/ml lysozyme.
7. Sonicator with a large sonication tip.
8. Wash buffer 1: 20 mM Tris–HCl pH 8.0, 23% (w/v) sucrose, 0.5 %(v/v) Triton X-100, 1 mM EDTA.
9. Wash buffer 2: 20 mM Tris–HCl pH 8.0, 1 mM EDTA.
10. Solubilization buffer: 6 M Guanidine HCl, 50 mM Tris–HCl pH 8.0, 1 mM DTT.
11. A serine protease inhibitor phenylmethanesulfonylfluoride or phenylmethylsulfonyl fluoride (PMSF): 100 mM stock in isopropanol.
12. Phenathroline: 1 M stock in DMSO.
13. Refolding buffer: 50 mM Tris–HCl pH 8.0, 1 mM $MgCl_2$, 5% (v/v) glycerol, 50 mg/L $CuSO_4$, 1 mM phenathroline, 0.1 mM PMSF (see Note 2).

14. Ion-exchange buffer A: 20 mM Tris–HCl pH 8.0.
15. Ion-exchange buffer B: 20 mM Tris–HCl pH 8.0, 1 M sodium chloride.
16. Size-exclusion buffer: 20 mM HEPES pH 7.5, 0.2 M sodium chloride.
17. FPLC equipment.
18. Q-sepharose ion-exchange column.
19. Superdex75prep size-exclusion column.

2.2. Expression and purification of Ligands in CHO Cells

1. CHO *lec* 3.2.8.1 cells stably transfected to express ICAM-5 D1D2 (6).
2. GMEM-MSX medium: To make 500 ml or 2 L of GMEM-MSX medium, add the following stock solutions as listed below using aseptic technique in a cell culture hood.

Sterile water (ml)	350	1,400
(a) 10XGMEM (ml)	50	200
(b) Sodium bicarbonate (ml)	18.1	72.4
(c) NEAA (ml)	5	20
(d) G+A (ml)	5	20
(e) Sodium pyruvate (ml)	5	20
(f) Nucleosides (ml)	10	40
(g) Pen-Step (ml)	5	20
(h) Dialyzed FCS (ml)	50	200
(i) L-MSX (ml)	0.125	0.5
Total (ml)	500	2,000

3. Stock solutions:
   (a) 10× Glasgow MEM without glutamine and without tryptose-phosphate broth (custom order to GIBCO or Sigma).
   (b) 7.5% sodium bicarbonate.
   (c) 100× Non-essential amino acids (NEAA).
   (d) 100× glutamic acid + asparigine (G+A): Dissolve 1,500 mg L-glutamic acid and 1,500 mg L-asparigine to 250 ml of distilled water and sterilize by filtration.
   (e) 100 mM sodium pyruvate.
   (f) 50× Nucleosides: Dissolve 175 mg adenosine, 175 mg guanosine, 175 mg cytidine, 175 mg uridine, 40 mg thymidine to 500 ml of autoclaved water, and sterilize by filtration.

(g) 100× Penicillin–Streptomycin at 5,000 units/ml.

(h) Dialyzed FCS: Heat inactivated at 56°C for 30–35 min (see Note 3).

(i) 100 mM L-MSX: Prepare 18 mg/ml solution in PBS. Sterilize by filtration and store at –20°C in 1-ml aliquots. Final concentration in medium is 25 μM.

4. FPLC equipment.
5. Cation exchange column SP.
6. Anion exchange column Mono Q.
7. Cation exchange column Mono S.
8. Size-exclusion column Superdex75 prep.

3. Methods

The αL I domain is expressed in *E. coli* as inclusion bodies, refolded, and purified to homogeneity. By contrast, the integrin ligand domain of ICAM-5 (ICAM-5D1D2) is expressed in CHO cells, as the protein is highly glycosylated and involves several disulfide bonds. A mutant cell line CHO lec 3.2.8.1 has four independent mutations in the N- and O-glycosylation pathways (12). N-linked carbohydrates produced by CHO *lec* 3.2.8.1 cells are all of the high mannose type, but different in the number of mannoses, ranging from Man_5 to Man_9. O-glycosylation is homogenous, with only a single GalNAc residue attached per site. When cultured in the presence of the alpha-glucosidase I inhibitor N-butyl-deoxynojirimycin (NB-DNJ), glycoproteins produced in CHO *lec* 3.2.8.1 cells are almost completely susceptible to Endo H digestion (13, 14). Endo H cleaves chitobiose, leaving a single N-linked N-acetylglucosamine per site, which is ideal for maintenance of protein solubility and special carbohydrate–protein interactions, such as between the first N-acetyl glucosamine residue and tryptophan. The property of CHO *lec* 3.2.8.1 makes it quite suitable for expression of proteins for structural studies. Therefore, CHO *lec* 3.2.8.1 will be our expression host for integrin IgSF ligands including ICAM-5 D1D2.

3.1. Expression and Purification of αL I Domain in E. coli

1. Inoculate a single colony of BL21(DE3) transformed with the expression vector into 30 ml of LB-ampicillin medium and shake at 37°C for 6–8 h.
2. The next day, transfer 30 ml of bacteria culture to 600 ml of rich medium, shake at 37°C for 3–4 h. When OD600 reaches to 1–1.2, add 1 mM IPTG to induce expression.
3. Three to four hours later harvest bacteria by centrifugation. Resuspend the pellet with 50 mM Tris–HCl pH 8.0 and centrifuge to harvest the pellet. Freeze pellet at –20°C for later use.

4. Suspend the pellet in 30 ml of lysis buffer, incubate it at 37°C for 10–15 min, and sonicate it with a large sonicator tip.
5. Harvest inclusion bodies by centrifuge (25,000 × *g* for 30 min).
6. Wash inclusion bodies with wash buffer 1 by sonication, centrifuge at 25,000 × *g* for 15 min and discard the supernatant. Repeat this step for five times.
7. Wash inclusion bodies with wash buffer 2 by sonication, centrifuge and discard the supernatant.
8. Resuspend inclusion bodies in solubilization buffer, stir at room temperature for 1–2 h.
9. Centrifuge at 25,000 × *g* for 30 min, filter and keep the supernatant. Measure the protein concentration at OD280 and adjust protein concentration to 0.5–1 mg/ml with solubilization buffer.
10. Refolding at 4°C overnight or longer by quick dilution into 19-fold of refolding buffer.
11. Centrifuge and filter with a 0.22 μm membrane to remove precipitant.
12. Concentrate the supernatant to small volume and filter it.
13. Purify the sample on a Q-sepharose ion-exchange column using a FPLC equipment. The supernatant is loaded to the column, washed with 2 column volumes (CV) of buffer A and eluted with a NaCl gradient from 0 to 30% of buffer B in 20 CV.
14. Collect and pool the peak fractions.
15. Concentrate the pooled fractions and further purify it on Superdex75 prep size-exclusion column using a FPLC equipment.
16. Collect and pool the peak fractions.
17. Desalt and change buffer to 20 mM HEPES pH 7.5, 50 mM sodium chloride and 5 mM magnesium chloride.
18. Concentrate the sample to >20 mg/ml. Flash freeze into liquid nitrogen and store at −80°C for later use of crystallization (see Note 5).

3.2. Expression and Purification of Ligands in CHO Cells

Our ICAM-5D1D2 construct does not contain any purification tags and we have no suitable antibodies against ICAM-5D1D2 to be used in affinity purification. However, the unique situation here is that the isoelectric point (pI) of ICAM-5 D1D2 is about 11, far above the pI value of most proteins (4–6). This property was exploited to purify this protein fragment using a series of ion-exchange columns.

1. Collect supernatant from CHO *lec* 3.2.8.1 cells stably transfected to express ICAM-5 D1D2. The cells were cultured in dishes, flasks or in roller bottles with GMEM-MSX medium. Roller bottles generally give higher expression due to larger surface area. We typically harvest the supernatant every week and add fresh medium into roller bottles (100–200 ml for a 2-L roller bottle) for long-term culture (see Note 4).
2. Concentrate the supernatant up to 10–20-folds and dialyze it against 50 mM Tris–HCl, pH 8.8.
3. Centrifuge to remove precipitants.
4. Load the concentrated supernatant to a SP cation exchange column pre-equilibrated with 50 mM Tris–HCl pH 8.8. Wash the column with 10 CV of 50 mM Tris–HCl pH 8.8. Elute ICAM-5D1D2 with 50 mM Tris–HCl pH 8.8, 1 M NaCl.
5. Collect and pool peak fractions containing ICAM-5D1D2. Dialyze it against 50 mM Tris–HCl pH 8.8.
6. Load the dialyzed sample to Mono Q anion exchange column pre-equilibrated with 50 mM Tris–HCl pH 8.8. ICAM-5D1D2 does not bind to the column at this pH. Collect the flow through fraction. Although ICAM-5D1D2 does not bind to Mono Q at this pH, contaminant proteins do and this step improves the purity.
7. Load the flow through fraction from Mono Q to Mono S pre-equilibrated with 50 mM Tris–HCl pH 8.8. Wash the column with 2 CV of 50 mM Tris–HCl pH 8.8 and elute with a NaCl gradient from 0 to 0.4 M in 20 CV. Collect and pool the peak fractions.
8. Concentrate the ICAM-5D1D2 fractions from Mono S and load to Superdex 75 prep column pre-equilibrated with 20 mM HEPES pH 7.5, 0.2 M NaCl. Collect and pool the peak fractions.
9. Concentrate the pooled fractions and change buffer by dialysis to 20 mM HEPES, pH 7.5, 50 mM NaCl. Aliquote, flash freeze into liquid nitrogen and store at −80°C for later use (see Notes 5 and 6).

4. Notes

1. Unless stated otherwise, all solutions should be prepared in water that has a resistivity of 18.2 MΩ cm and total organic content of less than five parts per billion. This standard is referred to as "water" in this text.
2. For wild-type I domain, $CuSO_4$ and phenathroline should not be included. $CuSO_4$ is used to facilitate oxidization of the

engineered disulfide-bond in the mutant locked high-affinity I domains, whereas phenathroline is used to inhibit metallo-protease activity. GSH and GSSH may be used instead of $CuSO_4$ to provide oxidization-reduction potential. $CuSO_4$ and phenathroline are less expensive.

3. Dialyzed FCS should be used especially when culturing CHO cells in the presence of L-MSX.
4. CO_2 is not needed for CHO *lec* cells cultured in roller bottles. The bottles can be placed on a rotating rack at a speed of 2–3 rpm in a 37°C warm room. Harvest the supernatant and add in fresh medium with an aseptic technique in a tissue culture hood. Parafilm can be used to wrap around the bottle cap to further reduce contaminant.
5. In the crystallization of ICAM-5 D1D2 with the integrin α_L I domain, ICAM-5 D1D2 was mixed with the I domain in equal molar ratio at a total concentration of 11 mg/ml. The mixture was used in crystallization screening and we obtained crystals from precipitant in a reservoir solution containing 0.1 M HEPES pH 7.5, 10% PEG8000, 8% ethylene glycol at room temperature. Crystals were optimized with a pH gradient from 7.0 to 8.0 and PEG8000 gradient from 5 to 15%. Later, ethylene glycol was replaced by glycerol with a gradient from 5 to 12%. Better crystals were obtained with 2 μl of protein mixed with 1 μl of reservoir solution (0.1 M HEPES pH 7.5, 7.5% PEG8000, 10% glycerol) at 4°C in 3 days.

   The crystals were harvested and soaked in 0.1 M HEPES pH 7.5, 15% PEG8000, 20% glycerol and 5 mM magnesium chloride. It is important to increase the concentration of PEG8000 from 7.5 to 15%; otherwise, the crystals dissolve slowly and this impairs the diffraction quality. After soaking in cryo-protectant, the crystals were flash frozen into liquid nitrogen for later data collection.
6. The diffraction data for crystals of ICAM-5D1D2 in complex with the integrin α_L I domain were collected at ID19 at Argonne National Laboratory and processed with HKL2000 (15). The scaled diffraction data (sca file) were fed into program dtrek2mtz in CCP4 (16) and converted into mtz file with 5% of the data added to FreeR column. The result of Mathew's coefficient calculation (17) showed that there was likely only one copy of the complex in the asymmetric unit with about 67% solution content. Self-rotation function did not show any pseudo-symmetry, and there was no pseudo-translation either. Thus, we figured that there should be only one copy of the complex in the asymmetric unit to look for.

Two homologous structures, ICAM-3 D1 in complex with high-affinity I domain (HA) (PDB code 1T0P) and ICAM-1 D1D2 with intermediate affinity I domain (IA) (PDB code

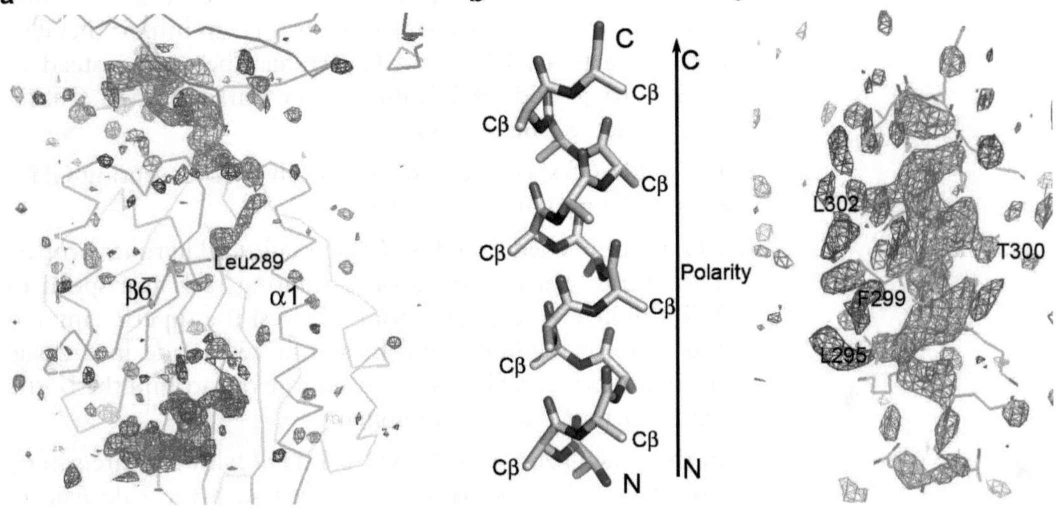

Fig. 1. Swung-out of the α7 helix of the I domain in complex with ICAM-5 D1D2. (**a**) Fo-Fc map showing the traces of β6–α7 loop and α7 helix. The positive map of Fo-Fc was shown as grey mesh. The I domain and ICAM-5 were shown as Cα traces. The density connecting to Leu289 showed an upward tracing for β6–α7 loop and α7 helix. And there was a break between Leu289 and the density at the lower part of I domain. (**b**) A typical α helix composed of alanines. All Cα–Cβ bonds of the helix pointed to the N terminus of the helix. (**c**) Polarity of the density for α7 helix showing a direction from down to up. The α7 helix was modeled into the density and was shown as Cα traces. Some of the residues were labeled showing a clear polarity from down to up.

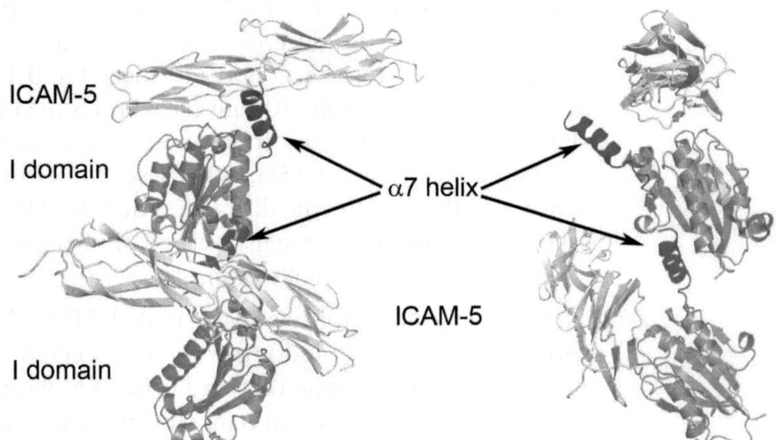

Fig. 2. Ribbon diagram of ICAM-5/I domain complex. Two symmetry-related ICAM-5/I domain complexes were shown with the C-terminal α7-helix of one I domain (at the lower position) inserted into a groove of the other I domain (at the upper position) and the α7-helix of upper I domain swung-out to insert into the third I domain, which was not shown in the figure.

1MQ8), were used as search models to solve the structure of ICAM-5D1D2 with I domain by molecular replacement. The structure of ICAM-1 D1D2 alone was also used (PDB code 1IC1). Phaser (18) from CCP4 package was used to accomplish molecular replacement with HA, ICAM-3D1, and ICAM1-D2 as search models.

The solution was refined in Refmac (19) in CCP4 suit by rigid body refinement followed by constrained refinement. By alternate model building with Coot (20) to fit density into ICAM-5 sequence and refinement with Refmac, there was not much difficulty in the tracing of ICAM-5. However, we did encounter some trouble in correctly tracing the I domain. In all previous integrin α I domain structures, α7 helix is located between β6 strand and α1 helix with a direction from top to down. At the beginning we tried to model the α7 helix of I domain in this direction and the density did not fit well with the residues of α7 helix. Furthermore, the β6–α7 loop did not have any reasonable density (Fig. 1a). When the symmetry-related molecules were checked, we noticed that α7 helix of one molecule might swing out and insert into a symmetry-related molecule in an upside-down fashion (Fig. 1a). Figure 1b shows a typical α helix, for which all the Cα–Cβ bond point to the N terminus of the helix. Upon carefully checking the density we noticed that most of the residues in α7 helix showed a polarity with α7 helix from down to up (Fig. 1c). The final model did have a swing-out of the α7 helix (Fig. 2).

References

1. Hynes, R. O. (2002) Integrins: bidirectional, allosteric signaling machines. *Cell* **110**, 673–687.
2. Springer, T. A., and Wang, J. H. (2004) The three-dimensional structure of integrins and their ligands, and conformational regulation of cell adhesion. *Adv. Protein Chem.* **68**, 29–63.
3. Luo, B. H., Carman, C. V., and Springer, T. A. (2007) Structural basis of integrin regulation and signaling. *Annu. Rev. Immunol.* **25**, 619–647.
4. Shimaoka, M., Xiao, T., Liu, J. H., Yang, Y., Dong, Y., Jun, C. D., McCormack, A., Zhang, R., Joachimiak, A. and Takagi, J. (2003) Structures of the alpha L I Domain and Its Complex with ICAM-1 Reveal a Shape-Shifting Pathway for Integrin Regulation. *Cell* **112**, 99–111.
5. Song, G., Yang, Y., Liu, J. H., Casasnovas, J. M., Shimaoka, M., Springer, T. A., and Wang, J. H. (2005) An atomic resolution view of ICAM recognition in a complex between the binding domains of ICAM-3 and integrin alphaLbeta2. *Proc. Natl. Acad. Sci. USA.* **102**, 3366–3371.
6. Zhang H., Casasnovas J.M., Jin M., Liu J.H., Gahmberg C.G., Springer T.A. and Wang J.H. (2008) An unusual allosteric mobility of the C-terminal helix of a high-affinity alpha L integrin I domain variant bound to ICAM-5. *Mol. Cell* **31**, 432–437.
7. Lee J.O., Rieu P., Arnaout M.A. and Liddington R. (1995) Crystal structure of the A domain from the alpha subunit of integrin CR3 (CD11b/CD18). *Cell* **80**, 631–638.
8. Xiao T., Takagi J., Coller B.S., Wang J.H. and Springer T.A. (2004) Structural basis for allostery in integrins and binding to fibrinogen-mimetic therapeutics. *Nature* **432**, 59–67.
9. Xiong, J.P., Stehle, T., Diefenbach, B., Zhang, R., Dunker, R., Scott, D.L., Joachimiak, A., Goodman, S.L., Arnaout, M.A. (2001) Crystal structure of the extracellular segment of integrin alphaVbeta3. *Science* **294**, 339–345.
10. Shimaoka M., Lu C., Palframan R.T., von Andrian U.H., McCormack A., Takagi J., Springer T.A. (2001) Reversibly locking a protein fold in an active conformation with a disulfide bond: integrin alphaL I domains with high affinity and antagonist activity in vivo. *Proc. Natl. Acad. Sci. USA.* **98**, 6009–6014.
11. Jin M., Song G., Carman C.V., Kim Y.S., Astrof N.S., Shimaoka M., Wittrup D.K., Springer T.A. (2006) Directed evolution to probe protein allostery and integrin I domains of 200,000-fold higher affinity. *Proc. Natl. Acad. Sci. USA.* **103**, 5758–5763.
12. Stanley P. (1989) Chinese hamster ovary cell mutants with multiple glycosylation defects for production of glycoproteins with minimal carbohydrate heterogeneity. *Mol. Cell. Biol.* **9**, 377–383.

13. Davis S.J., Davies E.A., Barclay A.N., Daenke S., Bodian D.L., Jones E.Y., Stuart D.I., Butters T.D., Dwek R.A. and van der Merwe PA. (1995) Ligand binding by the immunoglobulin superfamily recognition molecule CD2 is glycosylation-independent. *J. Biol. Chem.* **270**, 369–375.
14. Ikemizu S., Sparks L.M., van der Merwe P.A., Harlos K., Stuart D.I., Jones E.Y. and Davis S.J. (1999) Crystal structure of the CD2-binding domain of CD58 (lymphocyte function-associated antigen 3) at 1.8-A resolution. *Proc. Natl. Acad. Sci. USA.* **96**, 4289–4294.
15. Otwinowski Z.and Minor W. (1997) Processing of X-ray Diffraction Data Collected in Oscillation Mode. *Methods Enzymol.* **276**, 307–326.
16. Collaborative Computational Project, Number 4. 1994. The CCP4 Suite: Programs for Protein Crystallography. *Acta Cryst.* **D50**, 760–763.
17. Matthews B.W. (1968) Solvent content of protein crystals. *J. Mol. Biol.* **33**, 491–497.
18. McCoy A. J., Grosse-Kunstleve R. W., Adams P. D., Winn M. D., Storoni L.C. and Read R.J. (2007). Phaser crystallographic software. *J. Appl. Cryst.* **40**, 658–674.
19. Murshudov G.N., Vagin A.A. and Dodson E.J. (1997) Refinement of Macromolecular Structures by the Maximum-Likelihood method. *Acta. Cryst.* **D53**, 240–255.
20. Emsley, P., and Cowtan, K. (2004) Coot: model-building tools for molecular graphics. *Acta Crystallogr. D Biol. Crystallogr.* **60**, 2126–2132.

Chapter 9

Electron Microscopic Imaging of Integrin

Kenji Iwasaki

Abstract

Rotary-shadowed samples often used for electron microscopy do not preserve native integrin conformations. Negatively stained integrins – or, more desirably, unstained integrins in a cryo-condition – are now being used with sophisticated imaging techniques. Additionally, a single-particle analysis (SPA) of integrins is advanced by the recent determination of several crystal structures of integrins. Nevertheless the conformational flexibility of integrins limits the ability of SPA to image physiologic conformations. To solve this problem, we apply electron tomography to purified integrin, thereby obtaining high-quality three-dimensional (3-D) images that fit well to the atomic structures. We have also taken typical SPA approaches to obtain a 3-D reconstruction of integrin, using conditions that favor the bent conformation.

Key words: Integrin, Electron microscopy, Single-particle analysis, Tomography

1. Introduction

X-ray crystallography reveals the atomic structure of a protein in a conformation that stabilizes the protein's structure in the crystalline state. Although this provides abundant information that links the three-dimensional (3-D) structure of a protein to its primary structure as well as its functions, a single crystal structure cannot provide the information necessary for this task if the protein is intrinsically flexible. Comparison of multiple crystal structures of a protein obtained under different conditions is one way to reveal the pathways used by a protein to change conformation. More generally, however, electron microscopic (EM) analysis of the protein can serve as a very powerful approach for ascertaining the mechanism underlying structural dynamics. The most common method for visualizing relatively large molecules (~>100 kDa)

Motomu Shimaoka (ed.), *Integrin and Cell Adhesion Molecules: Methods and Protocols*,
Methods in Molecular Biology, vol. 757, DOI 10.1007/978-1-61779-166-6_9, © Springer Science+Business Media, LLC 2011

using EM is single-particle analysis (SPA) (1). However, since this method assumes that, under the same conditions, the majority of particles in the same solution bear the same conformation, it is not a suitable technology for visualizing the structure of flexible proteins. The best way to determine 3-D structures of proteins or protein complexes with conformational diversity using EM is the random conical tilt (RCT) method (2). Dube et al. (3) clearly showed that RCT is much more reliable than normal SPA in analyzing the structure of proteins with flexible domains. A third approach for visualizing purified proteins by molecular EM is electron tomography (ET). Unlike normal SPA and RCT that rely on image averaging of numerous particles, ET derives individual 3-D shapes of all particles present in a field, resulting in a gallery of true "single-particle" images. This technique has seldom been applied to purified proteins because the final structures are noisy, but its independence from image averaging, followed by classification into a limited number of classes, makes it an ideal method for analyzing proteins with high conformational flexibility, such as integrins. In this chapter, I describe practical aspects of ET observation of integrins, focusing on specimen preparation and image analysis. Normal SPA of negatively stained integrins using the atomic resolution structure as a template is also described. For the experimental as well as theoretical details of the RCT method, readers are referred to the recent original papers (4, 5).

2. Materials

2.1. Cell Culture and Lysis

1. RPMI medium: Supplemented with 20 mM HEPES, 10% fetal calf serum, 100 mg/l kanamycin sulfate (GIBCO/Invitrogen), and 2 g/l sodium hydrogen carbonate (special grade).
2. Lysis buffer: 20 mM Tris–HCl pH 7.4, 150 mM NaCl, 1 mM $MgCl_2$, 1 mM $CaCl_2$, 1 mM PMSF, and 1% Triton X-100. Store at 4°C.
3. Minimum Essential Medium Eagle α modification (α-MEM): Supplemented with 5% fetal calf serum and containing 1% (v/v) sodium pyruvate, 1% (v/v) GIBCO MEM non-essential amino acids 10 mM solution, and antibiotics (0.5 mg/ml neomycin, 5 μg/ml puromycin, 0.5% (v/v) of 1,000 units/ml penicillin, and 5 mg/ml streptomycin solution).
4. Human erythroleukemia (HEL) cells (ATCC).
5. Amicon YM30 membrane (Millipore, Billerica, MA).
6. Roller bottles with a 4,200 cm^2 expanded surface area (CORNING).

2.2. Column Chromatography for Purification

1. Buffer A: 20 mM Tris–HCl pH 7.4, 150 mM NaCl, 1 mM $MgCl_2$, 1 mM $CaCl_2$, and 0.1% Triton X-100. Store at 4°C.
2. GRGDSPK-Sepharose column: Heptapeptide GRGDSPK coupled to CNB-activated Sepharose 4B (GE Healthcare, WI). Store at 4°C.
3. Heparin-affinity column: The pre-packed Hi-Trap Heparin 5 ml (GE Healthcare) is easy to handle.
4. Concanavalin A-Sepharose: Pack 30 ml of Con A Sepharose (GE Health) resin into the Econo Column (Bio-Rad, Hercules, CA). Store at 4°C.
5. Buffer B: 20 mM Tris–HCl, pH 7.5, 1 mM $MgCl_2$, and 1 mM $CaCl_2$. Store at 4°C.
6. Elution buffer: 50 mM triethylamine, pH 11.5, 150 mM NaCl, and 1 mM $CaCl_2$. Store at 4°C.
7. Superdex S200 10/300GL (GE Healthcare UK Ltd., England).
8. AKTA-FPLC system (GE Healthcare).

2.3. EM

1. 2.0% (w/v) Uranyl acetate solution: Dissolve 100 mg of $UO_2(CH_3COO)_2 \cdot 2H_2O$ in 5 ml distilled water in an 15-ml tube. The tube is foil wrapped to protect the solution from light and is rotated overnight at 25°C. The solution is dispensed into Eppendolf tube (1.5 ml) and stored in light-resistant container at 4°C. To remove the insoluble crystals, filter the solution using a 0.22-μm filter.
2. A holey grid (QUANTIFOIL Micro Tools GmbH, Jena, Germany; Protochips Inc., Raleigh, NC; and JEOL, Tokyo, Japan): 200-mesh C-flat™ carbon copper grids (CF-2/2-2C; Protochips Inc.) are used to prepare carbon-coated grids for tomography.
3. Buffer C: 50 mM Tris–HCl, pH 7.5, 150 mM NaCl, 1 mM $MgCl_2$, and 5 mM $CaCl_2$.

3. Methods

3.1. Preparation of Integrin Samples

3.1.1. Purification of Full-Length Integrin αIIbβ3 for ET

Full-length integrin αIIbβ3 is purified from HEL cells. The HEL cell line was established from the peripheral blood of a patient with Hodgkin's lymphoma who later developed erythroleukemia (6). The platelet fibrinogen receptor, glycoprotein IIB (αIIb)/IIIa (β3) complex, is expressed on these cells and was identified (7, 8) and characterized (9). The following purification protocol for αIIbβ3 from HEL cells is very similar to the one described by Yamada et al. (10), but contains a few modifications.

1. Prepare a culture flask containing 250 ml of culture medium. Seed with HEL cells and culture for several days in a 5% CO_2 atmosphere.
2. Harvest the HEL cells by centrifugation, resuspend in 12.5 ml of lysis buffer per liter of culture solution, and lyse cells by stirring for 0.5–1 h at 4°C.
3. Collect the supernatant after pelleting the insoluble debris by centrifugation at 30,000 × *g* for 50 min.
4. Load the supernatant onto a tandemly arranged heparin-affinity column and concanavalin A-Sepharose, pre-equilibrated with a buffer A.
5. Disconnect the two columns and elute the glycoprotein from the conA-column with buffer A containing 0.5 M α-methyl-d-mannopyranoside.
6. Analyze the elution fractions on a 10% Tris-glycine SDS polyacrylamide gel, and by immunoblotting using both anti-gpIIb and anti-gpIIIa monoclonal antibodies after transfer to nitrocellulose membranes.
7. Concentrate the collected elution fractions by ultrafiltration using an Amicon YM30 membrane. This step also removes most solubilized sugars.
8. Load the samples onto a 30-ml GRGDSPK-Sepharose column pre-equilibrated with buffer A. Elute specifically bound integrin with buffer A containing 1 mg/ml GRGDSP peptide (11, 12).

3.1.2. Preparation of Recombinant αvβ3 Integrin Ectodomain Fragment for Normal SPA

The construct design of soluble recombinant αVβ3 is the same as that reported by Takagi et al. (13). Extracellular portions of the αV (residues 1–990) and β3 (residues 1–718) subunits were fused to 30-residue ACID and BASE peptides (14), respectively; each peptide contained one cysteine mutation (15). The constrained C-terminal interaction through the coiled coil clasp linked by a covalent bond between cysteine residues represents close apposition between the membrane proximal regions of both subunits.

1. Establish stable transfectant of soluble integrin αVβ3 in CHO lec 3.2.8.1 cells (16).
2. Add 300 ml fresh culture medium and antibiotics to a roller bottle. Seed ~1×10^8 cells into each bottle. Place the bottles in a Bellco roller bottle incubator at 37°C and rotate at 1 rpm. Culture supernatants are harvested every 7 days.
3. Add ammonium sulfate to the harvested cells up to 50% saturation. Stir the solution for 1 h at 4°C. Resuspend the pellet obtained by centrifugation with buffer B. Concentrated supernatants can be stored at −80°C until use.
4. Prepare a 2.5-ml antibody affinity column using anti-leucine zipper monoclonal antibody 2H11 (Chang et al. (14)). The

concentrated supernatants are loaded onto the column pre-equilibrated with buffer B, followed by 10 column volume washes and elution with elution buffer. Eluted fractions are immediately neutralized by addition of 1/10 volume of 1 M Tris–HCl, pH 7.2. Fractions containing integrin (>90% pure) are combined and concentrated to ~1 mg/ml and dialyzed extensively against buffer B.

3.2. Negative Staining of Integrin

The molecular weight of integrin is approximately 250 kDa. Although integrin is considered a large cell surface receptor, it is too small to be analyzed reliably by cryo-EM. The low contrast of the protein (which is composed of C, N, O, H, and S) within the vitreous ice phase tends to result in incorrect image analysis. In particular, low contrast images would increase the chance of misalignment and misclassification of conformationally flexible integrins during conventional SPA. Therefore, negative staining of specimens to increase image contrast is essential for successful 3-D structural analysis. In negative staining, a heavy metal solution, usually uranium, is used to cover the surface of the protein, allowing the metal compound to penetrate into the accessible regions around the protein particles. Heavy metal atoms are resistant to the electron beam and produce strong contrast, with a high ratio of amplitude contrast, which is mainly produced by electrons scattered outside of the aperture (17, 18). However, there are problems associated with this staining procedure. One is the deformation of the protein caused by air-drying, which usually results in a flattened shape. Another problem is partial staining. If regions of a protein particle are not covered by the stain solution, these regions do not contribute to the image. As a consequence, structural information of this uncovered "summit" region will not be reflected in the final calculations. The protocol for negative staining routinely used in our laboratory to minimize these artifacts is given below.

3.2.1. Preparation of Carbon-Coated Grids

Commercially available carbon-coated grids are widely used for negative staining. However, our laboratory prepares custom-made carbon grids by putting a thin carbon film on a holey grid (Fig. 1). Because the carbon membrane itself scatters electrons and thus decreases the contrast of proteins, the thinner the carbon layer the better, particularly for small proteins. On the other hand, the thinner the carbon membrane is, the more easily it is torn. When a mesh containing a large number of holes (e.g., a 1,000-mesh grid) is used, even a very thin carbon membrane can be positioned without being torn. It is not practical to use such fine mesh grids for tomography, however, because the effective area becomes very small due to blockade of the electrons by the grid bars when the specimen is tilted at large angles. Therefore, a layered carbon membrane on a holey carbon-coated grid is preferred.

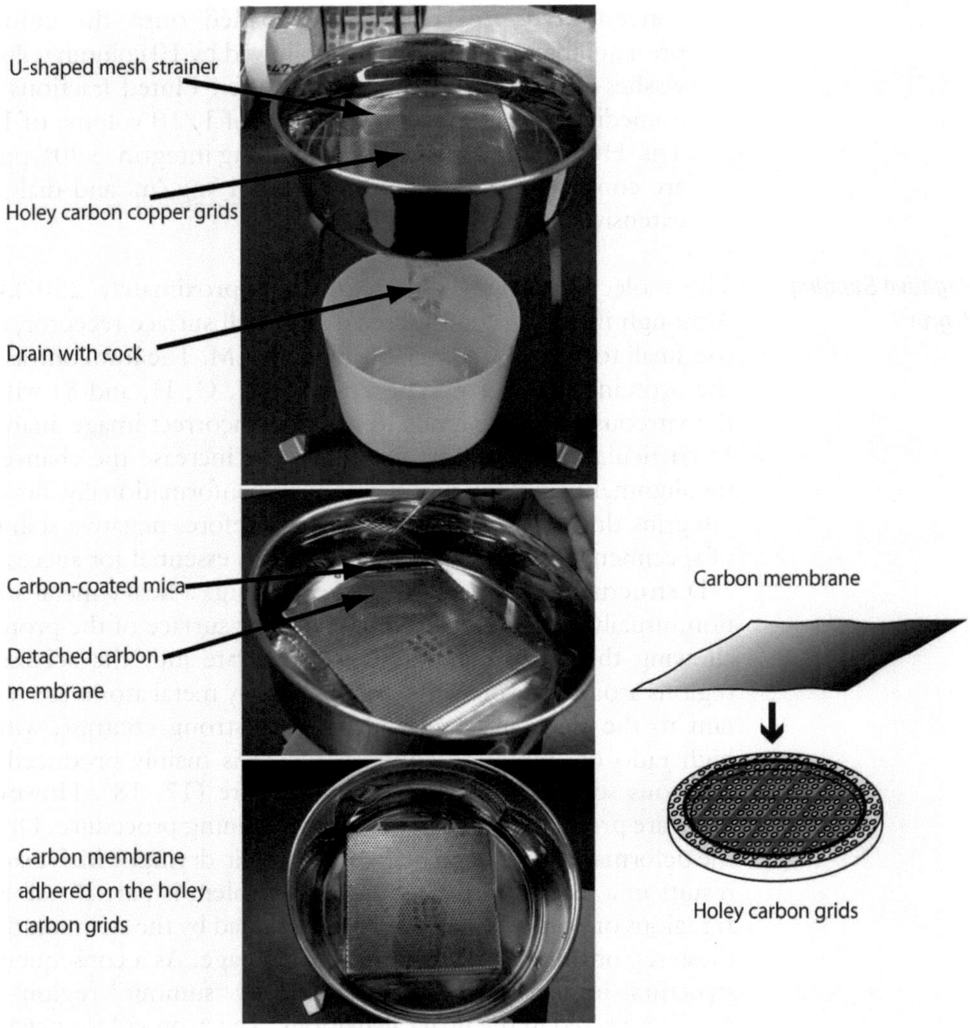

Fig. 1. The process of making carbon-coated holey carbon grids.

1. Evaporate carbon onto a freshly cleaved, unstepped mica surface. A carbon rod is used as the carbon source for the evaporation. The thickness of the carbon film is ~10 nm, which is measured very roughly by comparing its gray density with the known gray density of evaporated carbon on a tile purchased from JEOL.
2. Fill a small water bath equipped with a drain valve at the bottom with distilled water and submerge a U-shaped mesh strainer. Submerge the grids (typically 20–30 pieces at a time) in the mesh strainer. We use 200-mesh C-flat™ carbon copper grids (19).
3. Slide the carbon film off the mica surface into the water and let it float on the surface of the water. Drain the water slowly by carefully opening the valve. The floating carbon film is gradually deposited onto the grids as the water level decreases. Finally, dry the grids well.

3.2.2. Staining

1. Hydrophilize the grids by glow discharge.
2. Hold the edge of the grid using forceps that have no capillary action, and lock the forceps using an O-ring or slide-lock (DUMONT #5 INOX microdissection forceps with slide).
3. Place an aliquot (3 μl) of sample on the carbon-coated side of the grid and leave for 1 min.
4. Tear off a piece of Advantec No. 2 filter paper and touch the frayed edge of the paper to the edge of the grid. As soon as the excess sample solution is blotted onto the filter paper, place 3 μl uranyl acetate on the grid and leave for 30 s.
5. Remove excess stain solution using a piece of filter paper. Take care not to remove the liquid completely. There should be a liquid residue on the grid, as judged by the shiny appearance of the surface. This is a very important point with respect to preventing partial staining and flattening, which are notorious artifacts associated with negative staining (see Note 1).
6. Dry the grid overnight or longer in a desiccator.

3.3. Electron Tomography of Full-Length αIIbβ3

ET is normally used for visualizing fine structural elements within cells or tissues, such as organelles. Molecular resolution in situ is a challenge for those working in the cryo-ET field. However, few use ET for the 3-D reconstruction of purified molecules. Because the acquisition of a tilt series (i.e., multiple images of the same area) results in an accumulation of electron-induced damage in a specimen, it reduces the resolution of the image. Therefore, the RCT method is generally superior to ET for the visualization of purified molecules. The RCT technique is good for discerning conformational changes because it takes into account both the conformational and orientational variations as the source of different particle images, while SPA assumes all particles have an identical 3-D shape, and only takes into account orientational heterogeneity. The one big disadvantage of RCT, however, is the laborious task of data collection. In contrast, ET requires less time to collect data, which increases total throughput of analyses. ET also enables us to check the 3-D structure of one molecule without image averaging of different particles, which is essential for RCT or SPA. Although ET is not an ideal technique for small and globular particles, we took advantage of the unique shape of integrin and succeeded in analyzing integrin αIIbβ3 in a variety of conformations by ET. To accomplish this, reconstructing the whole area and extracting the particles is not a recommended strategy. During the acquisition of a tilt series, the carbon membrane may not retain its flatness due to damage by the electron beam. This makes it difficult to align the images if the specimen is considered as a rigid body. As a result, the tomogram of the whole area is calculated with a certain degree of error. Ideally, a local alignment of each particle should be performed. Most of these difficulties are overcome by using the following procedure (Fig. 2).

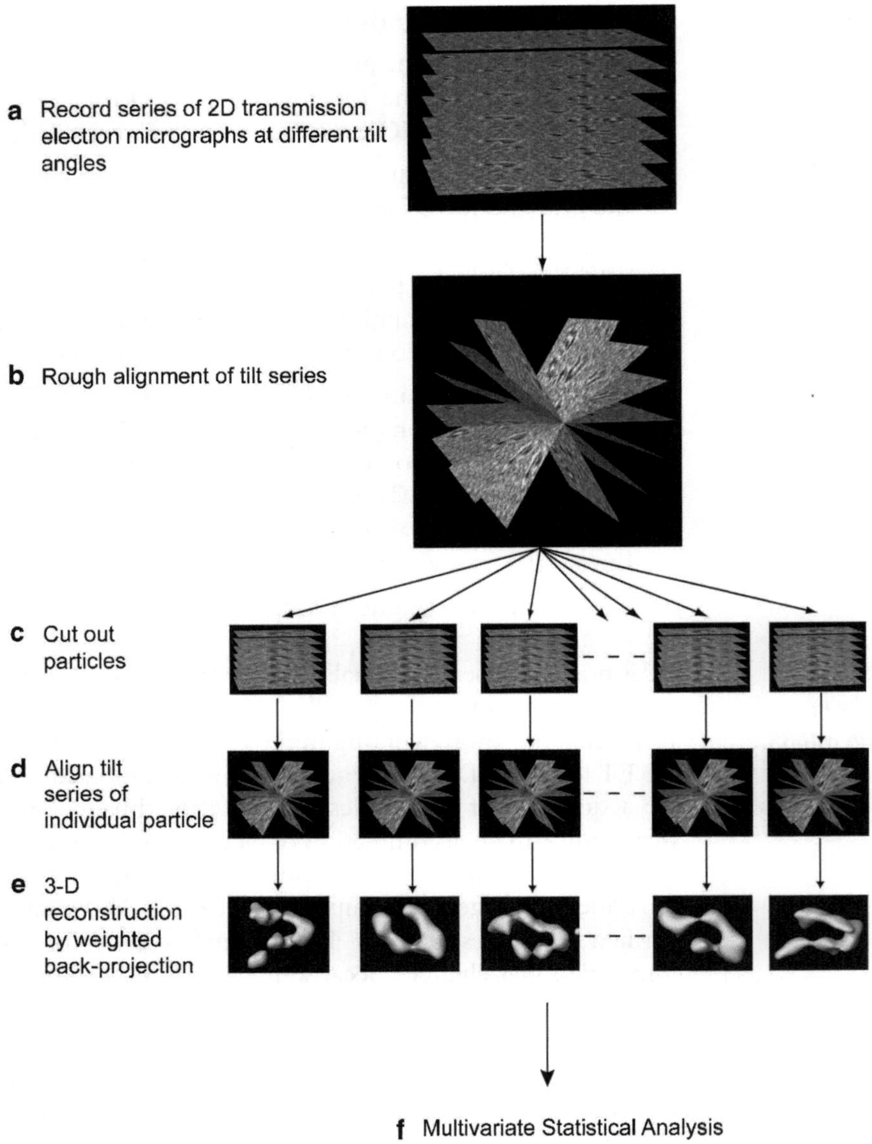

Fig. 2. Scheme for tomography of single particles (20). Sub-tilt series extracted from the original tilt series are realigned, which results in very accurate alignment of its particle image (Reproduced from Ref. (20) with permission from Elsevier Science).

3.3.1. Data Collection and Image Analysis

Observation using a low acceleration voltage is not suitable for high resolution data collection, but it can produce the high contrast images required to carry out accurate image processing. As negative stain EM cannot reach very high resolution, it is wise to use a low acceleration voltage in the range 80–100 kV to generate high contrast. Another hardware specification that requires serious consideration is the choice of detector type: either a charge coupled device (CCD) camera or photographic films. For tomography, sensitivity and detection quantum deficiency (DQE) is important because acquiring each image using a low dose of radiation via a tilt series is required to reduce radiation damage, even in the case of negative stained samples. A CCD camera, which has scintillators to transform electrons to photons, meets this requirement and is therefore the detector of choice. In addition, a CCD camera enables us to check the acquired image in near-real time. This is necessary for increasing the chances of success in acquiring an intact tilt series. Our first attempt at single-particle tomography of integrin used film to capture the images (20), which was very delicate operation because there was no way to know if all images were stored correctly until the films were developed.

Modulation transfer function (MTF), which represents how much contrast at a particular frequency is transferred from the object to the image (i.e., how much of the blurriness of the original signals is caused by the camera) is more important for high resolution techniques, such as 2-D crystallography and SPA (21). For instance, the choice of CCD camera type is not so important for tomography of tissue sections. For tomography of protein particles, however, we have to pay a great deal of attention to the MTF of the CCD camera. If you use CCD cameras with poor MTF, the obtained images become obscure, leading to blurred 3-D images. MTF is related to how much magnification should be selected (22). For example, when the CCD element size is 24 μm, nominal magnification is ×80,000 and postmagnification is 1.25, so the pixel size becomes 0.24 nm/pixel. Based on the sampling theorem proposed and proved by Hurry Nyquist and Claude E. Shannon (see Chapter 3 subheading 2.2 of Ref. (1)), the recoverable limit of the reconstructed object is 0.48 nm (Nyquist limit). Since the image is deteriorated by the camera and image manipulations during analysis, however, the resolution of the reconstructed image cannot reach this value. Thus, sampling size should be much smaller than half size of desired resolution. For example, in our laboratory, if the goal is 1.2 nm resolution, we choose a magnification that generates a pixel size of 0.24–0.3 nm (2/5–1/2 Nyquist frequency) on the CCD.

1. Set the acceleration voltage of an electron microscope to between 80 and 100 kV.
2. Take a tilt series from −60° to +60°. The step size for tilting should not be too small if an autotomography system incorporating a minimum dose system is not available. We select steps 2°–5° when recording a tilt series without a low dose system at room temperature. The step 5° work well, but it is possible to use smaller steps even if images are taken without using a low dose system. Magnification should be decided as described above.
3. Coarse alignment of a tilt series. We align the images by considering the protein molecules as a fiducial marker using SPIDER.

4. Extract the particles from each image. Make a sub-tilt series.
5. Precisely align the sub-tilt series. Coarse alignment of IMOD works well for this purpose (23).
6. Reconstruct each 3-D model from the sub-tilt series. Back-projection or the Simultaneous Iterative Reconstruction Technique (SIRT) (24) should be used.
7. Align the 3-D models. This step is the most difficult part. Whatever programs are used, it is very difficult to align these unsatisfactory models with each other without a visual check. The most convenient tool to implement this step is the visualization software Avizo (VSG, Inc., Burlington, MA). In Avizo, the command *AffineRegistration* enables us to fit one 3-D map onto another by defining the threshold of density used for calculation. It is also possible to define the initial position to fit the map. The fitting process can be visualized in real time. If the results of fitting are clearly wrong, it is possible to specify the starting position manually. Alternatively, one of the options of *AffineRegistration* (*Align principal axes*) can be used to align 3-D models more objectively. However, including a visual check in the alignment process is more effective in this case. After the execution of *AffineRegistration*, shift and rotate the coordinates of the 3-D maps. This process can also be performed easily using *ApplyTransform*.
8. Classify 3-D models. If the alignment of the 3-D models works, it is possible to classify the 3-D models. SPIDER software includes the commands for various multivariate data analyses. k-means clustering using factors produced by principal component analysis (see Ref. (1)) is useful for classifying 3-D models. Principal component analysis, which is the representative method in multivariate data analysis (for the difference between the terms "multivariate statistical analysis" and "multivariate data analysis," see Ref. (1)) is defined simply as follows: If there is data set $\{v_1, v_2, v_3 \ldots, v_i \ldots, v_n\}$, find the coefficients $\{a_1, a_2 \ldots, a_i \ldots, a_n\}$ that maximize the variance of u, where $u = \Sigma a_i v_i$, and $\Sigma a_i^2 = 1$. Then, analyze the data using u, which is called the "principal component." The coefficients are often called factors, and are the same as eigenvalues. The advantage of using principal component analysis is reduction of the dimension of data space. Since comparing images involves vast calculations, dimension-reduction is more important as the number of images increases. Indeed, multivariate data analysis (not principal component analysis but correspondence analysis) was introduced by van Heel and Frank to classify electron microscopic images (25). The classification of 3-D models obtained from tomography has been

undertaken by Walz et al. (26) using their own program. It is now possible to perform multivariate data analysis using the "CA S" and "CL KM" commands of SPIDER for correspondence analysis or principal component analysis and k-means clustering, respectively.

9. Average the 3-D models in each class. Ideally, the alignment parameters, that is, the rotation and the shift to align the 3-D models with each other, are applied to the sub-tilt series. Then, as an average model, one 3-D model is calculated from one data set merged from these aligned sub-tilt series. However, to simplify this, we average the aligned 3-D models in each class. The command AS in SPIDER or Arithmetic module in Avizo works well.
10. Fit the atomic coordinates obtained from crystal structures. Figure 3 shows the averaged 3-D map from tomograms of each particle. A fragment of the atomic coordinates of αIIbβ3 in the active state can be fitted. The coordinates were obtained by extracting head piece portions from the PDB code, 3FCS, in a bent form as reported by Zhu et al. (27) and replacing the β-subunit with 2VDL (28, 29), the hybrid domain which swings out. Movement of the hybrid domain

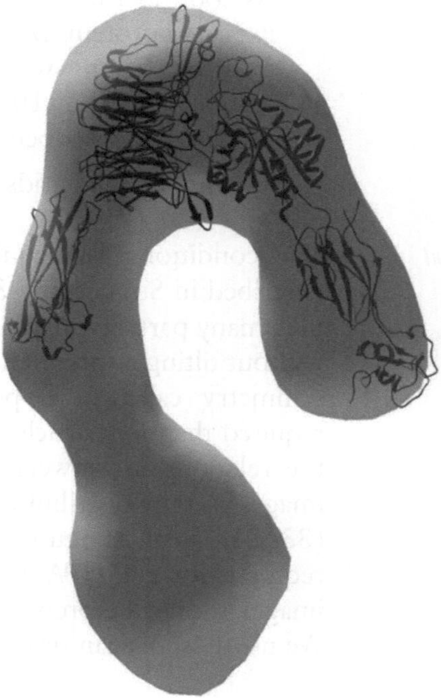

Fig. 3. 3-D averaged map obtained from tomograms of sub-tilt series of purified integrin images. The head piece regions of atomic coordinates of αIIbβ3 in active form are fitted onto the map. The hybrid domain of the swung-outward β-subunit is fitted well. All the processing was done using CHIMERA (40).

related to the integrin-affinity state was reported by Springer's group (13, 30, 31).

3.4. Single-Particle Reconstruction of αvβ3

As mentioned in the previous section, simple single-particle analysis of highly flexible integrin samples can easily lead to miscalculation and to incorrect structures. However, when integrins are imaged in a condition that favors a certain conformation, normal SPA would work. It is generally accepted that integrins in a low-affinity state assume a so-called "bent conformation," which is stable. Indeed, the crystal structure of αVβ3 was solved as a bent form (32). The following is the protocol for obtaining a low resolution 3-D structure of integrin in a low-affinity form.

3.4.1. Prepare Grids for EM

1. Superdex S200 10/300GL is an appropriate size-exclusion chromatography matrix for integrins. Use the column with an AKTA-FPLC system pre-equilibrated with a buffer containing buffer C at a flow rate of 0.5 ml/min.
2. Dilute the stock solution of αVβ3 to 0.12 mg/ml. Inject 60 μl (7.3 μg protein) into the sample loop and start the run at a flow rate of 0.5 ml/min. This concentration results in a peak absorbance height of 0.9 mAU (absorbance unit) at 280 nm on our AKATA-FPLC system.
3. As soon as the peak appears at the elution volume for the expected protein size, take the peak fraction and place on the grid for staining with 2% (w/v) uranyl acetate. Use a thin carbon film from 10 to 15 nm. Such a thin carbon film is prepared as described in Subheading 3.2.1 (see Note 2).
4. Vacuum dry the grids in a desiccator overnight.

3.4.2. Data Collection

The conditions for acquiring images are almost the same as described in Subheading 3.3.1. The primary difference is that in SPA many particle images with various orientations are collected without tilting a specimen. For proteins like integrin that have no symmetry (called Cl in point group symmetry), more images are required than for particles that have higher symmetry. Although the relationship between the theoretically required number of images and the distribution of orientations is described elsewhere (33, 34), a rough figure would be that 1,000 particle images are required for 2-D SPA and more than several thousand particle images should be prepared for 3-D SPA in the case of integrin. We use a "spot scan system" to collect images (35).

3.4.3. Image Processing

Software packages for SPA, SPIDER, and EMAN, are available for free, and IMAGIC-5 is commercially available. There are many other software packages (please refer to (36)). In our laboratory, depending on the advantage of each software, these three software packages are used in individual case.

2-D SPA. A complete description of the 2-D SPA process using IMAGIC-5 is described in the IMAGIC-5 manual that can be downloaded, and is briefly explained as follows:

1. Pick up particle images on the computer *boxer*, which is one the EMAN commands, enables us to do this interactively. One thousand particle images are enough to examine integrin structures.
2. Band-pass filter the particle image set using the *PRET-BOXED* command. In general, the highest and lowest frequency is set depending on the desired resolution and the size of particle, respectively.
3. Use the *CENTER-IMAGE* command to align images to the sum of total images.
4. Calculate principal components using the *MSA-RUN* command.
5. Classify particle images using principal components and calculate average images in each class. This process is implemented by *MSA-CLASSIFY* and *MSA-SUM*.
6. Use the averaged images as reference images to multiple reference alignments. This is executed by *M-R-A*. Go to 4.
7. Repeat from 4 to 6 until there is no change in the average image before and after. Figure 4 shows that αVβ3 is in the bent form even in the raw image. Collect particle images from

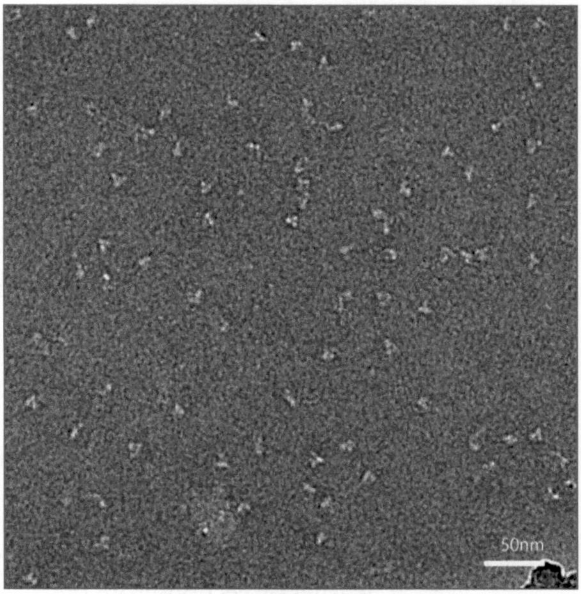

Fig. 4. CCD image of negative stained αVβ3. The contrast of the protein is white, which is surrounded by uranyl acetate with black contrast.

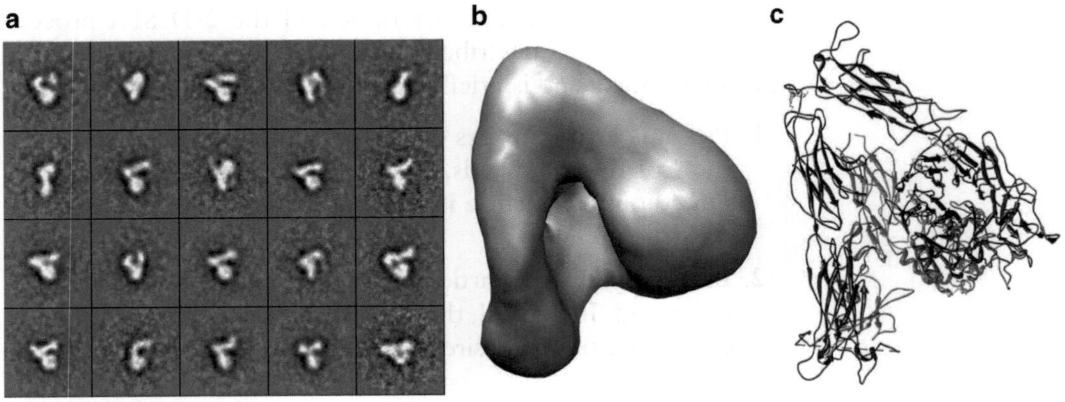

Fig. 5. (**a**) 2-D class average image of αVβ3 in a bent conformation; 1504 particle images were used. (**b**) 3-D map calculated from 19255 particle images by EMAN. *refine* was implemented for 20 cycles. (**c**) Crystal structure of αVβ3 in the same orientation as (**b**) (PDB CODE:1JV2).

many of these CCD images or digitized images from micrographs. Figure 5a shows the result of 2-D SPA. Improvement of the signal-to-noise ratio more clearly shows the outline of the structure.

3-D SPA. The following process is explained for the case of EMAN. We now often use SPIDER for 3-D SPA. The important point is that SPA needs an initial model. In most cases, we start from one model that assumes that there is a majority group with the homogenous structure close to the initial model. To resolve the bent conformation of integrin, this process is appropriate. Like extended integrin, when the structures are diverse, this strategy will generate wrong results.

In any observing system, the incoming signal is modulated by the instrument's function. The electron microscope is no exception. The distortion of the original mass distribution of the object caused by electron microscopes is described by contrast transfer theory. Since explanation of this theory goes beyond the purpose of this book, please refer to (1). However, one significant modulation, which is expressed by the phase contrast function (CTF), is related to the phenomenon expressed as the oscillation in the amplitude of the Fourier transform of the image contrast. The oscillation is dependent on the defocus value. Therefore, this process should be conducted for each micrograph or CCD image.

Typically, the 3-D SPA is composed of three main processes: projection matching, 2-D image averaging, and calculation of the 3-D map. The *refine* command automatically implements these processes sequentially. Once the 3-D map is obtained, *refine* calculates projections from the 3-D map and goes back to projection matching using these new references.

1. Box more than several thousand particle images from CCD images or micrographs using *boxer*.
2. Correct the phase of the images using *ctfit*.
3. Create an initial model at 2 nm resolution from the crystal structure of αVβ3 (PDB:1JV2) (32). This is done by *pdb2mrc*.
4. *refine* command defines the number of cycles of sequential image processing: (1) calculate projection images from the 3-D map; (2) classify 2-D images using projection matching; (3) align and average particle images in each class; (4) calculate the 3-D map. Repeat this cycle until the structure of the 3-D map does not change. See Fig. 5b for an example of a result following *refine*.

4. Notes

1. However carefully step 6 in Subheading 3.2.2. is performed, it is difficult to completely eliminate partial staining and flattening when the subjects are protein molecules. In such cases, adding *tobacco mosaic virus* (TMV) to the sample solution can dramatically improve the quality of the staining. TMV is relatively resistant to the flattening caused by air-drying due to its very rigid structure. Also, it is possible to check for the presence of deformation effects by examining the layer lines in the power spectrum computed from the images, caused by disturbance of the helical structure of the TMV envelope (37). Having TMV particles in the specimen has two benefits. First, the regions close to the TMV develop a deep stain, where flattening and partial staining of the target molecules is less likely to occur compared to areas far from the TMV. Therefore, inclusion of TMV particles in a sample at an appropriate density can result in a large area with desirable staining characteristics for the target particles. Second, the degree of magnification can be accurately calculated. In the transmission electron microscope, magnification is slightly changed from the set value when the focus is adjusted by the current of mainly objective lens. Ideally, the magnification should be measured at each image acquisition, which is not realistic. When TMV is contained in the image, accurate magnification can be calculated from the layer line at 2.3 nm obtained from the image by Fourier-transformation (38, 39). TMV particles are stable under most buffer conditions, enabling its universal use in widely different applications. Prior to addition to the sample solution, it is important to dialyze the TMV solution against the same buffer as used for the sample solution.

2. The most important point is that the negative staining must be performed immediately following sample separation by size-exclusion chromatography to obtain mono-dispersed particles on the carbon. This procedure worked for most protein samples (including non-integrin samples) we tested for obtaining mono-dispersed particles on the carbon grid, even though the mechanism behind this protocol is not known.

Acknowledgments

The author would like to thank Professor Junichi Takagi for critical reading of this manuscript and for his advice, and Emiko Mihara and Naoyuki Miyazaki for technical assistance. A part of this work was supported by "Nanotechnology Network Project of the Ministry of Education, Culture, Sports, Science and Technology (MEXT), Japan" at the Research Center for Ultrahigh Voltage Electron Microscopy, Osaka University (Handai multifunctional Nano-Foundry) and the CREST, JST.

References

1. Frank, J. (2006) *Three-dimensional electron microscopy of macromolecular assemblies*, 2nd ed., Oxford University Press.
2. Radermacher, M., Wagenknecht, T., Verschoor, A., and Frank, J. (1986) A new 3-D reconstruction scheme applied to the 50S ribosomal subunit of E. coli, *J. Microsc.* **141**, RP1-2.
3. Dube, P., Herzog, F., Gieffers, C., Sander, B., Riedel, D., Muller, S. A., Engel, A., Peters, J. M., and Stark, H. (2005) Localization of the coactivator Cdh1 and the cullin subunit Apc2 in a cryo-electron microscopy model of vertebrate APC/C, *Mol. Cell* **20**, 867–879.
4. Radermacher, M. (1988) Three-dimensional reconstruction of single particles from random and nonrandom tilt series, *J. Electron Microsc. Tech.* **9**, 359–394.
5. Shaikh, T. R., Gao, H., Baxter, W. T., Asturias, F. J., Boisset, N., Leith, A., and Frank, J. (2008) SPIDER image processing for single-particle reconstruction of biological macromolecules from electron micrographs, *Nat. Protoc.* **3**, 1941–1974.
6. Martin, P., and Papayannopoulou, T. (1982) HEL cells: a new human erythroleukemia cell line with spontaneous and induced globin expression, *Science* **216**, 1233–1235.
7. Tabilio, A., Rosa, J. P., Testa, U., Kieffer, N., Nurden, A. T., Del Canizo, M. C., Breton-Gorius, J., and Vainchenker, W. (1984) Expression of platelet membrane glycoproteins and alpha-granule proteins by a human erythroleukemia cell line (HEL), *EMBO J.* **3**, 453–459.
8. Papayannopoulou, T., Nakamoto, B., Yokochi, T., Chait, A., and Kannagi, R. (1983) Human erythroleukemia cell line (HEL) undergoes a drastic macrophage-like shift with TPA, *Blood* **62**, 832–845.
9. Bray, P. F., Rosa, J. P., Lingappa, V. R., Kan, Y. W., McEver, R. P., and Shuman, M. A. (1986) Biogenesis of the platelet receptor for fibrinogen: evidence for separate precursors for glycoproteins IIb and IIIa, *Proc. Natl. Acad. Sci. USA.* **83**, 1480–1484.
10. Yamada, T., Uyeda, A., Kidera, A., and Kikuchi, M. (1994) Functional analysis and modeling of a conformationally constrained Arg-Gly-Asp sequence inserted into human lysozyme, *Biochemistry* **33**, 11678–11683.
11. Pytela, R., Pierschbacher, M. D., Ginsberg, M. H., Plow, E. F., and Ruoslahti, E. (1986) Platelet membrane glycoprotein IIb/IIIa: member of a family of Arg-Gly-Asp-specific adhesion receptors, *Science* **231**, 1559–1562.
12. Pfaff, M., Gohring, W., Brown, J. C., and Timpl, R. (1994) Binding of purified collagen receptors (alpha 1 beta 1, alpha 2 beta 1) and RGD-dependent integrins to laminins and

laminin fragments, *Eur. J. Biochem.* **225**, 975–984.

13. Takagi, J., Petre, B. M., Walz, T., and Springer, T. A. (2002) Global conformational rearrangements in integrin extracellular domains in outside-in and inside-out signaling, *Cell* **110**, 599–511.
14. Chang, H. C., Bao, Z., Yao, Y., Tse, A. G., Goyarts, E. C., Madsen, M., Kawasaki, E., Brauer, P. P., Sacchettini, J. C., Nathenson, S. G., and et al. (1994) A general method for facilitating heterodimeric pairing between two proteins: application to expression of alpha and beta T-cell receptor extracellular segments, *Proc. Natl. Acad. Sci. USA.* **91**, 11408–11412.
15. Takagi, J., Erickson, H. P., and Springer, T. A. (2001) C-terminal opening mimics 'inside-out' activation of integrin alpha5beta1, *Nat. Struct. Biol.* **8**, 412–416.
16. Stanley, P. (1989) Chinese hamster ovary cell mutants with multiple glycosylation defects for production of glycoproteins with minimal carbohydrate heterogeneity, *Mol. Cell. Biol.* **9**, 377–383.
17. Rose, H. (1984) Infromation transfer in transmission electron microscopy, *Ultramicroscopy* **15**, 173–192.
18. Fitzgerald, L. A., Leung, B., and Phillips, D. R. (1985) A method for purifying the platelet membrane glycoprotein IIb-IIIa complex, *Anal. Biochem.* 151, 169–177.
19. Quispe, J., Damiano, J., Mick, S. E., Nackashi, D. P., Fellmann, D., Ajero, T. G., Carragher, B., and Potter, C. S. (2007) An improved holey carbon film for cryo-electron microscopy, *Microsc. Microanal.* **13**, 365–371.
20. Iwasaki, K., Mitsuoka, K., Fujiyoshi, Y., Fujisawa, Y., Kikuchi, M., Sekiguchi, K., and Yamada, T. (2005) Electron tomography reveals diverse conformations of integrin alphaIIbbeta3 in the active state, *J. Struct. Biol.* **150**, 259–267.
21. Hesse, J., Hebert, H., and Koeck, P. J. (2000) Evaluation of scanners and CCD cameras for high-resolution TEM of protein crystals and single particles, *Microsc. Res. Tech.* **49**, 292–300.
22. Ludtke, S. J., Booth, C. R., Serysheva, I. I., Chen, D.-H., and Chiu, W. (2005) Single Particle Reconstructions at Subnanometer Resolution from a JEOL 2010 F and a 4 k×4 k Gatan CCD Camera. *Microsc. Microanal.* **11(Suppl 2)**, 60–61.
23. Kremer, J. R., Mastronarde, D. N., and McIntosh, J. R. (1996) Computer visualization of three-dimensional image data using IMOD, *J. Struct. Biol.* **116**, 71–76.
24. Gilbert, P. (1972) Iterative methods for the three-dimensional reconstruction of an object from projections, *J. Theor. Biol.* **36**, 105–117.
25. van Heel, M., and Frank, J. (1981) Use of multivariate statistics in analysing the images of biological macromolecules, *Ultramicroscopy* **6**, 187–194.
26. Walz, J., Typke, D., Nitsch, M., Koster, A. J., Hegerl, R., and Baumeister, W. (1997) Electron Tomography of Single Ice-Embedded Macromolecules: Three-Dimensional Alignment and Classification, *J. Struct. Biol.* **120**, 387–395.
27. Zhu, J., Luo, B. H., Xiao, T., Zhang, C., Nishida, N., and Springer, T. A. (2008) Structure of a complete integrin ectodomain in a physiologic resting state and activation and deactivation by applied forces, *Mol. Cell* **32**, 849–861.
28. Xiao, T., Takagi, J., Coller, B. S., Wang, J. H., and Springer, T. A. (2004) Structural basis for allostery in integrins and binding to fibrinogen-mimetic therapeutics, *Nature* **432**, 59–67.
29. Springer, T. A., Zhu, J., and Xiao, T. (2008) Structural basis for distinctive recognition of fibrinogen gammaC peptide by the platelet integrin alphaIIbbeta3, *J. Cell Biol.* **182**, 791–800.
30. Carman, C. V., and Springer, T. A. (2003) Integrin avidity regulation: are changes in affinity and conformation underemphasized?, *Curr. Opin Cell Biol.* **15**, 547–556.
31. Takagi, J., Strokovich, K., Springer, T. A., and Walz, T. (2003) Structure of integrin alpha5beta1 in complex with fibronectin, *EMBO J.* **22**, 4607–4615.
32. Xiong, J. P., Stehle, T., Diefenbach, B., Zhang, R., Dunker, R., Scott, D. L., Joachimiak, A., Goodman, S. L., and Arnaout, M. A. (2001) Crystal structure of the extracellular segment of integrin alphaVbeta3, *Science* **294**, 339–345.
33. Glaeser, R. M. (1999) Review: electron crystallography: present excitement, a nod to the past, anticipating the future, *J. Struct. Biol.* **128**, 3–14.
34. Henderson, R. (1995) The potential and limitations of neutrons, electrons and X-rays for atomic resolution microscopy of unstained biological molecules, *Q. Rev. Biophys.* **28**, 171–193.
35. Brink, J., Chiu, W., and Dougherty, M. (1992) Computer-controlled spot-scan imaging of crotoxin complex crystals with 400 keV electrons at near-atomic resolution, *Ultramicroscopy* **46**, 229–240.
36. Smith, R., and Carragher, B. (2008) Software tools for molecular microscopy, *J. Struct. Biol.* **163**, 224–228.

37. Klug, A., and Berger, J. E. (1964) An Optical Method for the Analysis of Periodicities in Electron Micrographs, and Some Observations on the Mechanism of Negative Staining, *J. Mol. Biol.* **10**, 565–569.
38. Finch, J. T. (1964) Resolution of the Substructure of Tobacco Mosaic Virus in the Electron Microscope, *J. Mol. Biol.* **8**, 872–874.
39. Kendall, A., McDonald, M., and Stubbs, G. (2007) Precise determination of the helical repeat of tobacco mosaic virus, *Virology* **369**, 226–227.
40. Pettersen, E. F., Goddard, T. D., Huang, C. C., Couch, G. S., Greenblatt, D. M., Meng, E. C., and Ferrin, T. E. (2004) UCSF Chimera – a visualization system for exploratory research and analysis, *J. Comput. Chem.* **25**, 1605–1612.

Chapter 10

An NMR Method to Study Protein–Protein Interactions

Noritaka Nishida and Ichio Shimada

Abstract

Specific interactions between proteins are a fundamental process underlying the various biological events, such as cell–cell contacts, signal transduction, and gene expression. Therefore, the structural investigations of protein–protein interactions provide useful information for understanding these events. We describe an NMR method, termed the cross-saturation (CS) method, to determine the binding sites of protein complexes more precisely than conventional NMR methods. The CS method can determine the binding sites of a protein complex that undergoes fast exchange between the free and the bound states, regardless of the molecular size of the complex.

Key words: NMR, Interaction, Deuterium, Cross-saturation, Macromolecular complex

1. Introduction

Structure determination of protein complexes, using the X-ray crystallography or NMR methods, provides valuable information for understanding various biological events and for developing the pharmaceutical compounds for clinical use. However, attempts to crystallize protein complexes may not always be successful. Conventional NMR methods, using chemical shift perturbation, or HD exchange experiments, cannot determine the binding interface accurately, because the information obtained from these methods is ambiguous. We recently developed an NMR method to identify the binding interface, which utilizes the cross-saturation phenomena that occur at the interface of the complex (1). The CS method can determine the binding interface with high accuracy, compared with the conventional NMR methods, because the cross-saturation mainly depends on the distance between the protons in the complex. In addition, if

Motomu Shimaoka (ed.), *Integrin and Cell Adhesion Molecules: Methods and Protocols*,
Methods in Molecular Biology, vol. 757, DOI 10.1007/978-1-61779-166-6_10, © Springer Science+Business Media, LLC 2011

the CS method is applied to a complex that undergoes a fast exchange between the free and the bound states, then the binding sites can be determined regardless of the molecular size of the complex (2). This modified CS experiment, termed the transferred CS (TCS) method, is very useful to investigate interactions with extremely large and heterogeneous molecules, such as detergent-solubilized membrane proteins, or components of the extracellular matrix.

1.1. Principle of the Cross-Saturation Method

The goal of the CS experiment is to identify the binding interface of protein I, which forms a stable complex with protein II. As shown in Fig. 1, protein I is uniformly ^{2}H and ^{15}N labeled, while protein II is not isotopically labeled (see Note 1). Therefore, when an radio frequency (RF) pulse is applied at the frequency corresponding to the aliphatic resonances (0–4 ppm), only the protons belonging to protein II will be saturated. Although protein I, with its uniform ^{2}H and ^{15}N labeling, is not directly affected by the RF field, the saturation can be transferred from protein II to protein I through the interface of the complex, by a phenomenon called cross-saturation (Fig. 1). If the proton density of the protein I is sufficiently low, then the saturation transferred to protein I is limited to the interface. The effect of the cross-saturation can be detected by the reduction of the signal intensity of the ^{1}H-^{15}N HSQC (heteronuclear single quantum coherence) spectrum. Since the saturation is transferred from protein II to protein I, we will refer to protein I as the donor protein and protein II as the acceptor protein in the following sections.

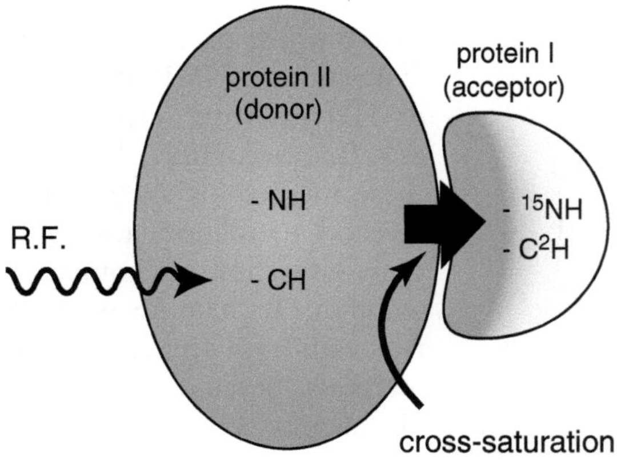

Fig. 1. Principle of the cross-saturation (CS) experiment. The band selective RF irradiation causes saturation of all of the protons of protein II, as shown in *gray*. On the contrary, protein I, which has no aliphatic protons, is not affected by the RF irradiation directly, but the saturation is transferred only at the binding sites for protein II, due to the cross-saturation phenomena.

2. Materials

1. LB medium: dissolve tryptone (10 g), yeast extract (5 g), and NaCl (5 g) in 1 L of Milli-Q water and adjust pH to 7.4.
2. M9 medium (D_2O) (see Note 2): dissolve Na_2HPO_4 (12.9 g), KH_2PO_4 (3.0 g), $^{15}NH_4Cl$ (1.0 g), NaCl (0.5 g), and ^2H glucose (2.0 g) in 1 L of D_2O (99.9%), and add the following stock solutions: 2 mL of 1 M $MgSO_4$, 100 μL of 10 mM $FeCl_3$, 1 mL of 1 mg/mL biotin, 100 μL of 50 mg/mL thiamine, 1 mL of 0.1 M $CaCl_2$, 100 μL of 0.5 M $ZnSO_4$, and 1 mL of 30 mg/mL antibiotics. Sterilize with a 0.22-μm filter. Use 100 mL for pre-culture and the remaining 900 mL for protein expression (see Note 3).
3. NMR buffer: 20 mM Na-phosphate (pH 6.0), 50 mM NaCl, 80%(v/v) D_2O, and 0.05% NaN_3.
4. Isopropyl β-D-1-thiogalactopyranoside (IPTG): dissolved in D_2O at 0.4 M.

3. Methods

3.1. Preparation of the Deuterated Acceptor Proteins

1. Inoculate 10 mL of LB medium (containing 50 μg/mL of antibiotics) with a single colony from a freshly plated transformation. Grow overnight in a 37°C shaker incubator.
2. Centrifuge the culture at 3,000 × *g* for 5 min. Remove the supernatant thoroughly, resuspend, and transfer to 100 mL of M9(D_2O) medium. Grow overnight in a 37°C shaker incubator.
3. Centrifuge at 3,000 × *g* for 5 min. Remove the medium thoroughly, resuspend, and transfer to 900 mL of M9(D_2O) medium. Grow in a 37°C shaker incubator.
4. Induce protein expression by adding 1 mL of 0.4 M IPTG when the OD at 600 nm reaches 0.6. The duration and the temperature of the protein expression may need to be optimized for each target protein (see Note 4).
5. Harvest the cells by centrifugation. Purify the proteins according to the established protocol.
6. Check the deuteration efficacy by mass spectrometry or ^1H NMR spectroscopy. More than 95% of deuteration for non-exchangeable protons is desirable.

3.2. CS Experiment and Data Analysis

1. Mix the purified acceptor protein with the donor protein at a 1:1 ratio and exchange the solvent to NMR buffer containing a higher (>80%) concentration of D_2O. Transfer to a 5-mm NMR tube.
2. The pulse scheme for the cross-saturation experiment is shown in Fig. 2. Pulse program is divided into two segments: the RF irradiation period, followed by the water-flip back HSQC spectroscopy or transverse relaxation optimized spectroscopy (TROSY) experiment (3) (see Note 5). During the irradiation period, the carrier frequency is set to 0–1 ppm, and the band selective RF field at the aliphatic proton frequency is applied using an adiabatic WURST-2 decoupling scheme (4) (see Note 6). Following the RF saturation, the carrier frequency is switched to water resonance, and the pulse train of a ^{1}H-^{15}N HSQC or TROSY experiment is implemented. After recording the free induction decay, the RF irradiation is repeated following an appropriate length of the delay (T_{adj}) for the recovery of the magnetization (see Note 7). In the CS experiment, a reference HSQC spectrum needs to be recorded without applying the RF irradiation. Therefore, two HSQC spectra with and without irradiation will be acquired alternately for each F1 increment.
3. The recorded NMR data are divided into the two HSQC (or TROSY) spectra with and without irradiation (see Note 8).

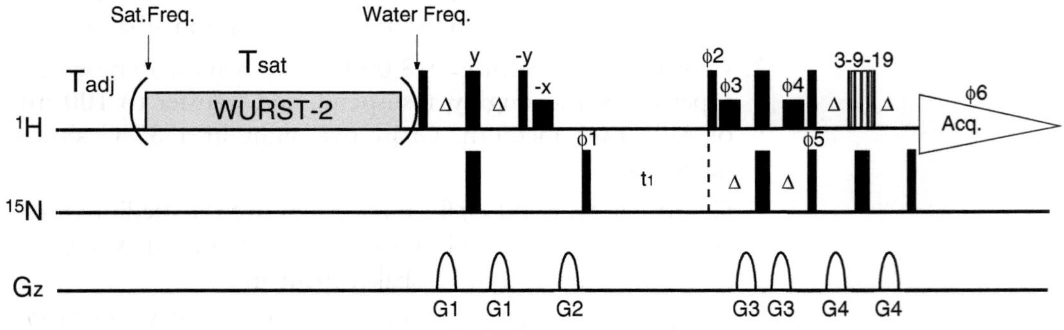

Fig. 2. Pulse scheme for the cross-saturation experiment. Unless otherwise specified, pulse phases are applied along the *x*-axis. *Narrow* and *wide bars* depict 90° pulses and 180° pulses, respectively. *Short solid bars* are water-selective 90° pulses. The line marked Gz indicates the duration and the amplitude of the sine-shaped pulsed magnetic field gradient applied along the *z*-axis: G1 = (600 μs, 7.5 G/cm); G2 = (1,000 μs, 10 G/cm); G3 = (600 μs, 14.5 G/cm); G4 = (600 μs, 20 G/cm). The delay, Δ, is 2.25 ms. The following phase cycling scheme was used: φ1 = {y, −y, x, −x}; φ2 = {−y}; φ3 = {−y}; φ4 = {−x}; φ5 = {−y}; φ6(receiver) = {y, −y}. In the ^{15}N (t1) dimension, a phase-sensitive spectrum is obtained by recording a second free induction decay (FID) for each increment of t1, with φ1 = {y, −y, −x, x}; φ2 = {y}; φ3 = {y}; φ4 = {−x}; φ5 = {y}; φ6 (receiver) = {−x, x}. In order to measure a spectrum with RF irradiation of an aliphatic region, a band-selective WURST-2 saturation scheme is applied during the T_{sat} period prior to the TROSY–HSQC scheme. In this case, the carrier frequency is switched to that for the irradiation before the WURST-2 saturation period (T_{sat}) and back to the water frequency just before the TROSY–HSQC scheme. The additional relaxation delay (T_{adj}) can be set to an appropriate value for obtaining a sufficient signal-to-noise ratio.

4. Perform a Fourier transformation of the time domain NMR data to obtain the frequency domain spectrum.
5. Measure the peak intensity of the spectra with or without RF irradiation (see Note 9).
6. Plot the peak intensity ratio according to the residue number, and map the residues showing a significant intensity reduction on the known structure of the acceptor protein.

3.3. Application of the CS Method: The FB and Fc Complex

The CS method was applied to the B domain of the protein A (FB) complexed with the Fc portion of human immunoglobulin G (IgG) (1). FB and Fc form a stable 1:1 complex with a total molecular mass of 64 kDa. The CS experiment was carried out using a sample containing 1.0 mM of uniformly ^{2}H- and ^{15}N-labeled FB as an acceptor protein, and 1.0 mM of the unlabeled Fc from the human IgG as a donor protein, dissolved in the NMR buffer with 10% H_2O/90% D_2O. The TROSY spectrum with RF irradiation exhibited the selective intensity reduction for some cross-peaks, compared with that without RF irradiation (Fig. 3a). This indicates that the saturation in the Fc fragment was transferred to the bound FB through the interface. A plot of the intensity ratio according to the residue number shows strong intensity reduction of the residues on helix I and helix II of FB, but not for those on helix III (Fig. 3b) (see Note 10). In addition, the residues with the strong signal reduction were localized on the side of the helix that is exposed to the solvent. When a solvent with a higher concentration of H_2O (90% H_2O/10% D_2O) was used, the residues on both sides of the helix exhibited similar levels of signal reduction, indicating that the saturation was diffused to the proximal protons (Fig. 3c) (see Note 11). A map of the residues with significant intensity reduction on the 3D structure of FB is shown in Fig. 3e. This result showed great accordance with the binding site defined by the X-ray crystal structure of the FB–Fc complex (Fig. 3d). It is clear that the CS method can determine the binding interface more rigorously, compared to the binding interface inferred from conventional NMR methods, such as chemical shift perturbation (Fig. 3f) or hydrogen–deuterium exchange (Fig. 3g) experiments (5). Our theoretical simulation suggested that the FB residues within about 7 Å from the interface would exhibit an intensity ratio of less than 0.5.

3.4. Transferred Cross-Saturation Method and its Application to the VWF-A3 Domain and Collagen

Due to the size limitation of the NMR spectroscopy, the CS method is difficult to apply to a protein complex with a molecular mass larger than 150 kDa. To overcome this limitation, we have developed a modified version of the CS experiment, termed the transferred cross-saturation (TCS) method (2). As shown in Fig. 4a, if the deuterated protein I, which is added in excess relative to protein II, undergoes a fast exchange process between the free

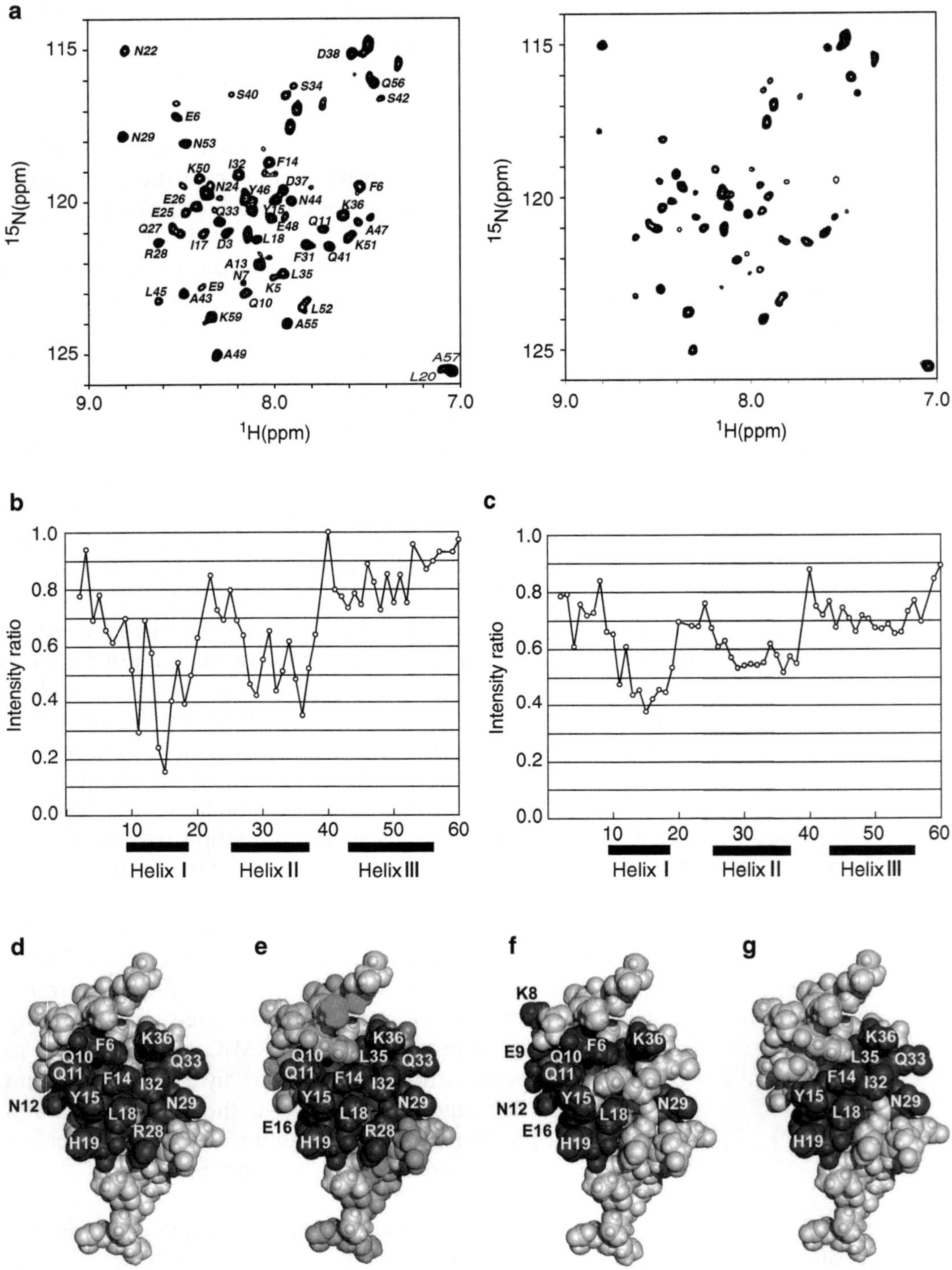

Fig. 3. Identification of the biding site of the FB–Fc complex using the CS experiment. (**a**) ^{1}H-^{15}N TROSY spectra observed for ^{2}H, ^{15}N-labeled FB complexed with the Fc fragment in 10% H_2O/90% D_2O without and with RF irradiation. (**b**–**c**) Plots of the intensity ratios of the cross-peaks of each amide signal in the CS experiment performed in the (**b**) higher (10% H_2O/90% D_2O) and (**c**) lower (90% H_2O/10% D_2O) concentration of D_2O in the solvent. (**d**–**g**) Comparison of the binding sites of the FB with the Fc fragment, defined by (**d**) X-ray crystallography, (**e**) the CS method, (**f**) the chemical shift perturbation experiment, and (**g**) the HD exchange experiment. For mapping of the CS experiment, the residues showing signal intensity ratios of less than 0.5 are colored *gray.*

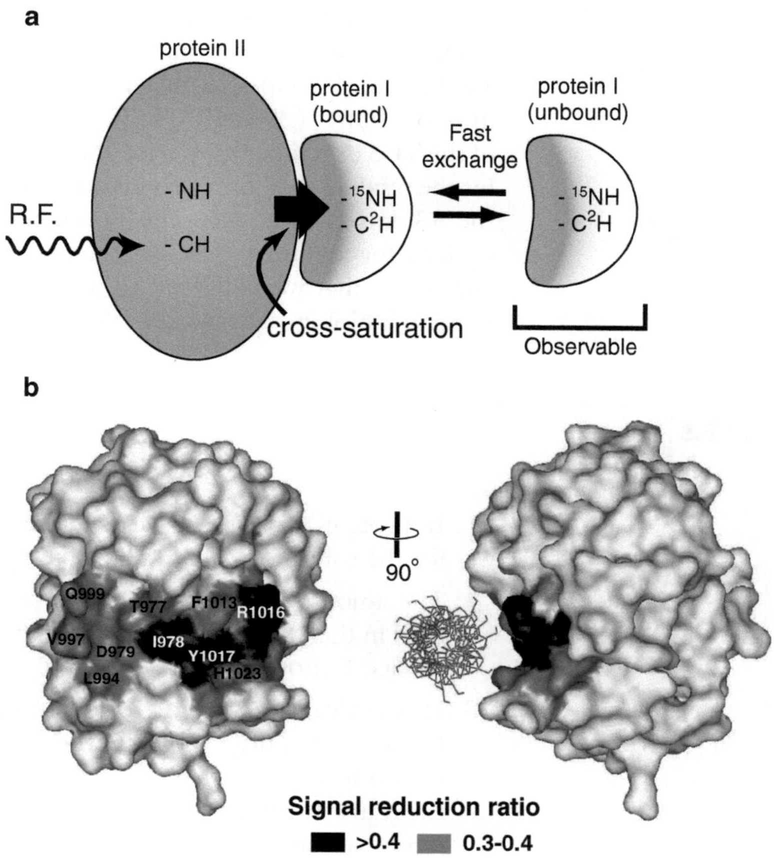

Fig. 4. Identification of the binding site of the VWF A3 domain for the collagen fibril using the TCS experiment.(**a**) Schematic representation of the transferred cross-saturation (TCS) experiments. (**b**) The surface representation of the VWF-A3. The residues with the signal reduction ratios of more than 0.3 are colored. On the *left panel*, the collagen triple helix is shown with a stick model, based on the docking model structure of the A3 domain and a collagen peptide.

and bound states, then it is expected that the effect of the cross-saturation that occurred in the bound state will be "transferred" to the NMR resonances in the free state. It should be emphasized that the NMR signal of protein I in the unbound state, not the bound state, will be observed in the TCS method. Therefore, the TCS method can be applied regardless of the total molecular mass of the complex.

We used the TCS method to investigate the interaction between collagen fibrils and their binding protein: the A3 domain of von Willebrand factor, which supports initial plate adhesion at the injured subendothelium (6). From the SPR experiment, the binding constant (Kd) between the A3 domain and the immobilized collagen type III was estimated to be 15 μM, with a dissociation rate (k_{off}) of >1/s, suggesting that the cross-saturation in the collagen-bound state can be transferred to the NMR signals of the A3 domain in the free state (see Note 12).

We carried out the TCS experiment using a sample containing 0.4 mM of uniformly ^{2}H- and ^{15}N-labeled A3 domain (Mr 20 kDa), and 0.04 mM of type III collagen from human placenta, dissolved in NMR buffer (20% H_2O/80% D_2O) (see Note 13). As a result of the TCS experiment, we were able to detect the signal reduction for the residues that form a continuous surface at the "front" face of the A3 domain (Fig. 4b). The TCS method has been successfully utilized for identifying the interfaces of various macromolecular complexes that were difficult to investigate by other methods in structural biology (reviewed in (7)).

4. Notes

1. In protein I, all protons are replaced with deuterium, except for the solvent-exchangeable amide protons.
2. The amount of glucose must be optimized. The expression level in the D_2O medium can be improved by adding ^{2}H,^{15}N-labeled Celtone powder.
3. It is critically important to use thoroughly dried flasks and beakers when preparing the M9(D_2O) media. Any contamination with H_2O will increase the amount of residual protons that cause the higher background signal reduction, due to the direct saturation of the acceptor protein.
4. In general, the growth speed of bacteria becomes slower in D_2O media. Thus, a longer induction period and a higher incubation temperature than those used for H_2O media may increase the yield.
5. For larger protein complexes (Mr >20 kDa), the TROSY spectrum gives rise to signals with a narrower line width than the HSQC spectrum.
6. The power of the RF pulse and the offset frequency need to be adjusted, so as not to irradiate the solvent signals directly.
7. The optimal length of the saturation and the recovery delay differ from one sample to another.
8. The "split" command executes this procedure in Bruker's Topspin software.
9. The resonance assignment of the amide signals on the HSQC spectrum needs to be established before the CS experiment.
10. A small signal reduction is observed for residues on helix III. This reduction is not due to the spin diffusion from helix I and II, but occurs because the residual protons in FB are saturated directly by the RF irradiation.

11. In the alpha helix structure, each amide proton is located within 4 Å, which is close enough to spread the saturation by spin diffusion.
12. In our theoretical simulations, the k_{off} value should be faster than 0.1/s to observe the effect of the transferred cross-saturation (8).
13. The molecular mass of one molecule of triple helical type III collagen is 300 K. Collagen type III was dissolved in 0.1 M acetic buffer (pH 4.0) and was dialyzed against the insoluble fibrillar collagen in phosphate buffer.

Acknowledgments

We thank Drs. H. Takahashi and T. Ueda for preparation of figures. This work was supported by grants from the Japan New Energy and Industrial Technology Development Organization (NEDO) and the Ministry of Economy, Trade, and Industry (METI).

References

1. Takahashi, H., Nakanishi, T., Kami, K., Arata, Y., and Shimada, I. (2000) A novel NMR method for determining the interfaces of large protein-protein complexes. *Nat. Struct. Biol.* 7, 220–223.
2. Nakanishi, T., Miyazawa, M., Sakakura, M., Terasawa, H., Takahashi, H., and Shimada, I. (2002) Determination of the interface of a large protein complex by transferred cross-saturation measurements. *J. Mol. Biol.* **318**, 245–249.
3. Pervushin, K., Riek, R., Wider, G., and Wuthrich, K. (1997) Attenuated T2 relaxation by mutual cancellation of dipole-dipole coupling and chemical shift anisotropy indicates an avenue to NMR structures of very large biological macromolecules in solution. *Proc. Natl. Acad. Sci. USA.* **94**, 12366–12371.
4. Kupce, E., and Wagner, G. (1995) Wideband homonuclear decoupling in protein spectra. *J. Magn. Res. B.* **109**, 329–333.
5. Gouda, H., Shiraishi, M., Takahashi, H., Kato, K., Torigoe, H., Arata, Y., et al. (1998) NMR study of the interaction between the B domain of staphylococcal protein A and the Fc portion of immunoglobulin G. *Biochemistry* **37**, 129–136.
6. Nishida, N., Sumikawa, H., Sakakura, M., Shimba, N., Takahashi, H., Terasawa, H., et al. (2003) Collagen-binding mode of vWF-A3 domain determined by a transferred cross-saturation experiment. *Nat. Struct. Biol.* **10**, 53–58.
7. Shimada, I., Ueda, T., Matsumoto, M., Sakakura, M., Osawa, M., Takeuchi, K., et al. (2009) Cross-saturation and transferred cross-saturation experiments. *Prog. Nucl. Magn. Reson. Spectrosc.* **54**, 123–140.
8. Matsumoto, M., Ueda, T., Shimada, I. (2010) Theoretical analysis of the transferred cross-saturation method. *J. Magn. Reson.* **205**, 114–124.

Chapter 11

Single-Molecule Methods to Study Cell Adhesion Molecules

Joonil Seog

Abstract

Single molecule techniques are used to characterize the biophysical properties of individual molecules in a mechanically well-controlled environment. The information obtained from direct force measurements can provide the dynamic adhesion forces of cell adhesion molecules, which may shed insights on molecular mechanisms of cellular adhesion. In addition, single-molecule techniques enable us to observe the detailed distributions of individual molecular behaviors that cannot be readily obtained from ensemble measurements. In this chapter, the protocols of using atomic force microscopy and optical tweezers to study cell adhesion molecules are presented.

Key words: Cell adhesion molecules, Atomic force microscopy, Optical tweezers, Single-molecule mechanics, Dynamic force spectroscopy, Nanomechanics

1. Introduction

Although classic biological methods successfully elucidated molecular mechanisms of biological machineries, the information that we obtained from conventional approaches are ensemble-averaged information. Single-molecule techniques are capable of probing individual molecule's behaviors by direct observation, providing new insights into the biological nanomachines (1–3). Atomic force microscopy (AFM), optical tweezers, biomembrane probe, and magnetic tweezer have been used to probe the effect of mechanical force on regulation of cellular adhesions, enzymatic activities, protein–DNA interactions, RNA polymerase–DNA interactions, and the mechanisms of motor proteins (4–10). In this chapter, we will present protocols to create functional surfaces and single-molecule constructs that can be applied to the various single-molecule techniques to study cellular adhesion mechanisms and their dynamic behavior in physiological condition.

Motomu Shimaoka (ed.), *Integrin and Cell Adhesion Molecules: Methods and Protocols*,
Methods in Molecular Biology, vol. 757, DOI 10.1007/978-1-61779-166-6_11, © Springer Science+Business Media, LLC 2011

When the adhesive interactions between biological molecules have been measured (11–13), the single bond rupture force was measured directly by separating interacting molecules which are usually fixed on the substrates. The bond rupture force measured at different loading rates produces information about kinetic barriers, a binding constant, binding mechanisms, and free energy (14, 15). In this chapter, we will briefly review the techniques of AFM and optical tweezers and focus on detailed protocols for single-molecule experiments using these techniques. Most of the protocols may be applicable to other single-molecule experimental methods.

1.1. Atomic Force Microscopy

AFM is capable of measuring single molecular interaction with sub-nanometer distance resolution. It has been widely used to probe DNA and polysaccharide single molecular mechanics, unfolding and refolding of proteins, and receptor–ligand interactions (5, 11, 16, 17). AFM employs a micro-machined soft, flexible cantilever with a sharp tip as a force transducer that deflects in response to the small forces between the cantilever tip and a sample surface (Fig. 1a). A laser beam is focused on the backside of the cantilever and reflected into a position-sensitive photodiode detector (Fig. 1a). The sample is placed on the top of the piezo actuator and by applying voltage to the piezo, the sample is approached and retracted to the cantilever tip. Depending on the interaction between the tip and the sample, the cantilever bends upward (repulsion) or downward (attraction). The deflection of the cantilever causes the changes in the laser spot, which in turn are recorded by photodiode and later converted to force vs. tip–sample separation distance (Fig. 1b).

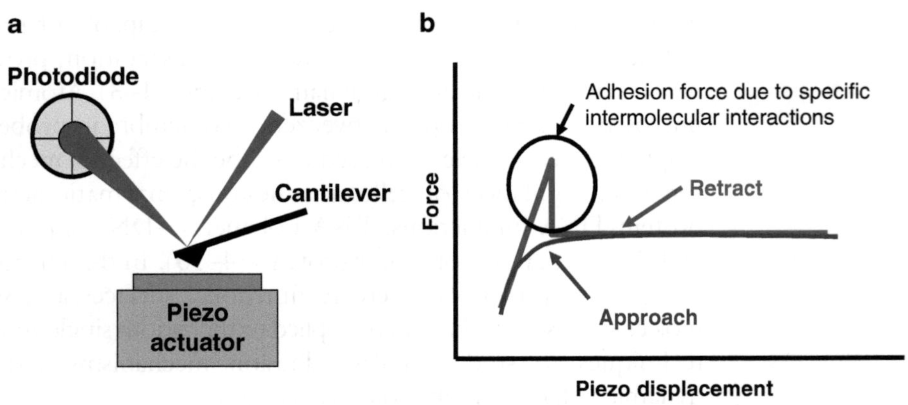

Fig. 1. An overview of AFM measurements. (**a**) A schematic of AFM setup. (**b**) A typical force profile of intermolecular force measurement using AFM. The AFM tip and the substrate are coated with two different molecules which are interacting with each other. As the tip approaches the surface (blue line), some repulsive force between the tip and the substrate is observed near the surface. Once the force reaches a certain setpoint, the piezo starts to retract. As the piezo is retracted (red line), the intermolecular bonding between two interacting molecules is ruptured, causing sudden drop in force.

The molecular level force measurement using AFM can shed a new light on the underlying design principles of biological systems. The force measurement between chondroitin 4-sulfate glycosaminoglycan molecules using molecular force probe showed that the highly charged nature of these molecules can account for up to half of the compressive modulus of the cartilage tissue (18). The single-molecule measurement of cardiac muscle titin demonstrated that N2B and PEVK regions in I band are mostly responsible for passive elasticity of titin in physiological strain range (19). In cell adhesion molecule experiments, the interaction force between P-selectin and P-selectin glycoprotein ligand-1 was directly measured, revealing a mechanism for regulating cell adhesion under mechanical stress (10). Single-molecule experiments on integrin were carried out using T-cell hybridoma, 3A9, and endothelial cell layers (20). More recently, the interaction force between LFA-1 and ICAMs was measured using cell-attached AFM tip and protein-coated substrates (21). In these experiments, cells or adhesion molecule functionalized substrates were used. The protocols for preparing functionalized substrates, single-molecule constructs, and experimental protocols for AFM will be provided in this chapter.

1.2. Optical Tweezers

Optical tweezers that has force resolution of less than pN were utilized to investigate single-molecule mechanics of DNA (22, 23). It has also been applied to the unfolding of RNA (24) and the titin protein (25) and to the mechanical unfolding and refolding of single globular protein of RNase H in a dynamic equilibrium (26), revealing an intermediate state between folded and unfolded state in a mechanically well-controlled environment. Recently, energy landscape of DNA and RNA (27, 28), mechanoswitching mechanism of von Willebrand factor (29), and dynamics of the RNA polymerase (30) were directly studied using optical tweezers.

The optical tweezer technique uses a tightly focused laser beam to trap a dielectric bead near the focal point (31). This optical trap is treated as a Hookean spring with constant spring stiffness, so if the trapped bead is moved out of its equilibrium position by an external force, the force can be calculated by multiplying a spring constant by the displacement. The optical tweezer has a higher force resolution than AFM due to its lower spring constant. The spatial resolution and thermal drift were improved significantly with the advent of dual laser beam trapping systems or back-focal-plane detection (32–34). The spatial resolution of the optical tweezer is now at the single base-pair level, making it ideal for probing biological mechanisms (8). The detailed information of the system setup is published elsewhere (35, 36).

In a typical experimental setup (Fig. 2a), the molecule of interest (yellow) is tethered between an optically trapped bead and a bead attached to a pipette tip on the piezoelectrically controlled stage,

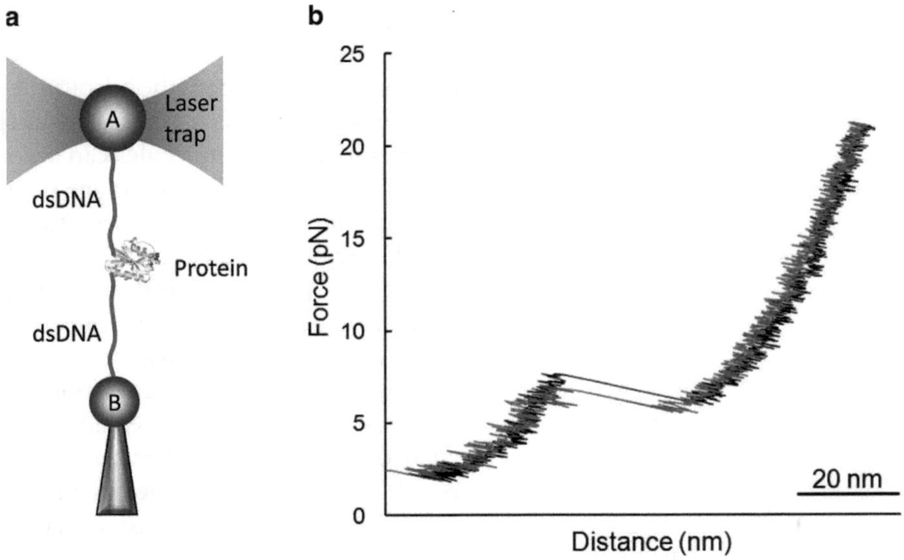

Fig. 2. Optical tweezer setup. (**a**) A laser with a Gaussian intensity distribution is focused by an objective lens, creating a tightly focused spot that traps a dielectric bead A near the laser focal point. Another bead B is attached to a pipette tip which is coupled to a mechanical piezo stage. The molecule of interest in this figure is a single protein domain connected to two ~560-bp double stranded DNA handles (dsDNA). The ends of the DNA handles are connected to beads A and B by specific binding interactions. (**b**) An example of a force vs. extension curve obtained by pulling α_L domain of integrin LFA-1. As the I domain is stretched, the force increases. At about 8 pN, it unfolds, showing a single transition during stretching. During relaxation, the unfolded I domain collapses at around 6 pN, forming a compact structure (unpublished data).

using two DNA handles. This setup dramatically reduces nonspecific interactions between the surfaces of the beads and enables us to study the behavior of single molecule reliably. As the bead B is moved away, the tension increases, and at a critical force level, the protein unfolds, showing an abrupt transition in the pulling curve (Fig. 2b). When the force is relaxed, the molecule recoils back at around 6 pN. This measurement can be done repeatedly, allowing rapid accumulation of high-quality data from one single molecule. Here, we present the protocol for single-molecule construct preparation for optical tweezer experiments.

2. Materials

2.1. Functionalized Surface for AFM Measurements

1. ~10 × 10 mm mica substrate (Ted Pella, Inc., Redding, CA).
2. 100 μM 3-aminopropyl-dimethylethoxysilane (APDES) (Fluorochem Ltd, Derbyshire, UK) in toluene (Fisher Scientific, Fair Lawn, NJ) (see Note 1).
3. 2 mM maleimide-poly(ethylene) glycol-*N*-hydroxysuccinimide (NHS) (MAL-PEG-NHS, molecular weight 5,000 Da, Creative PEGWorks, Winston Salem, NC) in a borate buffer at pH 8.6 (see Note 2).
4. Phosphate-buffered saline (PBS) solution.

5. Thiol-functionalized molecule of interest. For example, protein with engineered cysteine or DNA with thiol group. DNA with thiol group is prepared using a 5′ modified base (Integrated DNA Technologies, San Diego, CA) (see Note 3).
6. Silicon nitride (Si_3N_4) V-shaped cantilever (MLCT, Veeco, Santa Barbara, CA).
7. UV ozone cleaning system (UVOX, Germany).
8. Dithiothreitol (DTT) or tris (2-carboxyethyl) phosphine hydrochloride (TCEP) as a reducing agent (Sigma–Aldrich, Inc., St. Louis, MO).
9. Zeba spin columns (Pierce Biotechnology, Rockford, IL).

2.2. Construct with Mechanical Signatures

1. Human titin cDNA (GeneCopoeia, Rockville, MD).
2. DTT.
3. 10 mM of 2,2′-dithiodipyridine (DTDP) in dimethyl sulfoxide (DMSO).
4. 0.1 M sodium phosphate buffer, pH 5.5.
5. Thermal cycler.
6. QuickChange site-directed mutagenesis kit (Agilent Technologies, La Jolla, CA).
7. Gold-coated silicon wafer (SPI supplies/Structure Probe, Inc., West Chester, PA).
8. Zeba spin columns (Pierce Biotechnology, Rockford, IL).

2.3. Construct with DNA Handles

1. HiSpeed Plasmid Maxi kit (Qiagen, Valencia, CA).
2. 10 mM of DTDP in DMSO.
3. 5′ modified primers (Integrated DNA Technologies, San Diego, CA) for DNA handles: DNA handle A needs one thiol at one end and a biotin at the other end. DNA handle B needs a thiol group at one end and digoxigenin at the other end.
4. Isopropanol.
5. Equilibration buffer: 750 mM NaCl, 50 mM MOPS-NaOH, pH 7, 15% isopropanol (v/v), and 0.15% Triton X-100 (v/v).
6. Wash buffer: 1.0 M NaCl, 50 mM MOPS at pH 7, and 15% isopropanol (v/v).
7. Elution buffer: 1.25 M NaCl, 50 mM Tris–HCl at pH 8.5, and 15% isopropanol (v/v).
8. Microcon (YM-10, Millipore, Billerica, MA).
9. Micro Bio-Spin 6 Columns (Bio-Rad, Hercules, CA).
10. Ethylenediaminetetraacetic acid (EDTA).

11. Streptavidin-coated polystyrene bead and anti-digoxigenin antibody-coated polystyrene bead (Spherotech, Lake Forest, IL).

3. Methods

3.1. Functionalized Surface for AFM Measurements

To study the interactions between two biological molecules, the AFM tip and the substrate are functionalized with each biological molecule, respectively. The interaction forces between them are directly measured by brief contacting and then separating the surfaces. When the AFM tip is close to the substrate to probe intermolecular interactions, nonspecific interactions can also occur, and it is very important to distinguish specific interactions from nonspecific interactions. Using poly(ethylene oxide) (PEO) as a linker molecule between the surface and the molecule reduces nonspecific interactions. It also makes bond rupture occur far away from the surface, providing a characteristic rupture length defined by the length of the PEO (12) (see Note 4).

3.1.1. Substrate Functionalization

1. A ~10 × 10 mm mica substrate is freshly cleaved using a tape right before a functionalization.
2. The mica substrate is washed with toluene and then treated with 100 μM 3-aminopropyl-dimethylethoxysilane (APDES) in toluene for 1 h. After 1 h incubation, the mica is washed with toluene and then with distilled water for further modifications.
3. An amine-functionalized mica was incubated with 2 mM maleimide-poly(ethylene) glycol-*N*-hydroxysuccinimide (NHS) (MAL-PEG-NHS) in a borate buffer at pH 8.6 for 1 h. Amine group on the surface of mica reacts with the NHS group to form a covalent bond with PEG (see Note 5).
4. The PEG-functionalized mica is washed with distilled water briefly and then incubated with a biological molecule of interest with engineered or native cysteines. For example, protein with engineered cysteine or DNA with thiol group is ready to be reacted with maleimide end group of PEO to form a covalent bond. DNA with thiol group can be prepared using a 5′ modified base. For thiol-functionalized DNA, 10 μM of DNA in PBS was incubated on PEO-functionalized mica for 1 h at room temperature.
5. After incubation, the biomolecules-functionalized mica surface is gently washed three times with PBS and kept in wet state until it is used. The functionalized surface is recommended to be used when it is prepared, but it can be kept for a longer time period depending on the stability of the biomolecules.

3.1.2. AFM Tip Functionalization

1. The silicon nitride (Si_3N_4) V-shaped cantilever is irradiated with UV under oxygen for 30 min using UV ozone cleaning system.
2. A cantilever is washed with toluene and then 50 μl of 100 μM APDES in toluene is dropped onto the cantilever so that it is completely immersed in the solution. The cantilever is incubated for 1 h at the room temperature and then washed with toluene and distilled water.
3. 50 μl of 2 mM MAL-PEG-NHS in a borate buffer (pH 8.6) is dropped onto an amine-functionalized cantilever and incubated for 1 h. Amine group on the surface of the tip reacts with NHS group to form the PEG-modified AFM tip.
4. The PEG-modified tip is washed with distilled water briefly and then incubated with thiol-functionalized biomolecules. For example, 50 μl of 200 μM peptide with engineered cysteine in PBS is dropped onto PEG-modified tip and incubated for 1 h. Thiol group forms a covalent bonding with maleimide group, forming a biomolecule-functionalized AFM tip through a PEG linker.
5. The biomolecule-functionalized tip is rinsed using PBS to remove nonspecifically bound molecules and used immediately for force measurements (see Notes 6–8).

3.2. Construct with Mechanical Signatures

The mechanical properties of individual protein molecule were studied using AFM. In this method, the molecule of interest is first physisorbed on the substrate. Then AFM tip picks up a part of the molecule and pulls it at a specific pulling rate. Titin, muscle protein, was investigated in a great detail, revealing the role of the mechanical components at the single-molecule level (19). The mechanical anisotropy of protein, protein folding, and the effect of force on chemical reaction have been studied using the same method (37–41). Immunoglobulin 27th domain (I27) of the titin shows a very well-defined force profile when it is pulled by AFM (5). It extends about 28 nm and unfolds around 200 pN, which can serve as an excellent signature in the single-molecule experiment. Hence, to study the mechanical properties of a specific domain such as N2B or PEVK in the muscle protein, the domain of interest was flanked by two or more I27 domains (19, 42). When the whole construct was pulled, the mechanical behavior of the specific domain was easily identified by using I27 signature as an internal control in the force profile. The construct with I27 domains can be prepared as a fusion protein. Here, a protocol for preparing a single-molecule construct by connecting I27 domains to biomolecules through engineered cysteine residues is presented. Two I27 domains will be attached to the C- and N-terminal of the protein through disulfide bond formation in this example.

1. The I27 domain is subcloned from human titin cDNA by polymerase chain reaction (PCR) amplification. The two I27 domains will be cloned in a tandem manner using different restriction sites in the vector. One cysteine residue is introduced at the C-terminal of the two I27 domains to use thiol group in the cysteine for coupling two I27 domains to your molecule of interest.
2. Once two I27 domains with a cysteine residue are purified by a following standard protocol (e.g., using affinity column), purified product is reduced in 5 mM DTT by incubating for 1 h at room temperature. After reduction, Zeba spin column will be used to remove excess DTT.
3. The two reduced I27 domains (1 mg/ml in 0.1 M sodium phosphate at pH 5.5) with an engineered cysteine are reacted with DTDP. 10 mM of DTDP in DMSO stock solution is added in a small volume to achieve a few times excess of free SH groups (see Note 9).
4. After incubating the reaction mixture for an hour at the room temperature (Fig. 3a), excess DTDP is removed by a Zeba spin column. The DTDP-reacted I27 domains are ready to be coupled to biomolecules through a disulfide bond formation.
5. Two engineered cysteine residues are introduced in a molecule of interest using a single residue insertion. (e.g., two cysteine residues are added to N- and C-terminal of a protein using QuickChange site-directed mutagenesis kit).

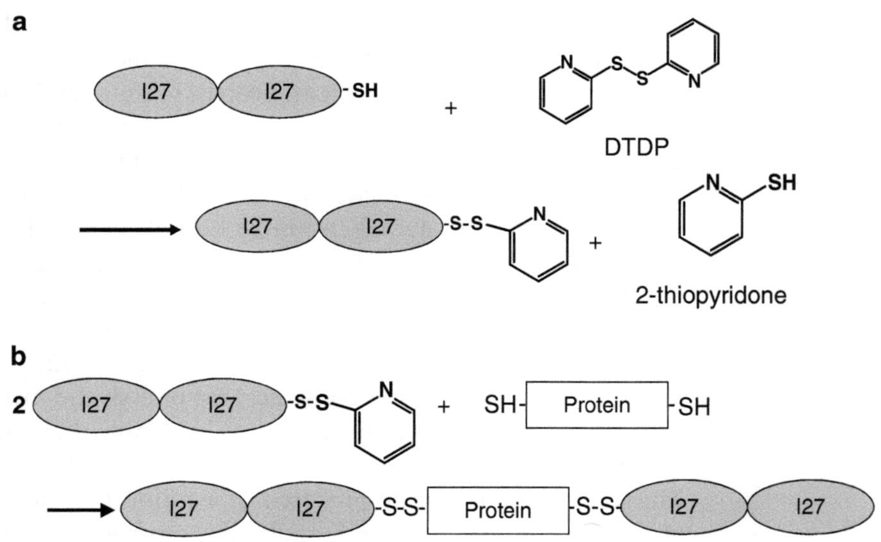

Fig. 3. Coupling of I27 domains. (**a**) The activation of thiol group in cysteine using 2,2′-dithiodipyridine. (**b**) Coupling of I27 domains with protein with two engineered cysteines at the C- and N-terminal. The activated thiol group in two I27 domains reacts with the thiol group in engineered cysteines of the protein. The multiple I27 domains will serve as a mechanical signature in a single-molecule experiment.

6. A molecule with two cysteines is mixed with the DTDP-activated group in titin I27 domains at a molar ratio of 4:1 (Fig. 3b). This reaction forms a disulfide bond between a molecule of interest and the titin I27 domains. Titin I27 domain will provide a signature of the single-molecule experiment when the whole molecule is pulled by AFM (43).
7. ~1 mg/ml single-molecule construct is diluted in 1:5 in PBS solution and then 6 μl of the diluted sample will be dropped onto a gold-coated silicon wafer that is pre-coated with 50 μl PBS buffer. Pipette up and down to mix the solution. The final concentration of the protein will be ~20 μg/ml.
8. After 5 min, the protein is adsorbed on the substrate. Si_3N_4 V-shaped cantilever tips with spring constant, k_c, 20 pN/nm is used. The pulling rate of the AFM tip is 400 nm/s and the maximum force exerted on the substrate is set to be 2–4 nN.
9. If the AFM tip picks up a construct during repeated pulling cycles, the characteristic I27 unfolding behavior is expected with a unique mechanical profile of the molecule of interest. Usually, only a few percent of the total force curves would show I27 unfolding pattern since picking up a molecule is a random process. The pick-up process can be automated so that AFM does the force pulling at the same spot for ten times and moves to a different spot.
10. The force vs. extension curves with more than three I27 unfolding behaviors will be analyzed using a worm-like chain model (22). Using this model, each peak of the force profile will be fitted to obtain the stretched distance between the transitions at fixed persistence length, 0.4 nm (see Notes 10–13).

3.3. Construct with DNA Handles

In optical tweezers experiment, DNA handle is used to position a molecule of interest far away from the surface (Fig. 2). This configuration significantly reduces chances for nonspecific interactions due to the presence of the surfaces. In addition, the overstretching region of DNA can serve as a good internal control of single-molecule experiment (22). Optical tweezers have less than one pN resolution which is much higher than that of AFM (~10 pN). This technique can be used to characterize reversible behaviors from the same molecule repeatedly. The lengths of the DNA handle that has been used are between ~500 and ~800 bp depending on the instrument setup (26, 33) (see Note 14).

3.3.1. DNA Handle Derivatization

1. Two DNA handles are prepared by PCR. One handle has one thiol at one end and a biotin at the other end (DNA handle A). The other handle has a thiol group at one end and digoxigenin at the other end (DNA handle B). These end functional groups are incorporated using 5′ modified primers.

The total volume of PCR is around 10 ml per DNA handle, but smaller volume can be used. 20 mM DTT is added in the PCR mixture to keep the thiol group reduced.

2. Equilibrate four HiSpeed Maxi Tip columns (HiSpeed Plasmid Maxi kit) by adding 10 ml of equilibration buffer and allow the column to drain by gravity flow.
3. Mix 10 ml of PCR product of each DNA handle with 100 ml of equilibration buffer. Allow the mixture of PCR product and the equilibration buffer to go through an equilibrated HiSpeed Tip column by gravity flow.
4. Wash the HiSpeed Maxi Tip column with 60 ml of wash buffer.
5. In the meantime, prepare 10 mM DTDP in DMSO.
6. Elute DNA handle with 10 ml of elution buffer. After elution, add 200 μl of 50 mM DTDP in DMSO to 10 ml of eluted sample. The final concentration of DTDP is 1 mM. Let the derivatization reaction go for an hour at room temperature (see Note 15).
7. Precipitate DNA handle by adding 7 ml of isopropanol. Mix and incubate at room temperature for 5 min.
8. Attach the QIAprecipitator module (HiSpeed Plasmid Maxi kit) to a syringe. Place the QIAprecipitator over a waste bottle, transfer the elute/isopropanol mixture into the syringe, and filter the eluate/isopropanol mixture through the QIAprecipitator using constant pressure.
9. Wash the QIAprecipitator with 2 ml of 70% ethanol three times.
10. Dry the membrane by blowing air and elute the DNA handle using 1 ml of distilled water (see Note 16).
11. Measure the concentration of each DNA handle by measuring the absorbance at 260 nm using UV spectrophotometer. Mix DNA handles A and B so that they are of equal molar concentration and then concentrate it to ~2,000 μg/ml (~5 μM) using Microcon centrifugal filter device.
12. Run the concentrated DNA handles through three Micro Bio-Spin 6 columns which are equilibrated using 0.1 M sodium phosphate buffer at pH 8.0 with 1 mM EDTA.
13. Measure a concentration of DNA handles at 260 nm again and then mix derivatized DNA handles with biomolecules with free cysteines at a molar ratio of 4:1. Typically, mix 16 μl of 5 μM DNA handles with 2 μl of 10 μM biomolecules. Let the DNA handle coupling reaction through disulfide bond formation go overnight at room temperature (see Note 17).

14. To connect the construct to the bead, a single-molecule construct with DNA handles is incubated with 2 μm anti-digoxigenin-coated polystyrene bead for 15 min at 4°C. Now the construct is bound to the bead through digoxigenin/anti-digoxigenin–antibody interaction. A few fold excess of DNA in a molar ratio is used (e.g., 5 μl of 1 pM bead is mixed with 5 μl of 3 pM single-molecule construct). The volume of the single molecule needs to be adjusted depending on coupling efficiency based upon SDS gel (see Note 18).
15. The other end of the construct is connected to the 3-μm streptavidin-coated polystyrene bead through biotin/streptavidin interaction. Now the construct is attached to both beads and is ready to be tested using the optical tweezers setup (see Note 19).

4. Notes

1. When APDES was used for amine functionalization of the mica surface, vaporization of APDES can also be used instead of immersing the substrate in the solution (44).
2. PEG was used as a linker since it reduces nonspecific bindings between AFM tip and a substrate (45). It also minimizes the nonspecifically adsorbed molecules. In addition, the linker provides characteristic rupture length defined by the length of the PEG (46).
3. It is recommended to reduce thiol-functionalized biomolecules since they may have protecting groups or have formed a disulfide bond. The disulfide bond can be reduced using DTT or TCEP in PBS to generate free thiol groups. Excess DTT or TCEP can be removed by using Zeba spin columns.
4. Molecular spring effect of PEO needs to be considered when dynamic force behaviors are studied using PEO as a linker (47). In this case, the actual loading rate is directly obtained from the slope in the force vs. distance curve rather than calculated using pulling rate and spring constant of the AFM tip.
5. The density of the available biomolecules attached on the tip can be controlled by mixing MAL-PEG-NHS with methoxy-PEG-NHS.
6. The adhesion force can be measured by contacting the functionalized substrate using the functionalized tip and retracting from the substrate. The contact force needs to be minimized to reduce the possibility of damaging the tip during the measurements. A contact force of 50–100 pN would be a good starting point.

7. Dwell time, the time that the tip spends while contacting the surface, can be varied to increase the chances of binding between the molecules.
8. For dynamic behavior of adhesions, pulling rate can be varied in a wide range (typically from 50 to 10,000 nm/s). The dependency of adhesion force on pulling rate variation can provide information on kinetic constants, transition state, and free energy (14, 15).
9. Notice that during the activation reaction, the release of the DTDP leaving group (pyridine-2-thione) can be monitored at 343 nm wavelength using UV spectrophotometer. In fact, the DTDP leaving group, unlike DTDP itself, absorbs light at 343 nm, with $\varepsilon_{343} = 7{,}651\ M^{-1}\ cm^{-1}$. It is suggested to follow the reaction in this way at least the first time to have an idea of the time course of the reaction.
10. The minimum number of I27 to be added per cysteine residue is two, and three signature peaks from titin I27 domain will guarantee that a molecule of interest is unfolded during the pulling. The mechanical behavior may show up before, in between, or after I27 peaks. For example, the single-molecule construct which is composed of domain 1 (D1) and domain 2 (D2) of intracellular adhesion molecule-1 (ICAM-1) flanked by two I27 titin domains is pulled by AFM (Fig. 4). The characteristic forces at which ICAM-1 D1 and D2 would unfold are unknown, but the characteristics of I27 domain unfolding behavior serve as a signature in single-molecule force profile due to its well-known unfolding force (~200 pN) and extension distance (~28 nm). With high extension resolution of the AFM, the characteristic extension length of D1 and D2 was readily identified from force vs. extension profile in Fig. 4.

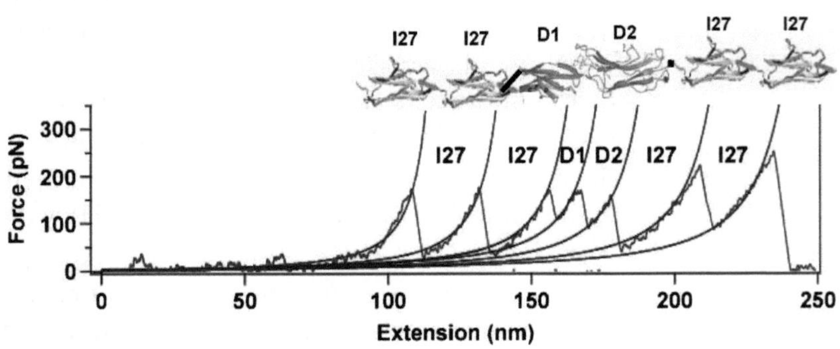

Fig. 4. The force profile that shows the unfolding of single-molecule construct of domains 1 and 2 of ICAM-1 flanked by two titin I27 domains. Each unfolding peaks were fitted with a worm-like chain model to obtain the extension distance between peaks. The delayed appearance of the first I27 unfolding peak is due to a long linker molecule that is not shown in the single molecule construct.

In Fig. 4, the well-known saw-tooth pattern from unfolding of titin I27 is clearly observed with additional new patterns from ICAM-1 domains with much shorter extension distances in the force profile. I27 unfolding was readily confirmed by fitting a worm-like chain model (22) with 27 nm stretched distance, which was very close to the reported value of 28.4 nm. The unfolding behavior of D1 and D2 was also noticed between the second and third I27 unfolding event. The two stretched distances from each domain were well fitted using 10 and 14 nm extension distance, which corresponds very well with the calculated distance of D1 and D2 based upon its limited stretching distance due to disulfide bond. Since there were four titin peaks observed in the profile, we can tell that the whole molecule was stretched by AFM. The magnitude of the unfolding force of I27 titin domain was about 197 pN, which was well correlated with the reported value (5).

The difference in folded and unfolded lengths of D1 and D2 was calculated from ICAM-1 structures and number of amino acids. Compared to I27 extension which is about 28 nm, the extension lengths for ICAM-1 D1 and D2 were shorter because one disulfide bond in each domain limits extension. For D1 of ICAM-1, the known N to C distance in the folded state is ~4.0 nm, and the calculated distance in the extended state is 14 nm, allowing 10 Å for disulfide-bonded cysteine-25 to cysteine-69 and 0.365 Å/residues 1–24 and 70–83. A net extension upon unfolding 10 nm (14 nm–4 nm) is thus calculated. For D2 of ICAM-1, a disulfide connects cysteine-108 to cysteine-159, and residues 84–107 and 159–185 are outside the disulfide, resulting in a calculated extended distance of 18 nm and a net extension of 14 nm.

To probe single-molecule mechanical behaviors, a similar model construct can be prepared. Instead of D1 and D2 of ICAM-1, any molecules can be inserted between two flanking titin I27 domains to probe the mechanical properties of a single biomolecule.

11. The nature of the intramolecular interactions can be studied by measuring forces by systematically varying pH values, salt concentrations, solvent conditions, and temperatures. For example, different contents of ethanol in aqueous solution were used to study the effect of the solvent quality on the strength of hydrophobic interactions (48).

12. Transition state and kinetic information of a structure can be obtained by direct measurement of mechanical strength using a single-molecule construct. In AFM experiments, force is linearly increased by moving AFM tip at a constant pulling speed. Under linearly increasing force, Evans et al. established

the following relationship between pulling rate, *r*, and the most probable unfolding or rupture force, *F** (15).

$$F^* = \frac{k_B T}{x} \ln\left(\frac{x}{k_o k_B T}\right) + \frac{k_B T}{x} \ln(r) \quad (1)$$

Equation 1 shows that the most probable force is a linear function of natural logarithm of pulling rate. The most probable unfolding or rupture force is obtained from the unbinding force histogram experimentally. When you plot F^* vs. ln(*r*), you will get a molecular level parameter, *x*, and a rate constant in the absence of force, k_o. The comparison of k_o between bulk measurement and single measurement was in good agreement, validating the model and the single-molecule experiment method (49, 50).

13. Some models of AFM are equipped with the "force clamp" capability (51). In this mode, an analog feedback control system is used to compare the deflection of the cantilever with a set point. If the feedback system senses the difference, it generates the error signal that is amplified using a proportional, integrative, and differential amplifier. The error signal then is fed back to the piezoelectric tube scanner, adjusting the deflection of the cantilever at the set point so that the applied force is "clamped" and maintained at a constant level. When the force is clamped at a specific force level, the unfolding will eventually occur and we can determine how long it takes before unfolding occurs as measured by a sudden increase in extension during unfolding. By analyzing the dwell time, the kinetic parameters can be obtained in a force-clamping experiment (52).
14. If the amount of the biomolecules such as proteins is enough, you may want to derivatize protein instead of DNA handles. Please refer to refs. (26, 53) for protein derivatization for DNA handle coupling.
15. The derivatization of DNA handles by DTDP can be monitored by measuring the absorbance by 2-thiopyridone at 343 nm using UV spectrophotometer.
16. After step 11, the derivatized DNA handle can be kept at −20°C for a few months for later uses.
17. The pH of the buffer used during coupling DNA handles may be adjusted to provide positive charges in the molecule of interest. The positively charged biomolecule may enhance coupling efficiency.
18. The result of the coupling chemistry can be checked by running the reaction solution on a 4% sodium dodecyl sulfate (SDS) polyacrylamide gel. Four bands may be observed: protein

with two DNA handles, two DNA handles coupled together, a protein with one DNA handle, and one DNA handle. The intensities of the bands can be correlated with the coupling efficiency of the reaction.

19. In certain optical tweezers system, the force-clamping mode may be available. This mode showed that lifetimes of unfolded and folded states of DNA and proteins can be directly measured (26–28).

Acknowledgments

The author would like to thank Adam Karcz, Jiwon Chung, Ciro Cecconi, Elizabeth Shank, and Ragan Robertson for their help in developing protocols. This work was supported by Nano-Bio Fund from the state of the Maryland.

References

1. Zlatanova, J., and van Holde, K. (2006) Single-molecule biology: what is it and how does it work? *Mol. Cell.* **24**, 317–329.
2. Bustamante, C., Macosko, J. C., and Wuite, G. J. (2000) Grabbing the cat by the tail: manipulating molecules one by one. *Nat. Rev. Mol. Cell Biol.* **1**, 130–136.
3. Schafer, D. A., Gelles, J., Sheetz, M. P., and Landick, R. (1991) Transcription by single molecules of RNA polymerase observed by light microscopy. *Nature* **352**, 444–448.
4. Strick, T. R., Allemand, J. F., Bensimon, D., Bensimon, A., and Croquette, V. (1996) The elasticity of a single supercoiled DNA molecule. *Science* **271**, 1835–1837.
5. Rief, M., Gautel, M., Oesterhelt, F., Fernandez, J. M., and Gaub, H. E. (1997) Reversible unfolding of individual titin immunoglobulin domains by AFM. *Science* **276**, 1109–1112.
6. Block, S. M., Goldstein, L. S., and Schnapp, B. J. (1990) Bead movement by single kinesin molecules studied with optical tweezers. *Nature* **348**, 348–352.
7. Khan, S., and Sheetz, M. P. (1997) Force effects on biochemical kinetics. *Annu. Rev. Biochem.* **66**, 785–805.
8. Herbert, K. M., La Porta, A., Wong, B. J., Mooney, R. A., Neuman, K. C., Landick, R., and Block, S. M. (2006) Sequence-resolved detection of pausing by single RNA polymerase molecules. *Cell* **125**, 1083–1094.
9. Evans, E., Ritchie, K., and Merkel, R. (1995) Sensitive force technique to probe molecular adhesion and structural linkages at biological interfaces. *Biophys. J.* **68**, 2580–2587.
10. Marshall, B. T., Long, M., Piper, J. W., Yago, T., McEver, R. P., and Zhu, C. (2003) Direct observation of catch bonds involving cell-adhesion molecules. *Nature* **423**, 190–193.
11. Florin, E. L., Moy, V. T., and Gaub, H. E. (1994) Adhesion forces between individual ligand-receptor pairs. *Science* **264**, 415–417.
12. Neuert, G., Albrecht, C., Pamir, E., and Gaub, H. E. (2006) Dynamic force spectroscopy of the digoxigenin-antibody complex. *FEBS Lett.* **580**, 505–509.
13. Merkel, R., Nassoy, P., Leung, A., Ritchie, K., and Evans, E. (1999) Energy landscapes of receptor-ligand bonds explored with dynamic force spectroscopy. *Nature* **397**, 50–53.
14. Dudko, O. K., Hummer, G., and Szabo, A. (2006) Intrinsic rates and activation free energies from single-molecule pulling experiments. *Phys. Rev. Lett.* **96**, 108101–108104.
15. Evans, E., and Ritchie, K. (1997) Dynamic strength of molecular adhesion bonds. *Biophys. J.* **72**, 1541–1555.
16. Noy, A., Vezenov, D. V., Kayyem, J. F., Meade, T. J., and Lieber, C. M. (1997) Stretching and breaking duplex DNA by chemical force microscopy. *Chem. Biol.* **4**, 519–527.
17. Rief, M., Clausen-Schaumann, H., and Gaub, H. E. (1999) Sequence-dependent mechanics of single DNA molecules. *Nat. Struct. Biol.* **6**, 346–349.

18. Seog, J., Dean, D., Rolauffs, B., Wu, T., Genzer, J., Plaas, A. H., Grodzinsky, A. J., and Ortiz, C. (2005) Nanomechanics of opposing glycosaminoglycan macromolecules. *J. Biomech.* **38**, 1789–1797.
19. Li, H., Linke, W. A., Oberhauser, A. F., Carrion-Vazquez, M., Kerkvliet, J. G., Lu, H., Marszalek, P. E., and Fernandez, J. M. (2002) Reverse engineering of the giant muscle protein titin. *Nature* **418**, 998–1002.
20. Zhang, X., Wojcikiewicz, E., and Moy, V. T. (2002) Force spectroscopy of the leukocyte function-associated antigen-1/intercellular adhesion molecule-1 interaction. *Biophys. J.* **83**, 2270–2279.
21. Wojcikiewicz, E. P., Abdulreda, M. H., Zhang, X., and Moy, V. T. (2006) Force spectroscopy of LFA-1 and its ligands, ICAM-1 and ICAM-2. *Biomacromol.* 7, 3188–3195.
22. Smith, S. B., Cui, Y. J., and Bustamante, C. (1996) Overstretching B-DNA: The elastic response of individual double-stranded and single-stranded DNA molecules. *Science* **271**, 795–799.
23. Bustamante, C., Bryant, Z., and Smith, S. B. (2003) Ten years of tension: single-molecule DNA mechanics. *Nature* **421**, 423–427.
24. Liphardt, J., Onoa, B., Smith, S. B., Tinoco, I. J., and Bustamante, C. (2001) Reversible unfolding of single RNA molecules by mechanical force. *Science* **292**, 733–737.
25. Kellermayer, M. S., Smith, S. B., Granzier, H. L., and Bustamante, C. (1997) Folding-unfolding transitions in single titin molecules characterized with laser tweezers. *Science* **276**, 1112–1116.
26. Cecconi, C., Shank, E. A., Bustamante, C., and Marqusee, S. (2005) Direct observation of the three-state folding of a single protein molecule. *Science* **309**, 2057–2060.
27. Woodside, M. T., Anthony, P. C., Behnke-Parks, W. M., Larizadeh, K., Herschlag, D., and Block, S. M. (2006) Direct measurement of the full, sequence-dependent folding landscape of a nucleic acid. *Science* **314**, 1001–1004.
28. Greenleaf, W. J., Frieda, K. L., Foster, D. A., Woodside, M. T., and Block, S. M. (2008) Direct observation of hierarchical folding in single riboswitch aptamers. *Science* **319**, 630–633.
29. Zhang, X., Halvorsen, K., Zhang, C. Z., Wong, W. P., and Springer, T. A. (2009) Mechanoenzymatic cleavage of the ultralarge vascular protein von Willebrand factor. *Science* **324**, 1330–1334.
30. Hodges, C., Bintu, L., Lubkowska, L., Kashlev, M., and Bustamante, C. (2009) Nucleosomal fluctuations govern the transcription dynamics of RNA polymerase II. *Science* **325**, 626–628.
31. Ashkin, A., Dziedzic, J. M., Bjorkholm, J. E., and Chu, S. (1986) Observation of a Single-Beam Gradient Force Optical Trap for Dielectric Particles. *Opt. Lett.* **11**, 288–290.
32. Moffitt, J. R., Chemla, Y. R., Izhaky, D., and Bustamante, C. (2006) Differential detection of dual traps improves the spatial resolution of optical tweezers. *Proc. Natl. Acad. Sci. USA.* **103**, 9006–9011.
33. Greenleaf, W. J., Woodside, M. T., Abbondanzieri, E. A., and Block, S. M. (2005) Passive all-optical force clamp for high-resolution laser trapping. *Phys. Rev. Lett.* **95**, 208102.
34. Nugent-Glandorf, L., and Perkins, T. T. (2004) Measuring 0.1-nm motion in 1 ms in an optical microscope with differential back-focal-plane detection. *Opt. Lett.* **29**, 2611–2613.
35. Smith, S. B., Cui, Y., and Bustamante, C. (2003) Optical-trap force transducer that operates by direct measurement of light momentum. *Meth. Enzymol.* **361**, 134–162.
36. Neuman, K. C., and Block, S. M. (2004) Optical trapping. *Rev. Sci. Instrum.* **75**, 2787–2809.
37. Wiita, A. P., Perez-Jimenez, R., Walther, K. A., Grater, F., Berne, B. J., Holmgren, A., Sanchez-Ruiz, J. M., and Fernandez, J. M. (2007) Probing the chemistry of thioredoxin catalysis with force. *Nature* **450**, 124–127.
38. Fernandez, J. M., and Li, H. (2004) Force-clamp spectroscopy monitors the folding trajectory of a single protein. *Science* **303**, 1674–1678.
39. Carrion-Vazquez, M., Li, H. B., Lu, H., Marszalek, P. E., Oberhauser, A. F., and Fernandez, J. M. (2003) The mechanical stability of ubiquitin is linkage dependent. *Nat. Struct. Biol.* **10**, 738–743.
40. Mickler, M., Dima, R. I., Dietz, H., Hyeon, C., Thirumalai, D., and Rief, M. (2007) Revealing the bifurcation in the unfolding pathways of GFP by using single-molecule experiments and simulations. *Proc. Natl. Acad. Sci. USA.* **104**, 20268–20273.
41. Dietz, H., and Rief, M. (2004) Exploring the energy landscape of GFP by single-molecule mechanical experiments. *Proc. Natl. Acad. Sci. USA.* **101**, 16192–16197.
42. Sandal, M., Valle, F., Tessari, I., Mammi, S., Bergantino, E., Musiani, F., Brucale, M., Bubacco, L., and Samori, B. (2008) Conformational equilibria in monomeric alpha-synuclein at the single-molecule level. *PloS. Biol.* **6**, e6.
43. Marszalek, P. E., Lu, H., Li, H., Carrion-Vazquez, M., Oberhauser, A. F., Schulten, K., and Fernandez, J. M. (1999) Mechanical

unfolding intermediates in titin modules. *Nature* **402**, 100–103.
44. Lohr, D., Bash, R., Wang, H., Yodh, J., and Lindsay, S. (2007) Using atomic force microscopy to study chromatin structure and nucleosome remodeling. *Method.* **41**, 333–341.
45. Alcantar, N. A., Aydil, E. S., and Israelachvili, J. N. (2000) Polyethylene glycol-coated biocompatible surfaces. *J. Biomed. Mater. Res.* **51**, 343–351.
46. Morfill, J., Blank, K., Zahnd, C., Luginbuhl, B., Kuhner, F., Gottschalk, K. E., Pluckthun, A., and Gaub, H. E. (2007) Affinity-matured recombinant antibody fragments analyzed by single-molecule force spectroscopy. *Biophys. J.* **93**, 3583–3590.
47. Guo, S., Ray, C., Kirkpatrick, A., Lad, N., and Akhremitchev, B. B. (2008) Effects of multiple-bond ruptures on kinetic parameters extracted from force spectroscopy measurements: Revisiting biotin-streptavidin interactions. *Biophys. J.* **95**, 3964–3976.
48. Walther, K. A., Grater, F., Dougan, L., Badilla, C. L., Berne, B. J., and Fernandez, J. M. (2007) Signatures of hydrophobic collapse in extended proteins captured with force spectroscopy. *Proc. Natl. Acad. Sci. USA.* **104**, 7916–7921.
49. Carrion-Vazquez, M., Oberhauser, A. F., Fowler, S. B., Marszalek, P. E., Broedel, S. E., Clarke, J., and Fernandez, J. M. (1999) Mechanical and chemical unfolding of a single protein: a comparison. *Proc. Natl. Acad. Sci. USA.* **96**, 3694–3699.
50. Fritz, J., Katopodis, A. G., Kolbinger, F., and Anselmetti, D. (1998) Force-mediated kinetics of single P-selectin/ligand complexes observed by atomic force microscopy. *Proc. Natl. Acad. Sci. USA.* **95**, 12283–12288.
51. Oberhauser, A. F., Hansma, P. K., Carrion-Vazquez, M., and Fernandez, J. M. (2001) Stepwise unfolding of titin under force-clamp atomic force microscopy. *Proc. Natl. Acad. Sci. USA.* **98**, 468–472.
52. Brujic, J., Hermans, R. I., Garcia-Manyes, S., Walther, K. A., and Fernandez, J. M. (2007) Dwell-time distribution analysis of polyprotein unfolding using force-clamp spectroscopy. *Biophys. J.* **92**, 2896–2903.
53. Cecconi, C., Shank, E. A., Dahlquist, F. W., Marqusee, S., and Bustamante, C. (2008) Protein-DNA chimeras for single molecule mechanical folding studies with the optical tweezers. *Eur. Biophys. J.* **37**, 729–738.

Part III

Imaging for Investigating Cell Adhesion and Migration

Chapter 12

Overview: Imaging in the Study of Integrins

Christopher V. Carman

Abstract

Integrins play critical adhesion and signaling roles during development, wound healing, immunity, and cancer. Central to their function is a unique ability to dynamically modulate their adhesiveness and signaling properties through changes in conformation, both homo- and heterotypic protein–protein interactions and cellular distribution. Genetic, biochemical and structural studies have been instrumental in uncovering overall functions, describing ligand and regulatory protein interactions and elucidating the molecular architecture of integrins. However, such approaches alone are inadequate to describe how dynamic integrin behaviors are orchestrated in intact cells. To fill this void, a wide array of distinct light microscopy (largely fluorescence-based) imaging approaches have been developed and employed. Various microscopy technologies, including wide-field, optical sectioning (laser-scanning confocal, spinning-disk confocal, and multiphoton), TIRF and range of novel "Super-Resolution" techniques have been used in combination with diverse imaging modalities (such as IRM, FRET, FRAP, CALI, and fluorescence speckle imaging) to address distinct aspects of integrin function and regulation. This chapter provides an overview of these imaging approaches and how they have advanced our understanding of integrins.

Key words: Integrin, Fluorescence, Microscopy, FRET, GFP, Conformation, Clustering, migration, Adhesion

1. Introduction

1.1. The Integrin Family of Adhesion Receptors

Integrins represent a large family of heterodimeric adhesion/signaling receptors composed of α and β subunits. In vertebrates, 18 α-subunits, and 8 β-subunits form 24 known ab pairs. These exhibit diverse cell expression patterns, ligand-binding properties, and coupling to cytoskeleton and signaling pathways (1). In this way, integrins mediate dynamic cell–cell, cell–extracellular matrix and cell–pathogen interactions and are critical for development, cell migration, phagocytosis, platelet adhesion, and immunological synapse formation.

Motomu Shimaoka (ed.), *Integrin and Cell Adhesion Molecules: Methods and Protocols*,
Methods in Molecular Biology, vol. 757, DOI 10.1007/978-1-61779-166-6_12, © Springer Science+Business Media, LLC 2011

1.2. Integrin Structure

Both integrin α and β subunits are type I transmembrane (TM) glycoproteins with large extracellular domains, single spanning TM domains and (with the exception of β4) short cytoplasmic domains (Fig. 1a). The overall topology of integrins, initially elucidated by electron microscopy studies, revealed a globular N-terminal ligand binding "head" domain (a critical α and β subunit interface) standing on two long C-terminal "legs" or "stalks," which connect to the transmembrane and cytoplasmic domains of each subunit (2) (Fig. 1a, right). X-ray crystallographic studies have revealed complex domain structure in the extracellular region of both subunits (3, 4). Moreover, such studies demonstrated a surprisingly compact three-dimensional overall topology consisting of legs that were severely bent at the so-called genu or knee generating a V-shaped conformation; In this structure, the head domain was closely juxtaposed to the membrane-proximal portions of the legs, which themselves were in close contact with each other (Fig. 1a, left) (3, 4). As discussed below, the significant differences between early EM and crystallographic studies suggested that integrins posses a propensity for dramatic conformational rearrangements.

1.3. Integrin Dynamics and Regulation

Among adhesion molecules, integrins are unique in their ability to dynamically regulate their adhesiveness through a process termed inside-out signaling or "priming," which in turn leads to ligand binding and signal transduction in the classical outside-in direction (5). As illustrated in Fig. 1, integrins display a wide array of interdependent dynamics (including changes in conformation, diverse protein–protein interactions, and cellular distribution) that collectively determine their adhesion and signaling properties. In terms of adhesion, the overall strength of cellular adhesiveness (i.e., "avidity") is governed by (1) the intrinsic *affinity* of the individual receptor–ligand bonds and (2) the number of receptor–ligand bonds formed (*valency*) (5). Whereas, affinity is dynamically modulated by conformational changes in integrins, valency is governed by the

Fig. 1. (continued) mediate diverse protein– protein interaction that include a wide variety of ligands (*left*), cytoplasmic cytosketelal/signaling molecules (*center*) and membrane proteins (e.g., IAP, uPAR, or tetraspanins) (*right*) proteins. (**d**) Cellular Redistribution. At a cellular level, integrins dramatically alter their distribution during adhesion and migration over ligand-bearing substrates. On the *left*, a cell is shown with even cell surface distribution. In the *center* is shown a cell spreading on a substrate bearing integrin ligands. Integrins accumulate at an increased density at the substrate interface. On the *right* is shown a polarized migrating cell in which integrins are particularly enriched behind the lead edge of migration. Additionally, intracellular vesicular trafficking of integrins is shown transporting integrins from the trailing to the leading edge. Integrin redistribution, can occur proactively, through the cytoskeleton or vesicular trafficking and/or via mass-action/diffusion. In turn, these properties are strongly influenced by affinity for ligand, microclustering, and cytoskeletal associations.

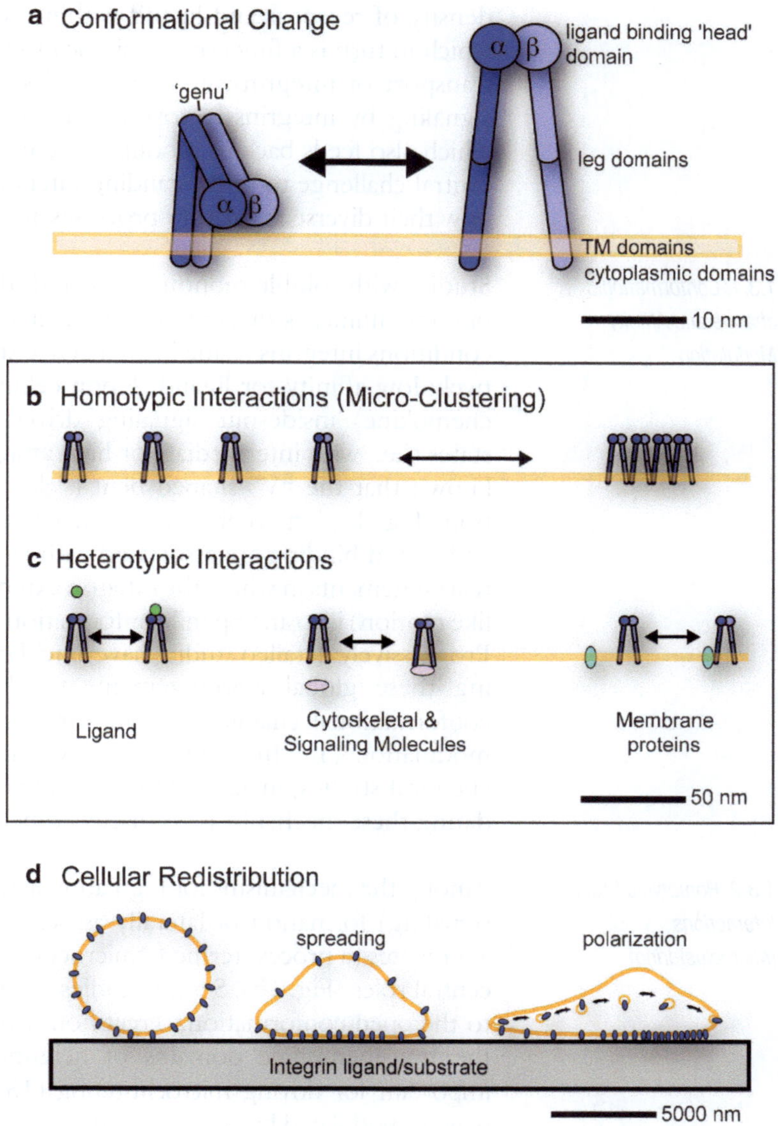

Fig. 1. Integrin Dynamics. Integrin regulation involves interdependent dynamic behaviors that are relevant at a range of spatial scales. (**a**) Conformational Change. At the molecular level, integrins undergo dramatic large-scale conformational changes, which are coupled to altered affinity for ligand, cytoskeletal proteins, and signaling molecules. The compact "V"-shaped low affinity/inactive conformer ("closed") is shown on the left. Note the extreme bend in the legs at the "genu," that the head domain is juxtaposed to the plasma membrane and that the α and β subunit membrane and cytoplasmic domains are closely associated. On the right is the extended ("open"), high-affinity conformer. Note that the head domain sits ~15–20 nm above the plasma membrane and that the membrane and cytoplasmic domains of the α and β subunit have separated. (**b**) Homotypic Interactions (Microclustering). Integrins undergo lateral associations with each other, which may be driven by direct interaction between individual integrin heterodimers, but also influenced by other protein interactions (see **c**, below) or association with membrane domains (e.g., lipid rafts). Such micron-scale clustering critically influences adhesiveness and signaling properties. (**c**) Heterotypic Interactions. Integrins

density of receptor and ligand within the zone of cell adhesion, which in turn is a function of microclustering, diffusion and active transport of integrins, as well as cell spreading (5). Outside-in signaling by integrins is governed by all of the above activities, which also feeds back to modulate these activities (6–10). Thus, a central challenge to understanding integrin function is elucidating how their diverse regulatory processes are integrated.

1.3.1. Conformational change and Affinity Modulation

Studies with soluble monomeric ligands demonstrate that ligand-binding affinity is dynamically modulated in integrins. In resting conditions integrins reside in an inactive state, which exhibits relatively low affinity for ligand. Upon cellular stimulation (e.g., by chemokine) inside-out signaling drives transition to activated states (i.e., with intermediate or high affinity for ligand). It is now known that the "V"-shaped bent ("closed") integrin conformation (Fig. 1a, left) represents the inactive state, whereas priming and ligand binding are associated with large-scale conformational rearrangements in which the integrin extends (with "switch-blade" like motion) into an "open" conformation (Fig. 1a, right) (11–14). Progressively detailed studies have elucidated mechanisms for linking these global rearrangements to the specific intradomain conformational changes that are directly responsible for affinity modulation (1). In addition to crystallographic, EM and biochemical studies, imaging approaches have been critical for elucidating these mechanisms, as discussed below.

1.3.2. Homotypic Protein Interactions (Microclustering)

Among the mechanisms for regulating avidity (as well as outside-in signaling) formation of laterally associated, micron-scale integrin aggregates (a process termed "microclustering") is thought to play central roles (Fig. 1b). Several studies suggest that, upon transition to the open conformation, direct homotypic associations between the transmembrane domains of neighboring integrins may be important for driving microclustering (15–18). These interactions may be modulated by association with the cytoskeleton (5, 7, 19–23) and/or association with lipid raft domains (24–27).

1.3.3. Heterotypic Protein Interactions

Integrins undergo dynamic heterotypic interactions with diverse proteins including extracellular ligand, membrane proteins (e.g., tetraspanins, IAP (CD47), uPAR (CD87), and Fcγ receptors (28, 29)), cytoskeletal adaptor proteins and signaling molecules (see Fig. 1c). These dynamics participate in both inside-out signaling/priming (to facilitate adhesion) and outside-in signals (that result from ligand binding) in interdependent ways. For example, binding of multivalent ligands can both be facilitated by integrin microclustering and help drive/stabilize microclusters. Lateral association with tetraspanin membrane proteins may further facilitate integrin microclustering (28). Furthermore, binding and unbinding of cytoskeletal adaptor proteins (e.g., talin, vinculin, paxillin, tensin) also functions both in promoting adhesion and in cellular responses to ligand binding (5, 7, 19–23). Finally, outside-in signaling responses in integrins are initiated by binding of cytoplasmic signaling molecules, such as FAK, ADAP, ILK, and RapL, as a consequence of both microclustering and ligand-stabilized conformational changes (30–34).

1.3.4. Cellular Redistribution

In addition to microclustering (discussed above), integrins dramatically modulate their cell surface distribution on a broader scale, particularly during processes of spreading, polarization, and migration on adhesive substrates and during phagocytosis and immunological synapse formation (32, 35–37) (Fig. 1d). Studies have shown dynamic changes in tethering to the cytoskeleton during activation and ligand binding alter integrin diffusivity and, thus, the propensity to redistribute to or from zones of contact with ligand (5, 7, 19–23, 38). In addition, active processes, such as vesicular trafficking of integrins (39, 40) and Rap1- and RapL-driven polarization of integrins to the lamellipodia (41, 42), represent important active modes of integrin redistribution (Fig. 1d).

2. Light Microscopy in the Study of Integrins

The previous section illustrates that integrin function/regulation involves a variety of discrete, yet interdependent, dynamics, each with distinct spatial and temporal scales. Thus, understanding integrin function/regulation requires a range of approaches suitable to the investigation of each of these dynamics. Though imaging through electron and scanning probe (e.g., atomic force) microscopy have utility in studying integrins, optical/light microscopy-based imaging (especially using fluorescence) has proven to be exceptionally powerful in this regard and is the focus of this chapter.

"Light microscopy" comprises an increasingly complex collection of approaches that can provide diverse types of information when applied to biological samples. Fundamentally, these approaches derive from the combination of specific microscope system designs (referred to here as "microscope technologies") and methods of using these systems (i.e., "imaging modalities"). In the following sections we will describe fundamental aspects of the major light microscope technologies and imaging modalities along with discussion how each have been applied to the study of integrins.

2.1. Nonfluorescence Light Microscopy

In its simplest form, light microscopy involves shining white light on a sample and using lenses to variously assess the light transmission, reflection, and diffraction as it interacts with the sample. Commonly used optical configurations include bright field, phase contrast and differential interference contrast (DIC), which are all broadly applicable to visualizing cells (43). In the context of integrin substrates, these techniques provide important, albeit indirect, measures of integrin function by allowing visualization of cell adhesion, spreading, polarization, and migration.

Interference reflection contrast microscopy (IRM) is a somewhat more specialized imaging technique, which is particularly well-suited for characterizing aspects integrin function. The basic principal of IRM is the imaging of interference patterns in reflected light created by thin (on the order of a fraction of the wavelength of light) spaces between materials, such as those between a coverslip and an opposing cell membrane (44). Several relatively simple optical configurations can be used to achieve IRM imaging on basic light, as well as laser-scanning confocal (see below), microscopes (44–46). IRM is particularly useful for imaging the distribution and dynamics of areas of cell–substrate adhesion where the close apposition gives rise to dark areas against a bright background of reflected light. Such dark spots are often correlated with clusters of integrins and integrin-associated proteins (e.g., vinculin and talin) marking specific adhesion structures, such as focal contacts, focal adhesions, podosomes, invadopodia and immunological synapses (37, 44–47).

Finally, single particle tracking (SPT) is a nonfluorescent method for direct visualization of diffusive dynamics of individual integrin molecules on the membrane surface (48). SPT uses optically dense (~1 μm) polystyrene beads coated with low density of antibody or ligand, which serve as probes for membrane proteins (48). When coupled to DIC imaging at high temporal resolution, SPT can extract information on the trajectory of integrin movement and diffusion coefficients (48). In this way, SPT has shown that the diffusion of the integrin LFA-1 is determined by both integrin conformation and cell activation status and that confinement of LFA-1 by cytoskeletal attachment regulates cell adhesion both negatively and positively (20, 21, 38).

2.2. Fluorescence Light Microscopy

By far the most broadly used imaging approaches to characterize the distribution dynamics and activity of integrins in intact cells and tissues are those based on fluorescence microscopy. Through use of appropriate tags/probes (see below) that label integrins (or related molecules of interest) fluorescence microscopy allows dynamic and noninvasive molecular imaging. As discussed in this section, these fundamental techniques offer flexible building blocks for diversely informative imaging modalities.

2.2.1. Fluorescence Defined

Fluorescent molecules (i.e., "fluorophores" or "fluorochromes") exhibit the property of being able to "borrow" photons of light for short periods of time before releasing them at a lower energy level (i.e., of a longer wavelength) (Table 1 and Fig. 2a–c). Ideal fluorophores exhibit a sharply defined wavelength range that they can absorb (i.e., excitation/absorption maxima). Upon photon absorption fluorophore electrons are raised to an excited state. After resonating for a period of time ("fluorescence lifetime";

Table 1
Fluorescence terminology

Term	Definition/comments
Absorption	The uptake of an incident photon by a fluorophore, which drives its electrons into an excited state
Auto-fluorescence	Endogenous fluorescence within a sample that may provide significant background "noise," particularly when using higher energy excitation (i.e., shorter wavelengths). In cells auto-fluorescence mostly comes from pyridinic (NADPH) and flavin coenzymes, as well as aromatic amino acids and lipo-pigments. Unhealthy/dyeing cells exhibit elevated auto-fluorescence. In tissues, extracellular matrix proteins (e.g., collagen and elastin) provide a significant source of auto-fluorescence
Brightness	The overall intensity of a fluorophore defined by its extinction coefficient, quantum yield, and photo-stability
Contrast	The degree to which specific fluorescence can be distinguished from reflected light and auto-fluorescence. Contrast is determined by fluorophore brightness and concentration, as well as, imaging parameters, such as intensity of excitation light source and exposure time
Emission	Photon release from an excited fluorophore during relaxation to the ground state
Emission maxima	The center point in a distribution curve of light wavelengths that are most intensely emitted by a specific fluorophore after excitation
Excitation	The elevation of a fluorophore to an excited electronic state as a result of absorbing a photon
Excitation Maxima	The center point in a distribution curve of wavelengths that are most efficiently absorbed by a fluorophore
Extinction coefficient	A measurement of how strongly a chemical species absorbs light at a given wavelength
Fluorochrome	See fluorophore
Fluorophore	A component of a molecule, which causes a molecule to be fluorescent. The amount and wavelength of the absorbed and emitted energy depend on both the fluorophore and the chemical environment of the fluorophore
Fluorescence	The process of by which certain molecules (i.e., fluorophores) absorb photons of a specific wavelength, become electronically "excited" and then "relax" to the ground state by emitting a photon of reduced energy (and longer wavelength)
Fluorescence lifetime	The average time the molecule stays in its excited state before emitting a photon
Photo-activation	The property of certain fluorophores to require absorption of specific wavelengths of light in order to produce photo-chemical conversion from a nonfluorescent to a fluorescent state
Photo-bleach	The photo-chemical destruction of a fluorophore as a result of photon absorption

(continued)

Table 1 (continued)

Term	Definition/comments
Photo-switching	The property of certain fluorophores to undergo photo-chemical conversions that alter their fluorescent properties (specifically absorption and emission maxima) as a consequence of photon absorption
Photo-stability	The relative resistance of a fluorophore to photo-bleaching
Photo-toxicity	A phenomenon known in live-cell imaging in which illumination of a fluorophore causes the cell damage and death. The main cause for photo-toxicity is the formation of oxygen radicals due to nonradiative energy transfer
Stokes shift	The difference between absorption and emission maxima for a given fluorophore
Quantum yield	The efficiency of the fluorescence process, defined as the ratio of the number of photons emitted to the number of photons absorbed
Quenching	Any process which decreases the fluorescence intensity of a given substance. A variety of processes can result in quenching, such as excited state reactions, energy transfer, complex-formation, and collisional quenching

typically on the scale of nanoseconds), the fluorophore electrons "relax" back to the ground state, releasing a photon in the process ("emission"). Emitted photons are of lower energy and, therefore, of longer wavelength than the absorbed photons. The difference in excitation and emission wavelengths is known as the Stokes Shift. The commonly used green fluorophores FITC, Alexa488, and GFP absorb blue photons (~490 nm wavelength) optimally and in turn releases lower energy green light (~520 nm) and, thus, have Stokes Shift of ~30 nm (Fig. 2b). The red fluorophores rhodamine, Cy3, and DsRed absorb yellow light (~530 nm) and emit red light (~570 nm; Stokes Shift is ~40 nm) (Fig. 2c). The principal of fluorescence microscopy is to take advantage of the Stokes Shift, using carefully designed wavelength-specific filters (or alternate techniques) to separate reflected excitation light from the emitted light. When fluorophores are used in conjunction with appropriate "probes" (see below) they enable noninvasive and dynamic imaging of molecules of interest in intact cells.

In practice, there are several fundamental features of fluorescence that need to be carefully considered/managed to produce effective imaging (Table 1). Extinction coefficient, quantum yield, photo-stability/photo-bleaching and photo-toxicity are among the most critical. Extinction coefficient describes the efficiency of photon absorption by a fluorophore, whereas quantum yield is the efficiency with which excited fluorophores ultimately emit photons. Together these parameters determine brightness. The process of fluorescence is usually coupled to oxidative destruction of the fluorophore

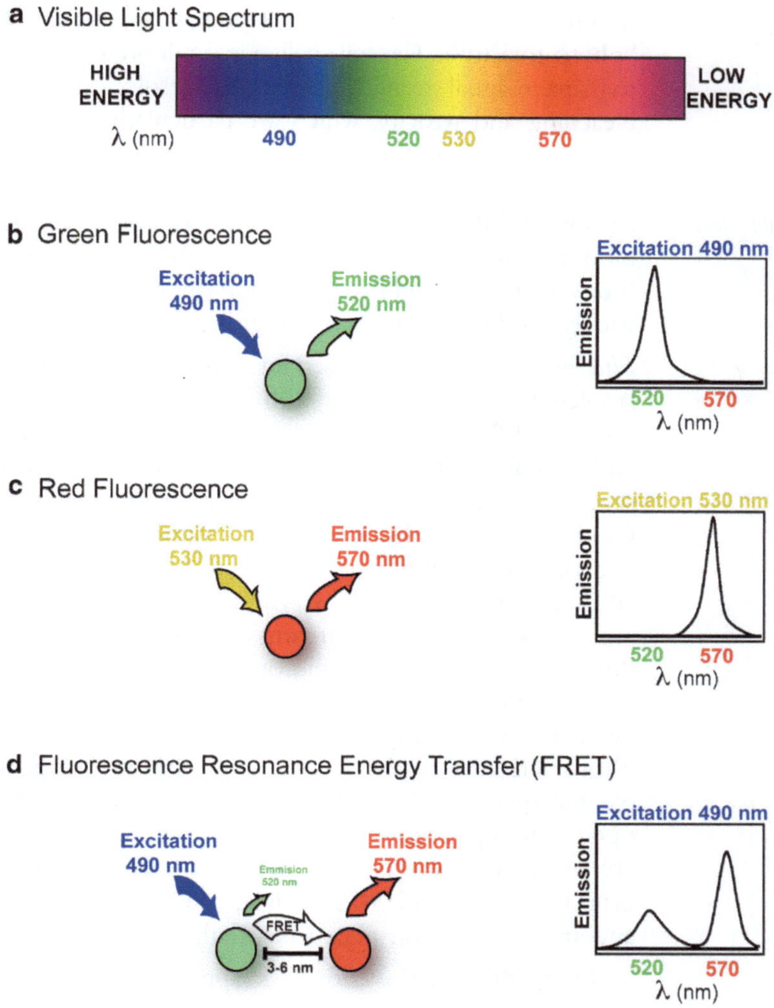

Fig. 2. Basics of Fluorescence and FRET. (**a**) Visible light spectrum displaying from left to right high to low energy (i.e., short to long wavelength). (**b**, **c**) The principle of fluorescence illustrated for two distinct fluorophores, which have properties roughly similar to common green (e.g., FITC, Alexa488, GFP; **b**) and red (e.g., Rhodamine, Cy3, DsRed; **c**) fluorophores. In each case, the emitted photons are shifted toward lower energy/longer wavelengths than those absorbed (i.e., exhibit a Stoke's Shift) due to energy dissipation during excited state "vibrations." (**d**) Fluorescence Resonance Energy Transfer (FRET). FRET is the nonradiative (that is without intermediate emission of a photon) transfer of excited state energy from on "donor" fluorophore to an "acceptor" fluorophore. FRET is a highly distance-dependent phenomenon and can be used as a "spectral ruler" to measure conformational changes or molecular interaction on the spatial scale of ~3–6 nm. Thus, when appropriately matched donor and acceptor come into close proximity and are exposed to light close the donor absorption maxima the donor emission becomes quenched and in turn the acceptor fluorophore undergoes "sensitized emission." In practice, FRET assays often monitor donor quenching, sensitized emission or donor fluorescence lifetime as a readout for FRET.

(photo-bleaching). Fluorophores exhibit vastly differing resistance to such destruction (i.e., photo-stability). Oxidative processes (e.g., reactive oxygen species (ROS) generation) that are coupled to absorption of photons both by fluorophores and endogenous "autofluorescent" molecules (see Table 1) can produce damage to proteins and other biomolecules that may substantially compromise cell health (i.e., cause cytotoxicity, a process specifically referred to as "photo-toxicity"). Central challenges in fluorescence microscopy are to manage signal strength, contrast and resolution, while minimizing bleaching and toxicity, which are particularly important for thick samples or those requiring repeated imaging over long durations.

2.2.2. Fluorescent Probes

Meaningful fluorescence microscopy requires specific labeling of biomolecules of interest with appropriate fluorescent molecules (i.e., via "fluorescent probes"). Fluorophores and tagging techniques widely used in the study of integrins are discussed below.

Fluorophores: Commonly used fluorophores can be classified as small (molecular weights ~300–500 kDa) organic dyes, fluorescent proteins (FPs), and quantum dots (QDs). Traditional fluorescein- and rhodamine-based dyes have been broadly used, but recently replaced by a wide spectral variety of new generation Alexa and Cy dyes, which exhibit superior brightness, photo-stability and hydrophilicity. These are available in broad formats for chemical conjugation to molecules of interest. Fluorescent proteins (FP) are ~30 kDa β-barrel domains surrounding a cyclized peptide fluorochrome. Since the discovery of green fluorescent protein (GFP) from the *Aequorea* jellyfish, a full pallet of FP colors have been discovered/engineered (49, 50). In addition, a variety of photo-activatable, photo-switchable, pH-sensitive, and reactive oxygen species (ROS) photo-synthesizing FPs have been developed that support advanced imaging approaches (49, 50). QDs are inorganic nano-crystals of ~10–30 nm in size that exhibit brightness that is ~10–100-fold greater than organic dyes or FPs and are virtually non-photo-bleachable (50).

Techniques to Tag Proteins: Basic approaches for tagging biomolecules of interest include immunolabeling and genetic tagging. Immunolabeling uses antibodies that have been covalently conjugated with either small organic dyes or QDs. Immunolabeling can be accomplished using primary antibodies directly conjugated to fluorophore or by combination of nonfluorescent primary antibody followed by fluorescent secondary antibodies. This indirect approach is widely used for its relative convenience (based on the broad commercial availability of conjugated secondary, but not most primary antibodies), but has limited flexibility for concomitant labeling of multiple proteins (based on the need for primary antibodies derived from distinct species). Whether, using a direct or indirect approach, a further concern relevant to live-cell imaging is the fact that antibodies are relatively large (~150 kDa) and bivalent molecules, which can induce protein cross-linking or otherwise alter target function through steric effects. This can be partially overcome by use of enzymatically generated monovalent antibody fragments (fAbs; ~50 kDa). The large extracellular epitope surfaces of integrins have made them easy targets for immunolabeling. Moreover, their profound conformational changes associated with activation have been exploited to generate antibodies specific for active conformations. In this way, researchers have been able to concomitantly monitor distribution and activation state.

Genetic tagging involves modifying the sequence of the protein of interest to directly incorporate a fluorescent tag (i.e., FP) or to incorporate a sequence that can facilitate the subsequent labeling with an exogenous fluorescent tag. FPs have become widely used in the study of integrins though fusing them (usually with use of several amino acid flexible linkers) to the C terminus of integrin α- and β-subunits (see Fig. 3). Additionally, many

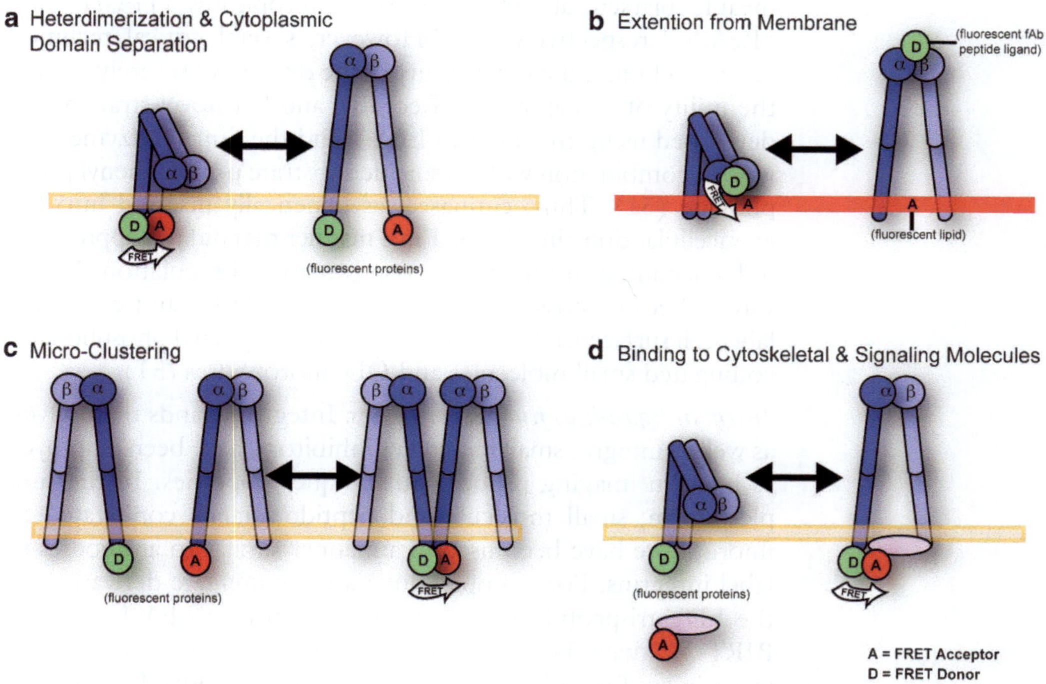

Fig. 3. FRET Approaches to Studying Integrins. FRET has been profoundly important for characterizing integrin regulation and employed in diverse ways to address distinct question. (**a**) Heterodimerization and Cytoplasmic Domain Separation. Integrin α and β subunits are differentially fused to FPs that can serve as FRET donors ("D") and acceptors ("A"). Strong FRET under basal conditions demonstrated that the when heterodimers form, α and β cytoplasmic domains are in close apposition. Strong loss of FRET upon integrin activation, ligand binding in the extracellular domain or talin binding in the cytoplasmic domain demonstrated the separation of the cytoplasmic domains in such settings. (**b**) Head Domain Extension from the Plasma Membrane. Fluorescent fAb fragments or peptide ligands that bind the head domain are used as FRET donors. Fluorescent lipids that intercalate into the plasma membrane are used as acceptors. Strong quenching of donor signal in the basal state that was relieved by activation or ligand binding demonstrated in intact cells that the head domain extends away from the plasma membrane during activation. (**c**) Microclustering. Similar to A, above, FP donors and acceptors are fused to the cytoplasmic domains. However, here donor and acceptor are both on either the α or the β subunit. In this experimental design significant FRET can only occur if individual heterdimers come into close apposition. (**d**) Binding of Cytoskeletal and Signaling Molecules. Donor and acceptor FPs are differentially fused to the cytoplasmic domain of one of the integrin subunits and a signaling/cytoskeletal protein of interest in order to monitor integrin binding to cytoplasmic proteins.

integrin-binding proteins such as talin, vinculin, paxillin, and focal-adhesion kinase (FAK) have similarly been labeled with FPs at their N- or C-termini. Thus, distribution dynamics of integrins and integrin interacting proteins can be monitored concomitantly in intact live cells (see Fig. 3).

The concept of indirect genetic tagging was first introduced by the addition of small peptide epitopes (e.g., "Myc," "Flag," and "HA" epitopes), to provide reactivity to well-characterized antibodies. This, "engineered" immunolabeling approach suggested potential for more diverse and flexible protein tagging strategies. The first such strategy to be realized involved addition of a 12 residue peptide sequence to intracellular proteins that allows for specific and high-affinity binding of the membrane permeable biarsenical green and red fluorophores "FlAsH" and "ReAsh," respectively (49). However, several critical technical issues, including the cytotoxicity of the dyes, have severely limited the utility of this approach. Recently, another novel strategy was developed using the integrin LFA-1 and the fungal enzyme cutinase in combination with its suicide substrate p-nitrophenyl phosphonate (51). Thus, cutinase was genetically inserted into the extracellular domain of LFA-1 in a manner that did not appreciably influence integrin function or conformational regulation. In this way, it became possible to efficiently, specifically, and covalently label cell surface LFA-1 integrins with p-nitrophenyl phosphonate-conjugated small molecules and QD fluorophores (51).

Integrin ligands as fluorescent probes: Integrin ligands themselves, as well as integrin small molecules inhibitors, have been employed as integrin imaging probes with unique properties. In the simplest form, small molecule and peptide ligands conjugated to fluorophore have been used in a manner similar to antibodies to label integrins. For example, fluorescein-conjugate derivatives of the LDV tri-peptide VLA-4 ligand or synthetic LFA-1 inhibitor BIRT-377 have both been used as integrin probes to monitor integrin conformation in the context of specific FRET assays (see below) (52, 53). Fluorescent ligands presented in the context of membrane can provide a readout for integrin behavior that is somewhat less direct, but in some ways more physiologic. An excellent example of this is the use of fluorescently conjugated ICAM-1 (a ligand for LFA-1) linked to a membrane tether (GPI) presented on the surface of a glass supported lipid planar bilayer as a readout for integrin rearrangements during formation of immunological synapse (37). With this approach, the initial integrin clustering and ligand binding by lymphocytes at the center of adhesive contacts and the progressive rearrangement into "peripheral supramolecular adhesion clusters" (pSMACs) during antigen recognition were defined (37). Moreover, VCAM-1(a ligand for the integrin VLA-4)- and ICAM-1-FP fusion proteins transfected into adherent cells (e.g., endothelial, epithelial, or CHO K1) provides a readout for integrin-driven ligand clustering, membrane reorganization, and topological alterations (including formation of "transmigratory cups," "docking structures," podosomes', and "adhesion rings") during migration of blood leukocytes (54–58).

2.2.3. Basic Wide-Field Fluorescence Microscopy

Wide-field fluorescence or "epifluorescence" microscopy is perhaps the most versatile, accessible, and affordable approach for conducting fluorescence imaging (see Table 2). Typical epifluorescence systems use a broad-spectrum excitation light source (e.g., mercury or xenon arc lamp) that is directed through an excitation filter to select desired excitation wavelengths (i.e., appropriate for a specific fluorophore in the sample). This light then passes through an objective lens and floods the entire sample imaging field with photons (hence the term wide-field). Reflected incident light and emitted fluorescent light pass back through the objective and then are then filtered by a combination of a dichroic mirror and emission filter (matched to the emission maxima of the fluorophore) before traveling to the detector (typically a charge-coupled device (CCD) camera).

Inherent advantages of wide-field imaging include sensitivity (and therefore the ability to use relatively low intensity excitation) along with relatively rapid image acquisition. Thus, for many (at least in vitro) systems wide-field microscopy is an excellent choice for dynamic imaging. As stated above, such systems are flexible and easily combined with most basic and advanced (e.g., FRET, FRAP, FLIP, etc.) fluorescence imaging modalities. Moreover, it is relatively straightforward to acquire multiple fluorescence and nonfluorescence channels in parallel over time. As a result, this basic technology has been used in an enormous number of studies to concomitantly image the localization dynamics of integrins and integrin-associated proteins during processes such as adhesion, migration, and phagocytosis. As discussed below, the most significant deficiency in wide-field imaging comes in the resolution of submicron range information (particularly when handling relatively thick samples) and in 3D reconstruction from serial images acquired along the z-axis.

2.2.4. Optical Sectioning Fluorescence Microscopy

An inherent limitation of traditional wide-field fluorescence imaging is the simultaneous excitation of fluorophores throughout the sample, both within and outside of the focal plane. As a result, out of focus light enters the detector and creates background "noise," which convolutes the signal coming from fluorophores within the focal plane, thereby limiting resolution. Optical sectioning microscopy aims to eliminate this effect in order to improve resolution in the x–y plane and facilitate 3D reconstruction of samples via serial sectioning along the z-axis. Several distinct technologies have been developed to achieve optical sectioning, each offering distinct advantages and limitations, as discussed below (Table 2).

Table 2
Optical microscopy technologies

Technology[a]	Advantage[b]	Disadvantage/limitation[c]	Primary use[d]
Wide-field fluorescence (epifluorescence)	Sensitive Extremely flexible to broad applications Highly affordable	*x*–*y* and *z* resolution compromised by out-of-focus light	Live-cell imaging Diverse imaging of integrin and integrin-associated protein dynamics Integrin FRET studies
Laser-scanning confocal	Robust and high quality 3D imaging Generally quite flexible to broad applications	Long scan times Relatively high photo-toxicity Relatively high photo-bleaching	Optical sectioning Analytical co-localization studies of integrin & interacting proteins 3D-localization of integrins/ligands during cell–cell or cell–substrate interactions Integrin FRET studies
Spinning-disk confocal	Rapid scan times Relatively low photo-toxicity and photo-bleaching	Modest sensitivity Modest depth of field	Optical sectioning Live-cell 4D imaging in vitro Integrin FRET studies
Mutliphoton	Excellent depth of field Limited photo-bleaching and photo-toxicity in thick samples	Long scan times Expensive	Optical sectioning Intravital 4D imaging Integrin function (adhesion & migration) in vivo Integrin FRET studies
TIRF	Outstanding axial resolution and contrast Relatively low photo-toxicity and photo-bleaching	Limited to the cell–substrate interface	Distribution dynamics of molecules at cell–substrate interface, especially adhesion structures Integrin FRET studies

Near-field scanning optical microscopy (NSOM)	Extremely high resolution (~20 nm *x*–*y*, ~5 nm *z*) Provide simultaneous optical & topological information	Low working distance and extremely shallow depth of field Limited to study of surfaces Long scan times Prone to artifacts Low sensitivity High skill required Highly specialized equipment	Imaging of nanoscale structures or single molecules Integrin diffusion dynamics Integrin organization into focal adhesions and raft domains
Far-field optical nanoscopy (FFON):4pi, STED, SPEM, PALM, STORM	Extremely high resolution (~20 nm *x*–*y*, ~80 nm *z*) "Normal" working distances	Long scan times Requires photo-activatable or photo-switchable fluorophores Low sensitivity High skill required Highly specialized equipment Requires significant computational power/time	Imaging of nanoscale structures or single molecules Dynamics of focal adhesions

[a] Major types of optical microscopy imaging technologies listed. Extensive permutations by addition/combinations with additional hardware and computational approaches exist that are not shown

[b] The major strength of each technology are listed

[c] The major weaknesses of each technology are listed

[d] Primary overall purpose and utility in the study of integrins

Laser-scanning fluorescence confocal microscopy (LSCM): Contrasting wide-field fluorescence, LSCM confocal microscope uses laser-mediated point illumination and a pinhole in an optically conjugate plane (i.e., "confocal") in front of the detector to eliminate out-of-focus light (59). Since only a single point in the sample is excited at a time, 2D or 3D imaging requires raster scanning over the sample. No doubt, LSCM remains the most robust and versatile optical sectioning technology and has excelled in resolving static 3D structures and conduct analytical co-localization studies. However, LSCM also suffers from a range of limitations including relatively long scan times, high-energy excitation, and limited sensitivity (owing to the "wasted" photons rejected by the pinhole). In practice, the resulting photo-bleaching, photo-toxicity, and low temporal resolution make LSCM a relatively poor choice for time-lapse imaging of rapid events or of those requiring extended durations.

Spinning-disk fluorescence confocal microscopy (SDCM): Spinning (or Nipkow) disk confocal microscopy relies on the same basic principals as LSCM. However, SDCM replaces a rastering laser/pinhole system with a pair of disks arrayed with a series of pinholes spinning in unison (60, 61). The advantage to SDCM is that of offers much fast, indeed video rate, sectioning, and lower excitation energy than LSCM, thereby enabling time-lapse 3D imaging of samples (i.e., 4D imaging). However, the quality of imaging is inferior to LSCM and sensitivity is low compared to wide-field microscopy.

Multiphoton fluorescence microscopy (MPFM): Though often referred to as a "confocal" technology, MPFM, in fact, achieves optical sectioning in a manner fundamentally distinct from truly confocal microscopes (see above). The principal, instead, is to only excite fluorophores within the focal plane. This is achieved using mode-locked pulsed lasers that deliver low energy, long wavelength (specifically ~twice that of the fluorophore absorption maxima) photons that get driven into high density within ~1 μm of the focal plane (62). At such density, individual fluorophores have the opportunity to concomitantly absorb two or more photons providing sufficient energy for excitation (That is, a fluorophore that can be excited by a single photon of ~490 nm can alternatively be excited by concomitantly absorbing two photons of ~980 nm (each of which have twice the wavelength and half of the energy of a 490 nm photon)). Outside the focal plane little light absorption or excitation takes place because the density of photons is too low. This allows for much deeper penetration and greatly reduced photo-bleaching and photo-toxicity in thick samples. Moreover, since only fluorophores within the focal plane emit photons, MPFM can be used with direct detection systems (lacking pinholes), which significantly improves sensitivity. Taken together, the features of MPFM are ideally suited for intravital imaging, where it has been exploited to characterize a variety of integrin-dependent processes such as leukocyte trafficking (63). It is important to note that for relatively thin samples, such as cultured cell monolayers, MPFM is largely similar to LSCM in terms of scan times, photo-toxicity, and photo-bleaching.

2.2.5. Total Internal Reflection Fluorescence Microscopy

Total internal reflection fluorescence (TIRF) microscopy is a unique technology that offers exceptionally high axial resolution (64). TIRF takes advantage of the fact that under conditions of total internal reflection (i.e., light striking an interface between materials of distinct refractive indices at sufficiently high incident angle) an evanescent energy wave is propagated perpendicular to the plane of the interface. The rapid exponential decay in energy intensity of this wave means that only fluorophores located very close to the cell–substrate interface can be excited, which effectively yields an axial resolution of ~100 nm (surpassing confocal or multiphoton optical sectioning by approximately five- to tenfold). That TIRF is inherently imaging only of the ventral cell–substrate interface makes it an ideal system for studying integrin substrate adhesion. Indeed, many studies (often coupled to other technologies) have exploited TIRF to characterized focal adhesion and podosome composition/dynamics, as well as integrin conformational states during adhesions (65–69).

2.2.6. Super-Resolution Optical Microscopy

Nanometer-scale resolution imaging has traditionally been the realm of electron, and more recently scanning probe (e.g., atomic force) microscopy (70). Resolution of optical microscopy has been characterized as being limited by diffraction (the process by which waves of light become distorted to create interference patterns as they pass through optical lenses). Thus, a single point source of light becomes blurred into a broader distribution of light intensities (that can be described by a point spread function; PSF), with maximal achievable resolution of ~200 nm in the $x-y$ plain and ~600 nm along the z-axis (values that are inherently dictated by the wavelength of light). A range of technologies, broadly characterized as "near-field" and "far-field," have overcome the diffraction limit and ushered in an emerging era of super/nano-resolution optical fluorescence microscopy (or "optical nanoscopy" (71)), which is already adding to our understanding of integrin-mediated functions.

Near-field scanning optical microscopy (NSOM/SNOM): NSOM was one of the first techniques to overcome the diffraction limit, and did so by exploiting the properties of evanescent waves. In NSOM, a nanometer-scale detector is placed at a very short distance (also on the scale of nanometers) to the specimen surface. With this technique, the resolution of the image is limited by the size of the detector aperture rather than the excitation wavelength, yielding lateral resolution of 20 nm and vertical resolution of 2–5 nm (72, 73). Furthermore, because of the extremely small near-field excitation volume (and therefore reduced background fluorescence from the cytoplasm), NSOM sensitivity extends to the level of single molecule detection. The downside, however, to this feature is that this technique is limited to imaging of the sample surface. NSOM is also generally quite requires highly specialized equipment/expertise and has, therefore, been applied only to a limited degree. Nonetheless, NSOM has provided insights to integrin behavior (73). Specifically, NSOM has been exploited to analyze the spatio-functional relationship between the integrin LFA-1 and lipid raft components (GPI-APs) on immune cells. Such studies revealed previously unappreciated nanoscale building blocks and hierarchical aggregation pathways driving activation/ligand-dependent integrin clustering (72, 73).

Far-field optical nanoscopy (FFON): Contrasting NSOM, FFON (as the name implies) allows nanoscale resolution at working distances similar to those for most other types of optical microscopy. FFON is not a single technique but rather a term that captures a collection of diverse approaches (such as, 4pi, STED, SPEM, PALM, and STORM) being developed for nanoscale optical imaging (71, 74). Among these, PALM (photo-activated localization microscopy (66, 67, 75)) and STORM (Sub-diffraction-limit imaging by stochastic optical reconstruction microscopy (76)) seem to show the most promise. These both take advantage of recent advances such as the development of photo-switchable fluorophores, high-sensitivity microscopes, and single particle localization algorithms to achieve resolution of several nanometers. At the moment, these approaches remain rather sophisticated requiring highly specialized equipment and expertise and additionally require substantial off-line computational time/power (Table 2) (71, 74). There are also a number of limitations in the experimental setup, such as need for photo-switchable fluorophores, that are nontrivial.

Despite the limitations, FFOM represents an exciting frontier in bio-imaging that has already shown utility in the study of integrins and adhesion biology. Among the first applications of FFOM was the use of PALM to visualize vinculin in focal adhesions and actin within a lamellipodium of fixed cells with 2–25 nm resolution (75). Shortly thereafter, PALM was used to investigate nanoscale dynamics within individual adhesion complexes (ACs) in living cells under physiological conditions (66, 67). In this study, AC dynamics were visualized (with 25 s temporal resolution and 60 nm spatial resolution) allowing measurement of the fractional gain and loss of individual paxillin and vinculin molecules as each AC evolved (66, 67).

2.3. Advanced Fluorescence Imaging Modalities

As discussed above diversity in imaging comes from the intersection of technologies and the means by which these are applied (i.e., "modalities"). Whereas the former emphasizes the role of hardware and fundamental capabilities of equipment (Table 2), the later emphasizes image acquisition/processing protocols and experimental design features such as choice of fluorescent probe and how these are introduced into the system and subsequently manipulated. The most rudimentary imaging modalities involve determining subcellular distribution of molecules and/or organelles of interest with respect to each other, often during dynamic processes such as adhesion and migration. A wide range of advanced imaging modalities have been developed that allow for discrete aspects of molecular behavior to be probed. Below, we focus on those most extensively exploited (or show important potential) for understanding integrins.

2.3.1. Fluorescence (or "Forster") Resonance Energy Transfer

FRET microscopy encompasses multiple techniques that allow for nanometer-scale conformational changes or intermolecular interactions to be probed with high spatial and (often) temporal resolution in intact cells (77, 78) (Fig. 2d). As described above (Subheading 2.2.1), fluorophores "borrow" photons of particular wavelengths for a short period of time before releasing photons of longer wavelengths (i.e., of lower energy). FRET takes advantage of the process described by Theodor Forster in which appropriately matched "donor" and "acceptor" fluorophores can directly pass energy on through "resonance" without intermediate release of photons when they are in close physical juxtaposition (i.e., ~2–10 nm apart) (Fig. 2d) (77, 78). As a result, the emitted donor wavelength is "quenched" by the acceptor resulting in "sensitized" fluorescence emission at the acceptor emission wavelength. In practice FRET can be measured by monitoring the ratio of donor and acceptor emission (ratiometric FRET), the quenching of the donor emission or the fluorescence lifetime of the donor.

Fluorescence lifetimes can be monitored by FLIM (fluorescence lifetime imaging microscopy), which makes use of pulsed lasers (such as those used in MPFM, above) to induce temporally discrete excitation events and then measures fluorescence lifetimes through so-called time domain or frequency domain methods (77, 78). FLIM-FRET takes advantage of the fact that fluorescence lifetime deceases proportionally to the efficiency of FRET. FLIM-FRET is generally regarded as the gold standard for FRET analysis (due to its relative freedom from artifacts, such as those associated with spectral cross-talk and donor/acceptor concentration), as well as its high temporal and spatial resolution. However, this method also requires the most sophisticated (and expensive) equipment, along with the highest degree of expertise.

Next to FLIM-FRET, the "donor de-quenching" (or "acceptor photo-bleach") method provides the best accuracy/freedom from artifacts and is compatible with most imaging technologies and experimental designs (i.e., with respect to types of biomolecules and probes that can be used as donor and acceptor) (77, 78). This method basically measures donor fluorescence before and after photo-bleaching of the acceptor (which effectively destroys the acceptor, thereby removing its quenching effect). The relative increase in the fluorescence in the "de-quenched" donor is proportional to FRET efficiency. The obvious downside to this methodology is that it is an endpoint measurement and, thus, is incompatible with dynamic monitoring within the same sample.

When used with appropriate spectral cross-talk correction, ratiometric FRET offers an easy option for dynamic measurements. However, this method is only appropriate in settings of specific "biosensors" engineered to contain donor and acceptor (typically Cyan FP (CFP) and Yellow FP (YFP)) in the same molecule. This ensures that the molar ratio of donor and acceptor in cells and any local volume is constant.

Finally, a variation of FRET termed BRET (Bioluminescence Resonance Energy Transfer) has been developed (79). This technique uses a bioluminescent luciferase to produce initial photon emission, serving effectively as a donor, compatible with green and yellow acceptors. The advantage of BRET is that no excitation light is required, thereby avoiding photo-bleaching and photo-toxicity. However, BRET is also constrained by kinetics of the luciferase enzymatic activity.

FRET studies of integrin conformation: cytoplamic domain separation: As discussed in Subheading 1.2, early EM and crystal structures suggested the hypothesis that integrin regulation involved large-scale (on the order of ~3–10 nm) conformational rearrangement that would specifically involve separation of the α and β subunits legs and extension of the head domain away from the plasma membrane (Fig. 1a). One of the first studies to test the validity of this hypothesis in intact cells used a FRET-based approach in which the C-termini of the α and β subunits were differentially fused to CFP and YFP to serve as FRET donor and acceptor, respectively (Fig. 3a). Using a photo-bleach FRET approach, it was demonstrated in live cells that, in fact, in the resting state the cytoplasmic domains of LFA-1 were close to each other and that they underwent significant spatial separation upon either activation or ligand binding (13). A similar FRET approach was subsequently employed in conjunction with mutagenesis in to demonstrate that the LFA-1 transmembrane domains play an important role in this conformational regulation (80). Additional studies using the cytoplasmic FRET approach with the integrin Mac1 demonstrate that constitutive heterodimerization of αM and β2 subunits is detectable in plasma membrane, peri-nuclear area, and Golgi in living cells (81) and that interaction of Mac1 in the extracellular domain with uPAR (glycosylphosphatidylinositol-linked urokinase-type plasminogen activator receptor) promotes conversion into the open conformation (82). Finally, the cytoplasmic FRET approach was applied to demonstrate specific zones in the leading edge where activation of the integrin VLA-4 is concentrated during lateral migration (65).

FRET studies of integrin conformation: head extension: As discussed above, in addition to leg separation, integrins were hypothesized to have a strongly bent extracellular domain on the cell surface in the resting state that undergoes large-scale extension during activation (Fig. 1a). To test this aspect of the conformational model, again FRET was applied (52). Here a fluorescein-conjugated peptide ligand that specifically binds to the VLA-4 (in the ligand-binding head domain) served as the FRET donor and a lipophilic dye, octadecyl rhodamine B, was incorporated into the plasma membrane to serve as a FRET acceptor (Fig. 3b) (52). These studies confirmed in intact cells that the head domain in resting cells lies close to the plasma membrane but is rapidly extended away from it upon activation (52) (83). A similar assay was developed for LFA-1 using a fluorescent derivative of the small molecule head domain-binding agonist BIRT-377 (53). Recent studies have applied Alexa488- and Cy3-conjugated fAb fragments (specific for the β subunit "I-like" domain and α subunit "calf" domain) to create more refined characterization of the conformational changes within the extracellular domain during activation of the platelet integrin αIIbβ3 (84).

FRET studies of integrin microclustering: To assess integrin microclustering variations on the cytoplasmic domain assays discussed above have been developed (17, 18, 79, 85). In such systems, rather than attaching donor and acceptor differentially to a and b subunits, donor and acceptor are placed either both on the α or both on the β subunit (Fig. 3c). Thus, individual integrin heterodimers will possess a single donor or acceptor and FRET will only occur when two of these come into close proximity. Such systems have been instrumental in determining mechanisms that microclustering. For example, the role of ligand in driving microclustering was demonstrated by the finding that PMA, cytochalasin-D and latrunculin, at concentrations that activate adhesion and diffusivity (20) do not promote BRET-detectable microclustering of αIIbβ3 (79) or FRET-detectable microclustering of LFA-1 (85), whereas addition of soluble multivalent ligands readily induces microclustering of these integrins. Additional studies suggest that the β2 transmembrane domains (and particularly residue Thr-686) play important roles in driving homomeric associations for integrins Mac1 (17) and LFA-1 (18).

In addition, a unique FRET-based assay to monitor integrin microclustering was developed in which extracellular YFP and dsRed (FRET donor and acceptor, respectively) each fused to β integrin transmembrane/cytopasmic domains were co-expressed with full-length integrin heterodimers (86). In this system, FRET increased during adhesion to ligand. Since the FRET reporters lack ligand-binding capability, these results support the idea that ligand-driven microclustering may be propagated, at least in part, through indirect mechanisms that are downstream of the actual ligand-binding event (e.g., conformational changes in heterodimers or outside-in signaling) (86).

FRET studies of integrin heterotypic protein binding: It is clear that interactions between integrin cytoplasmic domains and both the cytoskeleton and various signaling molecules are altered dynamically as part of both inside-out and outside-in signaling. Integrins also mediated lateral associations with other membrane proteins that may modulate function. In order to study the temporal and spatial dynamics of these interactions, FRET-based assays have been designed in which FP FRET donors and acceptors are differentially incorporated into the cytoplasmic domain of one of the integrin subunits and a target molecule (Fig. 3d). In this way, through use of a FLIM-FRET assay, it was shown that formation of an integrin-α4/14-3-3z/paxillin protein ternary complex mediates localized Cdc42 activity and accelerates cell migration (87). Similarly, FRET studies have been developed to measure β1-integrin lateral association with ErbB2 kinase (88) and cytoplasmic association with PKCα (89). Moreover, FLIM-FRET has been developed to quantify integrin receptor agonism using β1 integrin-GFP and effector-mRFP interactions as a readout (90). Thus, association of talin with β1 integrin and paxillin with α4 integrin was demonstrated to be dependent on both the ligand and receptor activation state, and sensitive to inhibition with small molecule RGD and LDV mimetics, respectively (90).

FRET studies of integrin signaling: FRET assays have also proven useful in understanding protein complexes and signaling events downstream of integrins using diverse experimental designs and biosensors (77). For example, biosensors that report activity of Rho family GTPases involved in integrin signaling (e.g., Rho, Rac, Cdc42, and Rap) have been used to understand localized signaling during integrin-dependent adhesion and migration (77). Additionally, FRET has been used to characterize signaling through integrin cytoplasmic domain-associated protein-1 (ICAP-1). ICAP-1 binds to both β1 integrin and the ROCK kinase (an effector of the RhoA GTPase) through separate domains (91). FRET between CFP-ICAP-1 and YFP-ROCK demonstrated that ICAP-1 could serve as a scaffold to drive indirect ROCK- β1 integrin interactions (within specific leading and trailing edge regions) that are important for properly orchestrated cell migration (91). Finally, the importance of temporal and spatial negative regulation of αIIbβ3 signaling for proper cell spreading was elucidated in part through FRET assays that report interactions between Src-YFP and CFP-Csk (positive and negative regulators, respectively, of signaling) (92).

2.3.2. Fluorescence Recovery After Photo-Bleaching

FRAP, and the related method FLIP (Fluorescence Loss in Photobleaching), are modes of measuring *en mass* diffusiveness of proteins within cells of within membranes of cells (Fig. 4). In FRAP, the movement of fluorescent species into a region acutely subjected to photo-bleaching (Fig. 4a) is monitored. In FLIP, a region is continuously subjected to photo-bleaching, while the depletion of fluorescence in a region outside the bleaching zone is monitored. A recent variation on these approaches involves the use of photo-activatable or photo-switchable FPs, which can be locally activated/switched within a limited zone in order to asses subsequent movement out of that zone. The two most important parameters that can be extracted from these methods are diffusion coefficients (rate at which fluorescent species move into or out of the monitored zone) and the mobile fraction (the percentage of fluorescent species that undergoes detectable movement) (Fig. 4b).

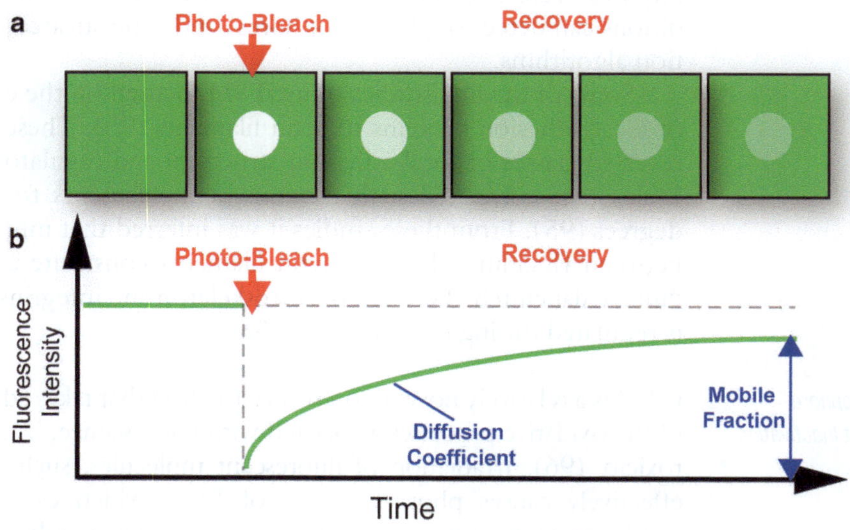

Fig. 4. Fluorescence Recovery After Photo-Bleaching (FRAP). FRAP is one of the most broadly used techniques to monitor molecular mobility of biomolecules, including integrin, in intact live-cell settings. (**a**) Basic FRAP approach. Green panels represent a cellular region containing fluorescent molecules (e.g., plasma membrane-bound integrin-GFP). The *first panel* on the *left* depicts baseline intensity. The *next panel* depicts the discrete photo-bleaching of a sub-region (most typically by an intense burst of laser excitation light). *Subsequent panels* show time-dependent recovery of fluorescence intensity within the bleached region. (**b**) FRAP analysis. The plot depicts a typical intensity profile during a FRAP experiment. The recovery curve provides a basis for calculating overall diffusion coefficients. The saturation point of the recovery curve provides an estimate for the overall fraction of the fluorescent molecules that exhibit detectable diffusion. When applied to integrins, inverse population, the immobile fraction, often represents to the proportion tethered to the cytoskeleton.

For integrins, diffusion rates/immobilization are critical determinant of their adhesive functions (5, 7, 19–23, 32, 35–38, 68). FRAP-based approaches have been widely employed to integrins in order to demonstrate the roles of inside-out signals, cytoskeletal tethering and integrin conformation in altering diffusivity and how this is related to adhesive dynamics (38, 85, 93).

2.3.3. Fluorescence Speckle Imaging

Among the methods to study molecular dynamics in cells FRAP-based studies (which measures whole population of molecules *en mass*) and single molecule fluorescence/SPT could be viewed as representing extremes. A novel methodology termed fluorescence speckle microscopy (FSM) might be viewed as representing an intermediate approach that offers unique advantages that are particularly useful in cytoskeletal and adhesion biology (60, 94). At high concentration, fluorescent tags illuminate cytoskeletal polymers uniformly. However, at low concentration (when percentage of tagged to endogenous molecules is ≤1%), random incorporation of tags produces a discontinuous pattern, which can then be used to determine whether the polymer is translocating or stationary. In practice, FSM conditions are achieved either by injecting very low concentrations of fluorescently tagged proteins or, for GFP-coupled proteins, adjusting conditions so that only small amounts of the labeled protein are expressed. Detection of speckles requires a sensitive microscopy system (which under optimal conditions can detect single fluorophores) and sophisticated correlation algorithms.

Some of the first studies to use FSM focused on the coupling of focal-adhesion proteins to actin filaments (95). These studies revealed classes of focal-adhesion structural and regulatory molecules that correlated with motions actin filaments to varying degrees (95). From these studies it was inferred that interactions between vinculin, talin, and actin filaments constitute a sort of "molecular clutch" between the cytoskeleton and integrins, which is regulated during cell migration (95).

2.3.4. Chromophore-Assisted Light Inactivation

CALI is a relatively novel fluorescence method that takes advantage of the oxidative processes associated with fluorescence, i.e., phototoxicity (96). Irradiation of fluorescent molecules, such as GFP, effectively causes photo-synthesis of ROS, which can damage nearby proteins, and especially the target molecule to which the FP may be fused. Thus, by fusion a target molecule of interest with an FP and then irradiating it with intense laser excitation one can selectively destroy the target in intact cells. Though initial studies have been conducted with GFP (see below), several newly identified/engineered FPs, such as Killer Red, show strongly enhanced (up to 1,000-fold over GFP) photo-synthesis of ROS suggesting even greater potential for future CALI applications (96).

Some of the first CALI studies were centered on understanding of focal adhesions (97). Here Swiss 3T3 cells expressing EGFP-alpha-actinin (but not those expressing EGFP-FAK) exhibited detachment of stress fibers from focal adhesions upon focused laser irradiation. CALI experiments also demonstrated a reduction of EGFP-α-actinin binding to the cytoplasmic domain of the β1 integrin subunit, but not to actin. Thus, CALI was able to provide unique demonstration that α-actinin is essential for the binding of microfilaments to integrins in focal adhesions (97).

2.3.5. Toward Imaging Force and Mechano-Transduction

An inherent aspect of integrin function is the ability to resist and transduce forces, as well as its ability to serve as "mechano-sensors" that transduce mechanical force information into biochemical signals ("mechano-transduction"). These features are increasing appreciated as central aspects of integrin regulation (98–101). Methods to assess responses to integrin-mediated mechano-transduction have largely been through whole sample biochemical (e.g., kinase and Rho-family GTPase activation assays) or gene expression analysis (98–101). However, since the very essence of mechano-transduction is the sensing of spatially and temporally discrete force application events, such whole sample methods are clearly missing critical information. Imaging modalities to better understand the spatial/temporal orchestration integrin mechano-transduction responses are only beginning to be developed and involves varied approaches (94, 98–101).

One intuitive approach to imaging integrin mechano-transduction is to simply monitor morphologic, protein distribution and signaling responses to application of a mechanical stimulus. In addition to various integrin/intergin-associated localization probes (discussed above), a wide range of FRET-based probes have been generated (that involve integrins (see above), vinculin, Myosinn II, N-WASP, many Rho-family GTPases and kinases) that have utility in the study of mechano-transduction (102). For example, a FRET biosensor for Src kinase activity was used in conjunction with locally applied pulling forces on integrins, via laser-tweezer traction or "magnetic twisting" of fibronectin-coated beads adhering to cell surfaces (103–105). These studies revealed the transmission of mechanically induced Src activation in a dynamic and directed process that relies on the cytoskeleton (103–105). In another study, cellular reorganization of focal adhesions and cytoskeleton was monitored in parallel with FRET-based RhoA, Rac1, and Cdc42 activity biosensors. These studies showed that application of cyclic stretching forces drives dramatic reorganization of focal adhesions and the actin cytoskeleton that correlated with strong increase in RhoA, but not Rac1 or Cdc42 (106).

Other approaches seek to visualize the genesis of traction forces. For example, Spatio-temporal image correlation spectroscopy

(STICS), uses correlation algorithms asses directed movement of fluorescently tagged proteins and generate high-resolution velocity maps of adhesion components during force generation. Maps for actin, α-actinin, α5-integrin, talin, paxillin, vinculin, and focal adhesion kinase revealed significant differences in the efficiency of the linkage between integrin and actin among different cell types and on the same cell type grown on different substrate densities during formation of adhesions and traction forces (107). Additionally, studies have developed correlative motion-sensing approaches to directly visualized force generation. High resolution traction force microscopy is one such method that uses fluorescent nano-beads introduced into deformable polyacrylamide gels (used as a cell adhesion substrate) which serve as markers for substrate deformation, the motion of which can be computationally transformed into force vector maps (108). Studies using combined traction force and fluorescent speckle microscopy, demonstrated that F-actin speed is a fundamental and biphasic regulator of traction force at focal adhesions during cell migration. That is, F-actin speed was inversely related to traction stress near the cell edge, where focal adhesions are formed and F-actin motion is rapid. In contrast, larger focal adhesion (where the F-actin speed is low) were marked by a direct relationship between F-actin speed and traction stress (109).

3. Summary and Perspective

Integrins are extremely dynamic adhesion/signaling receptors with diverse modes of regulation. Because integrin behavior is inherently coupled to temporally and spatially discrete organization patterns, imaging approaches capable of providing dynamic subcellular spatial resolution are absolutely essential to elucidating integrin function. Traditional and emerging imaging technologies/modalities have been employed diversely in an effort to address the many distinct aspects of integrin function. In this way, major components of integrin conformational regulation, micro-clustering, protein–protein interactions and cellular distribution dynamics have been elucidated in intact cell systems. Moving forward, continued development of multidimensional approaches (e.g., that coordinately report cell migration behavior, integrin localization, integrin conformation, and signaling processes) represents a feasible (and perhaps the only) way to understand how integrin functions are orchestrated in cells. In addition, continued application of existing, and development of novel, modalities to image mechano-transduction is now recognized as a particularly important goal in integrin adhesion biology. The rapid advances in imaging technologies and computational power/

approaches, along with the growing collection of fluorescent probes and biosensors, suggest, at least over the short term, that available imaging modalities for studying integrins will only be limited by the creativity of the investigator.

Acknowledgments

This work was supported by the Arthritis Foundation, American Heart Association, and Roche Organ Transplant Research Foundation.

References

1. Luo, B.-H., Carman, C. V., and Springer, T. A. (2007) Structural basis of integrin regulation and signaling, *Annu. Rev. Immunol.* **25**, 619–647.
2. Nermut, M. V., Green, N. M., Eason, P., Yamada, S. S., and Yamada, K. M. (1988) Electron microscopy and structural model of human fibronectin receptor, *EMBO J.* 7, 4093–4099.
3. Xiong, J.-P., Stehle, T., Diefenbach, B., Zhang, R., Dunker, R., Scott, D. L., Joachimiak, A., Goodman, S. L., and Arnaout, M. A. (2001) Crystal structure of the extracellular segment of integrin αVβ3, *Science* **294**, 339–345.
4. Xiong, J. P., Stehle, T., Zhang, R., Joachimiak, A., Frech, M., Goodman, S. L., and Arnaout, M. A. (2002) Crystal structure of the extracellular segment of integrin αVβ3 in complex with an Arg-Gly-Asp ligand, *Science* **296**, 151–155.
5. Carman, C. V., and Springer, T. A. (2003) Integrin avidity regulation: Are changes in affinity and conformation underemphasized?, *Curr. Opin. Cell Biol.* **15**, 547–556.
6. Miranti, C. K., and Brugge, J. S. (2002) Sensing the environment: a historical perspective on integrin signal transduction, *Nat. Cell Biol.* **4**, E83–90.
7. Constantin, G., Majeed, M., Giagulli, C., Piccib, L., Kim, J. Y., Butcher, E. C., and Laudanna, C. (2000) Chemokines trigger immediate β2 integrin affinity and mobility changes: differential regulation and roles in lymphocyte arrest under flow, *Immunity* **13**, 759–769.
8. Chan, J. R., Hyduk, S. J., and Cybulsky, M. I. (2001) Chemoattractants induce rapid and transient upregulation of monocyte alpha4 integrin affinity for vascular adhesion molecule 1 which mediates arrest: an early step in the process of emmigration, *J. Exp. Med.* **193**, 1149–1158.
9. Chan, J. R., Hyduk, S. J., and Cybulsky, M. I. (2003) Detecting rapid and transient upregulation of leukocyte integrin affinity induced by chemokines and chemoattractants, *J. Immunol. Meth.* **273**, 43–52.
10. Shimaoka, M., Kim, M., Cohen, E. H., Yang, W., Astrof, N., Peer, D., Salas, A., Ferrand, A., and Springer, T. A. (2006) AL-57, a ligand-mimetic antibody to integrin LFA-1, reveals chemokine-induced affinity up-regulation in lymphocytes, *Proc. Natl. Acad. Sci. USA.* **103**, 13991–13996.
11. Takagi, J., Petre, B. M., Walz, T., and Springer, T. A. (2002) Global conformational rearrangements in integrin extracellular domains in outside-in and inside-out signaling, *Cell* **110**, 599–611.
12. Takagi, J., Strokovich, K., Springer, T. A., and Walz, T. (2003) Structure of integrin α5β1 in complex with fibronectin, *EMBO J.* **22**, 4607–4615.
13. Kim, M., Carman, C. V., and Springer, T. A. (2003) Bidirectional transmembrane signaling by cytoplasmic domain separation in integrins, *Science* **301**, 1720–1725.
14. Xiao, T., Takagi, J., Wang, J.-h., Coller, B. S., and Springer, T. A. (2004) Structural basis for allostery in integrins and binding of ligand-mimetic therapeutics to the platelet receptor for fibrinogen, *Nature* **432**, 59–67.
15. Li, R., Babu, C. R., Lear, J. D., Wand, A. J., Bennett, J. S., and Degrado, W. F. (2001) Oligomerization of the integrin αIIbβ3: Roles of the transmembrane and cytoplasmic domains, *Proc. Natl. Acad. Sci. USA.* **98**, 12462–12467.

16. Li, R., Mitra, N., Gratkowski, H., Vilaire, G., Litvinov, S. V., Nagasami, C., Weisel, J. W., Lear, J. D., DeGrado, W. F., and Bennett, J. S. (2003) Activation of integrin αIIbβ3 by modulation of transmembrane helix associations, *Science* **300**, 795–798.
17. Fu, G., Wang, C., Wang, G. Y., Chen, Y. Z., He, C., and Xu, Z. Z. (2006) Detection of constitutive homomeric associations of the integrins Mac-1 subunits by fluorescence resonance energy transfer in living cells, *Biochem. Biophys. Res. Commun.* **351**, 847–852.
18. Vararattanavech, A., Lin, X., Torres, J., and Tan, S. M. (2009) Disruption of the integrin alphaLbeta2 transmembrane domain interface by beta2 Thr-686 mutation activates alphaLbeta2 and promotes micro-clustering of the alphaL subunits, *J. .Biol. Chem.* **284**, 3239–3249.
19. Jin, T., and Li, J. (2002) Dynamitin controls β2 integrin avidity by modulating cytoskeletal constraint on integrin molecules, *J. Biol. Chem.* **277**, 32963–32969.
20. Kucik, D. F., Dustin, M. L., Miller, J. M., and Brown, E. J. (1996) Adhesion-activating phorbol ester increases the mobility of leukocyte integrin LFA-1 in cultured lymphocytes, *J. Clin. Invest.* **97**, 2139–2144.
21. Peters, I. M., van Kooyk, Y., van Vliet, S. J., de Grooth, B. G., Figdor, C. G., and Greve, J. (1999) 3D single-particle tracking and optical trap measurements on adhesion proteins, *Cytometry* **36**, 189–194.
22. Smith, A., Carrasco, Y. R., Stanley, P., Kieffer, N., Batista, F. D., and Hogg, N. (2005) A talin-dependent LFA-1 focal zone is formed by rapidly migrating T lymphocytes, *J. Cell Biol.* **170**, 141–151.
23. Felsenfeld, D. P., Choquet, D., and Sheetz, M. P. (1996) Ligand binding regulates the directed movement of β1 integrins on fibroblasts, *Nature* **383**, 438–440.
24. Leitinger, B., and Hogg, N. (2001) The involvement of lipid rafts in the regulation of integrins function, *J. Cell Sci.* **115**, 963–972.
25. Hogg, N., Henderson, R., Leitinger, B., McDowall, A., Porter, J., and Stanley, P. (2002) Mechanisms contributing to the activity of integrins on leukocytes, *Immunol. Rev.* **186**, 164–171.
26. Shamri, R., Grabovinsky, V., Feigelson, S. W., Dwir, O., van Kooyk, Y., and Alon, R. (2002) Chemokine stimulation of lymphocyte α_4 integrin avidity but not leukocyte functional-associated antigen-1 avidity to endothelial ligands under shear flow requires cholesterol membrane rafts, *J. Biol. Chem.* **277**, 40027–40035.
27. Krauss, K., and Altevogt, P. (1999) Integrin leukocyte function-associated antigen-1-mediated cell binding can be activated by clustering of membrane rafts, *J. Biol. Chem.* **274**, 36921–36927.
28. Hemler, M. E. (2003) Tetraspanin proteins mediate cellular penetration, invasion, and fusion events and define a novel type of membrane microdomain, *Annu. Rev. Cell Devel. Bio.* **19**, 397–422.
29. Petty, H. R., Worth, R. G., and Todd, R. F. r. (2002) Interactions of integrins with their partner proteins in leukocyte membranes, *Immunol. Res.* **25**, 75–95.
30. Shattil, S. J., P. J. Newman. (2004) Integrins: dynamic scaffolds for adhesion and signaling in platelets, *Blood 104*, 1606–1615.
31. Grashoff, C., Thievessen, I., Lorenz, K., Ussar, S., and Fassler, R. (2004) Integrin-linked kinase: integrin's mysterious partner, *Curr. Opin. Cell Biol.* **16**, 565–571.
32. Ridley, A. J., Schwartz, M. A., Burridge, K., Firtel, R. A., Ginsberg, M. H., Borisy, G., Parsons, J. T., and Horwitz, A. R. (2003) Cell migration: integrating signals from front to back, *Science* **302**, 1704–1709.
33. Guo, W., and Giancotti, F. G. (2004) Integrin signalling during tumour progression, *Nat. Rev. Mol. Cell Biol.* **5**, 816–826.
34. Zhu, J., Carman, C. V., Kim, M., Shimaoka, M., Springer, T. A., and Luo, B. H. (2007) Requirement of alpha and beta subunit transmembrane helix separation for integrin outside-in signaling, *Blood* **110**, 2475–2483.
35. Plancon, S., Morel-Kopp, M. C., Schaffner-Reckinger, E., Chen, P., and Kieffer, N. (2001) Green fluorescent protein (GFP) tagged to the cytoplasmic tail of αIIb or β3 allows the expression of a fully functional integrin αIIbβ3: effect of β3GFP on αIIbβ3 ligand binding, *Biochem. J.* **357**, 529–536.
36. Laukaitis, C. M., Webb, D. J., Donais, K., and Horwitz, A. F. (2001) Differential dynamics of α5 integrin, paxillin, and α-actinin during formation and disassembly of adhesions in migrating cells, *J. Cell Biol.* **153**, 1427–1440.
37. Dustin, M. L. (2009) Supported bilayers at the vanguard of immune cell activation studies, *J. Struct. Biol.* **168**, 152–160.
38. Cairo, C. W., Mirchev, R., and Golan, D. E. (2006) Cytoskeletal regulation couples LFA-1 conformational changes to receptor lateral mobility and clustering, *Immunity* **25**, 297–308.
39. Lawson, M. A., and Maxfield, F. R. (1995) Ca^{2+}- and calcineurin-dependent recycling of

an integrin to the front of migrating neutrophils, *Nature* **377**, 75–79.

40. Tohyama, Y., Katagiri, K., Pardi, R., Lu, C., Springer, T. A., and Kinashi, T. (2003) The critical cytoplasmic regions of the αL/β2 integrin in Rap1-induced adhesion and migration, *Mol. Biol. Cell* **14**, 2570–2582.
41. Katagiri, K., Maeda, A., Shimonaka, M., and Kinashi, T. (2003) RAPL, a novel Rap1-binding molecule, mediates Rap1-induced adhesion through spatial regulation of LFA-1, *Nat. Immunol.* **4**, 741–748.
42. Shimonaka, M., Katagiri, K., Kakayama, T., Fujita, N., Tsuruo, T., Yoshie, O., and Kinashi, T. (2003) Rap1 translates chemokine signals to integrin activation, cell polarization, and motility across vascular endothelium under flow, *J. Cell Biol.* **161**, 417–427.
43. Centonze Frohlich, V. (2008) Phase contrast and differential interference contrast (DIC) microscopy, *J. Vis. Exp.* **6**, 844.
44. Verschueren, H. (1985) Interference reflection microscopy in cell biology: Methodology and applications, *J. Cell. Sci.* **75**, 279–301.
45. Bailly, M., Condeelis, J. S., and Segall, J. E. (1998) Chemoattractant-induced lamellipod extension, *Microsc. Res. Tech.* **43**, 433–443.
46. Neumeister, P., Pixley, F. J., Xiong, Y., Xie, H., Wu, K., Ashton, A., Cammer, M., Chan, A., Symons, M., Stanley, E. R., and Pestell, R. G. (2003) Cyclin D1 governs adhesion and motility of macrophages, *Mol. Biol. Cell* **14**, 2005–2015.
47. Linder, S. (2009) Invadosomes at a glance, *J. Cell Sci.* **122**, 3009–3013.
48. Ritchie, K., and Kusumi, A. (2003) Single-particle tracking image microscopy, *Meth. Enzymol.* **360**, 618–634.
49. Giepmans, B. N., Adams, S. R., Ellisman, M. H., and Tsien, R. Y. (2006) The fluorescent toolbox for assessing protein location and function, *Science* **312**, 217–224.
50. Wang, Y., Shyy, J. Y., and Chien, S. (2008) Fluorescence proteins, live-cell imaging, and mechanobiology: seeing is believing, *Annu Rev. Biomed. Eng.* **10**, 1–38.
51. Bonasio, R., Carman, C. V., Kim, E., Sage, P. T., Love, K. R., Mempel, T. R., Springer, T. A., and von Andrian, U. H. (2007) Specific and covalent labeling of a membrane protein with organic fluorochromes and quantum dots, *Proc. Natl. Acad. Sci. USA.* **104**, 14753–14758.
52. Chigaev, A., Buranda, T., Dwyer, D. C., Prossnitz, E. R., and Sklar, L. A. (2003) FRET detection of cellular alpha4-integrin conformational activation, *Biophys J.* **85**, 3951–3962.
53. Larson, R. S., Davis, T., Bologa, C., Semenuk, G., Vijayan, S., Li, Y., Oprea, T., Chigaev, A., Buranda, T., Wagner, C. R., and Sklar, L. A. (2005) Dissociation of I domain and global conformational changes in LFA-1: refinement of small molecule-I domain structure-activity relationships, *Biochemistry* **44**, 4322–4331.
54. Barreiro, O., Yanez-Mo, M., Serrador, J. M., Montoya, M. C., Vicente-Manzanares, M., Tejedor, R., Furthmayr, H., and Sanchez-Madrid, F. (2002) Dynamic interaction of VCAM-1 and ICAM-1 with moesin and ezrin in a novel endothelial docking structure for adherent leukocytes, *J. Cell Biol.* **157**, 1233–1245.
55. Carman, C. V., Jun, C. D., Salas, A., and Springer, T. A. (2003) Endothelial cells proactively form microvilli-like membrane projections upon intercellular adhesion molecule 1 engagement of leukocyte LFA-1, *J. Immunol.* **171**, 6135–6144.
56. Carman, C. V., and Springer, T. A. (2004) A transmigratory cup in leukocyte diapedesis both through individual vascular endothelial cells and between them, *J. Cell Biol.* **167**, 377–388.
57. Carman, C. V., Sage, P. T., Sciuto, T. E., de la Fuente, M. A., Geha, R. S., Ochs, H. D., Dvorak, H. F., Dvorak, A. M., and Springer, T. A. (2007) Transcellular diapedesis is initiated by invasive podosomes, *Immunity* **26**, 784–797.
58. Yang, L., Froio, R. M., Sciuto, T. E., Dvorak, A. M., Alon, R., and Luscinskas, F. W. (2005) ICAM-1 regulates neutrophil adhesion and transcellular migration of TNF-alpha-activated vascular endothelium under flow, *Blood* **106**, 584–592.
59. Paddock, S. W. (1999) An introduction to confocal imaging, *Meth. Mol. Biol.* **122**, 1–34.
60. Maddox, P. S., Moree, B., Canman, J. C., and Salmon, E. D. (2003) Spinning disk confocal microscope system for rapid high-resolution, multimode, fluorescence speckle microscopy and green fluorescent protein imaging in living cells, *Meth. Enzymol.* **360**, 597–617.
61. Graf, R., Rietdorf, J., and Zimmermann, T. (2005) Live cell spinning disk microscopy, *Adv. Biochem. Eng. Biotechnol.* **95**, 57–75.
62. Rubart, M. (2004) Two-photon microscopy of cells and tissue, *Circulation Res.* **95**, 1154–1166.
63. Zarbock, A., and Ley, K. (2009) New insights into leukocyte recruitment by intravital microscopy, *Curr. Top. Microbiol. Immunol.* **334**, 129–152.

64. Fish, K. N. (2009) Total internal reflection fluorescence (TIRF) microscopy, *Current Protoc. Cytometry* J. Paul Robinson, Ed. Chapter **12**, Unit12 18.

65. Hyun, Y. M., Chung, H. L., McGrath, J. L., Waugh, R. E., and Kim, M. (2009) Activated integrin VLA-4 localizes to the lamellipodia and mediates T cell migration on VCAM-1, *J. Immunol.* **183**, 359–369.

66. Shroff, H., Galbraith, C. G., Galbraith, J. A., and Betzig, E. (2008) Live-cell photoactivated localization microscopy of nanoscale adhesion dynamics, *Nat. Meth.* **5**, 417–423.

67. Shroff, H., White, H., and Betzig, E. (2008) Photoactivated localization microscopy (PALM) of adhesion complexes, *Curr. Protoc. Cell Biol.* Juan S. Bonifacino, Ed. Chapter **4**, Unit 4 21.

68. Vicente-Manzanares, M., Koach, M. A., Whitmore, L., Lamers, M. L., and Horwitz, A. F. (2008) Segregation and activation of myosin IIB creates a rear in migrating cells, *J. Cell Biol.* **183**, 543–554.

69. Liu, D., Bryceson, Y. T., Meckel, T., Vasiliver-Shamis, G., Dustin, M. L., and Long, E. O. (2009) Integrin-dependent organization and bidirectional vesicular traffic at cytotoxic immune synapses, *Immunity* **31**, 99–109.

70. Koster, A. J., and Klumperman, J. (2003) Electron microscopy in cell biology: integrating structure and function, *Nat. Rev. Mol. Cell Biol.* **Suppl**, SS6–10.

71. Hell, S. W. (2007) Far-field optical nanoscopy, *Science* **316**, 1153–1158.

72. de Lange, F., Cambi, A., Huijbens, R., de Bakker, B., Rensen, W., Garcia-Parajo, M., van Hulst, N., and Figdor, C. G. (2001) Cell biology beyond the diffraction limit: near-field scanning optical microscopy, *J. Cell Sci.* **114**, 4153–4160.

73. van Zanten, T. S., Cambi, A., Koopman, M., Joosten, B., Figdor, C. G., and Garcia-Parajo, M. F. (2009) Hotspots of GPI-anchored proteins and integrin nanoclusters function as nucleation sites for cell adhesion, *Proc. Natl. Acad. Sci. USA.* **106**, 18557–18562.

74. Henriques, R., and Mhlanga, M. M. (2009) PALM and STORM: what hides beyond the Rayleigh limit?, *Biotechnol. J.* **4**, 846–857.

75. Betzig, E., Patterson, G. H., Sougrat, R., Lindwasser, O. W., Olenych, S., Bonifacino, J. S., Davidson, M. W., Lippincott-Schwartz, J., and Hess, H. F. (2006) Imaging intracellular fluorescent proteins at nanometer resolution, *Science* **313**, 1642–1645.

76. Rust, M. J., Bates, M., and Zhuang, X. (2006) Sub-diffraction-limit imaging by stochastic optical reconstruction microscopy (STORM), *Nat. Meth.* **3**, 793–795.

77. Wang, Y., and Chien, S. (2007) Analysis of integrin signaling by fluorescence resonance energy transfer, *Meth. Enzymol.* **426**, 177–201.

78. Periasamy, A. (2001) Fluorescence resonance energy transfer microscopy: a mini review, *J. Biomed. Opt.* **6**, 287–291.

79. Buensuceso, C., De Virgilio, M., and Shattil, S. J. (2003) Detection of integrin αIIbβ3 clustering in living cells, *J. Biol. Chem.* **278**, 15217–15224.

80. Vararattanavech, A., Tang, M. L., Li, H. Y., Wong, C. H., Law, S. K., Torres, J., and Tan, S. M. (2008) Permissive transmembrane helix heterodimerization is required for the expression of a functional integrin, *Biochem. J.* **410**, 495–502.

81. Fu, G., Yang, H. Y., Wang, C., Zhang, F., You, Z. D., Wang, G. Y., He, C., Chen, Y. Z., and Xu, Z. Z. (2006) Detection of constitutive heterodimerization of the integrin Mac-1 subunits by fluorescence resonance energy transfer in living cells, *Biochem. Biophys. Res. Commun.* **346**, 986–991.

82. Tang, M. L., Vararattanavech, A., and Tan, S. M. (2008) Urokinase-type plasminogen activator receptor induces conformational changes in the integrin alphaMbeta2 headpiece and reorientation of its transmembrane domains, *J. Biol. Chem.* **283**, 25392–25403.

83. Chigaev, A., Waller, A., Amit, O., and Sklar, L. A. (2008) Galphas-coupled receptor signaling actively down-regulates alpha4beta1-integrin affinity: a possible mechanism for cell de-adhesion, *BMC Immunol.* **9**, 26.

84. Coutinho, A., Garcia, C., Gonzalez-Rodriguez, J., and Lillo, M. P. (2007) Conformational changes in human integrin alphaIIbbeta3 after platelet activation, monitored by FRET, *Biophys. Chem.* **130**, 76–87.

85. Kim, M., Carman, C. V., Yang, W., Salas, A., and Springer, T. A. (2004) The primacy of affinity over clustering in regulation of adhesiveness of the integrin {alpha}L{beta}2, *J. Cell Biol.* **167**, 1241–1253.

86. Smith, E. A., Bunch, T. A., and Brower, D. L. (2007) General in vivo assay for the study of integrin cell membrane receptor microclustering, *Anal. Chem.* **79**, 3142–3147.

87. Deakin, N. O., Bass, M. D., Warwood, S., Schoelermann, J., Mostafavi-Pour, Z., Knight, D., Ballestrem, C., and Humphries, M. J. (2009) An integrin-alpha4-14-3-3zeta-paxillin ternary complex mediates localised Cdc42 activity and accelerates cell migration, *J. Cell Sci.* **122**, 1654–1664.

88. Mocanu, M. M., Fazekas, Z., Petras, M., Nagy, P., Sebestyen, Z., Isola, J., Timar, J., Park, J. W., Vereb, G., and Szollosi, J. (2005) Associations of ErbB2, beta1-integrin and lipid rafts on Herceptin (Trastuzumab) resistant and sensitive tumor cell lines, *Cancer Lett.* **227**, 201–212.
89. Ng, T., Shima, D., Squire, A., Bastiaens, P. I., Gschmeissner, S., Humphries, M. J., and Parker, P. J. (1999) PKCalpha regulates beta1 integrin-dependent cell motility through association and control of integrin traffic, *EMBO J.* **18**, 3909–3923.
90. Parsons, M., Messent, A. J., Humphries, J. D., Deakin, N. O., and Humphries, M. J. (2008) Quantification of integrin receptor agonism by fluorescence lifetime imaging, *J. Cell Sci.* **121**, 265–271.
91. Stroeken, P. J., Alvarez, B., Van Rheenen, J., Wijnands, Y. M., Geerts, D., Jalink, K., and Roos, E. (2006) Integrin cytoplasmic domain-associated protein-1 (ICAP-1) interacts with the ROCK-I kinase at the plasma membrane, *J. Cell. Physiol.* **208**, 620–628.
92. Vielreicher, M., Harms, G., Butt, E., Walter, U., and Obergfell, A. (2007) Dynamic interaction between Src and C-terminal Src kinase in integrin alphaIIbbeta3-mediated signaling to the cytoskeleton, *J. Biol. Chem.* **282**, 33623–33631.
93. Lele, T. P., Thodeti, C. K., and Ingber, D. E. (2006) Force meets chemistry: analysis of mechanochemical conversion in focal adhesions using fluorescence recovery after photobleaching, *J. Cell. Biochem* **97**, 1175–1183.
94. Chown, M. G., and Kumar, S. (2007) Imaging and manipulating the structural machinery of living cells on the micro- and nanoscale, *Int. J. Nanomed.* **2**, 333–344.
95. Hu, K., Ji, L., Applegate, K. T., Danuser, G., and Waterman-Storer, C. M. (2007) Differential transmission of actin motion within focal adhesions, *Science* **315**, 111–115.
96. Bulina, M. E., Lukyanov, K. A., Britanova, O. V., Onichtchouk, D., Lukyanov, S., and Chudakov, D. M. (2006) Chromophore-assisted light inactivation (CALI) using the phototoxic fluorescent protein KillerRed, *Nat. Protoc.* 1, 947–953.
97. Rajfur, Z., Roy, P., Otey, C., Romer, L., and Jacobson, K. (2002) Dissecting the link between stress fibres and focal adhesions by CALI with EGFP fusion proteins, *Nat. Cell Biol.* **4**, 286–293.
98. Alon, R., and Dustin, M. L. (2007) Force as a Facilitator of Integrin Conformational Changes during Leukocyte Arrest on Blood Vessels and Antigen-Presenting Cells, *Immunity* **26**, 17–27.
99. Wang, N., Tytell, J. D., and Ingber, D. E. (2009) Mechanotransduction at a distance: mechanically coupling the extracellular matrix with the nucleus, *Nat. Rev. Mol. Cell Biol.* **10**, 75–82.
100. Puklin-Faucher, E., and Sheetz, M. P. (2009) The mechanical integrin cycle, *J. Cell Sci.* **122**, 179–186.
101. Cai, Y., and Sheetz, M. P. (2009) Force propagation across cells: mechanical coherence of dynamic cytoskeletons, *Curr. Opin. Cell Biol.* **21**, 47–50.
102. Sabouri-Ghomi, M., Wu, Y., Hahn, K., and Danuser, G. (2008) Visualizing and quantifying adhesive signals, *Curr. Opin. Cell Biol.* **20**, 541–550.
103. Wang, Y., Botvinick, E. L., Zhao, Y., Berns, M. W., Usami, S., Tsien, R. Y., and Chien, S. (2005) Visualizing the mechanical activation of Src, *Nature* **434**, 1040–1045.
104. Na, S., and Wang, N. (2008) Application of Fluorescence Resonance Energy Transfer and Magnetic Twisting Cytometry to Quantitate Mechano-Chemical Signaling Activities in a Living Cell, *Sci Signal* **26**, *pl1*.
105. Na, S., Collin, O., Chowdhury, F., Tay, B., Ouyang, M., Wang, Y., and Wang, N. (2008) Rapid signal transduction in living cells is a unique feature of mechanotransduction, *Proc. Natl. Acad. Sci. USA.* **105**, 6626–6631.
106. Goldyn, A. M., Rioja, B. A., Spatz, J. P., Ballestrem, C., and Kemkemer, R. (2009) Force-induced cell polarisation is linked to RhoA-driven microtubule-independent focal-adhesion sliding, *J.Cell Sci.* **122,** 3644–3651.
107. Brown, C. M., Hebert, B., Kolin, D. L., Zareno, J., Whitmore, L., Horwitz, A. R., and Wiseman, P. W. (2006) Probing the integrin-actin linkage using high-resolution protein velocity mapping, *J. Cell Sci.* **119**, 5204–5214.
108. Sabass, B., Gardel, M. L., Waterman, C. M., and Schwarz, U. S. (2008) High resolution traction force microscopy based on experimental and computational advances, *Biophys. J.* **94**, 207–220.
109. Gardel, M. L., Sabass, B., Ji, L., Danuser, G., Schwarz, U. S., and Waterman, C. M. (2008) Traction stress in focal adhesions correlates biphasically with actin retrograde flow speed, *J. Cell Biol.* **183**, 999–1005.

Chapter 13

Live Imaging of LFA-1-Dependent T-Cell Motility and Stop Signals

Andrew J. Wiemer, Sarah Wernimont, and Anna Huttenlocher

Abstract

T-cell motility is critical for leukocyte trafficking both in normal host defense and in pathologic conditions including chronic inflammatory disease. Despite progress in understanding the mechanisms of T-cell polarity and motility, we have limited understanding of the mechanisms that contribute to antigen-induced T cell arrest. Here, we describe methods to analyze leukocyte function antigen-1-mediated T-cell motility and T-cell receptor-induced stop signals using in vitro assays on two-dimensional surfaces. Specifically, methods for live time-lapse imaging of T cell random migration and arrest on ICAM-1-coated surfaces are described. Additionally, we detail methods for live imaging of T-cell motility within 3D substrates to analyze T cell–antigen-presenting cell (APC) interactions and APC-mediated stop signals.

Key words: Integrin, Leukocyte function-associated antigen-1 (LFA-1), Leukocyte, lymphocyte, T cell, Stop signal, Immune synapse, D10 T cell, Conjugation, Arrest, Microscopy

1. Introduction

T cells migrate between vascular and lymphatic systems surveying protein fragments presented by major histocompatibility complexes (MHCs) on antigen-presenting cells (APCs). Within lymph nodes, T cell receptor (TCR) recognition of an antigen–MHC complex induces the rapid arrest of T cells and the formation of a stable, long-lasting interaction between the T cell and APC (1). This interaction has been described as an immune synapse (2) and has been characterized by its prototypical "bulls eye" morphology consisting of a central region (cSMAC) where the T cell receptor, co-stimulatory molecules, and signaling molecules such

Motomu Shimaoka (ed.), *Integrin and Cell Adhesion Molecules: Methods and Protocols*,
Methods in Molecular Biology, vol. 757, DOI 10.1007/978-1-61779-166-6_13, © Springer Science+Business Media, LLC 2011

as protein kinase C theta (PKC-θ) cluster, and a peripheral region (pSMAC) that includes molecules such as the integrin leukocyte function-associated antigen-1 (LFA-1, alphaLbeta2) (3). Engagement of T cell LFA-1 with its ligand intercellular adhesion molecule-1 (ICAM-1) on the APC stabilizes this cell–cell interaction. Ultimately, stable immune synapse formation allows for TCR signaling and consequently proliferation, differentiation, and immune response generation (4).

While immune synapse formation has been extensively studied and many molecules that localize to it identified (5), less is known about the regulation and dynamics of a T cell's transition from a polarized migratory cell to a rounded cell that maintains long-lasting interactions with APCs. Engagement of TCR by antigen-bearing APCs induces proximal signaling and calcium flux, both of which are known to be sufficient for T cell arrest (6). Further downstream, elements including inside-out LFA-1 activation (7) and vav-mediated cytoskeletal reorganization (8) are also associated with this transition. However, other factors involved remain unclear despite recent progress in understanding TCR signaling and integrin activation. Few studies of TCR engagement and T cell–APC conjugation have investigated the cellular dynamics of these processes.

Both LFA-1-mediated T cell migration and TCR-induced arrest are relevant to several human disease states. For instance, leukocyte adhesion deficiency (LAD) is characterized by the absence of surface expression or defective function of integrins including LFA-1 in T cells (9–11). Additionally, several T cell subsets have been found to have abnormal association between T cells and APCs. For example, CTLA-4 expression on T cells has been shown to prevent T cell arrest in response to antibody-induced TCR signals and APCs (12, 13). Studies using two-photon imaging within lymph nodes have found that regulatory T cells ($CD4^+$ $CD25^+$) impair contact between effector $CD4^+$ T cells and antigen-bearing dendritic cells (14). Increased understanding of the mechanisms underlying T cell–APC interactions may help to identify new immunomodulatory drugs for pathologic conditions including chronic inflammatory disease (15).

Here, we describe two methods using time-lapse microscopy to study T cell arrest in response to TCR stimuli (Fig. 1). The first method describes antibody stimulation of the TCR in human peripheral blood T (HPBT) cells to induce migration arrest on substrate-coated surfaces. The second method uses T cells from TCR transgenic mice and a three-dimensional medium to observe T cells migrating to and arresting on antigen-loaded APCs.

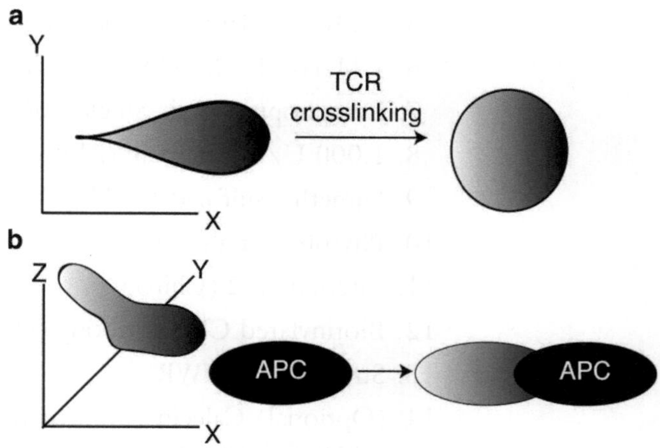

Fig. 1. Schematic representation of T cell arrest methods described. (**a**) T cell arrest following TCR stimulation on two-dimensional ICAM-1-coated surface, characterized by decrease in speed and shift from polarized to rounded morphology. (**b**) Three-dimensional migration of a polarized T cell prior to its arrest on surface of antigen-presenting cell.

2. Materials

2.1. General Equipment and Supplies

1. Tissue culture incubator at 37°C and 5% CO_2.
2. Automated or multichannel pipettes.
3. Aspirator: Preferably equipped for multiple-well aspiration.
4. Swinging bucket centrifuge.
5. Hemocytometer.
6. 15- and 50-mL conical centrifuge tubes.
7. 1.5-mL centrifuge tubes.

2.2. Live Imaging of Migrating Cells (DIC/ Fluorescence)

1. Microscope (Nikon).
2. Climate-controlled chamber (The Box).
3. Camera with time-lapse capability (Nikon).
4. Image acquisition software (Metamorph).
5. Image analysis software (Metamorph/Microsoft Excel).

2.3. HPBT Stop Signal

1. 21-g needles and 23 g × 12″ vacutainer needles.
2. 1- and 30-mL syringes.
3. Culture media: For HPBT cells, use RPMI-1640 supplemented with 10% heat-inactivated FBS, 10 mM HEPES, 1 mM pyruvate, 1× nonessential amino acids, and 1 μM β-mercaptoethanol; store at 4°C.
4. 1× PBS (without calcium and magnesium).

5. 2% BSA in 1× PBS (blocking buffer).
6. 1 M Tris–HCl, pH 9.5 (coating buffer).
7. Lymphoprep (Axis-Shield PoC AS).
8. 1,000 U/mL heparin (Abraxis).
9. Dimethyl sulfoxide (DMSO).
10. Phytohemagglutinin.
11. Interleukin-2 (Chiron).
12. Biotinylated CD3 antibody (UHCT1) (eBioscience).
13. Streptavidin (VWR).
14. (Optional) Calcein AM (Invitrogen): Suspend in DMSO at 100 μM. Store at –20°C. Light sensitive.
15. Optically clear, 384-well plates (BD Optilux) (see Note 1 on plate selection).
16. ICAM-1 (1 mg/mL): We use self-purified ICAM-1-Fc from CHO cells; however, ICAM-1 from a commercial source (e.g., R&D Systems) will work. Store at –80°C for long term, or store working aliquots at 4°C. Avoid multiple freeze-thaw cycles.

2.4. Supplies for T Cell Antigen Recognition in Real Time

1. Poly-L-lysine (0.01% solution).
2. Heat-inactivated fetal bovine serum (HI-FBS).
3. Culture media: For D10 T cell culture, use Clicks media (Irvine Scientific) supplemented with 10% HI-FBS, 10 mM HEPES, 1× penicillin–streptomycin, and 1× L-glutamine; store at 4°C.
4. PBS (without calcium and magnesium).
5. Hanks' buffered salt solution supplemented with 10% FBS and 10 mM HEPES.
6. Low-melt agarose.
7. IL-2 (Chiron).
8. Conalbumin (10 mg/mL stock prepared in D10 culture media).
9. PHK-26 (1 mM) plus diluent C or 5% dextrose (w/v in water).
10. 6-cm non-tissue culture-treated plate with an 18-mm hole in the center.
11. Norland optical adhesive.
12. 22-mm acid-washed glass coverslip.

3. Methods

3.1. Studying HPBT Stop Signal

Within lymph nodes, TCR recognition of an antigen–MHC complex induces rapid T cell arrest and formation of a stable immune synapse (1). There are several methodologies to monitor T cell arrest in vitro; the most widely used are transwell migration chambers and live-cell imaging (12, 16). Use of live imaging offers the advantage of reduced assay time, ability to perform more complicated kinetic analyses rather than endpoint studies, and resolution at the single cell or subcellular levels, all of which are not easily attainable with transwell migration assays. Advances in automated tracking software has greatly reduced time required for these analyses and will only increase the prevalence of live-cell imaging for studying cell migration and arrest.

Here, we describe a live-cell imaging method for assessing T cell random migration and the TCR-induced stop signal. Cells are plated onto an optically clear multi-well plate coated with ICAM-1 on which they adhere and migrate, and a stop signal is induced by cross-linking CD3 with a biotinylated antibody. This allows for observation and quantification of integrin-dependent T cell migration and arrest. Using T cells fluorescently stained with calcein allows for enhanced automated tracking. This methodology is straightforward and readily adaptable for studying specific protein functions via siRNA-induced knockdowns or small molecule inhibitors.

3.1.1. T Cell Purification (See Note 2 on T Cell Methodology)

1. Obtain 30 mL of blood using heparinized syringe from donor with informed consent.
2. Dilute whole blood to 60 mL with 1× PBS.
3. Overlay 30 mL of blood/PBS mixture onto 15 mL of Lymphoprep in a 50-mL conical tube.
4. Centrifuge for 30 min at 500×*g*.
5. Resuspend mononuclear cells in 30 mL of PBS.
6. Wash by centrifugation at 400×*g* for 10 min.
7. Resuspend the cells in 30 mL pre-equilibrated media and wash by centrifugation at 350×*g*. Repeat once.
8. Stimulate the cells with phytohemagglutinin (2 μg/mL).
9. Add fresh media containing IL-2 (50 U/mL) at 48 and 96 h post-draw.
10. Expand for 7–14 days, replacing with fresh media every 2–3 days.

3.1.2. Plate Coating

1. Dilute ICAM-1 to 2.5 μg/mL in coating buffer.
2. Add 20 μL to each well of the 384-well plate (see Note 3 on controls).
3. Incubate the plates at 37°C for 1 h (see Note 4 on coating methods).
4. Aspirate off the remaining liquid.
5. Add 30 μL blocking buffer to each well of the 384-well plate.

6. Incubate the plates at 37°C for 1 h.
7. Immediately prior to cell addition, aspirate off the remaining liquid. Do not allow the plates to dry completely.

3.1.3. Cell Preparation

1. Count the cells using a hemocytometer.
2. Pre-equilibrate media at 37°C and 5% CO_2.
3. Calculate cell volume needed to yield 50,000 cells per well of 384-well plate.
4. Collect the cells by centrifugation for 5 min at $350\times g$.
5. Wash the cells with an equivalent volume of pre-equilibrated media followed by centrifugation for 5 min at $350\times g$.
6. Suspend at a final concentration of 1×10^6 cells/mL.

3.1.4. (Optional) Calcein AM Labeling

1. Use calcein AM for subsequent fluorescence-based tracking (see Note 5) (Fig. 2) and incubate the cells from above with 1 μM calcein AM for 15 min.
2. Collect the cells by centrifugation for 5 min at $350\times g$.
3. Wash the cells with equivalent volume of pre-equilibrated media, followed by centrifugation for 5 min at $350\times g$.

3.1.5. Adhesion of Cells to ICAM-Coated Plates

1. Remove blocking buffer from wells.
2. Wash wells using 20 μL of 1× PBS.
3. Add 50 μL of cells to each well of 384-well plate to be used.

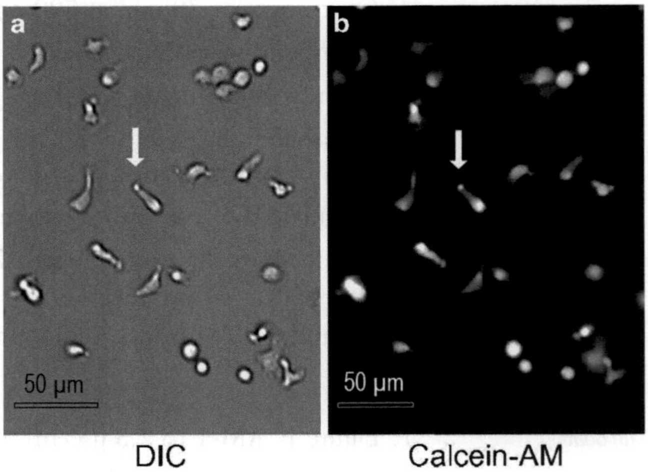

Fig. 2. HPBT cells migrating on ICAM-coated surface. DIC (**a**) and fluorescent (**b**) images of HBPT cells plated on a 384-well plate coated with 20 μL of ICAM (2.5 μg/mL). The uropod (*white arrow*) and overall cell morphology can be visualized clearly with either imaging method. Attached and migrating cells exhibit polarized morphology and diffuse staining, while floating non-adherent cells exhibit round morphology with intense staining.

4. Incubate for 30 min at 37°C and 5% CO_2 (see Note 6 on alternate adhesion protocols).
5. Wash 2× with 30 μL of pre-equilibrated culture media (see Note 7 on washing).
6. Optional. Marking bottom of each well with a razor or permanent marker will allow image alignment if stage drift occurs.

3.1.6. HPBT Cell Migration Stop Signal Following CD3 Cross-linking (Fig. 3)

1. Immediately prior to plating, incubate HPBT cells on ice for 5 min.
2. Add biotinylated CD3 antibody (UHCT1) or nonspecific control antibody (IgG) to cells at 500 ng/mL for 10 min on ice.
3. Plate 50 μL of HPBT cells as described above (Subheading 3.2.1).
4. Immediately following plating, cross-link by adding streptavidin at 1 μg/mL.
5. Incubate for 10 min at 37°C prior to imaging.

3.1.7. Live Imaging of Migrating Cells (DIC/Fluorescence)

1. Place the plate on microscope stage with 10× lens in closed system that has been preheated to 37°C. For longer experiments, use CO_2-controlled chamber or non-bicarbonate buffering system to avoid pH fluctuations.
2. Examine the live image and select a stage position that contains between 50 and 200 cells. A smaller cell number may not provide statistical power, while too many cells will increase tracking difficulty.
3. Focus cells and record stage position for each well (see Note 8 on focus).
4. Capture every 30 s for 15-min time frame (see Note 9 on imaging times and Note 10 on intervals).

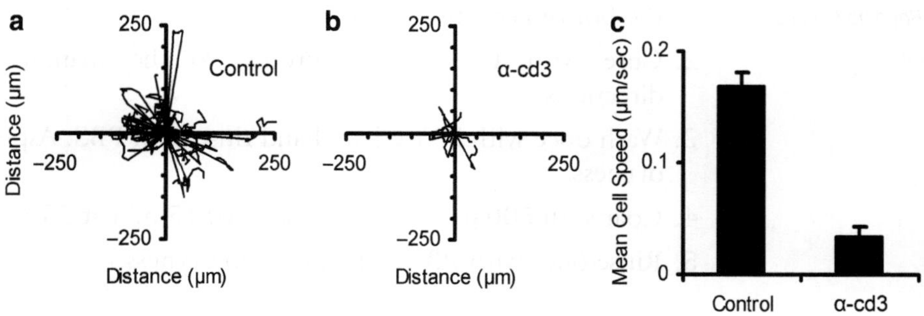

Fig. 3. CD3 cross-linking induces a HPBT cell migration stop signal. HPBT cells were expanded in culture. Cells were coated on ice for 10 min with 500 ng/mL of biotin-labeled anti-CD3 (UCHT1) (**b**) or control antibody (**a**). Cells were plated and TCR signaling induced by cross-linking with 500 ng/mL streptavidin. Cells were imaged for 15 min at 45-s intervals. Cells were tracked and directionality and speed of migration were assessed (**c**).

3.1.8. Tracking Cell Migration

1. Load images into software.
2. Select box around the cells to be tracked. Cell selection is important; we usually highlight and track any cell that appears in the first frame, or randomly select 30–40 cells from the image center. Cells are tracked until they move out of the image or, less frequently, detach. Floating, touching, or dividing cells may be excluded.
3. Data logged to a spreadsheet. Most software will allow for exportation of average speeds and X, Y position which can be sorted by frame or object.

3.2. Antigen Recognition in Real Time

Since its molecular characterization (3), studies of the immune synapse have been traditionally conducted under static, fixed conditions and have successfully identified many molecular players involved in its formation. More recently, studies using lipid bilayers, total internal reflection fluorescence (TIRF) microscopy, and recombinant ICAM-1 and MHC-I molecules have demonstrated how dynamic the process of immune synapse formation can be (17). Rather than being a static structure, the immune synapse is constantly breaking and reforming as the T cell continues it interactions with the APC. These dynamics of T cell–APC interactions are not accounted for in static, fixed systems.

Here, we describe a method for observing T cell antigen recognition in real time. After loading with antigen, APCs are fixed onto a glass-bottomed plate and overlaid with T cells migrating in a solution containing low-melt agarose and FBS. This permits observation of integrin-independent T cell migration and arrest with antigen-loaded APCs in real time. Using T cells expressing fluorescently tagged proteins, polarization of immune synapse components may be observed in real time (Fig. 4). This method has previously been used concurrently with calcium-sensitive dyes to study chemokine modulation of T cell attachment to APCs (18).

3.2.1. Preparation of Glass-Bottomed Plates for Imaging

1. Using an optical adhesive, secure a 22-mm glass coverslip to the bottom of a 6-cm plate.
2. Cure with UV light according to the manufacturer's directions.
3. Wash once with 70% ethanol and rinse with PBS. Aspirate to dryness.
4. Coat with 500 μL of poly-L-lysine for 15 min at 25°C.
5. Rinse once with PBS and aspirate to dryness.

3.2.2. Preparation of Antigen-Presenting Cells (CH12 Cells)

1. CH12 cells (ATCC) are maintained at 5×10^5 cells/mL in Clicks media (see Note 11).
2. At 18 h prior to experimentation, CH12 cells are resuspended in fresh Clicks media supplemented with 250 μg/mL conalbumin. As a control, CH12 may be left untreated with antigen.

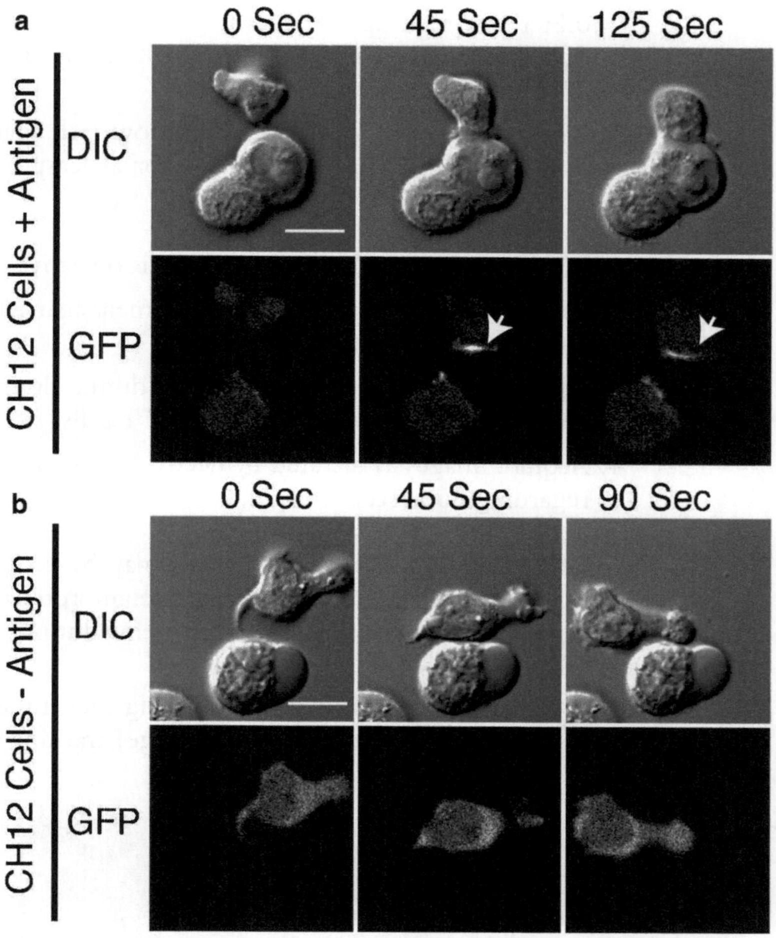

Fig. 4. Polarization of PKC-θ during T cell recognition of antigen in real time. (**a**) DIC and GFP images showing T cell arrest on CH12 cell (antigen-presenting cell) loaded with antigen. PKC-θ polarizes to the synapse rapidly following initial contact. (**b**) DIC and GFP images showing T cell contacting and not arresting upon contact with CH12 cells not bearing antigen. In the absence of conjugation, PKC-θ does not polarize. Scale bar represents 10 μm.

3. Optional. Dye CH12 cells with fluorescent membrane label PKH26. Pellet the cells and aspirate media completely. Resuspend in 250 μL 5% dextrose (or Diluent C) supplemented with 2 μM PKH26. Incubate at 25°C for 2 min. Quench by adding 250 μL FBS. Wash 1× with PBS and 1× with media and resuspend at 5×10^5 cells/mL in Clicks media.
4. Plate 2.5×10^5 cells onto a poly-L-lysine-coated cover slip and incubate at 37°C and 5% CO_2 for 15 min.
5. Wash once with pre-equilibrated HBSS with 10% FBS and HEPES and resuspend in 500 μL for imaging.

3.2.3. Preparation of T Cells

1. To prepare low-melt agarose migration media, combine HBSS supplemented with 10% FBS and 1 mM HEPES with

0.25% low-melt agarose. Heat gently in a microwave with frequent inversion until agarose is melted and allow to cool to 37°C (see Notes 12 and 13).

2. Resuspend 3×10^5 T cells in 2.5 mL low-melt agarose migration media and incubate at 37°C for at least 15 min until ready to image.

3.2.4. Imaging

1. In a climate-controlled chamber, focus objective on APC.
2. Overlay APCs with T cells in low-melt agarose migration media using a transfer pipette.
3. Optional: To prevent evaporation during longer movies, overlay 1 mL of mineral oil on top of T cells (see Note 14).
4. Acquire images as dictated by needs. See Note 15 for caveat regarding fluorescent imaging.

3.2.5. Analysis

1. Following acquisition, still images may be processed using image acquisition software such as Metamorph to show localization of fluorescently labeled proteins during conjugation (Fig. 4).
2. T cells may be tracked (Fig. 5) during their time of contact with APC using Metamorph or ImageJ tracking software to

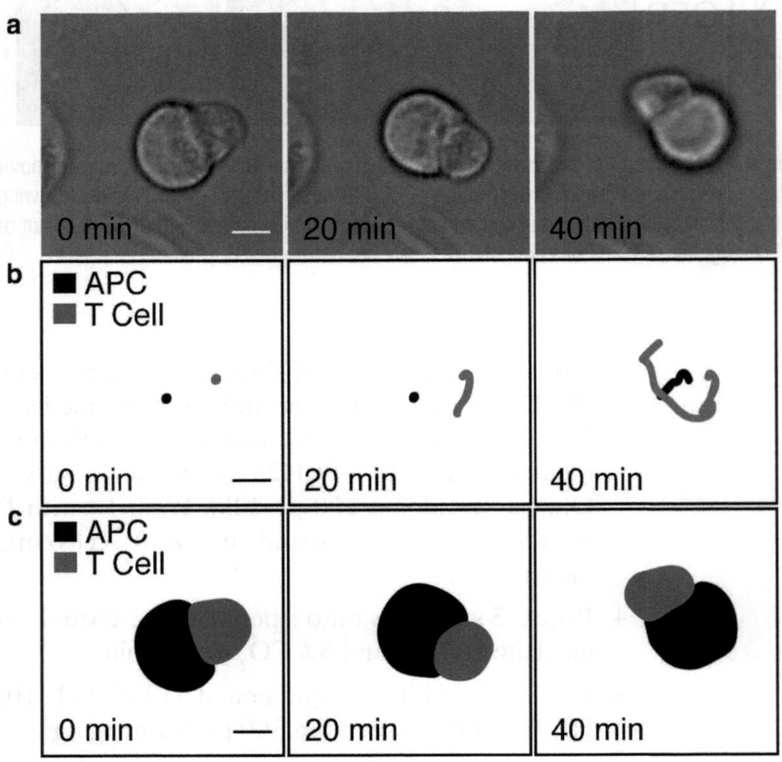

Fig. 5. Analysis of T cell–APC conjugates. (**a**) DIC images of T cell–APC conjugates over time. (**b**) Tracks of T cell and APC movement over time. (**c**) Morphometric analysis of T cell–APC interactions over time. Scale bar represents 10 μm.

show change in position of T cell relative to APC over time (Fig. 5b). Additionally, positions of T cell relative to APC over time may be assessed by morphometric analysis (Fig. 5c).

4. Notes

1. Plate selection is an important step for studying migration. Plates vary significantly between manufacturers and lots, so it is best to use one plate type consistently. Any size between 6 wells and 384 wells will work in this assay. We prefer to use BD Optilux 96- or 384-well plates. Optical clarity of the plates should be determined prior to start of the assay by visualization of test cells. High-protein binding plates traditionally used in ELISA applications give enhanced results in integrin adhesion assays, but are frequently not optimized for microscopy and may not provide a truly clear surface.
2. HPBT cell purification methods can vary significantly; therefore, we recommend using one method consistently and do not recommend changing your established laboratory protocol. Some factors that may affect this assay include HPBT cell age, subset of interest, amount of IL-2 stimulation, and cell concentration. One should also keep in mind species specificity of integrin/substrate interactions (e.g., mouse ICAM-1 is required for studying migration of mouse cells).
3. It is best to use several controls when optimizing assay conditions. In particular, the tissue culture treatment process is optimized to enhance cell adhesion. Therefore, it is important to include empty wells and blocked-only wells. Where available, integrin-blocking or integrin-activating antibodies should be used as positive controls.
4. Coating for 1 h at 37°C is preferred due to speed, although overnight coating at 4°C has been reported and may allow lower adhesion molecule concentrations. We recommend using one coating method for consistent results. Integrin-binding site numbers can be determined by ELSIA if required. Cell adhesion molecule coating concentrations are typically between 1 and 25 μg/mL, and maximal speeds are obtained at intermediate adhesion molecule levels (19).
5. Longer staining with higher calcein AM concentrations will increase signal-to-noise ratio; however, this may cause toxicity. Fluorescently tagged proteins have also been used as a way to standardize fluorescence and avoid potential toxicity from fluorescent dyes such as calcein AM. When performing fluorescent image acquisition in live cells, it is important to add a control well for optimizing exposure time and focus

parameters, as migration can be sensitive to overexposure. If auto-fluorescence is apparent, one may use a phenol red-free imaging media.

6. A wide variety of adhesion protocols for HPBT cells have been reported (20, 21), and the optimal protocol depends on cell type and stimulus. We find for HPBT cells that centrifugation on a plate centrifuge at forces <500 × *g* allows for simultaneous adhesion without altering migration rates.
7. Even gentle washing of adherent cells results in shear forces that can increase assay variability (22). This effect can be visualized by microscopy following washing steps by appearance of empty spaces on the surface and can be quantified by fluorescence measurement after each wash step. The variability is most pronounced on smaller well sizes. Therefore, in 384-well plates, it is important to aspirate and wash in one corner to avoid reducing the available imaging area.
8. It is desirable to use a microscope and software with automatic focus. This allows stage positions to be programmed prior to start of assay and avoids spending a significant amount of time obtaining focus.
9. We have observed HPBT cell migration up to 24 h following plating, although longer time frames would probably work. For longer incubations, it would be recommended to overlay with mineral oil, or to use a climate control box with the ability to maintain proper CO_2 concentration and humidity. Media that contains HEPES buffer is more resistant to changes in pH during the course of the experiment than those relying on bicarbonate buffering.
10. Imaging intervals larger than 30 s will work; however, as the interval gets larger, automatic tracking programs become less useful. For other cell types that migrate slower, larger intervals can be more readily used. If, upon data analysis, frames appear too frequently, alternating frames can be readily removed from the stack.
11. Do not allow CH12 cells to exceed a density of 1×10^6 cells/mL. Doing so causes them to lose antigen presentation capability.
12. For preparation of low-melt agarose migration media, use a polypropylene tube that has sufficient space (generally 3× the volume of product) in case of overheating. Generally, when heating in the microwave, we heat in 10-s increments with frequent inversion to mix reagents and check for dissolution of low-melt agarose. Be careful not to overheat since proteins will aggregate and T cells will not migrate. Additionally, allow to return to 37°C so that T cells are not damaged by heat. We always make fresh migration media on the day of experimentation.

13. One caveat to using low-melt agarose migration media is that FBS is providing adhesive proteins. Variability in FBS lots may contribute to variability in the assay, so we recommend using single lots of FBS for experiments.
14. We have successfully imaged T cell–APC interactions for greater than 5 h. For longer acquisition periods, we always overlay with mineral oil to prevent evaporation. Additionally, to reduce phototoxicity from fluorescence exposure, we decrease the frequency of fluorescent acquisition to less than one image every 4 min.
15. For fluorescent imaging, be careful to not overexpose cells. Doing so can damage cells and inhibit migration and conjugation.

Acknowledgments

This work was supported by NIH NIAID R01 to A. H. (AI068062). Postdoctoral support was provided to A.W. by the University of Wisconsin Institute on Aging Training Grant (NIH T32AG000213-17), Sanjay Asthana Principle Investigator. We would also like to thank Max Krummel and Anne Ridley for providing useful advice during method development.

References

1. Mempel, T. R., Henrickson, S. E., and Von Andrian, U. H. (2004) T-cell priming by dendritic cells in lymph nodes occurs in three distinct phases. *Nature* **427**, 154–159.
2. Norcross, M. A. (1984) A synaptic basis for T-lymphocyte activation. *Ann. Immunol. (Paris)* **135D**, 113–134.
3. Monks, C. R., Freiberg, B. A., Kupfer, H., Sciaky, N., and Kupfer, A. (1998) Three-dimensional segregation of supramolecular activation clusters in T cells. *Nature* **395**, 82–86.
4. Huppa, J. B., Gleimer, M., Sumen, C., and Davis, M. M. (2003) Continuous T cell receptor signaling required for synapse maintenance and full effector potential. *Nat. Immunol.* **4**, 749–755.
5. Dustin, M. L. (2009) The cellular context of T cell signaling. *Immunity* **30**, 482–492.
6. Negulescu, P. A., Krasieva, T. B., Khan, A., Kerschbaum, H. H., and Cahalan, M. D. (1996) Polarity of T cell shape, motility, and sensitivity to antigen. *Immunity* **4**, 421–430.
7. Rose, D. M., Alon, R., and Ginsberg, M. H. (2007) Integrin modulation and signaling in leukocyte adhesion and migration. *Immunol. Rev.* **218**, 126–134.
8. Burkhardt, J. K., Carrizosa, E., and Shaffer, M. H. (2008) The actin cytoskeleton in T cell activation. *Annu. Rev. Immunol.* **26**, 233–259.
9. Springer, T. A., Thompson, W. S., Miller, L. J., Schmalstieg, F. C., and Anderson, D. C. (1984) Inherited deficiency of the Mac-1, LFA-1, p150,95 glycoprotein family and its molecular basis. *J. Exp. Med.* **160**, 1901–1918.
10. Mathew, E. C., Shaw, J. M., Bonilla, F. A., Law, S. K., and Wright, D. A. (2000) A novel point mutation in CD18 causing the expression of dysfunctional CD11/CD18 leucocyte integrins in a patient with leucocyte adhesion deficiency (LAD). *Clin. Exp. Immunol.* **121**, 133–138.
11. Hogg, N., Stewart, M. P., Scarth, S. L., Newton, R., Shaw, J. M., Law, S. K., and Klein, N. (1999) A novel leukocyte adhesion deficiency caused by expressed but nonfunctional beta2 integrins Mac-1 and LFA-1. *J. Clin. Invest.* **103**, 97–106.
12. Schneider, H., Downey, J., Smith, A., Zinselmeyer, B. H., Rush, C., Brewer, J. M.,

Wei, B., Hogg, N., Garside, P., and Rudd, C. E. (2006) Reversal of the TCR stop signal by CTLA-4. *Science* **313**, 1972–1975.
13. Rudd, C. E. (2008) The reverse stop-signal model for CTLA4 function. *Nat. Rev. Immunol.* **8**, 153–160.
14. Tadokoro, C. E., Shakhar, G., Shen, S., Ding, Y., Lino, A. C., Maraver, A., Lafaille, J. J., and Dustin, M. L. (2006) Regulatory T cells inhibit stable contacts between CD4+ T cells and dendritic cells in vivo. *J. Exp. Med.* **203**, 505–511.
15. Cantor, J. M., Ginsberg, M. H., and Rose, D. M. (2008) Integrin-associated proteins as potential therapeutic targets. *Immunol. Rev.* **223**, 236–251.
16. Bromley, S. K., Peterson, D. A., Gunn, M. D., and Dustin, M. L. (2000) Cutting edge: hierarchy of chemokine receptor and TCR signals regulating T cell migration and proliferation. *J. Immunol.* **165**, 15–19.
17. Sims, T. N., Soos, T. J., Xenias, H. S., Dubin-Thaler, B., Hofman, J. M., Waite, J. C., Cameron, T. O., Thomas, V. K., Varma, R., Wiggins, C. H., Sheetz, M. P., Littman, D. R., and Dustin, M. L. (2007) Opposing effects of PKCtheta and WASp on symmetry breaking and relocation of the immunological synapse. *Cell* **129**, 773–785.
18. Friedman, R. S., Jacobelli, J., and Krummel, M. F. (2005) Mechanisms of T cell motility and arrest: Deciphering the relationship between intra- and extracellular determinants. *Semin. Immunol.* **17**, 387–399.
19. von der Mark, K., Schober, S., and Goodman, S. L. (1999) Integrins in cell migration. *Methods Mol. Biol.* **129**, 219–230.
20. Simonson, W. T. N., Franco, S. J., and Huttenlocher, A. (2006) Talin1 regulates TCR-mediated LFA-1 function. *J. Immunol.* **177**, 7707–7714.
21. Nolz, J. C., Nacusi, L. P., Segovis, C. M., Medeiros, R. B., Mitchell, J. S., Shimizu, Y., and Billadeau, D. D. (2008) The WAVE2 complex regulates T cell receptor signaling to integrins via Abl- and CrkL-C3G-mediated activation of Rap1. *J. Cell Biol.* **182**, 1231–1244.
22. Park, J. Y., Arnaout, M. A., and Gupta, V. (2007) A simple, no-wash cell adhesion-based high-throughput assay for the discovery of small-molecule regulators of the integrin CD11b/CD18. *J. Biomol. Screen.* **12**, 406–417.

Chapter 14

Monitoring Integrin Activation by Fluorescence Resonance Energy Transfer

Craig T. Lefort, Young-Min Hyun, and Minsoo Kim

Abstract

Aberrant integrin activation is associated with several immune pathologies. In leukocyte adhesion deficiency (LAD), the absence or inability of β_2 integrins to undergo affinity upregulation contributes to recurrent infectious episodes and impaired wound healing, while excessive integrin activity leads to an exaggerated inflammatory response with associated tissue damage. Therefore, integrin activation is an attractive target for immunotherapies, and monitoring the effect of agents on integrin activation is necessary during preclinical drug development. The activation of integrins involves the structural rearrangement of both the extracellular and cytoplasmic domains. Here, we describe methods for monitoring integrin conformational activation using fluorescence resonance energy transfer (FRET).

Key words: FRET, Integrin, Cell adhesion, LFA-1, Mac-1, VLA-4, Fluorescence microscopy, Flow cytometry

1. Introduction

Integrins are heterodimeric transmembrane receptors that mediate cell–cell and cell–extracellular matrix interactions (1). Integrins play critical roles in a wide range of processes, including embryonic development and the coordinated immune response (1, 2). The activation state of an integrin is regulated by its conformation, similar to many other membrane proteins and proteins involved in signaling (3). Integrin receptors are not fixed in a particular conformation; rather, they equilibrate between a compact, bent structure with low affinity for ligand, and an extended, high-affinity conformation, with several apparent intermediate conformational states (3). In contrast to the low-affinity integrin, the activated integrin exhibits spatially separated cytoplasmic tails and an extracellular headpiece that is extended away from the plasma membrane (Fig. 1).

Motomu Shimaoka (ed.), *Integrin and Cell Adhesion Molecules: Methods and Protocols*,
Methods in Molecular Biology, vol. 757, DOI 10.1007/978-1-61779-166-6_14, © Springer Science+Business Media, LLC 2011

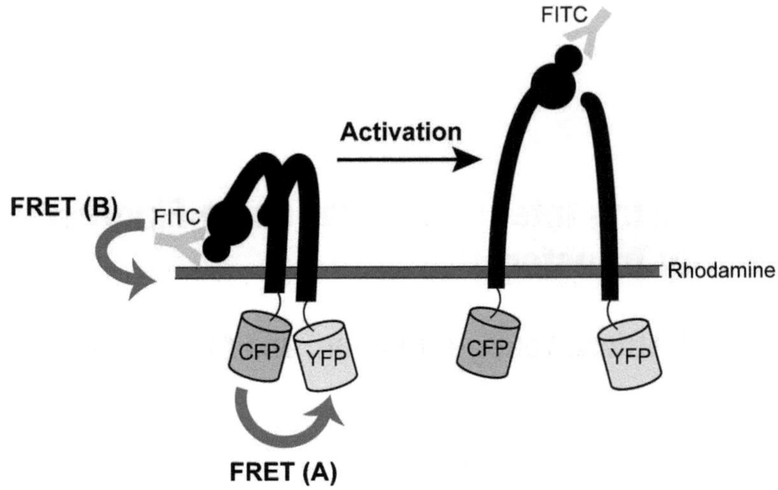

Fig. 1. Experimental systems used for FRET analysis of integrin activation. (**a**) Loss of FRET between CFP- and YFP-tagged integrin cytosolic tails indicates spatial separation during integrin activation. (**b**) Loss of FRET between FITC-conjugated anti-integrin I domain antibodies and the membrane dye ORB indicates extracellular domain extension during integrin activation.

The structural changes that occur in the integrin molecule during conversion from the inactive to the active form can be observed and quantified using fluorescence resonance energy transfer (FRET). FRET is a spectroscopic phenomenon in which energy emitted from a fluorophore (donor) is transferred to a second fluorophore (acceptor) in close proximity (<10 nm) and with overlapping spectra, resulting in the emission of energy from the acceptor. While initial studies using electron microscopy demonstrated the changes in conformation between the resting and activated integrin (4), fluorescence microscopy and FRET provide a means to study the structure, and therefore function, of integrins on the living cell surface. In addition, FRET is a powerful method because it has the sensitivity for measuring dynamic integrin activation with nanometer resolution. There are several modalities for measuring FRET signals, including acceptor photobleaching FRET, sensitized emission FRET, and ratiometric FRET. Here, we describe three different FRET methods for studying integrin activation (Fig. 1).

2. Materials

2.1. Acceptor Photobleaching FRET

1. K562 human leukemia cell line (ATCC).
2. RPMI 1640 media supplemented with 10% fetal bovine serum (FBS), 100 μg/ml streptomycin, and 100 U/ml penicillin.

2.1.1. K562 Cell Culture and Transfection

3. α_M-mCFP and β_2-mYFP cDNA: Generated using the mammalian expression vectors pECFP-N1 and pEYFP-N1 (ClonTech), respectively (see Note 1).
4. G418 sulfate.

2.1.2. K562 Cell Treatment and FRET Measurement

1. L-15 media (Gibco) containing 2 mg/ml D-glucose.
2. Nikon Eclipse TE2000-E microscope coupled to a QuantEM:512SC CCD camera (Photometrics).
3. Imaging filters (Chroma): YFP (HQ500/20×, Q515LP, HQ535/30M), CFP (D436/20×, 455DCLP, D480/40M), and YFP photobleach (D535/50×, Dichroic Full Mirror, Metal Slug).

2.1.3. Data Analysis

1. NIS-Elements software (Nikon).

2.2. Sensitized Emission FRET

2.2.1. GD25 Cell Culture and Transfection

1. GD25 cells (5).
2. DMEM supplemented with 10% FBS, 100 μg/ml streptomycin, and 100 U/ml penicillin.
3. α_4-mCFP and β_1-mYFP, generated using the mammalian expression vectors pECFP-N1 and pEYFP-N1 (ClonTech), respectively (see Note 1).
4. Amaxa transfection system, including nucleofector kit V (Lonza).

2.2.2. GD25 Cell Treatment and FRET Measurement

1. Trypsin-EDTA.
2. L-15 media containing 2 mg/ml D-glucose.
3. Delta T dish, 0.17 mm.
4. Human VCAM-1-Ig recombinant fusion protein (R&D Systems).
5. Nikon Eclipse TE2000-E microscope coupled to a QuantEM:512SC CCD camera (Photometrics) with a dual-view image splitter (Photometrics).
6. CFP/YFP dual-band filter set (Chroma).
7. Filter wheel with excitation filters (431–441 nm/490–510 nm).
8. White-light TIRF illuminator (Nikon).
9. Perfect focus unit (Nikon) and vibration isolation system (Technical Manufacturing Corp.).

2.2.3. Data Analysis

1. AutoQuant and AutoDeblur software (Media Cybernetics).

2.3. Ratiometric FRET

2.3.1. Neutrophil Isolation

1. 1-Step Polymorphs cell separation media (Accurate Chemical).
2. Neutrophil isolation buffer (NIB): Hanks' balanced salt solution (HBSS) without calcium chloride or magnesium chloride buffered with 10 mM HEPES, pH 7.4, and 0.1% bovine serum albumin (BSA).

3. Hypotonic erythrocyte lysis buffer (HELB): 10× phosphate-buffered saline (PBS) without calcium chloride or magnesium chloride, pH 7.4, diluted 1:100 in water to make a final concentration of 0.1× PBS.
4. Normalization buffer: 10× PBS without calcium chloride or magnesium chloride, pH 7.4, diluted 1:3 in water to make a final concentration of 4× PBS, and containing 0.4% BSA.

2.3.2. Neutrophil Treatment and Labeling

1. Neutrophil labeling buffer (NLB): HBSS containing calcium chloride, magnesium chloride, and 0.1% BSA (Sigma).
2. Fluorescein isothiocyanate (FITC)-conjugated monoclonal antibody (mAb) directed against the Mac-1 integrin head domain (see Note 4).

2.3.3. Flow Cytometry Data Acquisition

1. Octadecyl rhodamine B (ORB; Sigma): Prepared as a 200 mM stock solution in DMSO.

3. Methods

3.1. Acceptor Photobleaching FRET

Acceptor photobleaching FRET represents a simple end-point technique for analyzing the spatial relationship of donor and acceptor fluorophores. In this method, FRET is measured by determining the fluorescence intensity of the donor before and after photodestruction of the acceptor. In the protocol below, we describe a technique we have used to measure the separation of the α- and β-subunit cytoplasmic tails of the integrin Mac-1 ($\alpha_M\beta_2$) in response to activation (Fig. 1). Monomeric forms of CFP and YFP are genetically fused to the carboxy-terminus of the α_M- and β_2-subunits, respectively. Cells expressing α_M-mCFP and β_2-mYFP are analyzed by fluorescence microscopy. The proximity of the Mac-1 cytoplasmic domains is reported as FRET efficiency, or the extent to which the acceptor fluorophore quenches donor fluorescence. Using this technique, it has been shown that activation of both LFA-1 (6) and Mac-1 (7) leads to the separation of the integrin α and β cytoplasmic tails.

3.1.1. K562 Cell Culture and Transfection

1. K562 cells are maintained in RPMI media containing FBS and penicillin/streptomycin.
2. K562 cells are transfected with α_M-mCFP and β_2-mYFP by electroporation and maintained under selection with 1 mg/ml G418 sulfate.
3. Stable K562 transfectants are sorted twice by immunofluorescence with anti-α_M clone ICRF44 and then seeded at a single cell per well in 96-well plates to obtain homogeneous and stable clones.

3.1.2. K562 Cell Treatment and FRET Measurement

1. K562 cells stably expressing α_M-mCFP and β_2-mYFP are maintained in RPMI media containing FBS, penicillin/streptomycin, and G418 sulfate.
2. Wash K562 cells and then resuspend in L-15 media supplemented with 2 mg/ml D-glucose.
3. Allow K562 cells to settle for 5–10 min on coverslips mounted on the microscope stage.
4. Treat K562 cells with integrin activation agonists/antagonists.
5. Acquire initial CFP and YFP images using a 60× oil immersion, 1.40 NA objective lens (see Note 2).
6. Expose K562 cells to YFP photobleach excitation wavelength for 3 min, without using any neutral density filters.
7. Acquire post-photobleach CFP and YFP images, using settings identical to those used for the initial image capture (exposure time, neutral density filter, etc.).

3.1.3. Data Analysis

1. For each image, analyze only the membrane region on each cell of interest.
2. Measure the mean CFP fluorescence intensity of the region of interest on each individual cell before and after acceptor photobleaching using Nikon NIS-Elements (or similar) software.
3. Repeat the analysis above for YFP to verify >90% destruction of the initial YFP signal after photobleaching.
4. Calculate FRET efficiency (E) with the following formula:

$$E = 1 - (F_{\mathrm{CFP}}(d)_{\mathrm{Pre}} / F_{\mathrm{CFP}}(d)_{\mathrm{Post}})$$

where $F_{\mathrm{CFP}}(d)_{\mathrm{Pre}}$ and $F_{\mathrm{CFP}}(d)_{\mathrm{Post}}$ are the mean CFP emission intensity of pre- and post-photobleach images (8).

3.2. Sensitized Emission FRET

In the previous FRET method, acceptor photobleaching FRET, the donor fluorescence was monitored to determine the extent of energy transfer and, thus, distance between donor and acceptor. In sensitized emission FRET, the fluorescence of the acceptor is monitored during excitation of the donor. Unlike acceptor photobleaching FRET, the sensitized emission FRET method allows for measurement of energy transfer over a time course. Sensitized emission FRET is also called the three-cube method, as the necessary measurements for data acquisition and signal correction are performed with three different filter sets on a fluorescence microscope. In the protocol below, we outline a method for monitoring the activation state of the integrin VLA-4 ($\alpha_4\beta_1$) over time in an adherent cell. We have coupled this FRET method with TIRF microscopy to analyze specifically integrins in close contact with an adhesive substrate. Similar to the acceptor photobleaching method above, we describe an experimental setup in which the donor (CFP) and acceptor (YFP) are genetically fused to the cytoplasmic tail of the integrin α- and β-subunits. Using this method, we have shown that VLA-4 activation is sustained at the leading edge of GD25 cells and migrating T lymphocytes (9).

3.2.1. GD25 Cell Culture and Transfection

About 5×10^5 GD25 cells are transfected with 2 μg α_4-mCFP and β_1-mYFP using Amaxa nucleofector kit V (protocol T-24), and the cells are then cultured overnight in DMEM containing FBS and penicillin/streptomycin.

3.2.2. GD25 Cell Treatment and FRET Measurement

1. Detach adhered GD25 cells from the plate with 0.25% trypsin-EDTA and resuspend the cells in L-15 medium supplemented with 2 mg/ml D-glucose.
2. Transfer the cells to a Delta T dish coated with 100 μg/ml VCAM-1 and allow to settle for 10 min at 37°C.
3. Acquire CFP and YFP dual-emission images with CFP and YFP excitations using a 0.5-s exposure time. Acquire images every 10 s for 30 min through a 100× oil immersion, 1.49 NA objective lens.

3.2.3. Data Analysis

Subtract the background for each fluorescence image and perform FRET efficiency calculations with the AutoQuant imaging algorithm of AutoDeblur (see Note 3).

3.3. Ratiometric FRET

The ratiometric FRET method for measuring energy transfer from donor to acceptor fluorophore utilizes the dependence of FRET on the donor–acceptor ratio. As the concentration of the acceptor fluorophore increases relative to the donor concentration, so does the probability that an acceptor is available for energy transfer. In the protocol below, we describe a method for measuring the spatial proximity of the Mac-1 integrin extracellular domain headpiece to the plasma membrane. The donor fluorophore is a FITC-conjugated mAb directed against the integrin, and the acceptor fluorophore is a rhodamine dye (ORB) that incorporates into the plasma membrane. FRET is measured by holding the donor concentration constant and varying the concentration of ORB. The fluorescence intensity of the donor and acceptor is measured using flow cytometry. By plotting the FITC mean fluorescence versus rhodamine mean fluorescence, we obtain a curve whose slope indicates the extent of FRET and, thus, the relative distance between donor and acceptor for a large population of integrin molecules. Using this method, we have shown that certain agonists, such as formyl peptides, and activating mAbs induce the extension of the Mac-1 headpiece away from the plasma membrane (7).

3.3.1. Neutrophil Isolation

1. After allowing the reagents to reach room temperature, gently layer 3 ml of heparinized whole blood obtained from healthy human donors on top of 3 ml 1-Step Polymorphs in an 8-ml round-bottom polystyrene tube.
2. Centrifuge the tubes for 45 min at 500 × *g*.
3. After centrifugation, the blood components will be separated into distinct layers. The top cell layer contains mononucleated cells, including monocytes and lymphocytes. The second cell layer, separated from the top cell layer by a small amount of clear solution, contains the neutrophils, or polymorphonuclear leukocytes. Erythrocytes are centrifuged to the bottom of the tube, though there is typically a small amount of erythrocyte contamination in the neutrophil layer. Discard the lymphocyte layer and collect the neutrophil layer, adding 1 ml of cells to 1 ml of NIB in each new 8-ml round-bottom polystyrene tube. Each tube of layered blood should provide 0.5–1 ml volume for the neutrophil layer.
4. Add 4 ml of NIB to each tube, bringing the volume to 6 ml, and then centrifuge the tubes at 300 × *g* for 7 min.
5. Discard the supernatant. At this point, cells from multiple tubes can be combined into a single tube. Gently resuspend the cell pellet and wash once with 6 ml of NIB. At this point, cells from multiple tubes can be combined into a single tube. Centrifuge the tubes at 300 × *g* for 7 min.
6. Discard the supernatant. Gently resuspend the cell pellet. Lyse contaminating erythrocytes by adding 4.5 ml of HELB and gently inverting the tubes several times over 30 s. Immediately add 1.35 ml of normalization buffer and gently invert the tubes several times. Centrifuge the tubes at 300 × *g* for 7 min.
7. Discard the supernatant. Gently resuspend the cell pellet and wash once with 6 ml of NIB. Count cells using a hemocytometer. You should be able to purify approximately $1–2 \times 10^6$ cells/ml whole blood. Centrifuge the tubes at 300 × *g* for 7 min.

3.3.2. Neutrophil Treatment and Labeling

1. Discard the supernatant. Gently resuspend the cell pellet in NLB at a concentration of 10×10^6 cells/ml.
2. In 1.5-ml microfuge tubes, treat 2×10^6 neutrophils per experimental group with integrin activation agonists/antagonists. Place the tubes on ice.
3. Label the cells with a FITC-conjugated mAb directed against the extracellular Mac-1 integrin head domain for 1 h. Retain a control group of cells that are unlabeled for use in determining

base level signal and signal compensation during flow cytometry acquisition.

4. Wash the cells three times with NLB containing the appropriate experimental agonists/antagonists used during the treatment step. During the washes, centrifuge the tubes at 300 × *g* for 5 min at 4°C.

3.3.3. Flow Cytometry Data Acquisition

1. For each experimental group, resuspend cells (2×10^6) in 2 ml of NLB and aliquot into four flow cytometry tubes in 0.5-ml volume, keeping the tubes on ice.
2. For each set of four tubes in an experimental group, add 0, 0.25, 0.5, or 1.0 μl of the 200 mM ORB stock. This gives final ORB concentrations of 0, 100, 200, and 400 nM. Mix well and incubate the tubes for 20 min on ice before acquisition on flow cytometer (see Notes 5 and 6). Acquire data for at least 10,000 cells.

3.3.4. Data Analysis

1. The relationship between the donor FITC fluorescence and the concentration of ORB in the membrane depends on the distance between the donor and the membrane surface, *L*, to the inverse fourth power (10):

$$F_{\mathrm{DA}} = F_{\mathrm{D}} / (1 + S[\mathrm{ORB}]),$$

$$S = \pi R_0^{\,6} / 2L^4,$$

   where F_{DA} is the donor (FITC) fluorescence in the presence of acceptor (rhodamine), F_{D} is the donor fluorescence in the absence of acceptor, (ORB) is the concentration of acceptor in the cell membrane (assumed to be proportional to the ORB fluorescence), and *S* is the slope factor, which depends on the Forster radius (R_0) and distance between donor and acceptor (*L*). The relative change in spatial separation of the donor and acceptor between two experimental conditions, called the distance ratio, can be calculated as follows:

$$L_2 / L_1 = (S_1 / S_2)^{1/4}.$$

2. Using flow cytometry data analysis software, such as FlowJo (Tree Star), gate on the neutrophil population. Obtain mean fluorescence intensity (MFI) values for both the donor (FITC) and the acceptor (rhodamine) for all samples.
3. For each experimental group, plot the data: acceptor MFI (*x*-axis) and donor MFI (*y*-axis). Use the curve-fitting tool in your graphing software to obtain a curve fit of data points in each experimental group according to the equation: $y = b/(1 + ax)$. In the derived equation for the curve, *b* should have

a value similar to your initial donor MFI (F_D), and *a* represents your slope factor, *S*, and should be on the order of 0.001–0.01.

4. Notes

1. Leu-221 in pECFP-N1 and pEYFP-N1 was replaced with Lys to produce the monomeric mutant (11).
2. For all CFP/YFP fluorescence image acquisitions, a 1/8 neutral density filter and 0.5-s exposure time is used.
3. For the sensitized emission FRET measurement, cross talk between CFP and YFP that occurs with the CFP/YFP dual-band filter set (CFP_{Ex}/YFP_{Em}) needs to be calculated and accounted for. This is done using GD25 cells transfected with α_4-mCFP/wt β_1 or α_4-mYFP/wt β_1.
4. We have used FITC-conjugated anti-CD11b clone ICRF44 mAb (Ancell) to observe Mac-1 extension on neutrophils. This is a mAb directed against the ligand-binding I domain in the headpiece of the α-subunit of Mac-1.
5. To complete the experiment in a reasonable amount of time and thus avoid variance in the signal from degradation of the FITC-conjugated mAbs, we found it is best to stagger the addition of ORB to each tube by 3 min. This allows 3 min for data acquisition to be performed for each tube on the flow cytometer at the 20-min time point of incubation with ORB.
6. In our experience, we have found that a negligible amount of compensation is needed for FL-2, the acceptor channel, whereas a significant amount of compensation is required for FL-1, the donor channel. The user should determine the appropriate compensation values using single-labeled cells.

References

1. Hynes, R. O. (2002) Integrins: bidirectional, allosteric signaling machines. *Cell* **110**, 673–687.
2. Evans, R., Patzak, I., Svensson, L., De Filippo, K., Jones, K., McDowall, A., and Hogg, N. (2009) Integrins in immunity. *J. Cell. Sci.* **122**, 215–225.
3. Luo, B. H., Carman, C. V., and Springer, T. A. (2007) Structural basis of integrin regulation and signaling. *Annu. Rev. Immunol.* 25:619–647.
4. Takagi, J., Petre, B. M., Walz, T., and Springer, T. A. (2002) Global conformational rearrangements in integrin extracellular domains in outside-in and inside-out signaling. *Cell* **110**, 599–511.
5. Wennerberg, K., Lohikangas, L., Gullberg, D., Pfaff, M., Johansson, S., and Fassler, R. (1996) Beta 1 integrin-dependent and -independent polymerization of fibronectin. *J. Cell Biol.* **132**, 227–238.
6. Kim, M., Carman, C. V., and Springer, T. A. (2003) Bidirectional transmembrane signaling by cytoplasmic domain separation in integrins. *Science* **301**, 1720–1725.

7. Lefort, C.T., et al. 2009. *J. Immunol.* **183**, 6460–6468.
8. Miyawaki, A., and Tsien, R. Y. (2000) Monitoring protein conformations and interactions by fluorescence resonance energy transfer between mutants of green fluorescent protein. *Methods Enzymol.* **327,** 472–500.
9. Hyun, Y. M., Chung, H. L., McGrath, J. L., Waugh, R. E., and Kim, M. 2009. Activated integrin VLA-4 localizes to the lamellipodia and mediates t cell migration on VCAM-1. *J. Immunol.* **183**, 359–369.
10. Holowka, D., and Baird, B. (1983) Structural studies on the membrane-bound immunoglobulinE-receptorcomplex.1.Characterization of large plasma membrane vesicles from rat basophilic leukemia cells and insertion of amphipathic fluorescent probes. *Biochemistry* **22**, 3466–3474.
11. Zacharias, D. A., Violin, J. D., Newton, A. C., and Tsien, R. Y. (2002) Partitioning of lipid-modified monomeric GFPs into membrane microdomains of live cells. *Science* **296**, 913–916.

Chapter 15

High-Resolution Fluorescence Microscopy to Study Transendothelial Migration

Christopher V. Carman

Abstract

Immune system functions rely heavily on the ability of immune cells (i.e., blood leukocyte) to traffic throughout the body as they conduct immune surveillance and respond to pathogens. A monolayer of vascular endothelial cells (i.e., the "endothelium") provides a critical, selectively permeable barrier between two principal compartments of the body: the blood circulation and the tissue. Thus, knowledge of the basic mechanisms by which leukocytes migrate across the endothelium (i.e., undergo "transendothelial migration"; TEM) is critical for understanding immune system function. Cultured endothelial cell monolayers, used in combination with isolated blood leukocytes, provide a basis for highly useful in vitro models for study of TEM. When used in conjunction with high spatial and temporal resolution imaging approaches, such models have begun to reveal complex and dynamic cell behaviors in leukocytes and endothelial cells that ultimately determine TEM efficiency. In this chapter, we provide protocols for setting up a basic in vitro TEM system and for conducting high-resolution dynamic live-cell and three-dimensional fixed-cell imaging of TEM.

Key words: Fluorescence, Microscopy, Leukocyte, Endothelium, Diapedesis, Transmigration, GFP, Imaging, Antibody, Lymphocyte

1. Introduction

In order to satisfy their roles in immune surveillance and pathogen elimination, blood leukocytes (e.g., lymphocytes, monocytes, dendritic cells, and neutrophils) must continuously traffic throughout the body (1). Such trafficking can be broken into two major phases: (1) movement within the vascular and lymphatic circulation and (2) migration within tissues. The vascular and lymphatic circulatory systems are lined by an endothelial cell monolayer (the "endothelium") that grows on an abluminal layer

Motomu Shimaoka (ed.), *Integrin and Cell Adhesion Molecules: Methods and Protocols*,
Methods in Molecular Biology, vol. 757, DOI 10.1007/978-1-61779-166-6_15, © Springer Science+Business Media, LLC 2011

of extracellular matrix (the basement membrane) and forms organized intercellular adherens, tight and gap junctions (2–4). In this way, the endothelium functions as a (selectively permeable) barrier between the circulation and the tissues. Thus, the movement of leukocytes into (intravasation) or out of (extravasation) the circulation requires the explicit crossing of the endothelial barrier (i.e., "diapedesis" or "transendothelial migration"; TEM). Thus, TEM is a critical and rate-limiting component of leukocyte trafficking and, in turn, of both normal immune system function and immune-mediated/inflammatory pathology. Understanding the regulatory mechanism of this process is, therefore, of great biomedical significance (1, 5).

TEM is important in diverse settings in vivo with a wide range of immediate purposes (1). For example, all leukocytes begin their life cycle by intravasation from the bone marrow, T and B lymphocytes constitutively enter and exit various lymphoid organs (including the thymus, lymph nodes, Peyer's patches, spleen, and tonsils) during maturation and immune surveillance processes, and diverse leukocytes (e.g., granulocytes, monocytes, memory/effector lymphocyte subsets, and NK cells) traffic into inflamed peripheral tissues in response to infection or injury.

Whereas TEM during intravasation has just begun to be studied (6), TEM during extravasation has been extensively characterized, revealing a discrete "five-step" cascade (Fig. 1). Extravasation begins with the accumulation of circulating leukocytes

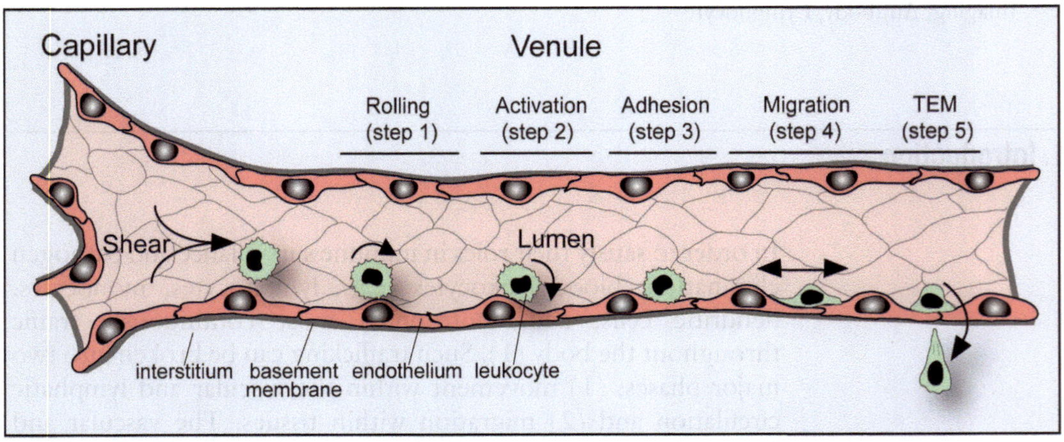

Fig. 1. The "Five-Step" Extravasation Cascade. Extravasation of leukocytes (*green*) across vascular structures (exemplified here as a postcapillary venule; *pink*) is a multistep process. In Step 1, leukocytes undergo transient rolling-type interactions with the endothelium that are mediated predominantly by selectins. This facilitates chemokine-dependent activation (Step 2) and firm adhesion or "arrest" (Step 3), which is mediated by the binding of leukocyte integrins (e.g., LFA-1, Mac1, and VLA4) to endothelial cell-adhesion molecules (e.g., ICAM-1, ICAM-2, and VCAM-1). Subsequently, leukocytes spread, polarize, and migrate laterally over the surface of the endothelium, probing for a site to penetrate the endothelium (Step 4). Finally, leukocytes cross the endothelial barrier (i.e., undergo transendothelial migration (TEM) or "diapedesis"; Step 5), either para- or trans-cellularly, and enter the interstitium.

on the luminal surface of the endothelium through a "classic" three-step adhesion and activation cascade (7–9). Initially, leukocytes undergo transient rolling interactions mediated by the selectin family of adhesion molecules (Step 1), which facilitates sensing of, and signaling responses to, chemokines presented on the surface of the endothelium (Step 2). This in turn triggers high-affinity interaction of lymphocyte integrin receptors (e.g., LFA-1, Mac1, and VLA-4) with their endothelial ligands (e.g., ICAM-1, ICAM-2, and VCAM-1), resulting in firm adhesion of leukocytes (Step 3) (10, 11). Subsequently, leukocytes undergo cytoskeleton-mediated spreading, polarization, and integrin-mediated lateral migration on the luminal surface of the endothelium (Step 4). This process confers leukocytes with the ability to search out sites permissive for endothelial barrier penetration (12, 13). Finally, leukocytes formally breach and transmigrate across the endothelium (Step 5), a process referred to specifically as "transendothelial migration" or "diapedesis" (though these terms are also often employed more loosely to refer to the entire five-step cascade).

Until recently, only one basic pathway, the "para-cellular" route, for diapedesis was widely recognized. Para-cellular diapedesis involves cooperative efforts on the part of both the leukocyte and the endothelium to disassemble the inter-endothelial junctions locally in order to form a para-cellular gap that will allow leukocyte transmigration (5, 14–17). In fact, however, their is a large body of literature (including some of the very first studies to investigate the route of diapedesis directly (18–20) and nearly 50 subsequent studies recently reviewed in detail (21)) that demonstrates the coexistence in vivo of the para-cellular route along with a second pathway termed that trans-cellular route, whereby leukocytes pass directly through individual endothelial cells via the formation of a trans-cellular pore. As a consequence of recent studies characterizing trans-cellular diapedesis in vitro for the first time, the mechanisms for this pathway have begun to be elucidated and its physiologic relevance more broadly appreciated (22). The relative roles, in distinct in vivo settings, of the para- and trans-cellular pathways remain important open questions.

Classic adhesion and TEM (e.g., transwell/Boyden chamber) assays coupled to various function perturbation approaches have been instrumental in identifying adhesion, chemoattractant, and signaling molecules that are important for TEM. However, a complete understanding of the cell biological mechanisms and orchestrated dynamics of TEM requires imaging approaches.

Dynamic live-cell imaging has been particularly instrumental in elucidating the dynamics of leukocyte lateral migration and initiation of diapedesis events. For example, recent studies of using fluorescent membrane markers in endothelial cells have uncovered a novel form for invadosome-like protrusive structures (23–25) formed dynamically by leukocytes that serve a role in migratory pathfinding by effectively probing the endothelial surface mechanically (26–29).

High-resolution three-dimensional (3D) confocal imaging of fixed samples has been critical for defining adhesion receptor distribution within, and detailed topology of, leukocyte–endothelial interactions. For example, this approach was essential for the discovery and characterization of novel ICAM-1-, VCAM-1-, and actin-enriched endothelial structures termed "transmigratory cups" or "docking structures" (30–33). Studies in diverse in vitro settings that have included a range of leukocyte types, endothelial models (including HUVECs, as well as human dermal, lung, and brain microvascular endothelium and lymphatic endothelium), and activation stimuli have demonstrated that transmigratory cups play critical roles in adhesion and directional guidance during diapedesis (34–43). A key feature of these structures is their 3D architecture formed by vertical microvilli-like projections that surround adherent and transmigrating leukocytes (31–33).

It remains clear that many additional mechanistically important features of leukocyte–endothelial interactions remain to be elucidated in order to refine our understanding of TEM. The following protocols describe the details for setting up one specific in vitro model system (designed to recapitulate effector lymphocyte homing to sites of inflammation in the microvasculature) and using this system to conduct both dynamic wide-field and fixed-sample confocal imaging of TEM. As pointed out in the extensive notes provided under Subheading 4, these basic approaches can be readily modified to incorporate diverse endothelium, leukocytes, and migratory stimuli, effectively allowing modeling of wide-ranging leukocyte–endothelial interactions.

2. Materials

2.1. Preparation of Endothelium

1. Phosphate-buffered saline (PBS).
2. Hanks balanced salt solution (HBSS).
3. EGM-2 MV, endothelial cell culture medium: Prepare complete EGM-2 MV media (Invitrogen) by thawing and adding all components of the "Bullet kit" (see Note 1).
4. Fibronectin (FN) stock solution (50×): Prepare a 1 mg/ml stock solution by dissolving 5 mg of sterile lyophilized FN (Invitrogen) in 5 ml sterile PBS (adjusted to pH 6.5 with ~50 ml 1 M sodium citrate, pH 3.0). Aliquot 100 ml to 15-ml conical tubes and store at −80°C.
5. Trypsin solution.

6. Trypsin inhibitor.
7. Falcon (Fisher, Hampton, NH) T25, T75, 6-well and 24-well cell culture plates (see Note 2).
8. 12-mm No. 1 thickness circular coverslips.
9. Bioptechs Delta-T cell-culture dishes (Bioptechs, Butler, PA).
10. Amaxa HMVEC-L Nucleofector kit and electroporator (Invitrogen).

2.2. Preparation of Effector Lymphocytes

1. Citrate solution: Dissolve 25 g sodium citrate and 8 g citric acid in 500 ml PBS. Sterile filter (0.2 μm) and store at room temperature for not more than 6 months.
2. Dextran solution: Dissolve 20 g dextran-500 into 500 ml MilliQ-H_2O (6% w/v). Sterile filter (0.2 μm) and store at room temperature for not more than 6 months.
3. Histopaque 1077.
4. Gelatin solution: Prepare fresh by dissolving 1 g gelatin into 50 ml PBS. Incubate at 37°C for ~30 min with occasional vortexing. Sterile filter (0.2 μm).
5. PBS with Ca/Mg: To 50 ml of PBS, add 50 ml of 1 M $CaCl_2$ and 50 ml of 1 M $MgCl_2$. Use a 60-ml syringe with 0.2-μm filter to sterilize.
6. RPMI base media: RPMI, 10% FCS, 1% glutamine, and 1% penicillin/streptomycin (see Note 1).
7. RPMI-phytohemagglutin (PHA) media: RPMI base media with 1 μg/ml PHA (see Note 1).
8. RPMI-interleukin-2 (IL-2) media: RPMI base media with 10 ng/ml IL-2 (R & D Systems, Minneapolis, MN) (see Note 1).

2.3. Live-Sample Dynamic Imaging of TEM

1. Live-cell imaging buffer (Buffer-A): Prepare fresh phenol red-free HBSS (see Note 3), supplemented with 20 mM HEPES, pH 7.4, and 0.5% (v/v) human serum albumin (HSA). Pre-warm to 37°C.
2. Fixative solution (3.7% formaldehyde): Combine 2.5 ml of 37% formaldehyde solution with 22.5 ml PBS (see Notes 4 and 5).
3. Tumor necrosis factor-α (TNF-α) (Invitrogen): Prepare a 100 μg/ml stock solution by dissolving 10 μg of TNF-α in 100 μl sterile filtered 5 mM Tris–HCl, pH 8.0. Aliquot single use volumes (~2–4 μl) in sterile PCR tubes and store at −80°C.
4. Disposable 5-ml transfer pipettes.
5. Delta-T live-cell imaging culture dishes (Bioptechs).

2.4. Immunofluorescence Staining and Confocal Imaging of TEM

1. Fixative solution (3.7% formaldehyde): Combine 2.5 ml of 37% formaldehyde solution with 22.5 ml PBS (see Notes 4 and 5).
2. Block solution (5% w/v NFDM): Dissolve 2.5 g NFDM in 50 ml PBS (see Notes 4 and 5).
3. Permeabilization solution: Prepare a 10% v/v stock solution of Triton X-100 in water to be stored at 4°C (this can be kept for ~1 year). Dilute 50 μl of 10% Tx-100 into 10 ml of Block solution (0.05% Tx-100 final) (see Note 6).
4. Primary and secondary antibodies (see Notes 7–9): See Subheading 3.4.2 for details on preparing Antibody Stain Solutions.

3. Methods

The methods outlined below describe (1) the preparation of primary human microvascular endothelial cell monolayers; (2) isolation and culture of primary human effector lymphocytes; (3) dynamic live-cell imaging of lymphocyte TEM; and (4) immunofluorescence staining and high-resolution, fixed-cell confocal imaging of lymphocyte TEM. The experimental system used throughout this protocol is designed to model the process of effector lymphocyte homing to microvascular sites of peripheral tissue inflammation (see Note 10).

3.1. Preparation of Endothelium

The following protocols are designed as a guideline for preparation of confluent primary endothelial monolayers to serve as basic in vitro models of the vasculature. The procedures described herein focus on primary human lung (HLMVECs) and dermal (HDMVECs) microvascular endothelial cells (see Note 11), with emphasis on generating optimal endothelial cell health and recapitulation of physiologic vascular phenotype and function.

3.1.1. Sterilizing Coverslips

For some of the protocols described below, endothelial cells will be grown on glass-coated coverslips placed in cell culture dishes. In a cell culture hood, use forceps to place a single 12-mm circular coverslip into each of the desired number of wells of a 24-well cell culture plate. With the cover off, place each plate in the back of the hood and sterilize by leaving the UV light on for 1–2 h (see Notes 12 and 13).

3.1.2. Fibronectin Coating of Plates

1. Thaw a fresh 100 μl aliquot of fibronectin (FN) stock and dilute to 1× (20 μg/ml) with 5 ml PBS (see Note 14).
2. Add enough 1× FN to culture flasks or dishes to cover surfaces completely (see Note 15). Use the following recommendations:

Culture vessel	Minimum volume (ml)
T75 flask	4
100-mm dish	3
T25 flask	2
6-well plate	1
60-mm dish	1
Bioptechs Delta-T4 dish	0
24-well plate w/coverglass	0.3–0.5

3. Store the dish in culture hood (without UV lamp on) for 1–2 h at room temperature or overnight at 4°C.
4. Aspirate FN immediately before adding medium. It is not necessary to wash plates to remove residual FN.

3.1.3. Starting HLMVECs/HDMVECs

1. Coat a T25 flask with FN, as described in Subheading 3.1.2.
2. Remove FN and add 5 ml complete medium.
3. Pre-incubate in 37°C cell culture incubator.
4. Thaw a vial of frozen HLMVECs or HDMVECs in a 37°C water bath with occasional gentle agitation for ~2–3 min. Proceed until a barely visible piece of ice remains. Immediately transfer cells to T25 flask containing pre-warmed media. Gently swirl and place in incubator at 37°C.
5. Change the media after ~4–6 h. Continue to change media approximately every 48 h (or when media becomes slightly yellow) until the cells reach ~90–95% confluency.

3.1.4. General Splitting and Expansion of Endothelial Cells

1. Grow cells to ~90–95% confluency. This may take 2–5 days. If media becomes yellow, it should be replaced with fresh, pre-warmed complete media as often as necessary until confluency is reached.
2. For splitting, remove media and rinse with HBSS.
3. Remove HBSS and replace with minimum volume of fresh 1× trypsin (0.5 ml for T25 or 1.5 ml for T75). Gently swirl to cover all surfaces with trypsin. Incubate at room temperature for ~3–6 min.
4. Monitor the detachment of the cells from the plate using a low-power light microscope. An occasional gentle tap of the side of the flask can be applied to help determine the extent of detachment (see Note 16).
5. When majority of cells appear rounded or detached, add 1 volume (i.e., equal to the trypsin volume added) of 1× trypsin inhibitor and gently triturate trypsin/trypsin inhibitor solution over the surfaces of the flask to detach all cells.

6. Count endothelial cells with a hemocytometer: At ~90% confluency, T25 flask will yield ~1–1.5 and a T75 flask ~3–4.5 million cells, respectively.
7. Adjust concentration to 0.5 million cells/ml by adding pre-warmed complete EGM-2 MV media.
8. Transfer aliquots of cells to the appropriate FN-coated dishes. For general maintenance/expansion of endothelial cultures, cells should be seeded at ~0.5 million cells (1 ml) for T25 and ~1.5 (3 ml) million cells for T75, and grown in 5 or 10 ml total EGM-2 MV media, respectively (see Note 17).
9. Gently swirl and place the plates in the incubator.
10. Change the media within 6–12 h of plating. Media should be changed approximately every 48 h or as needed, according to media color. Repeat this process approximately every 2–4 days until the cells reach 90–95% confluency (see Note 18).

3.1.5. Preparative Plating of Endothelial Cells for Imaging Experiments

1. FN coat the requisite number of Delta-T dishes (for live-cell imaging as described in Subheading 3.3) or 24 wells containing coverglass (for fixed-cell confocal imaging, as described in Subheading 3.4) as described above (Subheading 3.1.2).
2. Starting with endothelial maintenance flasks that are ~90–95% confluent (and have been passaged less than 6 times) trypsinize, count, and adjust the endothelial cell density to 0.5 million/ml, as in Subheading 3.1.4.
3. To 24 wells, add ~50–100 μl of 0.5 million cells/ml (i.e., 25,000–50,000) to 0.5 ml media. To Bioptechs Delta-T dishes, add ~200–300 μl of 0.5 million cell/ml to 1 ml media.
4. Place in cell culture incubator (see Note 19).
5. Change media within 4–6 h and again at 12–16 h.

3.1.6. Endothelial Cell Transfection

For many experiments (especially live-cell fluorescence imaging), expression of fluorescent protein-tagged probes is extremely useful, if not required. Endothelial cells are highly resistant to transient transfection by cationic lipid-based approaches. Amaxa Nucleofection, using an Amaxa Electroporator and HMVEC-L Nucleofector kit, however, allows for highly efficient (typically, 30–60%) transfection of HLMVECs and HDMVECs with many fluorescent probes (see Note 20). The following is protocol modified from those provided by the manufacturer (Invitrogen/Lonza).

1. Prepare T25 or T75 flasks (as needed) of HLMVECs or HDMVECs to a final density of 90–95% confluency (see Notes 21–23).
2. Prepare Delta-T plates and/or 24-well dishes containing coverglass by FN coating as described above (Subheading 3.1.2).

3. Add 1 or 0.5 ml of complete EGM-2 MV culture media to Delta-T or 24 wells, respectively, and pre-incubate/equilibrate plates in a humidified 37°C/5% CO_2 incubator.
4. Harvest and count endothelial cells, as described in Subheading 3.1.4.
5. Centrifuge the required number of cells (0.5 millions cells per sample) at 200×*g* for 5 min at room temperature.
6. Resuspend the cell pellet carefully in 100 μl room temperature Nucleofector solution per sample.
7. Combine 100 μl of cell suspension with 1–5 μg of DNA. Transfer the cell/DNA suspension into certified cuvette; the sample must cover the bottom of the cuvette without air bubbles.
8. Close the cuvette with the cap. Insert the cuvette with cell/DNA suspension into the Nucleofector cuvette holder of the Amaxa electroporator and apply electroporation program S-005.
9. Take the cuvette out of the holder once the program is finished.
10. Add ~500 μl of the pre-equilibrated culture media to the cuvette and gently remove the cell suspension from the cuvette using the plastic transfer pipettes provided in the Nucleofection kit.
11. For experiments using Delta-T dishes, partition the cell suspension from one reaction equally between two dishes containing pre-warmed media. For experiments using 24 wells, partition one reaction equally into four separate wells (see Note 24).
12. Incubate the cells in a humidified 37°C/5% CO_2 incubator and change media 4–6 h, and again at 12–16 h post-transfection.

3.2. Preparation of Effector Lymphocytes

The following protocols are designed as a guideline for isolation and culture of primary human Th1 effector-like lymphocytes (see Notes 25 and 26) for investigation of the cell biological mechanisms of TEM, and are optimized for maximal retention of viability and physiologic characteristics.

3.2.1. Acquiring Blood Sample

1. Place a fresh sheet of bench paper on the benchtop. Gather an elastic tourniquet, 21-gauge butterfly needle/tubing, Band-Aids, and ethanol wipes on bench paper.
2. Prepare blood donor form and complete informed consent process (see Note 27).
3. Sterilely transfer 10 ml of dextran and 7 ml of citrate solution to a sterile 10-cm tissue culture dish in the hood. Use the dish lid to prop the dish at a slight angle.

4. Uncap a fresh sterile 60-ml syringe and draw up the dextran/citrate mixture. Invert the syringe, expel excess air, and recap the syringe until blood drawing.
5. A trained phlebotomist should attach a butterfly needle/tubing to the dextran/citrate-containing syringe and commence blood draw via venipuncture (see Note 28). Continue drawing until the syringe is filled.
6. Remove the needle from the subject and *using extreme caution*, replace cap on the needle. Then gently invert the syringe 3–4 times to mix the blood with the dextran/citrate.
7. Affix the inverted syringe to the surface of the hood with tape and let stand for 30 min (Fig. 2). The red blood cells (RBCs) will sediment into a dark red lower phase, and the serum,

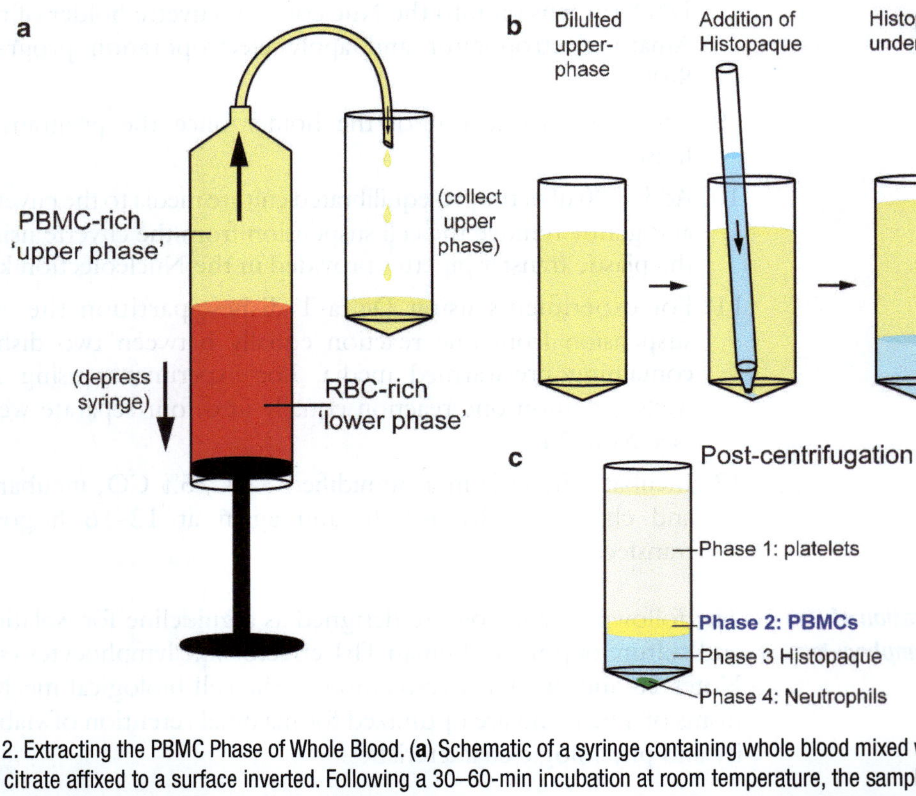

Fig. 2. Extracting the PBMC Phase of Whole Blood. (**a**) Schematic of a syringe containing whole blood mixed with dextran and citrate affixed to a surface inverted. Following a 30–60-min incubation at room temperature, the sample will spontaneously separate into an upper PBMC-rich phase and a lower RBC-rich phase. Collection of the upper phase is accomplished by cutting the tubing connecting the syringe to the butterfly needle and placing the free end of the tube into a 50-ml collection tube, followed by slow depression of the syringe. (**b**) After the "upper phase" is diluted with PBS, Histopaque is underlaid by slowly pipetting into the bottom of the tube (see *center panel*). After addition, a sharp interface between the "upper phase" and the Histopaque should be evident (see *right panel*). (**c**) Following centrifugation, the sample will have separated into four phases: (1) a large (25–30 ml) milky upper phase, composed largely of blood platelets; (2) a dense white cloud of PBMCs termed the "buffy coat" concentrated at the original blood sample–Histopaque interface; (3) an underlying ~7-ml phase of relatively clear Histopaque; and (4) a pellet containing neutrophils and residual RBCs. Collection of PBMC is accomplished by careful aspiration of phase 1 with a pipette, followed by collection of phase 2 without perturbation of phase 4.

white blood cells (lymphocytes, monocytes, and neutrophils), and platelets will form a pale yellow upper phase.

3.2.2. Isolation of Peripheral Blood Mononuclear Cells

1. Sterilely transfer 20 ml of Histopaque-1077 to a 50-ml test tube, let stand at room temperature, and set aside.
2. In the hood, drench the tubing connecting the butterfly needle and syringe, as well as a pair of scissors (open and close scissors during this process to ensure all cutting surfaces are sterilized).
3. Carefully cut the tubing about 3 in. from the tip of the syringe and place the newly cut end of the tubing in a sterile 50-ml conical test tube (Fig. 2a). Transfer the free needle to an appropriate "sharps disposal" waste container immediately.
4. Collect the yellow upper phase of the blood sample by slowly depressing the syringe causing this phase to be expelled into the 50-ml conical test tube (Fig. 2a). Proceed at a modest pace; it should take ~1–2 min to collect the majority of the upper phase (30–35 ml). Stop collection when ~1–2 ml of this phase remains.
5. Briefly set aside the collected upper phase.
6. Detach the syringe from the surface of the hood and expel the RBCs into a 200-ml beaker containing 50–100 ml bleach. Let stand for 5 min. Dispose in sink. Rinse excessively with running water.
7. Split the volume of the upper blood phase equally into two sterile 50-ml conical tubes (usually ~15–17 ml each).
8. Add sterile PBS to each tube to bring the volume to ~38–40 ml. Cap each and invert several times.
9. Using a 10-ml pipette gently underlay 10 ml of room temperature Histopaque-1077 in each of the two test tubes (Fig. 2b). A sharp interface should form between the clear Histopaque on the bottom and the yellow upper phase of the blood on top.
10. Transfer to a benchtop swinging bucket centrifuge. Spin at $600 \times g$ for 30 min at room temperature with no brake applied during deceleration.
11. Collect the tubes and place in a cell culture hood. The samples should now contain four phases: (1) a large (25–30 ml) milky upper phase, composed largely of blood platelets; (2) a dense white cloud of peripheral blood mononuclear cells (PBMCs) termed the "buffy coat" concentrated at the original blood sample–Histopaque interface; (3) an underlying ~7-ml phase of relatively clear Histopaque; and (4) a pellet containing neutrophils and residual RBCs (Fig. 2c).

12. Sterilely remove and discard the upper phase into a waste beaker containing bleach.
13. Next, without touching the underlying pellet, remove the PBMCs/buffy coats (in ~10 ml volume each for each tube) and transfer each one to a separate sterile 50-ml tube.
14. Add bleach to the remaining material and discard.
15. To the extracted buffy coat, add ~40 ml of sterile PBS. Cap and invert several times.
16. Centrifuge at 280 × *g* in a swinging bucket table-top centrifuge for 10 min at room temperature with brake on low.
17. Discard the platelet-rich supernatants and resuspend the PBMCs-rich pellets in 5 ml sterile PBS via trituration.
18. Add 45 ml PBS to each tube and centrifuge at 180 × *g* for 10 min with brake on low.
19. Discard the supernatant and resuspend PBMC pellets from both tubes in a total of 12 ml of PBS containing 1 mM calcium and 1 mM magnesium.

3.2.3. Preparation of Effector Lymphocytes

1. To each of three T75 flasks, aliquot ~10–15 ml of sterile gelatin (Subheading 2.2, step 5), swirl to coat bottom surface, and incubate at 37°C for 30 min.
2. Remove gelatin and rinse each flask once with 10 ml sterile PBS.
3. Aliquot 4 ml of PBMCs (Subheading 3.2.2) to each of the three flasks. Incubate at 37°C for 30–60 min to allow adhesion of monocytes to the gelatin-coated surface.
4. Pre-warm 15 ml of RPMI-PHA media in a separate T75 flask.
5. Upright the gelatin-coated, PBMC-containing T75 flasks and gently remove the monocyte-depleted lymphocytes contained in the liquid at the bottom.
6. Pool the lymphocyte from all three flasks in a 15-ml conical test tube and centrifuge at 230 × *g* for 3 min.
7. Resuspend pellet in 15 ml of pre-warmed RPMI-PHA media, transfer to a T75 flask, and incubate at 37°C overnight.
8. After 24 h of growth in RPMI-PHA, add 15 ml of additional pre-warmed RPMI-PHA media and transfer lymphocytes to T150 flask.
9. After 72 h of growth in PHA, centrifuge lymphocytes (230 × *g*, 3 min) in two 15-ml test tubes.
10. Remove as much supernatant as possible without perturbing the cell pellet.

11. Resuspend each pellet in 15 ml of pre-warmed RPMI-IL-2 media (see Note 29), transfer to T75 flasks, and place in cell culture incubator.
12. Expand cells by adding 15 ml of fresh RPMI-IL-2 media within ~48 h.
13. Continue to expand/split cells in RPMI-IL-2 every 24–48 h as needed (based on media color) thereafter (see Note 30). The resulting effector lymphocytes can be used for TEM experiments after 3 days in IL-2 and can be continuously cultured and used for further experiments for an additional ~5 days thereafter.

3.3. Live-Sample Dynamic Imaging of TEM

The following protocols are designed as a guideline for setting up and performing live-cell imaging studies of leukocyte diapedesis that will provide a high degree of topological detail. Specifically, these methods are designed for high-resolution, live-cell dynamic imaging using an inverted wide-field light microscope. In this case, the protocol is written for an AxiovertS200 microscope outfitted with a 63×, 1.4 NA PlanApo objective coupled to a CCD camera, automated fluorescence filter cube turret, and Axiovision time-lapse image acquisition software (see Note 31). Selected fluorescent probes (to be introduced into endothelium) will include actin-GFP and membrane-RFP (mRFP), which will be used in concert with differential interference contrast (DIC) bright-field imaging (see Notes 32–34).

3.3.1. Preparing Endothelium

1. Co-transfect primary HLMVECs with actin-GFP and membrane-RFP via Amaxa Nucleofection and plate onto Delta-T live-cell imaging culture plates, as described in Subheading 3.1.6 (see Notes 32–34).
2. Change media within 4 h and again between 12 and 16 h post-transfection (see Note 35). Imaging experiments should be performed 48–72 h post-transfection.
3. At 4–12 h prior to the imaging, replace media with fresh media containing 50 ng/ml TNF-α-transfected endothelium (see Notes 36 and 37). Once TNF-α is added, endothelial cell proliferation largely ceases and, as endothelial cells become activated, stress fibers and contractility develop that appose endothelial monolayer integrity. One must be sure that at the time of initiating imaging experiments that the monolayers are intact with well-formed intercellular junctions free of gaps (see Note 38).

3.3.2. Basic Setup

1. To initiate an imaging experiment, prepare Buffer A and fixative solution (Subheading 2.3) and turn on/set up the imaging system.

2. Thoroughly clean the imaging objective with ethanol and objective lens paper.
3. Only after all buffers are prepared and imaging equipment/acquisition software setup are confirmed, take a sample of cultured lymphocytes and determine the density by counting with a hemocytometer (see Note 39).
4. Remove a Delta-T plate containing transfected and activated HLMVECs from cell culture incubator. Using a disposable transfer pipette, rapidly remove media and rinse once by adding ~1 ml of pre-warmed Buffer A. Aspirate Buffer A and then add 0.5 ml of Buffer A.
5. Prepare the underside of the Delta-T dish by successive cleaning with water and then with ethanol-soaked Kimwipes (Fisher).
6. Add fresh objective oil and mount Delta-T dish on the heating stage adaptor and immediately turn on to equilibrate to 37°C (will take ~2–3 min) (Fig. 3a).

3.3.3. Live-Cell Imaging

1. Bring the objective into contact and use bright-field imaging to find the focal plane.
2. Switch to fluorescence imaging and identify a field of interest using the ocular lenses (see Note 40).
3. Identify fields in which brightly fluorescent-positive transfected endothelial cells are present that appear healthy with well-formed intercellular junctions.
4. Switch imaging to CCD camera and adjust acquisition parameters (e.g., exposure time, detector gain, and binning). Be sure that mean fluorescence signal intensity in each channel falls between 25 and 75% of the dynamic range of the detector (see Note 41).
5. Remove a volume of lymphocytes from culture equivalent to 0.5–1 million cells. Transfer to a 15-ml conical tube and centrifuge at 200 × *g* for 3 min. Aspirate media and gently but thoroughly resuspend lymphocytes in 20–40 ml of Buffer A.
6. Begin live-cell imaging of endothelial cells by successive capture of sets of DIC, green fluorescence, and red fluorescence images with an acquisition interval of ~5–30 s per set (see Note 42).
7. Acquire 30–60 s of baseline imaging of endothelium.
8. During acquisition, apply ~5 ml of concentrated lymphocytes to the center of Delta-T imaging field by inserting the tip of a small volume (P-5 or P-20) pipette into the media close to the center of the objective and ejecting slowly.
9. Image for ~30 s to allow for cells to settle into the imaging field. For an average imaging system using a 63× objective, ~10–20

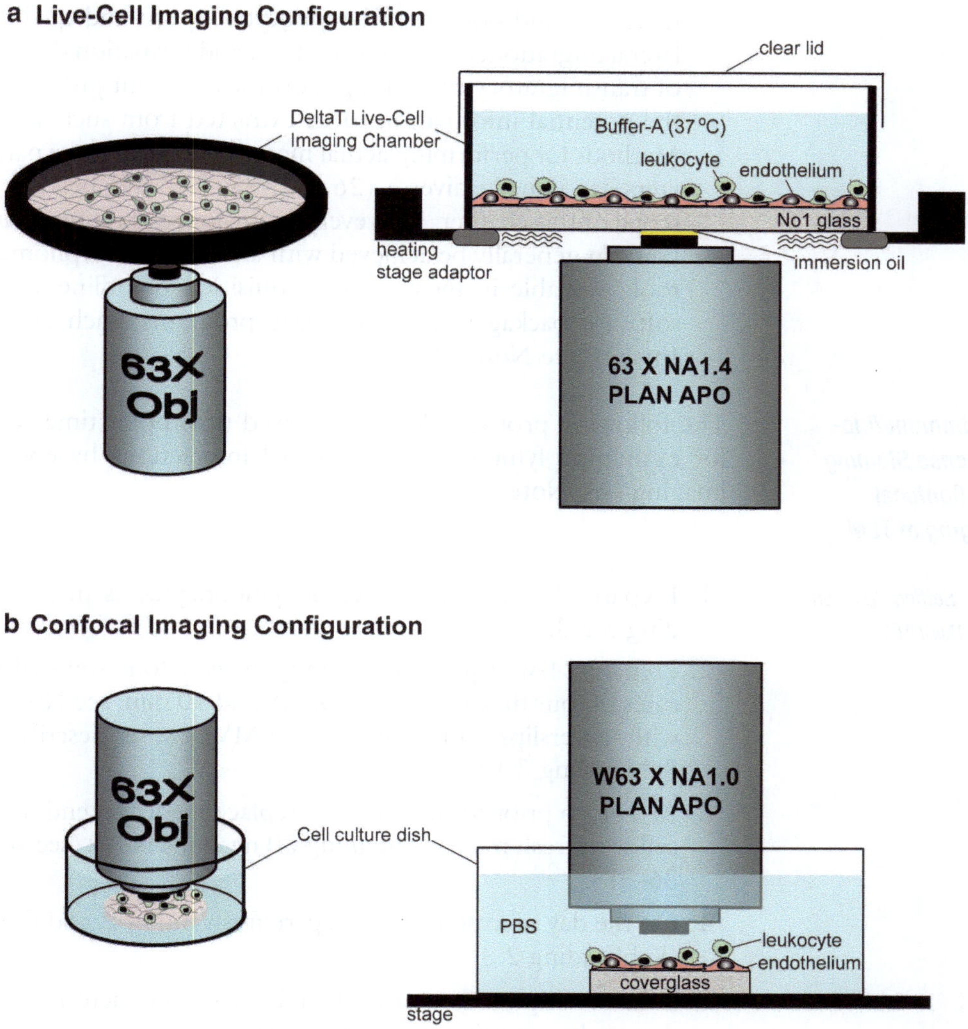

Fig. 3. Live-Cell and Confocal Imaging Configurations. (**a**) Live-Cell Imaging. A monolayer of endothelial cells (*pink*) grown on a Delta-T live-cell imaging chamber and immersed in Buffer A is mounted on a heating stage adaptor and equilibrated to 37°C. Lymphocytes (*green*) are added to the top of the chamber and lymphocyte-endothelial interactions are imaged live via time-lapse epifluorescence/wide-field microscopy. (**b**) Confocal Imaging. Endothelial cells grown on 12-mm circular cover glass in 24 wells are incubated with overlaid lymphocytes in Buffer A for various times, followed by fixation and immunofluorescence staining. The cover glass is then transferred to 6-well or 60-mm culture dishes containing PBS and imaged by confocal microscopy using an upright microscope coupled to a water-dipping objective.

cells per field is optimal (see Note 43). If the lymphocyte number is too low, repeat steps of lymphocyte addition and analysis until an appropriate density is reached. Continue imaging over the course of ~20–60 min (see Notes 44–47).

10. Analyze results. Potential analysis of resulting videos is almost limitless. The design of these imaging experiments will yield high spatial and temporal resolution dynamic of lymphocyte and endothelial morphological changes and behavior during TEM. Diverse parameters such as rate/efficiency of TEM,

spreading and polarity of lymphocytes, speed and quality of lateral migration, route of diapedesis, and formation dynamics of transmigratory cups and podosome represent just some of the potential information to be extracted from such studies. Methods for performing actual measurements of these parameters are equally diverse (26, 32, 33, 44) and beyond the scope of this chapter. However, the majority of required analyses can generally be achieved with a few basic morphometric tools available in most image acquisition or offline analysis software packages (including free programs, such as NIH Image) (see Note 48).

3.4. Immunofluorescence Staining and Confocal Imaging of TEM

The following protocol describes a fixed time point time course for examining lymphocyte–endothelial interactions by confocal imaging (see Note 47).

3.4.1. Setting Up Fixed Time Point TEM

1. Prepare IL-2-cultured effector lymphocytes as in Subheading 3.2.3.
2. Prepare 24-well plates containing at least four wells (duplicates of four time points: 5, 10, 20, and 40 min; see Note 49) with coverslips and confluent HLMVECs, as described in Subheading 3.1.5.
3. At 4–12 h prior to the imaging, replace media on endothelial cell with fresh media containing 50 ng/ml TNF-α (see Notes 36–38).
4. On the day of experiment, prepare fresh Buffer A and fixative (Subheading 2.3).
5. Take a sample of cultured lymphocytes and determine the density by counting with a hemocytometer.
6. Transfer a lymphocyte culture volume equivalent to at least 3×10^5 lymphocytes/sample (e.g., 2.4×10^6 for the eight HLMVECs well prepared above) to a 15-ml conical tube and centrifuge at $200\times g$ for 3 min. Remove the supernatant and resuspend the lymphocyte pellet in Buffer A at a concentration of 6×10^5 lymphocytes/ml.
7. To begin the TEM experiment, remove the endothelial cells from the cell culture incubator, (proceeding one sample at a time) successively aspirate media, and replace with 0.5 ml of Buffer A/lymphocyte suspension (3×10^5 lymphocytes/sample) and replace in 37°C incubator (see Note 50).
8. After appropriate incubation times, noted above, aspirate Buffer A and add sufficient fixative solution to cover the sample completely (e.g., in 24-well plate, ~300–500 μl; in a 6-well plate, ~1–1.5 ml; and in a Delta-T dish, ~0.5–1.0 ml). Incubate at room temperature for 5–10 min (see Note 51).
9. Rinse the sample 3× with PBS (see Note 52).

3.4.2. Sample Staining

The following protocols are designed as a guideline for immunofluorescence labeling of isolated adherent cells for the purposes of confocal fluorescence microscopy (see Note 53). In this procedure, staining will be for endothelial ICAM-1, lymphocyte LFA-1, and total sample F-actin, a combination previously established to be highly informative for understanding the topology of leukocyte–endothelial interactions (26, 32, 33) (see Note 54). See Notes 55–58 for general tips on handling procedures for sample staining.

1. Aspirate PBS. Add 0.5 ml permeabilization solution per well (see Note 59).
2. Incubate at room temperature for 5–10 min.
3. Rinse the sample 1× with PBS. Aspirate PBS.
4. Add 0.5 ml Block Solution per well and incubate at room temperature for 10–15 min (see Note 60).
5. To prepare stain solution, aliquot a volume of Block Solution equivalent to 0.3 ml × the total number of samples to be stained and then add directly conjugated antibodies (see Notes 61 and 62) to human ICAM-1 (IC1/11-488) and LFA-1 (TS2/4-Cy3) at 10 μg/ml (see Note 63) and phalloidin-647 stock solution at 10 μl/ml, as recommended by the manufacturer (Invitrogen).
6. Centrifuge stain solution (at ~10,000 × *g* in a microfuge for 2 min) to pellet any antibody/stain aggregates. Recover the supernatant (see Note 64).
7. Aspirate the block solution from samples and replace with aliquots of stain solution. Incubate at room temperature for 30 min in the dark (see Note 65).
8. Remove the stain solution from samples and collect into a clean Eppendorf or 15-ml conical tube (see Note 66).
9. Wash the samples three times with PBS (allowing ~30–60-s incubations for each) (see Note 67). Aspirate the PBS from samples.
10. Add fixation solution and incubate at room temperature for 3–5 min (see Note 68).
11. Rinse three times with PBS.

3.4.3. Confocal Imaging

1. This protocol is designed for use with an upright Zeiss LSM510 laser-scanning confocal microscope used with a 63× water-dipping objective (see Note 69). To accommodate the width of a water-dipping objective, transfer samples on 13-mm glass coverslips from 24-well plates to 6-well plates or 60-mm dishes (containing ~4 ml of PBS) using fine tweezers and a "hook-tipped" syringe needle (see Notes 70–72). The samples can be either stored at 4°C for imaging at a later time

(see Note 73) or transferred to the microscope to initiate imaging immediately. If the samples are stored at 4°C, allow to warm to room temperature before initiating imaging (see Note 74).

2. To begin imaging, place the 6-well plate containing samples on the microscope stage. Be sure that the sample-containing cover glasses are lying in the center of each well and then lower the 63× dipping objective into one well (Fig. 3b).
3. Use the bright-field imaging mode and ocular lenses to locate the focal plane.
4. Switch to epifluorescence and inspect the sample via the ocular lenses (see Note 40).
5. Select a field of interest.
6. Switch to laser-scanning mode.
7. Using the fast-scan mode and working in the widest (i.e., non-zoomed or -cropped, see Note 75) field possible, individually adjust laser power and gain of each channel to optimize signal such that, ideally, the specific signal intensity reaches at least ~25% and not more than 75% of the dynamic range of the detectors.
8. Using manual focus controls, quickly scan through the *Z*-axis and identify the upper and lower limits of sectioning (typically all of the information should be contained within a thickness of ~15 mm).
9. Select the *Z*-axis section thickness in the range of ~0.2–1.0 μm (see Note 76).
10. Finally, zoom/crop the imaging field to the specific region of interest (see Note 75) and conduct scan (see Note 77). In addition to having very bright and specific fluorescence signal in the sample, acquiring high-resolution 3D imaging requires iterations of scanning and making adjustments to acquisition parameters (see Notes 76 and 77). The objective is to obtain maximal *x*–*y*- and *z*-axis resolution, without appreciable photobleaching.
11. Analyze images. The resulting optical sections can be subjected to extensive analysis (e.g., Pearson's co-localization analysis), orthogonal viewing, and digital 3D rendering to extract information from the sample, and ultimately better understand the mechanisms for TEM (see Note 78). The precise protocols to be employed will depend on the experimental questions (see Note 78).

4. Notes

1. Store at 4°C and use for no more than ~3 months.
2. Though vendors can make a difference in some cases, generally, similar cell-culture pretreated polystyrene culture plates from diverse vendors are acceptable.
3. Phenol-free RPMI can be used instead of HBSS.
4. Always prepare this solution fresh, i.e., same day of use.
5. Total volumes will change based on sample needs.
6. While 0.5% Tx-100 works well for many applications, it may be necessary to vary this value, typically in the range of ~0.01–0.1%, to produce optimal results (i.e., that balance sufficient permeabilization with minimal sample disruption).
7. Antibody staining can use two main strategies: (1) Primary antibodies that are directly conjugated to fluorochrome or (2) unconjugated primary antibody followed by secondary antibody that is fluorochrome conjugated.
8. Non-antibody fluorescent dyes (e.g., Phalloidin-488 (actin stain), ToPro-3 (nuclear stain), and Lyso-tracker-Red (lysosomal stain)) may also be used in combination with antibody staining. Protocol details will vary according to the manufacturer's instructions.
9. Special notes on fluorophores: (a) For imaging, new generation fluorophores (i.e., Alexa and Cy dyes) are vastly superior to more traditional fluorescein (e.g., FITC) and rhodamine (e.g., TRITC, texas red) derivatives. (b) UV wavelength dyes (e.g., DAPI) are fine for epifluorescence microscopy, but may not be compatible with laser-scanning confocal microscopy, as most systems are not equipped with the appropriate UV lasers for exciting such fluorophores. Thus, alternatives that fall in the green, red, or far-red categories (e.g., TopPro3, Invitrogen) must be used. (c) Far-red dyes (e.g., Cy5 and Alexa 647) are not visible to the naked eye and should only be used when a third color is needed. (d) Generally speaking, ideal fluorophores for imaging are as follows: Green channel – Alexa 488, Cy2, eGFP, and eYFP; red channel – Alexa 546, Cy3, and mCherry; and far-red channel – Cy5, Alexa 647, and mPlum(FP).
10. Though focused on one specific context, in general these protocols offer a widely applicable platform of approaches relevant for modeling of broad leukocyte–endothelial interactions.

11. These protocols are also generally applicable to diverse human (e.g., umbilical vein, coronary artery, heart microvascular, and bladder microvascular) and mouse (e.g., heart, lung, and skin microvascular) endothelium, though with various cell-specific tweaks.
12. Alternatively, glasses can be flame sterilized, by dipping each one briefly in ethanol followed by flaming with a Bunsen burner inside the cell culture hood and then placing each in cell culture plate wells. This requires more manipulation initially, but ultimately can save some time. Anecdotal experience suggests that some endothelia may grow preferentially on flamed glass.
13. Mouse primary endothelium is relatively more challenging to grow on glass compared to human.
14. HBSS can be substituted.
15. When using 24-well plates with circular coverglass, attention must be paid to the complete coverage of the glass. Initially, the glass is more hydrophobic than the plastic and will repel the FN solution or even cause the cover glass to float on top. Agitation of the plates will help to begin the coating process, making the glass increasingly hydrophilic.
16. Do not excessively swirl or knock the plate during this stage; while it may help to remove cells, it will also cause cell damage and aggregation.
17. Seeding endothelial cells at significantly lower densities may lead to low viability and extremely slow growth.
18. It is not recommended to use endothelial cells beyond ~6 passages, as phenotypic drift may seriously compromise the quality/physiologic relevance of the endothelial monolayers formed.
19. When preparing monolayers to be used in experiments, seeding density should be sufficiently high that following initial attachment and spreading (i.e., with ~4–6 h of plating), confluency will be ~80%. This will allow ~100% confluency to be reached by ~12–16 h. Cells should subsequently be cultured further for a minimum of 24 h (ideally 48–72 h) before any experimental manipulations. This regime allows intercellular junctions to form and at least a minimal amount of differentiation toward an integrated vascular monolayer to occur.
20. Even with the Amaxa approach, some constructs may be expressed poorly due to issues of stability, toxicity, or aspects of construct design (e.g., promotor type used).
21. Transfection results are highly dependent on health of the endothelial cells. Ideally, cells should be passaged 3–4 days prior to transfection and be at relatively early total passage number (i.e., p2–p4).

22. At 90% confluency, T25 and T75 flasks typically yield ~1.5 million and ~4 million cells, respectively.
23. A single Nucleofection reaction required 0.5 million endothelial cells.
24. Viability will vary depending on type and health of the endothelium and the plasmids transfected.
25. From the same starting point of isolated primary human blood, one can readily also isolate/culture naïve lymphocytes, memory lymphocytes, neutrophils, monocytes, and NK cells for similar use in in vitro TEM studies.
26. Generally, similar protocols can be used for derivation of mouse Th1 and memory lymphocytes, neutrophils, and monocytes from spleen with some variations.
27. An alternative strategy to drawing blood specifically for experiments is to purchase "blood leukopacks" from nearby hospitals, blood banks, or commercial vendors (e.g., Cambrex). Leukopacks are bags of fresh human blood cells collected from normal peripheral blood by automated apheresis in anticoagulant followed by density gradient centrifugation. Each leukopack contains a mixture of monocytes, lymphocytes, platelets, plasma, and red cells. This not only offers convenience over setting up IRB-approved human subject protocols for drawing blood, but also tends to be an expensive option. Protocols to work with leukopacks and quality of resulting cell preparations are generally similar to those of blood drawn freshly in house.
28. Details of appropriate blood drawing techniques and safety protocols are beyond the scope of the current chapter. Those interested in learning more on this topic should consult the appropriate human subjects research administrators within their host institutions.
29. RPMI-IL15 media (RPMI base media supplemented with 10 μg/ml of IL-15; R & D Systems) can be used as an alternate here to produce a relatively more memory, rather than effector, T cell phenotype.
30. Based on FACs analysis, resulting cells are typically ~97% CD3+, ~50% CD4+, ~50% CD8, ~95% CD45RO+, and CD56- and thus represent a mixture of CD4 and CD8 effector lymphocytes. One can readily further purify subpopulations using immuno-magnetic bead sorting (e.g., MACS, Miltenyi Biotec, Germany) if desired.
31. Generally, similar types of image acquisition can be achieved using inverted scopes/CCD camera setup from diverse vendors.

32. Membrane-tagged fluorescent proteins (e.g., mYFP and mRFP) are expressed with high efficiency (~50–70%) and at high levels in endothelial cells and provide a crisp delineation of individual cells and an extremely sensitive readout for endothelial topological changes during interactions with leukocytes (26). Actin-GFP is expressed somewhat less efficiently in endothelium and provides important additional topological information and more general insights into endothelial cytoskeletal responses to interactions with leukocytes (26). DIC imaging provides exquisite detail of all cell surface dynamics and complements the fluorescence information.
33. Co-transfection of diverse combinations of markers can be employed to address distinct questions. Particularly interesting markers for TEM studies include fluorescent protein fusion contructs of ICAM-1, VCAM-1, PECAM-1, VE-cadherin, tubulin, caveolin, and Rho family GTPases. Further advanced imaging can be obtained by combining topological markers (e.g., mRFP) with ratiometric- or FRET-based biosensors that readout localized signaling activities.
34. The protocols described here do not include any specific fluorescent markers for the lymphocytes, as primary blood leukocytes remain extremely poorly transfectable, by virtually all known approaches. The combination of DIC and endothelial mRFP which can serve as an indirect readout for leukocyte topological dynamics (26) can provide a great deal of information on leukocyte morphology. However, when desired, fluorescence can be introduced to leukocytes by staining with fluorescent cytoplasmic (e.g., BCECF or SNARF) or membrane (e.g., DiI) dyes or with fluorescent-conjugated antibody fAb fragments that target leukocyte cell surface markers (e.g., LFA-1), although these strategies are not optimal.
35. Amaxa transfection is extremely cytotoxic. Only a fraction of the input cells will survive and effectively adhere to the culture plates. To ensure that these grow as healthy monolayers, it is important to remove the dead cells and debris by gentle swirling and exchanging the media.
36. The current protocols model inflammatory leukocyte–endothelial interactions. As such, the endothelium must be activated to upregulate the expression of adhesion and chemoattractant molecules appropriately, which drive leukocyte adhesion and TEM. TNF-α arguably represents the most potent and broadly relevant endothelium-activating cytokine. However, alternative activation by other cytokines (such as IL-1β or interferon-γ) or by distinct stimuli (such as hypoxia, thrombin, or lipopolysacharide) may be substituted when modeling distinct inflammatory settings.

37. It is also important to note that a key challenge to studying leukocyte TEM in vitro is the formation and maintenance of an intact endothelial monolayer. In the absence of this, any apparent endothelial barrier crossing activities by leukocytes (i.e., TEM) is meaningless. Thus, it is important to start with conditions that allow for confluent monolayers to form and begin to differentiate within 12–16 h of transfection/plating.
38. The process of creating transfected and activated endothelial monolayers with well-formed intact junctions is an empirical one. If monolayers are not well formed, modifications (such as increasing the input cell density, culturing for longer periods before addition of TNF-α, or reduction in either the concentration or treatment time (e.g., 4 h instead of 12 h) of TNF-α application) should be implemented systematically until optimal monolayer conditions are identified.
39. It is critical to minimize the time interval between removal of cells from culture incubators and initiation of imaging. Thus, it is important to first fully prepare all aspects of the experiment and image acquisition. All live-cell time-lapse image systems are equipped with software that drive multidimensional image acquisition. Confirm parameter setup before starting experiment.
40. Effort should be made to work quickly and minimize the exposure of a sample to cytotoxic fluorescence excitation sources. Apply neutral density filters to attenuate excitation source to the minimal usable levels and turn excitation sources off (i.e., close relevant shutters) immediately whenever a pause in imaging activities is required.
41. The basic principle to be applied is to use the minimal amount of excitation energy required to provide good quality brightness and contrast. Generally speaking, exposure times should be set in the range of 200–800 ms. If high-quality images are possible at less than 100 ms, it is likely that the intensity of excitation light source being applied is greater than necessary and should be attenuated (e.g., using a neutral density filter). Alternatively, when required exposure times are longer than 1 s, aberrations due to the relative speed of cell dynamics may appear. Thus, in such settings, it may be necessary to increase the detector gain and/or binning.
42. As with exposure intensity/duration (see Note 41), acquisition rate/frequency must be selected carefully. This selection should be based on the minimum required temporal resolution and duration of imaging, as defined by the speed and duration of the specific behaviors being studied, respectively. Each exposure of the sample to excitation light sources induces some degree of phototoxic damage that hinders

cell viability and photobleaching, which reduces quality of fluorescence signals. The total amount of exposure (intensity and duration of each exposure multiplied by the total number of exposures, which in turn is defined by the frequency of acquisition multiplied by the total duration of imaging) that a given sample can tolerate before cell viability and/or fluorescence signal falls below acceptable levels is finite. *Thus, one should bear in mind that every imaging experiment works on a limited "photon budget."* High-quality, reliable results can be obtained only when all parameters are managed such that the experiment remains well within these limits by "spending" the minimal amount of photons necessary to obtain the required results.

43. Successful and robust imaging analysis requires a carefully balanced number of input lymphocytes in the field of interest. Too few cells may mean that a time-consuming live-cell imaging experiment may only yield one or several leukocyte–endothelial cell interactions to evaluate. Too many cells will make it extremely difficult to discriminate individual interactions, thus precluding meaningful analysis. Moreover, excess amounts of lymphocytes may also overwhelm and, thereby, compromise the health of the endothelium.

44. For high-quality imaging, it is absolutely critical to pay careful attention to the focal plane throughout the course of the experiment.

45. The current model system using effector lymphocytes and activated endothelium should produce completion of TEM by the majority of lymphocytes within ~30–45 min. If for any reason experimental objective requires imaging for much longer durations, careful consideration should be given to issues of evaporation and CO_2/pH. Clear plastic covers for Delta-T plates (Bioptechs) that do not interfere with imaging can offer some protection from evaporation. However, for truly long-term experiments, humidified, CO_2-equilibrated, 37°C environmental control boxes that fit around the entire microscope stage are highly recommended.

46. In many cases, it may be desirable to end the imaging experiment by fixation via removal of Buffer A followed by addition of fixative solution and incubation at room temperature for 5 min, followed by rinsing three times with PBS. This allows for the possibility of performing additional staining of samples and high-resolution, 3D confocal, fixed-cell end-point image analysis (Subheading 3.4).

47. It is important to note that this approach does not include physiological fluid shear forces as normally experienced in blood circulation. One can readily modify these experiments using parallel wall flow chambers (e.g., Biotechs FCS2 chamber)

to include shear force application and determine whether this parameter is important to the phenomena being measured (26, 32, 33). However, since these techniques are invariably more technically challenging and time consuming, it is recommended to develop results first with a static system before progressing to shear systems. Additionally, it must be noted that acute application of fluid shear forces to an in vitro model system will not necessarily make it more physiologic as a whole. While these forces may recapitulate those felt suddenly by leukocytes as they transition from free flow to firm adhesion on endothelium, the acute onset of shear felt by the endothelium itself is entirely non-physiologic (under normal condition, endothelium exhibits a high degree of long-term shear adaptation) and induces aberrant calcium flux, stress fiber formation, and intercellular junction weakening.

48. It must be stressed that the strength of the imaging approach described throughout this protocol is in delineation of fine morphological dynamics and attempting to correlate them with overall TEM behavior. In this way, cell biological mechanisms governing TEM may be uncovered and characterized. Thus, it is up to the investigator's power of observation to first identify and then rigorously characterize/quantify specific morphological dynamics, or changes in such dynamics in response to experimental conditions. This approach provides potential as both an exploratory tool and an assay system. However, the realization of this potential is a highly empirical process, in which measurement protocols often need to be custom developed.

49. These time points are selected to provide representative snap shots of behaviors at major phases of TEM. Extensive experience with this model shows that a small fraction of lymphocytes begin diapedesis within 5 min, whereas the majority complete the process by 30–40 min (26, 32, 33). Time points can/should be freely adjusted to suite experimental questions or distinct model systems.

50. It is recommended that the addition of cells for each time point be staggered (i.e., adding lymphocytes to the longest time point samples first and the shortest ones last) to allow for fixation of all samples together in one step to avoid issues noted below (see Note 51). Alternatively, one can plan a priori to plate each time point in separate 24-well plates.

51. It is possible to over-fix a sample, which may result in damaging of antibody epitopes or fluorescent proteins. This fixative should never be used at 37°C or for more than 30 min.

52. Samples can be transferred to 4°C and stored at this point if desired. Staining performed within several weeks often shows no appreciable difference from immediate staining.

However, for first time analysis, it is highly recommended that samples be stained and imaged immediately.

53. While general principles and procedures are similar, staining of suspension cells, whole-mount tissues, or frozen sections requires modifications. Moreover, immunofluorescence staining for flow cytometry also has distinct procedures and requirements.
54. This basic protocol is also generally applicable to broad combinations of stains (e.g., antibodies to other cell surface or intracellular proteins) and stains for other markers (e.g., lysotracker and mitotracker to mark the lysosome and mitochondrial compartments, respectively).
55. This protocol covers basic procedures common to most immunofluorescence staining needs. However, aspects of this protocol are written specifically for use with an upright microscope coupled to a water-dipping objective. For this, it is recommended that adherent cells be grown in 24-well plates containing 13-mm circular cover glasses.
56. Never let the samples dry. When aspirating and replacing, solutions always move quickly to minimize the amount of time that the sample is exposed to air. If working with a large number of samples, do not perform steps on all samples at once. That is, do not aspirate media or buffer off of 20 samples and then replace buffer in all 20. Rather, quickly aspirate buffer from one (or a few) samples and replace buffer, then move to the next sample (or a small group of samples).
57. Balance speed and efficiency of buffer changes with gentleness. Harsh pipetting will lead to damage or even loss of sample. Use of disposable transfer pipettes (which have relatively large diameter opening compared with pipette tips) allows for rapid delivery of buffer with minimal shear force. Additionally, add buffers to the side of the sample well rather than directly onto the center of the sample.
58. Once fluorophores are introduced (which may mean *a priori* if samples have been transfected to express a fluorescent protein), care should be taken to minimize sample exposure to light (especially direct sunlight). Perform incubation steps inside of a laboratory bench drawer or under a sheet of aluminum foil.
59. Permeabilize samples only when necessary (i.e., intracellular staining is being conducted). If staining only includes extracellular/cell surface markers, it is best not to perform a permeabilization step, as this almost always increases the background signal.
60. BSA, HSA, or FCS may be preferable to NFDM in some cases.

61. Direct antibody conjugation of primary antibodies can readily, quickly, and cost effectively be accomplished with as little as ~100 μg of protein using commercially available Alexa (Invitrogen) and Cy (Amersham) dye micro-labeling kits. Direct conjugation offers greatly increased flexibility (i.e., freedom from issues of selecting primary antibodies derived from distinct species in order to be able to use non-cross-reacting secondary antibodies), convenience (one antibody stain step instead of two, see Note 62), and generally cleaner, more specific staining results.
62. Traditional methods using nonconjugated primary antibodies in conjunction with fluorescent-conjugated secondary antibodies can similarly be used in this protocol by preparing primary antibodies as indicated, but without phalloidin-647, followed by washing, using a second antibody stain (i.e., containing secondary, for example, goat anti-mouse-488 or rat anti-rabbit-Cy3, antibodies; Phalloidin-647 or other non-antibody stains, such as ToPro3, should also be included in this step), and a wash step.
63. It is best to determine optimal concentrations empirically, but 10 μg/ml works well for most antibodies.
64. It is normal to see a significant amount of NFDM pelleted.
65. Longer incubations may sometimes be used/necessary to increase the level of signal. However, this may also increase the nonspecific background staining of the sample as well.
66. Stored at 4°C in the dark, stains can often be reused several times over a period of weeks or months without appreciable loss in stain quality, thereby offering substantial conservation of precious reagents and convenience. However, the shelf-life of a given premade stain solution will vary greatly and must be determined empirically.
67. In some cases, additional washes and/or extended incubations may be useful in reducing nonspecific background staining.
68. This step covalently links stains to the sample and dramatically increases the stability of the staining, allowing appropriately stored samples to be repeatedly imaged over the course of weeks or months. However, some dyes (e.g., Phalloidin-488 and ToPro3) do not seem to be efficiently "fixed" into the sample and may leach out over time. These, therefore, may require a fresh restain immediately before viewing samples that have been stored.
69. Use of an upright microscope and water-dipping objective offers several advantages over oil immersion objectives. The sample preparation technique used for a water-dipping objective is substantially easier than that used for traditional

coverslip mounting (see Notes 70 and 72) and also offers dramatically more stable samples (see Note 72). Additionally, the ability to image the sample directly through an aqueous medium (e.g., PBS) reduces *Z*-axis aberrations, which are strongly promoted by imaging through media of mixed refractive index (e.g., oil, glass, and aqueous sample).

70. First, carefully press the pointed edge of the needle against a hard benchtop at ~15° angle and slowly increase the angle to ~70° while applying pressure. This will result in a "hooked" appearance at the end of the needle that can be used to lift one edge of the coverslip and then raise it until the tweezers can be used to grab and transfer it. Be careful not to allow the sample to be exposed to air for more a couple of seconds and to maintain the coverslip with the samples surface up. It is highly recommended that this approach be practiced with blank coverslips before transferring "real" samples.
71. Samples can also be grown directly on 6-well or Delta-T plates. However, this will require significantly greater numbers of cells and large amounts of (usually precious) staining reagents.
72. If a water-dipping objective is not available, coverslips can also be mounted using traditional techniques to be used with an oil immersion objective mounted on either upright or inverted confocal systems. To do so, place a drop of mounting medium (e.g., Slow Fade, Invitrogen) on a cover glass, then flip the cover glass (sample-side down) on to the medium, and then use nail polish to seal the edges. Be extremely cautious to ensure that the nail polish is completely dry before attempting to image.
73. To store samples for subsequent imaging sessions, make sure that 3–4 ml of PBS remains in each well. Cover the plate with adhesive sealing strips (Fisher) and then with aluminum foil and store at 4°C. Samples stored in 6 wells covered by PBS and protected from light often retain high-quality signal for weeks or months, whereas mounted coverslips may retain reliable, good quality signal typically for only several days.
74. If samples are in the process of warming up during image acquisition, significant drift in focal plane can occur during scanning. This will be particularly problematic if 3D serial sectioning is being attempted.
75. The more zoomed or cropped the field is, the more intense the total laser energy being absorbed per unit area. Thus, the continuous scanning required to adjust parameters will have a higher tendency to photobleach the sample when it is highly zoomed. It is, therefore, recommended to set parameters first using as wide a field as possible and then crop/zoom to smaller regions of interest as the last step before collecting the final scan.

76. Ideally, all scans would be conducted in the lower range to provide optimal 3D information. However, as noted for live-cell imaging (see Note 42), we always need to be aware of the limited "photon budget" when working with fluorescent samples. The smaller the section thickness, the more scans will be required to complete the Z-sectioning. When using a laser-scanning confocal system, it is very easy to photobleach the sample as image acquisition is taking place, such that signal intensity in the sections obtained toward the end of the series may be substantially diminished (or lost all together) compared to those at the beginning. Thus, section thickness and/or laser intensity will often need to be compromised to avoid photobleaching issues. Ultimately, this is an empirical process. After several rounds of trail and error, optimal laser power and section thickness for a given sample can be accomplished.
77. An important tool/parameter for high-quality confocal imaging is the multiple-line scan feature or Kalman filter. In short, this feature causes the laser to raster repeatedly over the same line multiple times, as set by the user. Basic algorithms provide a type of recursive filtration that can dramatically enhance signal to noise. However, this comes with an increase in laser energy input to the sample and thus, increased photobleaching. Again, ultimate imaging quality requires balancing the number of line scans applied, the laser intensity, and the *Z*-axis section thickness.
78. No set analytical procedure exists. As with the dynamic live-cell imaging described in Subheading 3.3 (see Note 48), the analysis to be conducted with confocal serial Z-sections is highly dependent on the questions being asked in any given setting and experiment. Wide-ranging offline image-processing software packages exist (including many free packages) with flexible capabilities.

Acknowledgments

This work was supported by the Arthritis Foundation, American Heart Association, and Roche Organ Transplant Research Foundation.

References

1. von Andrian, U. H., and Mackay, C. R. (2000) T-cell function and migration. Two sides of the same coin. *N. Engl. J. Med.* **343**, 1020–1034.
2. Bazzoni, G., and Dejana, E. (2004) Endothelial cell-to-cell junctions: molecular organization and role in vascular homeostasis. *Physiol. Rev.* **84**, 869–901.

3. Pepper, M.S., and Skobe, M. (2003) Lymphatic endothelium: morphological, molecular and functional properties. *J. Cell Biol.* **163**, 209–213.
4. Baluk, P., Fuxe, J., Hashizume, H., Romano, T., Lashnits, E., Butz, S., Vestweber, D., Corada, M., Molendini, C., Dejana, E., and McDonald, D. M. (2007) Functionally specialized junctions between endothelial cells of lymphatic vessels. *J. Exp. Med.* **204**, 2349–2362.
5. Ley, K., Laudanna, C., Cybulsky, M. I., and Nourshargh, S. (2007) Getting to the site of inflammation: the leukocyte adhesion cascade updated. *Nat. Rev. Immunol.* 7, 678–689.
6. Huttenlocher, A., and Poznansky, M. C. (2008) Reverse leukocyte migration can be attractive or repulsive. *Trend. Cell Biol.* **18**, 298–306.
7. Springer, T. A. (1994) Traffic signals for lymphocyte recirculation and leukocyte emigration: the multistep paradigm. *Cell* **76**, 301–314.
8. Luscinskas, F. W., Kansas, G. S., Ding, H., Pizcueta, P., Schleiffenbaum, B. E., Tedder, T. F., and Gimbrone, M. A., Jr. (1994) Monocyte rolling, arrest and spreading on IL-4-activated vascular endothelium under flow is mediated via sequential action of L-selectin, beta 1-integrins, and beta 2-integrins. *J. Cell Biol.* **125**, 1417–1427.
9. Butcher, E. C. (1991) Leukocyte-endothelial cell recognition: Three (or more) steps to specificity and diversity. *Cell* **67**, 1033–1036.
10. Carman, C. V., and Springer, T. A. (2003) Integrin avidity regulation: are changes in affinity and conformation underemphasized? *Curr. Opin. Cell Biol.* **15**, 547–556.
11. Luo, B. H., Carman, C. V., and Springer, T. A. (2007) Structural basis of integrin regulation and signaling. *Annu. Rev. Immunol.* **25**, 619–647.
12. Schenkel, A. R., Mamdouh, Z., and Muller, W. A. (2004) Locomotion of monocytes on endothelium is a critical step during extravasation. *Nat. Immunol.* **5**, 393–400.
13. Phillipson, M., Heit, B., Colarusso, P., Liu, L., Ballantyne, C. M., and Kubes, P. (2006) Intraluminal crawling of neutrophils to emigration sites: a molecularly distinct process from adhesion in the recruitment cascade. *J. Exp. Med.* **203**, 2569–2575.
14. Luscinskas, F. W., Ma, S., Nusrat, A., Parkos, C. A., and Shaw, S. K. (2002) Leukocyte transendothelial migration: a junctional affair. *Semin. Immunol.* **14**, 105–113.
15. Muller, W. A. (2003) Leukocyte-endothelial-cell interactions in leukocyte transmigration and the inflammatory response. *Trend. Immunol.* **24**, 327–334.
16. Muller, W. A. (2001) Migration of leukocytes across endothelial junctions: some concepts and controversies. *Microcirculation* **8**, 181–193.
17. Burns, A. R., Smith, C. W., and Walker, D. C. (2003) Unique structural features that influence neutrophil emigration into the lung. *Physiol. Rev.* **83**, 309–336.
18. Williamson, J. R., and Grisham, J. W. (1960) Leucocytic emigration from inflamed capillaries. *Nature* **188**, 1203.
19. Williamson, J. R., and Grisham, J. W. (1961) Electron microscopy of leukocytic margination and emigration in acute inflammation in dog pancreas. *Am. J. Pathol.* **39**, 239–256.
20. Marchesi, V. T., and Gowans, J. L. (1964) The Migration of Lymphocytes through the Endothelium of Venules in Lymph Nodes: an Electron Microscope Study. *Proc. R. Soc. Lond. B. Biol. Sci.* **159**, 283–290.
21. Sage, P. T., and Carman, C. V. (2009) Settings and Mechanisms for Trans-cellular Diapedesis. *Front. Biosci.* **4**, 5066–5083.
22. Carman, C. V., and Springer, T. A. (2008) Trans-cellular migration: cell-cell contacts get intimate. *Curr. Opin. Cell Biol.* **20**, 533–540.
23. Linder, S., and Aepfelbacher, M. (2003) Podosomes: adhesion hot-spots of invasive cells. *Trend. Cell Biol.* **13**, 376–385.
24. Buccione, R., Orth, J. D., and McNiven, M. A. (2004) Foot and mouth: podosomes, invadopodia and circular dorsal ruffles. *Nat. Rev. Mol. Cell Biol.* **5**, 647–657.
25. Yamaguchi, H., Wyckoff, J., and Condeelis, J. (2005) Cell migration in tumors. *Curr. Opin. Cell Biol.* **17**, 559–564.
26. Carman, C. V., Sage, P. T., Sciuto, T. E., de la Fuente, M. A., Geha, R. S., Ochs, H. D., Dvorak, H. F., Dvorak, A. M., and Springer, T. A. (2007) Transcellular diapedesis is initiated by invasive podosomes. *Immunity* **26**, 784–797.
27. Hidalgo, A., and Frenette, P. S. (2007) Leukocyte podosomes sense their way through the endothelium. *Immunity* **26**, 753–755.
28. Carman, C. V. (2009) Mechanisms for transcellular diapedesis: probing and pathfinding by "invadosome-like protrusions". *J. Cell Sci.* **122**, 3025–3035.
29. Gerard, A., van der Kammen, R. A., Janssen, H., Ellenbroek, S. I., and Collard, J. G. (2009) The Rac activator Tiam1 controls efficient T-cell trafficking and route of transendothelial migration. *Blood* **113**, 6138–6147.

30. Wojciak-Stothard, B., Williams, L., and Ridley, A. J. (1999) Monocyte adhesion and spreading on human endothelial cells is dependent on Rho-regulated receptor clustering. *J. Cell Biol.* **145**, 1293–1307.
31. Barreiro, O., Yanez-Mo, M., Serrador, J. M., Montoya, M. C., Vicente-Manzanares, M., Tejedor, R., Furthmayr, H., and Sanchez-Madrid, F. (2002) Dynamic interaction of VCAM-1 and ICAM-1 with moesin and ezrin in a novel endothelial docking structure for adherent leukocytes. *J. Cell Biol.* **157**, 1233–1245.
32. Carman, C. V., Jun, C. D., Salas, A., and Springer, T. A. (2003) Endothelial cells proactively form microvilli-like membrane projections upon intercellular adhesion molecule 1 engagement of leukocyte LFA-1. *J. Immunol.* **171**, 6135–6144.
33. Carman, C. V., and Springer, T. A. (2004) A transmigratory cup in leukocyte diapedesis both through individual vascular endothelial cells and between them. *J. Cell Biol.* **167**, 377–388.
34. Barreiro, O., Yanez-Mo, M., Sala-Valdes, M., Gutierrez-Lopez, M. D., Ovalle, S., Higginbottom, A., Monk, P. N., Cabanas, C., and Sanchez-Madrid, F. (2005) Endothelial tetraspanin microdomains regulate leukocyte firm adhesion during extravasation. *Blood* **105**, 2852–2861.
35. Millan, J., Hewlett, L., Glyn, M., Toomre, D., Clark, P., and Ridley, A. J. (2006) Lymphocyte transcellular migration occurs through recruitment of endothelial ICAM-1 to caveola- and F-actin-rich domains. *Nat. Cell Biol.* **8**, 113–123.
36. Nieminen, M., Henttinen, T., Merinen, M., Marttila-Ichihara, F., Eriksson, J. E., and Jalkanen, S. (2006) Vimentin function in lymphocyte adhesion and transcellular migration. *Nat. Cell Biol.* **8**, 156–162.
37. van Buul, J. D., Allingham, M. J., Samson, T., Meller, J., Boulter, E., Garcia-Mata, R., and Burridge, K. (2007) RhoG regulates endothelial apical cup assembly downstream from ICAM1 engagement and is involved in leukocyte trans-endothelial migration. *J. Cell Biol.* **178**, 1279–1293.
38. Riethmuller, C., Nasdala, I., and Vestweber, D. (2008) Nano-surgery at the leukocyte-endothelial docking site. *Pflugers Arch.* **456**, 71–81.
39. Barreiro, O., Zamai, M., Yanez-Mo, M., Tejera, E., Lopez-Romero, P., Monk, P. N., Gratton, E., Caiolfa, V. R., and Sanchez-Madrid, F. (2008) Endothelial adhesion receptors are recruited to adherent leukocytes by inclusion in preformed tetraspanin nanoplatforms. *J. Cell Biol.* **183**, 527–542.
40. Kanters, E., van Rijssel, J., Hensbergen, P. J., Hondius, D., Mul, F. P., Deelder, A. M., Sonnenberg, A., van Buul, J. D., and Hordijk, P. L. (2008) Filamin B mediates ICAM-1-driven leukocyte transendothelial migration. *J. Biol. Chem.* **283**, 31830–31839.
41. Cayrol, R., Wosik, K., Berard, J. L., Dodelet-Devillers, A., Ifergan, I., Kebir, H., Haqqani, A. S., Kreymborg, K., Krug, S., Moumdjian, R., Bouthillier, A., Becher, B., Arbour, N., David, S., Stanimirovic, D., and Prat, A. (2008) Activated leukocyte cell adhesion molecule promotes leukocyte trafficking into the central nervous system. *Nat. Immunol.* **9**, 137–145.
42. Dittmar, S., Harms, H., Runkler, N., Maisner, A., Kim, K. S., and Schneider-Schaulies, J. (2008) Measles virus-induced block of transendothelial migration of T lymphocytes and infection-mediated virus spread across endothelial cell barriers. *J. Virol.* **82**, 11273–11282.
43. Rohlena, J., Volger, O. L., van Buul, J. D., Hekking, L. H., van Gils, J. M., Bonta, P. I., Fontijn, R. D., Post, J. A., Hordijk, P. L., and Horrevoets, A. J. (2009) Endothelial CD81 is a marker of early human atherosclerotic plaques and facilitates monocyte adhesion. *Cardiovasc. Res.* **81**, 187–196.
44. Beltman, J. B., Maree, A. F., and de Boer, R. J. (2009) Analysing immune cell migration. *Nat. Rev. Immunol.* **9**, 789–798.

Chapter 16

Multiphoton Intravital Microscopy to Study Lymphocyte Motility in Lymph Nodes

Thomas T. Murooka and Thorsten R. Mempel

Abstract

Intravital microscopy (IVM) allows for the direct in vivo visualization of dynamic biological processes in their physiological context at high spatial and temporal resolution. Novel nonlinear optical imaging modalities, most prominently multiphoton microscopy, have extended the spectrum of cellular functions amenable to IVM investigation to include migration and cell–cell interactions occurring deep inside the highly light-scattering environments of solid tissues, which had so far been inaccessible to conventional microscopy techniques. This has led to important new insights into immune cell behavior at steady state, as well as their change in behavior during an immune response. Here, we describe in detail a technique that allows for the monitoring of lymphocyte motility in the lymph nodes of mice at the single cell level using multiphoton intravital microscopy (MP-IVM).

Key words: Multiphoton intravital microscopy, In vivo imaging, Popliteal lymph node, Lymphocyte motility, Migration, Instantaneous velocity, Turning angles, Confinement ratio, Mean displacement

1. Introduction

Cellular behavior in vivo is determined by a multitude of extrinsic factors that are impossible to replicate in their full complexity outside of the living organism. In order to audit and refine hypotheses based on in vitro studies, in vivo experiments are required. MP-IVM is a powerful imaging modality that allows for the direct dynamic observation of biological processes in their physiological tissue environment at cellular and subcellular resolution. The optical principles underlying multiphoton microscopy hereby allow for deeper optical penetration into tissues, compared to conventional fluorescence microscopy techniques, as well as for prolonged observation while limiting phototoxicity and

Motomu Shimaoka (ed.), *Integrin and Cell Adhesion Molecules: Methods and Protocols*,
Methods in Molecular Biology, vol. 757, DOI 10.1007/978-1-61779-166-6_16, © Springer Science+Business Media, LLC 2011

photobleaching (1, 2). The study of the migratory behavior and function of immune cells is particularly amenable to this technology (3, 4).

Here, we describe in detail a technique that allows for the visualization of naïve T-lymphocyte migration in the popliteal lymph nodes (popLNs) of mice and quantitative analysis of their motility. The technique involves the isolation of T cells from the spleen and LN of donor mice, their labeling with commercially available fluorescent dyes, and intravenous re-injection into recipient mice. The popLN is then surgically exposed to image the dynamic cellular behavior of the tagged lymphocytes by MP-IVM. This method can be utilized to assess the effects of experimental interference with integrin functions, such as through transgene expression, gene silencing by siRNA, or small molecules, on lymphocyte motility in vivo.

2. Materials

2.1. Preparation and Injection of Lymphocytes to Track In Vivo

1. Spleen and lymph node cells from donor mice: Donor and recipient mice are of the same genetic background (e.g., C57BL/6).
2. Immunomagnetic cell purification kit for pan T cells, $CD4^+$ T cells, or $CD8^+$ T cells (e.g., mouse $CD8^+$ T cell negative selection kit from Miltenyi Biotec, Bergisch Gladbach, Germany).
3. Cell tracker dyes: 5-(6)-carboxyfluorescein diacetate succinimidyl ester (CFSE) and 5-(and-6)-(((4-chloromethyl) benzoyl) amino) tetramethylrhodamine (CMTMR, CellTracker Orange, both from Molecular Probes/Invitrogen, Carlsbad, CA).
4. PBS with 1% FCS.

2.2. Surgical Anesthesia

1. Ketamine HCl solution: Prepared at 100 mg/mL.
2. Xylazine HCl solution: Prepared at 20 mg/mL.
3. NaCl 0.9%: Used for injection.

2.3. Microsurgical Preparation of the Mouse Popliteal Lymph Node for Intravital Microscopy

1. Custom-built microscope stage (see Note 1, Fig. 1a, j).
2. Small animal clipper (e.g., Pocket Pro, Wahl, Sterling, IL).
3. Depilation crème.
4. Cotton tip applicators.
5. 2 × 2 in. gauze sponges.
6. Surgical instruments: Straight "tough cut" iris scissors; Vannas spring scissors with 3–5 mm blades; Dumont #5 Forceps standard tip, straight, 0.1 × 0.6 mm, Inox, 11 cm total length; Dumont

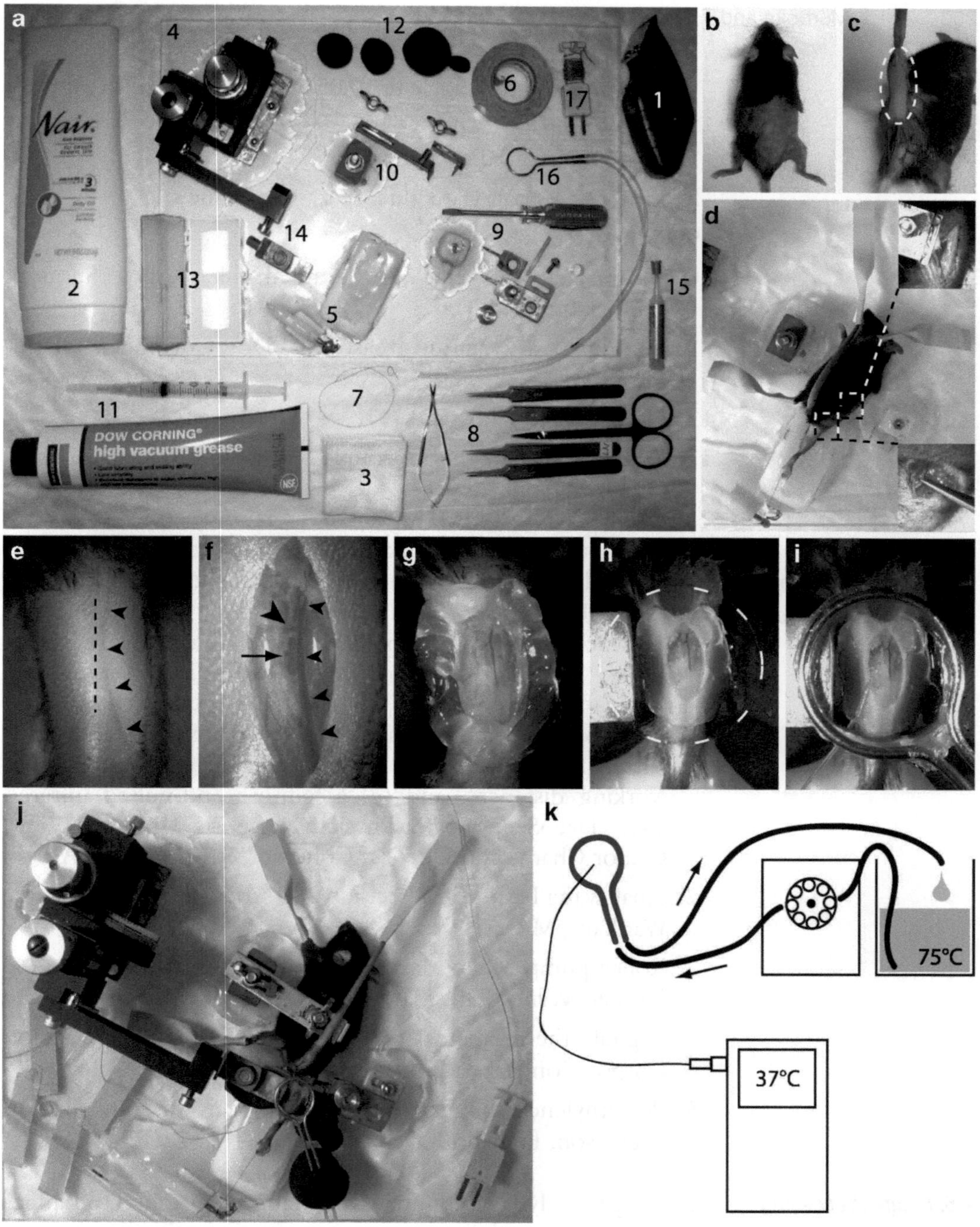

Fig. 1. Equipment and procedure of the popLN preparation for MP-IVM. (**a**) 1: Clipper; 2: Depilation crème; 3: Gauze sponges; 4: Custom-built microscope stage; 5: Pedestal for positioning of right hindleg; 6: Adhesive tape; 7: Suture; 8: Surgical instruments; 9: Dissembled holding clamp for immobilization of trochanter major; 10: Dissembled holding clamp for immobilization of spine; 11: Vacuum grease loaded in syringe; 12: Plasticine; 13: Coverglasses; 14: Mounting device for coverglass and micromanipulator; 15: Crazyglue; 16: Circular metal tube for temperature control; 17: Thermocouple. (**b**) Shaved areas of the experimental animal. (**c**) Depilated area. (**d**) Positioning of experimental animal prior to the surgical preparation. *Upper insert*: Site for spine clamp application. *Lower insert*: Site for hip fixation. (**e–i**) Steps of the popLN exposure and creation of a closed chamber. *Dashed line*: position of skin incision. *Small arrowheads*: outline of lateral marginal vein. *Large arrowhead*: position of a characteristic branch of the marginal vein that assists in locating the popLN. *Arrow*: position of popLN. **g**: Ring of vacuum grease. (**h**) Coverglass glued to mounting device (highlighted by a *circular dashed line*). (**i**) Circular metal tube sealed to coverglass. (**j**) Fully assembled preparation. (**k**) Schematic of the temperature control system.

#5 Forceps; and Biologie tip, 0.05 × 0.02 mm, Inox, 11 cm total length (e.g., from Fine Science Tools, Foster City, CA).

7. 18 mm #1 round coverglasses.
8. Ethyl-2-cyanoacrylate glue. (e.g., Crazyglue).
9. Adhesive tape.
10. Suture material (e.g., 5-0 braided silk, Roboz Surgical, Gaithersburg, MD).
11. High-viscosity vacuum grease (e.g., Dow Corning, Midland, MI).
12. Plasticine modeling clay.
13. Miniature K type thermocouple (e.g. 5SC-TT-K-40-72 from Omega, Stamford, CT).

2.4. Multiphoton Intravital Microscopy of the Mouse Popliteal Lymph Node

1. Upright multiphoton microscope equipped with (a) at least two (ideally three to four) non-descanned PMT detectors (e.g., Ultima V, Prairie technologies, Middelton, WI; TriMScope, LaVision Biotec, Bielefeld, Germany; TCS SP5, Leica, Wetzlar, Germany, LSM 710 NLO Zeiss, Jena, Germany; or Fluoview FV1000MPE, Olympus, Center Valley, PA), (b) a high numerical aperture objective lens (Olympus XLUMPLFL20XW 20×, 0.95 NA, water immersion, 2 mm working distance), and (c) a femtosecond-pulsed infrared laser (DeepSee, Spectra-Physics/Newport, Mountain View, CA; or Chameleon Ultra II, Coherent, Santa Clara, CA).
2. Small water bath (e.g., Precision 280 series, Thermo Scientific, Waltham, MA).
3. Roller pump (e.g., Masterflex L/S with Easyload II, Cole Parmer, Vernon Hills, IL).
4. Digital thermometer (e.g., model 710, BK Precision Corporation, Yorba Linda, CA).
5. Polyethylene tubing (e.g., Intramedic PE160, Becton Dickinson, Franklin Lakes, NJ).

2.5. Image Processing and Quantitative Analysis of Lymphocyte Motility

1. Image analysis computer workstation with high graphics processing capability.
2. Image analysis software that allows processing and analysis of 4-dimensional (3-D and time) data sets (e.g., Volocity, Improvision, Coventry, England; or Imaris, Bitplane, Zurich, Switzerland): As a more cost-effective alternative to these commercial products, the freeware program ImageJ with its wide array of freely available plugins (http://rsb.info.nih.gov/ij/) can substitute for some of the most relevant functions of image processing and analysis. More complex computational analyses of lymphocyte migration can be performed through scripts written in Visual Basic (Microsoft, Redmond, WA) or in Matlab (Mathworks, Natick, MA).

3. Methods

3.1. Preparation and Injection of Lymphocytes to Track In Vivo

1. Spleen and LNs from donor mice are harvested and T cells isolated using negative immunomagnetic selection. Resuspend cells at 10^6 cells/mL in PBS with 1% FCS.
2. Stain 10^6 cells with CFSE at 2.5 μM for 15 min in a 37°C waterbath. For direct comparison of cell motility between two different populations, stain the second population (e.g., treated with a small molecule inhibitor) with CMTMR at 10 μM for 15 min at 37°C. To stop the staining reaction, centrifuge cells through a cushion of 2 mL of 100% fetal calf serum, which binds residual protein-binding dyes such as CFSE and CMTMR. Resuspend cells in ~200 μL PBS. Alternative to staining with organic fluorescent dyes, cells from mice which express fluorescent proteins (such as enhanced green fluorescent protein, EGFP, and it color variants) globally or specifically in T cells can be used.
3. Inject lymphocytes intravenously into recipient mice through the lateral tail vein and image the popLN 6–24 h post-injection. For cells which show impaired recruitment to LNs after intravenous injection (a process which depends of $\alpha_L\beta_2$ integrin function), proportionally larger populations of cells relative to the control population should be injected and allowed more time to accumulate in LNs. The density of cells for both populations during image acquisition should be comparable. Be aware of possible selection effects imposed by the recruitment step (e.g., naïve and memory T cell may be differentially impaired in LN recruitment by inhibitor treatment), which may bias subsequent comparative analysis.

3.2. Surgical Anesthesia

Mice are anaesthetized by initial intraperitoneal injection of a mixture of Ketamine, Xylazine, and NaCl 0.9% (to achieve a dose of 150 mg/kg body weight for Ketamine and 10 mg/kg for Xylazine). A surgical plane of anesthesia is achieved if the mouse does not react to firm pinching of the footpad. Repeat injections during the surgical procedure and later on during the imaging session can be carried out intramuscularly to achieve slower release and more even plasma levels of the drugs.

3.3. Microsurgical Preparation of the Mouse Popliteal Lymph Node for Intravital Microscopy

1. The rig mouse leg, flank, and back are shaved using clippers (Fig. 1b) and depilation crème briefly (<30 s) applied to the dorsal aspect of the right leg, avoiding maceration of the skin (see Note 2). Thoroughly wipe off the crème using moist gauze and repeat if required (Fig. 1c). The goal is to remove all hair from this area since fragments of hair remaining in the imaging field produce undesirable autofluorescence.

Shaving the left upper leg facilitates intramuscular injections when the mouse is under the microscope.

2. The mouse is positioned on the microscope stage using adhesive tape attached to the left hindleg, the left foreleg, and the tip of the tail and by applying tension to the right hindleg using a sling of suture attached to one of the toes (Fig. 1d).
3. The tip of the right trochanter major (a prominent bony pivot of the upper leg bone) is identified through palpation and exposed though a 1 cm skin incision in the flank. The attached connective tissue is teased away and the first holding clamp applied (Fig. 1d). During this, the right leg should be in a rotational position so that the popliteal fat pad, which is visible though the skin, is facing straight upward.
4. The dorsal skin is incised over a distance of about 1 cm to expose the upper lumbar spine and a small fraction of its circumference freed of the attached musculature. The second holding clamp is attached to the spine's dorsal aspect (Fig. 1d).
5. Using the iris scissors, the skin overlying the popLN is longitudinally incised over a length of about 1 cm, using the lateral marginal vein as a landmark structure (Fig. 1e). Through blunt dissection, using the fine forceps, some the connective and fat tissue to the left of the lateral marginal vein is gently teased away and the popLN is readily exposed (Fig. 1f). During all times, the LN should be kept moist by applying NaCl 0.9% (see Note 3).
6. Plasticine blocks are molded onto the leg from both sides at the level of the popLN. Applying gentle pressure from both sides pushes the popLN into a slightly elevated position, making it more accessible without the need for extensive tissue dissection (Fig. 1g). Then a ring of vacuum grease is applied to the dry skin around the incision (Fig. 1h). The resulting basin is filled with NaCl 0.9%.
7. A clean coverglass glued to a micropositioning device mounted on the stage is lowered onto the popliteal fossa to seal off the pool of NaCl solution and thus create a closed chamber, taking care not to inclose air bubbles (Fig. 1h). The coverglass can gently touch the surface of the popLN, but excessive pressure will perturb its blood perfusion and should be avoided.
8. A second ring of vacuum grease is applied to the rim of the coverglass and used to seal a piece of metallic tubing made from 16G needle tips that is shaped to an incomplete circle to align with the coverglass (Fig. 1i). The metal tubing rests on another block of plasticine and is attached to flexible plastic tubing that will later be connected to the circular water heating system (Fig. 1j, k) (see Note 4).

9. The thermocouple is positioned between the mouse leg and the coverglass in close proximity to the popLN (Fig. 1i, j).
10. The stage is transferred to the intravital microscope. The heating element is connected to the roller pump and the waterbath and the pump turned on. The thermocouple is plugged into the digital thermometer.

3.4. Multiphoton Intravital Microscopy of the Mouse Popliteal Lymph Node

1. Through the eyepieces of the microscope the popLN can be distinguished from the surrounding tissue structures by its ovoid shape and its green autofluorescence under mercury arc lamp illumination (see Note 5). Negatively contrasted blood vessels can be assessed for robustness of perfusion. Any residual movement of the specimen that is noticeable at this point will preclude acquisition of high-resolution 3D image stacks and should be corrected by readjustment of the preparation under the dissecting microscope.
2. Lymphocyte migration is highly temperature dependent (5). Since the water immersion objective lenses used for MP-IVM function as heat sinks, the ambient popLN temperature must be actively maintained close to 37°C. Perfusing the circular metal tube overlying the coverglass with warm water is a cost-effective and convenient approach, since the temperature can be quickly adjusted by changing the flow rate (Fig. 1k).
3. The laser source is tuned to an appropriate wavelength (e.g., 800 nm for CFSE and CMTMR, 930 nm for EGFP and DsRed) and the microscope equipped with appropriate filter combinations to detect green and orange-red fluorescence emission (e.g., 525/50, and 620/100 nm bandpass filters). If a third channel is available, it can be used to record backscattered second harmonic signals from collagen fibers, which emanate from the tissue with exactly half the illumination wavelength (Emission filters centered on 400 or 460 nm should be used at 800 or 920 nm laser wavelength, respectively). Collagen fibers serve as convenient anatomical landmarks for orientation in the tissue and also for any post-acquisition image correction for small tissue drifts. The green channel will show fluorescence of CFSE-labeled or EGFP-expressing T cells, whereas the red channel will show florescence of CMTMR-labeled or DsRed-expressing lymphocytes. Some autofluorescence will typically be detected in all channels.
4. To test the viability of the preparation, the baseline migratory behavior of labeled lymphocytes is briefly recorded. To this end, a field of view containing the deep T-cell area is chosen (typically >200 μm away from the LN capsule) and a stack of optical sections (e.g., 16 sections spaced 4 μm apart in *z*, corresponding to 60 μm of depth) repetitively recorded (e.g., at 15 s intervals) for 15 min without optical zoom

(when using a 20× lens this typically results in a field size of 500–700 μm). Most commercial image acquisition software allows subsequent browsing through the acquired image sequence along the time axis to visually assess cell motility. At this point attention should also be paid to the stability of the preparation. In particular, a high frequency jitter caused by the respiratory movement of the mouse or the pulsation of a nearby arterial vessel, as well as a gradual drift of the specimen need to be corrected by readjustment of the preparation under the dissection microscope. At this point, the optimal microscope settings (laser power, PMT gain, offset, etc.) to obtain balanced signals of CFSE and CMTMR is determined and loaded.

5. Once a suitable location with large numbers of fluorescently labeled cells is found, begin image acquisition. Note that T-cell motility is not uniform in all regions of the LN, but is lower in superficial location close to the B cell follicles. To limit the potential bias introduced by selecting a certain region, record from different location of the T-cell area. Individual 30–60 min recordings usually provide a sufficient amount of cell tracks for a meaningful statistical analysis. Ideally, a continuous recording of several hours should be generated, but because of practical limitations, typically multiple sequential recordings have to be obtained. The recorded volume should be as large as possible to allow for tracking of individual cells of long periods, but again practical limitations, e g., on file size, must be considered (see Note 6).

3.5. Image Processing and Quantitative Analysis of Cell Migration Behavior

As outlined in Fig. 2, data analysis can be divided into at least two steps: (1) Cell tracking and (2) Track analysis. The first step can nowadays be performed in a semi-automated fashion using specialized software packages, such as Volocity or Imaris. These applications not only help to convert time-series of stacks of optical sections into maximum intensity-projected or three-dimensionally rendered movies, but also provide image segmentation and object tracking capabilities to extract the *x*, *y*, and *z*-spatial coordinates of individual migrating cells over time ("cell tracking"). Upon automated cell tracking, it is important to visually inspect each track and delete those that track accumulations of fluorescence that are not associated with live cells. Computational tools have been developed to help identify other tracking artifacts (6). Alternative to automated tracking, the open source-program ImageJ (http://www.rsbweb.nih.gov/ij/) provides a cost-free tool for slightly more laborious, computer-assisted manual tracking of migrating lymphocytes.

Information on the calibrated (in μm instead of pixels) spatial position in the *x*, *y*, and *z* dimension for all individual time points of measured T-lymphocyte tracks are exported as tab-delimited *.txt-files.

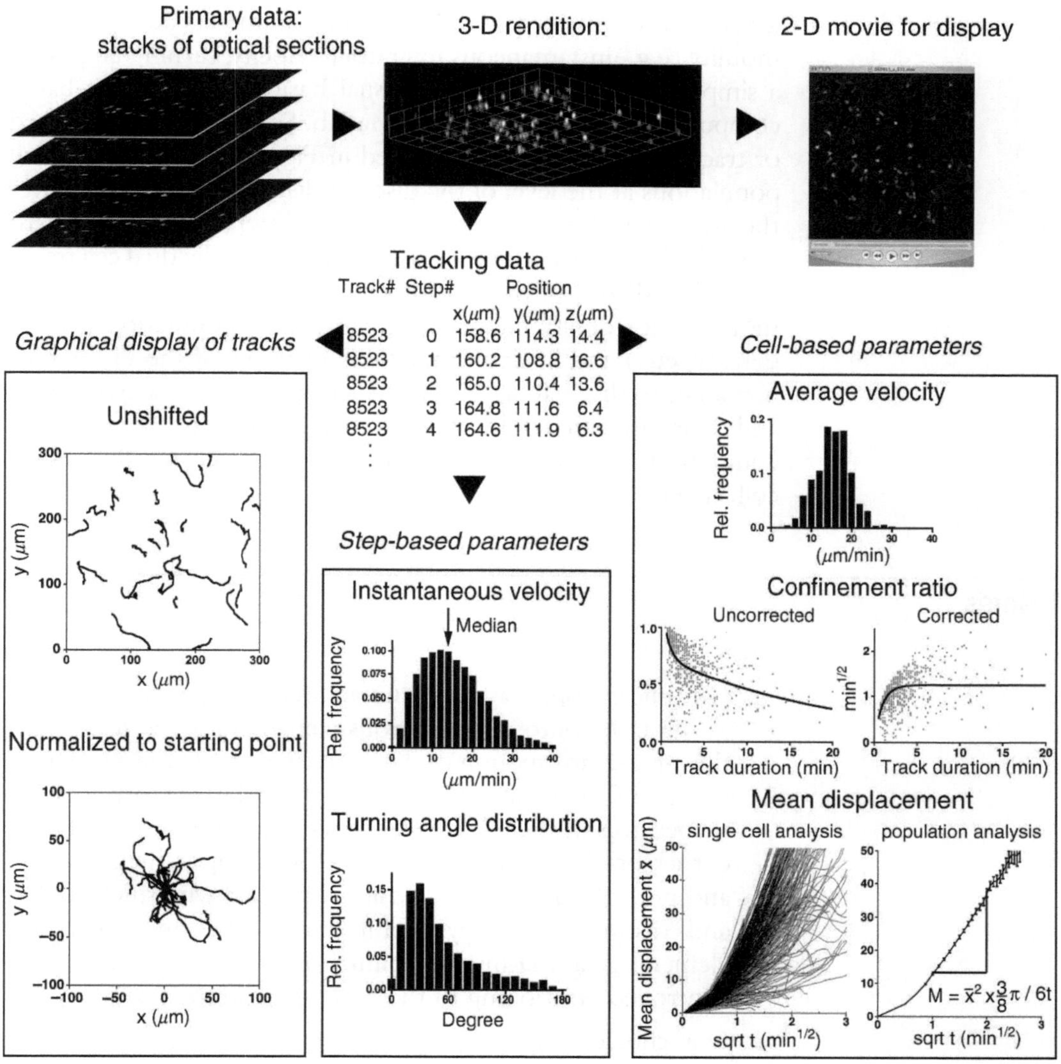

Fig. 2. Steps of the analysis process to obtain quantitative data on cell motility from 4-dimensional data sets. Stacks of optical sections are rendered into 3-D data sets through software that extrapolates the information contained in the sections to "fill in" the space between the individual sections. Typically, orthogonal views of these 3-D renditions are exported as movies for display purposes. Virtual cell centroids or centers of mass of cells in the 3-D volume over time are determined and cell tracks generated. The resulting positional information (*x*, *y*, *z* coordinates, time) is the basis for all subsequent computational analysis. For an intuitive assessment of cell motility, tracks can be displayed graphically. Step-based parameters, on the one hand, provide information on speed and directionality of migration on the population level, and are less prone to analytical bias than cell-based parameters. Cell-based parameters, on the other hand, are more powerful to detect heterogeneity within cell populations. Confinement (chemotactic index, meandering index) is described by the ratio of net displacement and the total path length of a track. To correct for varying path lengths, which affect the confinement ratio, a correction factor comprised of the square root of the track duration, can be introduced ("Corrected confinement ratio") (6). Mean displacement (MD) plots can be used to calculate the motility coefficient for cell populations following biased random walk behavior (7). Use of the mean displacement (instead of mean square displacement) requires the introduction of a correction factor (3/8 for 3-D measurements) when calculating the motility coefficient M from the slope of the MD plot.

These can then be computed to obtain parameters describing cell motility (e.g., instantaneous migratory velocity, turning angels) in a simple Excel spreadsheet, in Visual Basic, or in a matrix-based computing environment such as Matlab. Several useful step-based or track-based parameters are listed in Fig. 2. These describe cell populations at the level of the discrete displacements of cells over the arbitrarily chosen measurement time interval, without regard to track identity (step-based), or at the level of individual cell tracks (track-based). All of these parameters differ in the aspects of cell motility they describe (speed, directionality, population heterogeneity, etc.) and in their limitations in terms of the biases they introduce to the analysis. For a detailed review of these parameters and their susceptibility to analytical bias, see ref. (6). We routinely compute a large panel of parameters and interpret the results collectively.

4. Notes

1. Our microscope stage is manufactured using a plexiglas plate as the base, onto which various elements for the fixation of the mouse are mounted. These elements for the most part consist of materials available in any hardware store that are shaped to suit their particular function. We found small steel corner brackets, angle style steel shelf support pegs, screws and nuts of various calibers, silicone-based windshield sealer, and casting resin very useful items. The only expensive element is a 3D-micromanipulator, which allows for the controlled positioning of the coverglass on the popLN.
2. Our choice of the popLN for MP-IVM studies over other skin-draining LNs was based on two observations: (1) The quantity of footpad-injected reagents that drains to the LN can be controlled more accurately with the popLN than with other skin-draining LNs; (2) Due to the popLN's remote location relative to the animal's trunk, the respiratory movements, which would inevitably cause motion artifacts during the slow image acquisition achieved by laser-scanning microscopy, could be shielded off effectively without applying pressure to the surrounding tissues. Possibly as a consequence of this, we consistently detected robust lymph flow in the popLN model.
3. During initial attempts with this procedure, sparing the afferent lymph vessels during the surgical dissection may be a technical challenge. This can be helped by footpad injection of a dye (e.g., Evans blue) in order to outline the draining lymphatic vasculature. Typically, at least three afferent lymph

vessels can be identified entering the popLN during the procedure. Blunt dissection (insertion of closed forceps or scissors into the tissue followed by their opening) minimizes bleeding. If a persistent bleeding still occurs, it can usually be stopped by grabbing and applying pressure on the vessel with forceps for ~30 s. Perfusion of the popLN occurs through perforating, deeper arteries and veins and is therefore not affected by damage to superficial blood vessels.

4. At this step it is important that the metal tube rests strain free on the coverglass. Otherwise slow release of tension (e.g., during warming and softening of the plasticine block) can lead to a gradual drift of the specimen, precluding acquisition from the same depth of the LN.
5. If a conventional epifluorescence unit is unavailable, locating the popLN can be helped by drawing a circle around it on the dry coverglass with a fine marker pen. Oblique brightfield epi-illumination with a handheld flashlight then allows viewing of the specimen though the eyepieces of the microscope. Despite poor image contrast the drawn circle helps to identify the popLN.
6. The duration of the recording is limited by: (1) The requirement to obtain access to the animal in order to maintain surgical anesthesia, (2) Evaporation of the immersion water, (3) Restrictions of either the acquisition software or the computer hardware on the maximum size of the recorded data files. Besides the total length of the recording, the file size can typically be adjusted through the pixel resolution, the number of *z* planes, and the cycle time. We typically use a pixel resolution of 256×256, 16 optical sections, and a cycle time of 15 s, since this provides us with a good compromise between file size, image detail, and the ability to follow individual motile cells for sufficient time to obtain meaningful data.

References

1. Denk, W., Strickler, J.H., Webb., W.W. (1990) Two-photon laser scanning fluorescence microscopy. *Science* **248**, 73–76.
2. Helmchen, F., Denk, W. (2005) Deep tissue two-photon microscopy. *Nat. Methods* **2**, 932–940.
3. Cahalan, M.D., Parker, I., Wei, S.H., Miller, M.J. (2002) Two-photon tissue imaging: seeing the immune system in a fresh light. *Nat. Rev. Immunol.* **2**, 872–880.
4. Pittet, M.J., Mempel, T.R. (2008) Regulation of T-cell migration and effector functions: insights from in vivo imaging studies. *Immunol. Rev.* **221**, 107–129.
5. Miller, M.J., Wei, S.H., Parker, I., Cahalan, M.D. (2002) Two-photon imaging of lymphocyte motility and antigen response in intact lymph node. *Science* **296**, 1869–1873.
6. Beltman, J.B., A.F. Maree, and R.J. de Boer. (2009) Analysing immune cell migration. *Nat. Rev. Immunol.* **9**, 789–798.
7. Sumen, C., Mempel, T.R., Mazo, I.B., Von Andrian, U.H. (2004) Intravital microscopy; visualizing immunity in context. *Immunity* **21**, 315–329.

Part IV

Signaling Through Cell Adhesion Molecules

Chapter 17

Overview of Integrin Signaling in the Immune System

Tatsuo Kinashi

Abstract

It has been well established that integrins mediate cell–cell and cell–matrix adhesion and play crucial roles in the immune system such as leukocyte–endothelium interactions, immune synapse formation, and effector functions. Since the discovery that integrins undergo dynamic changes of adhesive activities in response to external stimuli, intensive studies have been conducted to elucidate the signaling events that control the activation of integrins (inside-out signaling) and signaling events from the induced integrin-dependent adhesion (outside-in signaling). The molecular characterization of these signaling pathways highlights the importance of integrins as bidirectional signaling receptors. The characteristics of integrin signaling are best exemplified in the immune system. This chapter highlights the recent studies of intracellular signaling pathways that regulate integrins in immunological contexts.

Key words: Integrin, Inside-out signaling, Outside-in signaling, Leukocytes, Chemokines, T-cell receptor (TCR), B-cell receptor (BCR), Small GTPase, Tyrosine kinase, Adaptor protein

1. Introduction

The immune system is supported by highly dynamic and contact-dependent processes, in which dynamic changes of integrin play crucial roles during homeostasis and disease. It was first reported almost 20 years ago that a transient change of LFA-1 adhesiveness occurred upon TCR crosslinking (1). Since then, activation of leukocyte integrins (upregulation of integrin adhesiveness) by external stimuli has been extensively studied. Integrins expressed on leukocyte are inactive in a resting state, but rapidly increase adhesiveness upon stimulation with antigens, chemokines, and cytokines. Basically, integrin adhesiveness is regulated by ligand-binding affinity (conformation) and cell surface clustering (valency). By definition, intracellular signaling events that increase integrin adhesiveness before ligand engagement is termed inside-out signaling. Conversely, adhesion strengthening and subsequent

Motomu Shimaoka (ed.), *Integrin and Cell Adhesion Molecules: Methods and Protocols*,
Methods in Molecular Biology, vol. 757, DOI 10.1007/978-1-61779-166-6_17, © Springer Science+Business Media, LLC 2011

morphological and functional changes following ligand engagement are categorized as outside-in signaling. Distinguishing between inside-out and outside-in signaling is often difficult as the two events often occur sequentially and rapidly in the immune cells. However, in recent years studies with gene-targeting and knock-down by short interference RNA (siRNA) techniques have revealed a number of signaling molecules involved in integrin activation and signaling processes. Since integrin-related adhesive events are too broad to cover in a limited space, this chapter will focus on two representative integrin activation (inside-out) signaling pathways, antigen receptors and chemokines, and also cover outside-in signaling in the immune system for readers interested in signaling aspects of integrin regulation.

2. Chemokine Signaling to Integrins

2.1. Leukocyte–Endothelial Interactions

During immunological surveillance naive T cells re-circulate between peripheral blood and lymphoid tissues. Upon entry into lymphoid tissues, T cells attach to a specialized post-capillary venule, the high endothelial venule (HEV), through a process known as the lymphocyte adhesion cascade involving sequential steps of selectin-dependent rolling, chemokine-triggered activation, and integrin-dependent adhesion (2–4) (Fig. 1). The activation of neutrophils and monocytes by inflammatory chemokines induces a transition from rolling to arrest adhesion by increasing the adhesiveness of integrins for their cognate ligands expressed on inflamed endothelium. This three-step model has been generally accepted during homeostasis and inflammation with some modifications, depending on the cellular contexts and molecules involved. For example, selectin-carbohydrate engagements can transmit signals (5, 6) and activate integrins (7). β2 integrins (LFA-1, Mac-1) and α4 integrins (α4β1, α4β7) can mediate rolling as well as arrest (8–10).

Chemokines play a major role in integrin activation, resulting in binding to integrin ligands on the endothelium. In the classical adhesion cascade, when the inside-out signaling is inhibited, leukocytes continue rolling on the endothelium. The combination of gene-targeting and siRNA technologies to deplete target proteins and in vitro flow adhesion assays reproducing L-selectin-dependent rolling to chemokine-induced arrest (11) has revealed a post-adhesion strengthening step after the initial lymphocyte arrest. Integrin activation can be also measured using activation state-specific antibodies that recognize extended, high-affinity open conformations, although this method does not necessarily guarantee the role of the signaling molecule in arrest adhesion. As described below, recent studies on signaling molecules have shown

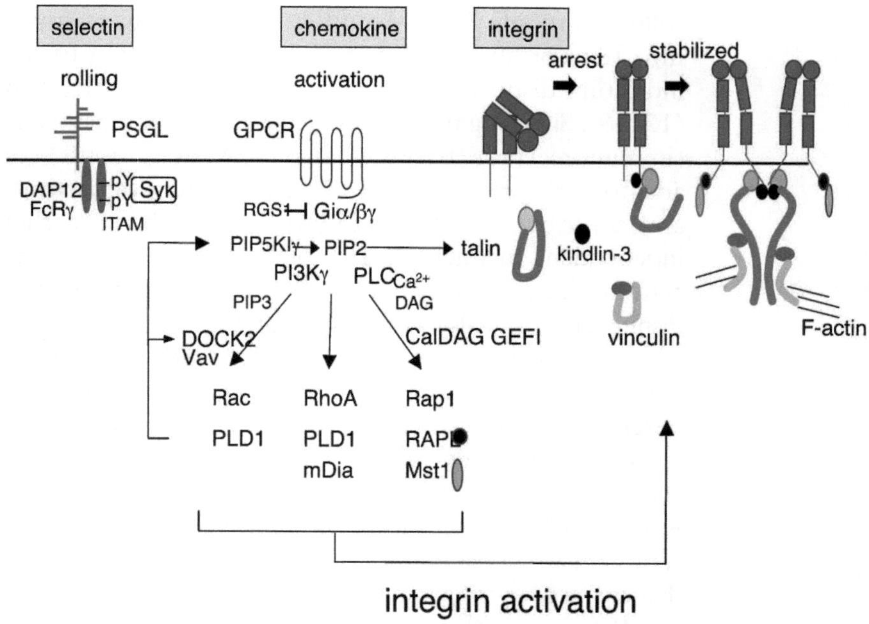

Fig. 1. Inside-out signaling in the leukocyte adhesion cascade. During selectin-dependent rolling on the endothelium, endothelial chemokines trigger inside-out signaling and leads to integrin-mediated arrest. The major intracellular signaling molecules are indicated. Signaling through binding to selectins activate integrins, depending on Syk and ITAM-containing FcRγ and DAP12. Chemokine activates the Gi family heterotrimeric G proteins, which leads to the activation of small GTPases (Rap1, Rac, RhoA) and the induction of arrest. Integrins exist in an inactive, bent form, and undergo conformational changes (extended intermediate affinity, extended high affinity) and clustering. Transition from rolling to arrest is followed by the adhesion-strengthening step. Exact signaling events controlling conformational changes during the adhesion cascade are not clear, but small GTPases are required for the initial arrest and RAPL/Mst1 are required in the adhesion-strengthening step. Talin and kindlin-3 directly bind to the β integrin tail and links with the actin cytoskeleton. The exact step involving talin and kindlin-3 is not clear but could be the initial arrest or stabilized step. Vinculin further associates with talin to connect the actin cytoskeleton to integrin.

that selectins are involved in integrin activation. Furthermore, adhesion stabilization is an important post-adhesion step that strengthens leukocyte attachment to the endothelium. Thus, from the viewpoint of the signaling pathways, the classical three-step model stands to be revised to dissect the true arrest process (12).

2.2. Inside-Out Signaling for Leukocyte Arrest

It is well established that endothelial chemokines induce arrest from rolling through the binding of leukocyte integrins to immunoglobulin superfamily proteins (ICAM-1, VCAM-1) expressed by HEV and inflamed endothelial cells. Chemokines do not distribute in a gradient fashion, but are localized on the apical surface of HEV, mostly through binding to glycosaminoglycan (13). In vitro flow assays have demonstrated that the transition from rolling to arrest can be mediated by immobilized chemokines on the endothelium, but not using soluble chemokines (14, 15). Lymphocyte arrest mediated by L-selectin-dependent rolling in response to chemokines occurs within a second, whereas neutrophils rolling through PSGL-1 and E-selectin requires much longer time (tens or hundreds of seconds) before coming to arrest (16), indicating distinct cell-dependent modalities in arrest induction (17, 18). Signaling through Syk (spleen tyrosine kinase) and ITAMs (immunoreceptor tyrosine-based activation motifs)-containing adaptors triggered by E-selectin engagement can activate LFA-1 and contribute to slow rolling of neutrophil in a chemokine-independent manner (7, 19). Since selectins do not appear to activate integrins in lymphocytes (18), neutrophils may employ distinct arrest signals in addition to those triggered by chemokines.

Leukocyte arrest by chemokine and chemoattractant is abrogated by pertussis toxin, which inhibits the Gi/o family of heterotrimeric G-proteins (2, 20, 21). Lymphocytes and neutrophils predominantly express Giα2 and Giα3. Giα2-deficient B cells exhibited poor attachment to HEV, whereas deficiency of RGS1, a GTPase-activating protein, increased attachment (22). Although arrest was not directly investigated in that study, it is likely that chemokine-triggered Giα2 played the major role in B cell arrest on HEV. It is unclear whether T cells also utilize Giα2 exclusively. Activation of pertussis toxin-sensitive Gi proteins resulted in release of α and βγ subunits, leading to inhibition of adenyl cyclase, and activation of phospholipase C (PLC), PI3 kinase, Ras/Rho family of small GTPases, tyrosine kinases, and MAP kinases.

Previous studies suggested a critical role for PLC in lymphocyte arrest, as the PLC-specific inhibitors inhibit arrest mediated by LFA-1 (23) and VLA-4 (24). Although PLCβ2 and PLCβ3 are required for chemotaxis of T cells (25), PLCβ2 and PLCβ3 double-deficient mice do not have a defect in arrest adhesion of lymphocytes and neutrophils (26). Further studies are needed to examine PLC isoforms required for arrest adhesion. A key target of PLC activity for integrin activation triggered by chemokines is CalDAG-GEF-I (RasGRP2), a Ca^{2+} and diacylglycerol-dependent guanine exchange factor for the small GTPase Rap1. A CalDAG-GEFI deficiency in mice caused defective β1, β2, and β3 integrin activation in platelets and leukocytes (23, 27–29). Interestingly, patients with a rare adhesion deficiency syndrome termed LAD-III, have a splice junction mutation in the CalDAG-GEFI gene (29).

The small GTPase Rap1 has received much attention as a cell adhesion regulator (30, 31). Rap1 is a potent activation signal for β1, β2, and β3 integrins, and enhances cellular adhesion in both immune and non-immune cells. Lymphocyte Rap1, which is rapidly activated by chemokines and cognate antigens, increases the adhesiveness of integrins to their ligands by modulating affinity and avidity (clustering), induces polarized cell shape, and facilitates cell migration (32) (see Chapter 24). Amino acid sequence identity of Rap1a and Rap1b is approximately 95% and functionally interchangeable regarding integrin activation (33, 34). Most tissues express both Rap1a and Rap1b to varying degrees.

Targeted deletion of the *rap1a* or *rap1b* genes resulted in impaired activation of lymphocyte (α4β1, β2) and platelet integrins (αIIbβ3), respectively (35–38). While both VLA-4 and LFA-1 were affected by Rap1 deficiency in mice, human peripheral T cells exhibited impaired LFA-1, but not VLA-4 by inhibition or depletion of Rap1, suggesting species distinct activation of integrins (18). Double deficiency of Rap1a and Rap1b has not yet been reported. In one study, Chat-H (Cas and Hef-1-associated signal transducer in hematopoietic cells) was shown to be required for chemokine-induced activation of the Rap-1 and integrin-mediated adhesion by using lentivirus-mediated RNA interference. Localization of Chat-H and its binding partner CasL (Crk-associated substrate in lymphocytes) to the plasma membrane were required for T-cell migration (39).

RAPL was identified as a Rap1-GTP-binding protein by yeast two-hybrid screening. RAPL is highly expressed in immune cells and it mediates Rap1 functions on integrins through association with the αL cytoplasmic subunit (40). Targeted deletion of the *rapl* gene impaired chemokine-induced lymphocyte adhesion and trafficking to the peripheral lymph nodes (41). The detailed analysis of arrest adhesion in vivo and in vitro indicated that $RAPL^{-/-}$ lymphocytes are defective in post-arrest adhesion strengthening, resulting in unstable attachment to HEV (Ebisuno, submitted). RIAM was also identified as a Rap1-GTP-binding protein, and was implicated as a Rap1 effector which activates LFA-1 (42). RIAM was also required for localization of Rap1 in the plasma membrane (42), as well as talin recruitment (43). Knockdown of human RAPL, but not RIAM, in T cells impaired chemokine-induced adhesion by VLA-4 (44). Thus, RIAM does not appear to be involved in inside-out signaling triggered by chemokines, instead it plays a role in TCR-induced adhesion (see below). Mst1 (Stk4), a ste20-like kinase, was identified as a RAPL-associated molecule. RAPL regulated Mst1 kinase activity and subcellular localization in response to chemokines (45). Mst1-deficient mice exhibited hypoplastic lymphoid organs due to defective lymphocyte trafficking. Mst1 deficiency did not affect L-selectin-mediated rolling and LFA-1-dependent arrest by chemokines, but impaired stable attachment to ICAM-1 under physiological shear flow (11). Similar phenotypes of Mst1 knockout mice are also reported by other groups (46, 47).

Chemokine-triggered RhoA and Rac1 activation in human primary T cells increased LFA-1 affinity and adhesion under shear flow (48–50), whereas Cdc42 was inhibitory on integrins (50). Importantly, PLD1 was indicated as a common downstream effector of RhoA and Rac1 on LFA-1 affinity modulation and attachment to HEV. Furthermore, inhibition of PLD1 led to decreased phosphatidyl inositol-4,5 biphosphate (PIP_2), a lipid involved in talin activation and actin reorganization (50).

Although further studies are necessary to establish a link between PLD1 and integrins, the Bolomini-Vittori study provides compelling evidence of the importance of RhoA and Rac1 in integrin activation of human T cells.

The Rho effector in straight actin filament generation, mDia1, was required for efficient T cell homing and chemotaxis in mice, since these processes were impaired in mDia1$^{-/-}$ T cells, but not B cells (51). Interestingly, Wiscott–Aldrich syndrome protein (WASP) was found to be downregulated in mDia1$^{-/-}$ T cells. However, WASP is dispensable for integrin-dependent adhesion (52), suggesting that mDia1 deficiency is responsible for impaired trafficking. Although this study did not examine arrest adhesion directly, it suggests an important role of these regulators on the actin cytoskeleton in T cell trafficking in mice. DOCK2, a Rac-GEF, plays an indispensable role in actin polymerization and chemotaxis of lymphocytes (53). DOCK2$^{-/-}$ B cells, but not T cells, were defective for attachment to HEV due to impaired activation of LFA-1 and α4 integrins (54). DOCK2 binds to PIP_3 (phosphatidylinositol 3,4,5-triphosphate) and phosphatidic acid, resulting in recruitment to the plasma membrane mediating the GTP-GDP exchange reaction of Rac (55). Thus, PI3 kinase and PLD are thought to be upstream of DOCK2. PI3Kγ deficiency abrogated the initial PIP_3-dependent membrane recruitment of DOCK2, but not later PLD-dependent recruitment (55). It is unclear whether Rac, Rho, and Rap1 crosstalk. Interestingly, it was reported that PLD1 was expressed on the same vesicular compartment in Jurkat T cells as Rap1, and Rap1 recruitment to the plasma membrane required PLD1 (56). It is therefore possible that spatial regulation through lipids, vesicle transport and cytoskeleton wire a crosstalk of small GTPases, and play important roles in integrin activation triggered by chemokines.

3. Antigen Receptor Signaling to Integrins

TCR-triggered LFA-1-dependent adhesion to ICAM-1 is a hallmark of inside-out signaling (1). A specialized antigen-dependent adhesion complex, known as the immunological synapse (IS) or SMAC (supramolecular activation cluster), were generated between T cells and antigen-presenting B cells or planar lipid membranes (57, 58) (Fig. 2). In mature IS an external LFA-1 ring surrounds central TCR clusters in the contact zone, which could facilitate relatively weak antigen binding by the TCR (59). Advances in imaging techniques has revealed that T-cell interactions with antigen-presenting cells (APC) is a dynamic process, rather than a long-lived stable adhesion, in which T cells crawl on, or contact briefly with APC, followed by stable adhesion resulting

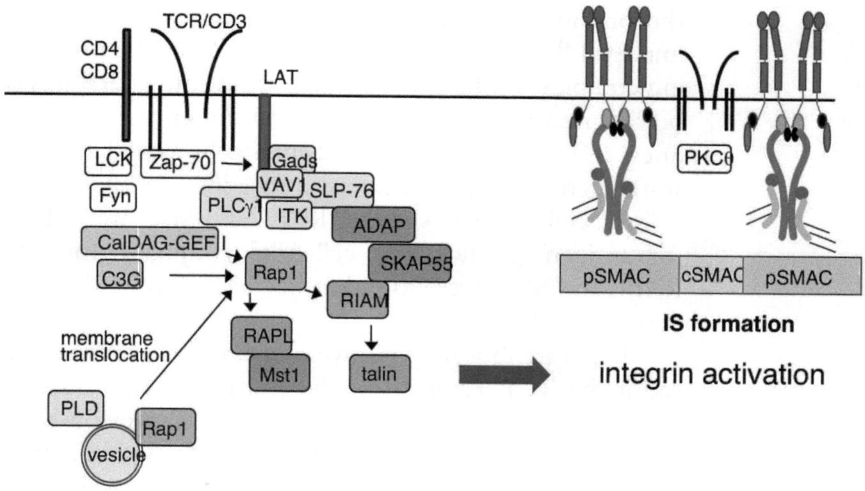

Fig. 2. TCR-mediated inside-out signaling pathways. Following TCR ligation, tyrosine kinases (LCK, FYN, ZAP-70) trigger the early signaling events, resulting in the LAT-VAV1-ITK-SLP76 complex formation, and leads to PLCγ1 activation. Rap1 activation occurs via Rap1-specific exchange factors, CalDAG-GEF I or C3G. Translocation of Rap1 to the plasma membrane occurs through vesicular transport that requires PLD activity. The binding of SLP76 to the ADAP/SKAP55 complex activates integrins in a Rap1-dependent manner via the RIAM-talin or RAPL-Mst1 pathways, leading to immunological synapse (IS) formation. In the mature IS, integrins (LFA-1) are redistributed to the peripheral SMAC (pSMAC) surrounding the TCR cluster in the central SMAC (cSMAC).

in T-cell clusters and proliferation (60, 61). These findings led to important issues of post-adhesion regulation, such as how spatial regulation of LFA-1 could be achieved upon TCR triggering, what makes a switch between mobile and stable contacts, or what are consequences of changes of adhesive behaviors on antigen recognition and response (62). These processes involve both inside-out and outside-in signaling, and are likely influenced by chemokines. To dissect the complicated process, it is crucial to examine the signaling pathways downstream of the TCR and clarify their direct effects on integrins in defined cellular contexts.

The early signaling of TCR engagement is an activation step of Src protein tyrosine kinases (Lck and Fyn) that phosphorylate the ITAM motifs of the CD3 complex, resulting in recruitment of the Syk kinase family member Zap70. Recruitment of Zap70 leads to a cascade of phosphorylation events and an assembly of the multimolecular signaling complex, which is responsible for propagating the TCR-triggered signal into diverse distal signaling pathways. The important targets of Zap70 are the adaptor proteins LAT (linker for activation of T cells) and SLP-76 (Src homology 2 domain-containing leukocyte protein of 76 kDa). These adaptors nucleate a complex composed of a TEC-family kinase ITK, a Rac/Cdc42 exchange factor VAV1, and the p85 subunit of PI3-kinase (63). ITK is important for the formation of the LAT-VAV1-SLP-76 complex and for PLCγ1 phosphorylation. Disruption of the proximal signaling complex, including VAV1, PI3K, ITK, impaired β1 and β2 integrins (52, 64–68). The proximal signaling complex results in the PLCγ1 activation, which was shown to be required for Rap1 activation and integrin activation in T-cell lines (69). Indeed, Rap1a-deficient T cells were defective in TCR-stimulated integrin clustering and adhesion (37). The studies using T cell lines showed that RAPL and Mst1 were also required for antigen-dependent T cell-APC conjugate formation, and RAPL and Mst1 were found to be colocalized with LFA-1 in the peripheral SMAC (40, 45).

The search for the selective signaling pathway to integrin activation by TCR came from the study of ADAP (adhesion and degranulation promoting adaptor protein, previously known as Fyb or SLAP-130) (70, 71). ADAP is expressed in hematopoietic cells, but not in B cells. ADAP associates with SLP76 after TCR ligation. $ADAP^{-/-}$ T cells showed impaired TCR-ligation-induced β1 and β2 integrin-dependent adhesion, although calcium flux and MAPK activation occurred normally. ADAP deficiency impaired TCR-ligation-induced LFA-1 clustering, but did not affect TCR clustering itself or assembly of the actin cytoskeleton. ADAP deficiency also resulted in reduced proliferation and IL-2 production. SKAP55 (Src-kinase associated phosphoprotein of 55 kDa, also known as SKAP1), constitutively associates with ADAP. When overexpressed, SKAP55 augmented LFA-1-mediated adhesion of primary T cells and IS formation (72). Interestingly, ADAP deficiency downregulated SKAP55 expression due to protein instability (73, 74), which raised an issue whether the phenotype of ADAP deficiency could be caused by ADAP or SKAP55. Since $SKAP55^{-/-}$ T cells expressed ADAP normally, and displayed the defects of integrin as well as cytokine production and proliferation, as seen in $ADAP^{-/-}$ T cells, SKAP55 is likely the effector for integrin activation (73). However, SKAP55, in the absence of ADAP, was insufficient to drive antigen-dependent T-cell–APC conjugate formation, suggesting that the ADAP/SKAP55 module is required for efficient IS formation (75).

The connection of the ADAP/SKAP55 module with Rap1 was suggested by a study showing that this module was required for the plasma membrane localization of Rap1 (74). RIAM appears to play an important role in this process. RIAM is constitutively bound to SKAP55 and associated with Rap1 after TCR ligation, resulting in Rap1 movement to the membrane (76). It is not clear how RIAM associates Rap1-GTP and SKAP55 through the RA (Ras association) domain. Although physiological relevance of the ADAP/SKAP55/RIAM complex in T cells has not yet been reported, in an heterologous cell system the Rap1/RIAM complex activated αIIbβ3 by forming a complex containing talin (43), suggesting the possibility that this signaling module directly associate and activate integrins through talin-regulated actin cytoskeleton.

Since requirement of ADAP and SKAP55 for T-cell adhesion was reduced when stronger TCR ligation was applied (73, 77), alternative signaling pathways must operate to activate integrins. As mentioned above, membrane localization and activation of Rap1 triggered by TCR ligation required PLD1, suggesting a role of vesicular transport in Rap1 activation and T-cell adhesion (56). In line with this, the RAPL-Mst1 pathway was shown to be present in vesicular compartments containing Rap1 and LFA-1 in resting lymphocytes (45). The other pathway connecting Rap1 activation with integrins is the serine/threonine kinase protein kinase D1 (PKD1) (78). After TCR ligation, PKD1 associated with Rap1 through the PH domain, and PKD1 further associated with β1 integrin through the carboxy-terminal end. The interaction of PKD1 and Rap1 did not require the kinase activity of PKD1. Since PKD1 also associated with Rap1-GEF, C3G, this could lead to Rap1 activation. It is unknown whether the same pathway exists for β2 integrin.

When B cells recognize antigens on the surface of APC, such as follicular dendritic cells, B cells develop an IS with a central cluster of BCR surrounded by a ring of LFA-1 and ICAM-1, similar to that seen in the T-cell IS (79). These adhesive responses may be beneficial for B cells to extract surface-bound antigens presented on APC (80). In B cells, after phosphorylation of ITAMs in the BCR, Syk phosphorylates the SLP76 homolog SLP-65, which then associates with BTK (Bruton's tyrosine kinase), leading to PLCγ2 phosphorylation and activation. The inside-out signaling pathway in B cells utilizes similar signaling components involved in T cells. VLA-4 mediated adhesion by BCR ligation required the protein kinases, Lyn and Syk, PI3 kinase, Btk, and PLCγ2 (81). Rap small GTPases also play an important role in B-cell adhesion (82, 83). Rap1a- or Rap1b-deficient B cells were defective in integrin-dependent adhesion as well as B-cell development (37, 38). Rap1 and Rap2 were activated in BCR-triggered adhesion through integrins, and play an important role in firm attachment and cell spreading (82, 83). Rac2, but not Rac1, was also shown to be critical for ICAM-1-dependent adhesion of B cells upon BCR ligation. Rac2 deficiency resulted in defective Rap1 activation in B cells, suggesting the importance of Rap1 activation downstream of Rac2. Regarding downstream signaling of Rap1, ADAP, and SKAP55 are not expressed in B cells. Although the B-cell ADAP homolog has yet to be determined, the SKAP55 homolog SKAP55R (also known as SKAP-HOM) is expressed in B cells, and the loss of SKAP55R caused reduced adhesion to ICAM-1 and fibronectin in B cells upon BCR ligation (84). Rap1 or Rap2 may utilize SKAP55R for B-cell adhesion, but the signaling pathways upstream and downstream of SKAP55R remain undetermined.

Overall, the inside-out signaling during the formation of the IS in T and B cells is beginning to emerge. Enhanced adhesion between lymphocytes and APC leads to increased contact areas, facilitating antigen recognition of T cells and antigen extraction of B cells for T cell-help. Thus, IS formation should be coordinately regulated by both inside-out and outside-in signaling. How these signaling pathways could translate into dynamic aspects of IS formation leading to productive immune responses awaits further investigation.

4. Outside-In Signaling

The signaling following ligand-engaged integrins are referred to as outside-in signaling (85). The outside-in signaling was known to be associated with integrin clustering (85) and also with separation of the cytoplasmic tails of α and β integrin subunits (86). Although separation of the integrin cytoplasmic tails by chemokines could be detected by a sensitive method using Fluorescence Resonance Energy Transfer analysis (FRET) (87), this study demonstrated that integrin activation could occur without tail separation whereas outside-in signaling does not. Since leukocytes require inside-out signals from antigen receptors and surface receptors in response to cytokines and chemokines for integrin activation, it is often difficult to separate outside-in signaling events specific to integrins. Nonetheless, it is clear that there are general (e.g., firm attachment and cell spreading) and specific (e.g., degranulation and respiratory burst) cellular responses that require outside-in signaling. These responses are likely regulated by an intricate network of protein and lipid kinases, small GTPases, their regulators and adaptor proteins. The literature on these molecules in integrin signaling is vast and beyond the scope of this section. Below is a brief summary of recent developments in this field.

Studies using neutrophils revealed that the deficiency of Src family kinases (Hck, Fgr, and Lyn) abrogated neutrophil degranulation and ROS (reactive oxygen species) production as well as firm adhesion and cell spreading following β2 and β3 integrin engagement, while the inside-out signaling to activate integrins was not affected (88–90). Similarly, platelets deficient for Src, Hck, Fgr, and Lyn exhibited complete defects in cell spreading on fibrinogen (91). Syk deficiency abrogated β1, β2, and β3 integrin signaling events in neutrophils and macrophages (92). Despite numerous studies indicating the importance of Src and Syk protein kinases in outside-in signaling, the molecular mechanism by which integrins activate these kinases has been elusive. Studies using FRET detected an association of β3 integrin and Src (93). These studies suggest that integrin clustering following ligand engagement could bring together associated Src kinases resulting in their activation and the initiation of the downstream signaling (91). More recently, it has become clear that two ITAM-containing adaptors FcRγ and DAP12 (DNAX activation protein-12) are requisitely involved in outside-in signaling from β2 and β3 integrins (94–96). In these cases, the binding of the Syk SH2 domain to the ITAMs phosphorylated by Src kinases is critical for outside-in signaling from integrins in neutrophils, platelets, and osteoclasts. Thus, as is the case with the antigen receptors, ITAM signaling is also induced by cell–cell and cell–extracellular matrix interactions that are mediated by integrins. However, it remains unclear how ITAM-containing signaling adaptors could associate with integrins.

Similar to antigen receptors, the SLP76/Vav/PLCγ complex was induced by integrin ligation. Disruption of the complex resulted in inhibition of the outside-in signaling (97). Deficiency of all three Vav family members severely inhibited cdc42, Rac, RhoA, and β2 integrin signaling, including defective firm cell attachment, cell spreading, and oxidative burst of neutrophils (98). Vav-deficient neutrophils failed to activate PLCγ2 (98). In the myeloid lineage, SLP76 deficiency severely inhibited outside-in signaling from β2 and β3 integrins (94, 99). ADAP-deficient neutrophils exhibited defects in adhesion-induced superoxide production (97). The fact that outside-in signaling molecules often overlap with inside-out signaling molecules highlights the importance of the cell-specific signaling, and also suggests in some cases that inside-out and outside-in signaling is a continuous process. For example, both selectin-triggered inside-out signaling to LFA-1 (19) and outside-in signaling from β2 and β3 integrins in neutrophils (95) occurs via ITAMs.

5. Talin and Kindlin-3

There are two FERM (protein 4.1, ezrin, radixin and moesin) domain-containing proteins, talin and kindlin-3 (100), that are critical for inside-out and outside-in signaling. Talin is a ubiquitous 270 kDa protein that links integrins to the actin cytoskeleton. It is composed of a head containing the integrin-binding FERM domain and a rod that binds to the F-actin binding protein vinculin, and also contains a second binding site of integrins. The binding of talin to the cytoplasmic tail region of β integrins induces a conformational change which increases its affinity, resulting in the development of focal adhesion through integrin clustering (101). Kindlin-3 is a hematopoietic-specific member of the kindlin family that contains a FERM domain and a PH domain (102, 103). Kindlin-3 also binds to β cytoplasmic regions, although distinct from those involved in talin binding. The critical importance of kindlin-3 in integrin regulation has been demonstrated by gene-targeting studies in mice as well as loss of function mutations of kindlin-3 in the human hereditary disease LAD-I variant or III, in which β1, β2, and β3 are defective (104–108). Kindlin-3$^{-/-}$ mice exhibited hypoplastic lymphoid organs and succumbed to fatal anemia and severe bleeding due to defective platelet aggregation as neonates. Studies using kindlin-3-deficient neutrophils have demonstrated that kindlin-3 is essential for firm attachment and spreading via β2 integrins. Furthermore, kindlin-3$^{-/-}$ neutrophils were defective in chemokine-triggered arrest through β1 and β2 integrins without affecting rolling. Interestingly, genes encoding CalDAG-GEF1 and kindlin-3 are located in proximity on chromosome 11. In some LAD-III subjects, mutations were found in both loci. However, cDNA encoding kindlin-3, but not CalDAG-GEF1, could rescue defective LFA-1-dependent adhesion of immortalized lymphocytes derived from this subject (106), demonstrating that the mutation of kindlin-3 is responsible for the defect. Human peripheral blood lymphocytes deficient for kindlin-3 exhibited relatively mild defects of VLA-4, compared to a total loss of LFA-1 adhesiveness, suggesting differential requirement of kindlin-3 for distinct integrin species (109).

The regulation of talin activation has been extensively studied. NMR studies revealed an autoinhibitory interaction between the carboxy terminus and the PTB (phospho tyrosine-binding)-like F3 subdomain of the FERM domain. Thus, talin is maintained in an inactive state without stimulation in leukocytes. The mechanism of talin activation, especially in leukocytes, when stimulated with chemokines and the antigen receptors, is unclear. Previous studies suggest that cleavage of the talin head from the rod by the protease calpain, or binding of the FERM domain to PIP_2 could disrupt the autoinhibitory interaction, leading to integrin binding of talin. Talin also binds through the FERM domain to the PIPKIγ (phosphatidylinositol 4-phosphate 5 kinase type I isoform γ) that produces PIP_2, which could increase local concentrations of PIP_2 at the contact sites, thereby augmenting adhesion stability. Since talin and kindlin-3 bind to distinct sites of the β cytoplasmic tails, some cooperation between talin and kindlin-3 is expected. Further studies are needed to clarify how the inside-out signaling triggered by chemokines and the antigen receptors leads to activation of talin and kindlin-3, which in turn coordinately regulate integrin activation as well as outside-in signaling to exert cellular functions.

6. Concluding Remarks

A wide variety of intracellular signaling molecules involved in inside-out and outside-in signaling have been identified and examined in detail under physiological conditions in vitro and in vivo. These studies have identified signaling molecules in arrest as well as the adhesion-strengthening step under shear flow. As seen in leukocytes, inside-out and outside-in signaling often overlap and it is difficult to separate them. Apparently, the inside-out pathways leading to integrins are likely different in cell types (e.g., T and B lymphocytes, neutrophils, platelets). Nonetheless, Rap small GTPases and talin/kindlin-3 have now been recognized as key players in intracellular signaling to and from integrins. Elucidation of the molecular mechanisms of the signaling pathways involving these molecules will shed light on integrin dynamics in the immune system, such as immune cell trafficking and surveillance, antigen response, or immune cell functions.

References

1. Dustin, M. L., and Springer, T. A. (1989) T-cell receptor cross-linking transiently stimulates adhesiveness through LFA-1, *Nature* **341**, 619–624.
2. Springer, T. A. (1995) Traffic signals on endothelium for lymphocyte recirculation and leukocyte emigration, *Annu. Rev. Physiol.* **57**, 827–872.
3. Butcher, E. C., and Picker, L. J. (1996) Lymphocyte homing and homeostasis, *Science* **272**, 60–66.
4. Butcher, E. C., Williams, M., Youngman, K., Rott, L., and Briskin, M. (1999) Lymphocyte trafficking and regional immunity, *Adv. Immunol.* **72**, 209–253.
5. Abbal, C., Lambelet, M., Bertaggia, D., Gerbex, C., Martinez, M., Arcaro, A., Schapira, M., and Spertini, O. (2006) Lipid raft adhesion receptors and Syk regulate selectin-dependent rolling under flow conditions, *Blood* **108**, 3352–3359.
6. Urzainqui, A., Serrador, J. M., Viedma, F., Yanez-Mo, M., Rodriguez, A., Corbi, A. L., Alonso-Lebrero, J. L., Luque, A., Deckert, M., Vazquez, J., and Sanchez-Madrid, F. (2002) ITAM-based interaction of ERM proteins with Syk mediates signaling by the leukocyte adhesion receptor PSGL-1, *Immunity* **17**, 401–412.
7. Zarbock, A., Lowell, C. A., and Ley, K. (2007) Spleen tyrosine kinase Syk is necessary for E-selectin-induced alpha(L)beta(2) integrin-mediated rolling on intercellular adhesion molecule-1, *Immunity* **26**, 773–783.
8. Berlin, C., Bargatze, R. F., Campbell, J. J., von Andrian, U. H., Szabo, M. C., Hasslen, S. R., Nelson, R. D., Berg, E. L., Erlandsen, S. L., and Butcher, E. C. (1995) alpha 4 integrins mediate lymphocyte attachment and rolling under physiologic flow, *Cell* **80**, 413–422.
9. Chan, J. R., Hyduk, S. J., and Cybulsky, M. I. (2001) Chemoattractants induce a rapid and transient upregulation of monocyte alpha4 integrin affinity for vascular cell adhesion molecule 1 which mediates arrest: an early step in the process of emigration, *J. Exp. Med.* **193**, 1149–1158.
10. Salas, A., Shimaoka, M., Kogan, A. N., Harwood, C., von Andrian, U. H., and Springer, T. A. (2004) Rolling adhesion through an extended conformation of integrin alphaLbeta2 and relation to alpha I and beta I-like domain interaction, *Immunity* **20**, 393–406.
11. Katagiri, K., Katakai, T., Ebisuno, Y., Ueda, Y., Okada, T., and Kinashi, T. (2009) Mst1 controls lymphocyte trafficking and interstitial motility within lymph nodes, *EMBO J.* **28**, 1319–1331.
12. Ley, K., Laudanna, C., Cybulsky, M. I., and Nourshargh, S. (2007) Getting to the site of inflammation: the leukocyte adhesion cascade updated, *Nat. Rev. Immunol.* **7**, 678–689.
13. Miyasaka, M., and Tanaka, T. (2004) Lymphocyte trafficking across high endothelial venules: dogmas and enigmas, *Nat. Rev. Immunol.* **4**, 360–370.
14. Alon, R., and Feigelson, S. (2002) From rolling to arrest on blood vessels: leukocyte tap dancing on endothelial integrin ligands and chemokines at sub-second contacts, *Semin. Immunol.* **14**, 93–104.
15. Shamri, R., Grabovsky, V., Gauguet, J. M., Feigelson, S., Manevich, E., Kolanus, W., Robinson, M. K., Staunton, D. E., von Andrian, U. H., and Alon, R. (2005) Lymphocyte arrest requires instantaneous induction of an extended LFA-1 conformation mediated by endothelium-bound chemokines, *Nat. Immunol.* **6**, 497–506.

16. Kunkel, E. J., Dunne, J. L., and Ley, K. (2000) Leukocyte arrest during cytokine-dependent inflammation in vivo, *J. Immunol.* **164**, 3301–3308.

17. Ley, K. (2002) Integration of inflammatory signals by rolling neutrophils, *Immunol. Rev.* **186**, 8–18.

18. Alon, R., and Ley, K. (2008) Cells on the run: shear-regulated integrin activation in leukocyte rolling and arrest on endothelial cells, *Curr. Opin. Cell Biol.* **20**, 525–32.

19. Zarbock, A., Abram, C. L., Hundt, M., Altman, A., Lowell, C. A., and Ley, K. (2008) PSGL-1 engagement by E-selectin signals through Src kinase Fgr and ITAM adapters DAP12 and FcR gamma to induce slow leukocyte rolling, *J. Exp. Med.* **205**, 2339–2347.

20. Warnock, R. A., Askari, S., Butcher, E. C., and von Andrian, U. H. (1998) Molecular mechanisms of lymphocyte homing to peripheral lymph nodes, *J. Exp. Med.* **187**, 205–216.

21. Campbell, J. J., Hedrick, J., Zlotnik, A., Siani, M. A., Thompson, D. A., and Butcher, E. C. (1998) Chemokines and the arrest of lymphocytes rolling under flow conditions, *Science* **279**, 381–384.

22. Han, S. B., Moratz, C., Huang, N. N., Kelsall, B., Cho, H., Shi, C. S., Schwartz, O., and Kehrl, J. H. (2005) Rgsl and Gnai2 regulate the entrance of B lymphocytes into lymph nodes and B cell motility within lymph node follicles, *Immunity* **22**, 343–354.

23. Ghandour, H., Cullere, X., Alvarez, A., Luscinskas, F. W., and Mayadas, T. N. (2007) Essential role for Rapl GTPase and its guanine exchange factor CalDAG-GEFI in LFA-1 but not VLA-4 integrin mediated human T-cell adhesion, *Blood* **110**, 3682–3690.

24. Hyduk, S. J., Chan, J. R., Duffy, S. T., Chen, M., Peterson, M. D., Waddell, T. K., Digby, G. C., Szaszi, K., Kapus, A., and Cybulsky, M. I. (2007) Phospholipase C, calcium, and calmodulin are critical for alpha4beta1 integrin affinity up-regulation and monocyte arrest triggered by chemoattractants, *Blood* **109**, 176–184.

25. Bach, T. L., Chen, Q. M., Kerr, W. T., Wang, Y., Lian, L., Choi, J. K., Wu, D., Kazanietz, M. G., Koretzky, G. A., Zigmond, S., and Abrams, C. S. (2007) Phospholipase cbeta is critical for T cell chemotaxis, *J. Immunol.* **179**, 2223–2227.

26. Li, Z., Jiang, H., Xie, W., Zhang, Z., Smrcka, A. V., and Wu, D. (2000) Roles of PLC-b2 and -b3 and PI3Kg in chemoattractant-mediated signal transduction, *Science* **287**, 1046–1049.

27. Bergmeier, W., Goerge, T., Wang, H. W., Crittenden, J. R., Baldwin, A. C., Cifuni, S. M., Housman, D. E., Graybiel, A. M., and Wagner, D. D. (2007) Mice lacking the signaling molecule CalDAG-GEFI represent a model for leukocyte adhesion deficiency type III, *J. Clin. Invest.* **117**, 1699–1707.

28. Crittenden, J. R., Bergmeier, W., Zhang, Y., Piffath, C. L., Liang, Y., Wagner, D. D., Housman, D. E., and Graybiel, A. M. (2004) CalDAG-GEFI integrates signaling for platelet aggregation and thrombus formation, *Nat. Med.* **10**, 982–986.

29. Pasvolsky, R., Feigelson, S. W., Kilic, S. S., Simon, A. J., Tal-Lapidot, G., Grabovsky, V., Crittenden, J. R., Amariglio, N., Safran, M., Graybiel, A. M., Rechavi, G., Ben-Dor, S., Etzioni, A., and Alon, R. (2007) A LAD-III syndrome is associated with defective expression of the Rap-1 activator CalDAG-GEFI in lymphocytes, neutrophils, and platelets, *J. Exp. Med.* **204**, 1571–1582.

30. Bos, J. L., de Rooij, J., and Reedquist, K. A. (2001) Rapl signaling: Adhering to new models, *Nat. Rev. Mol. Cell Biol.* **2**, 369–377.

31. Kooistra, M. R., Dube, N., and Bos, J. L. (2007) Rapl: a key regulator in cell-cell junction formation, *J. Cell Sci.* **120**, 17–22.

32. Kinashi, T., and Katagiri, K. (2005) Regulation of immune cell adhesion and migration by regulator of adhesion and cell polarization enriched in lymphoid tissues, *Immunology* **116**, 164–171.

33. Katagiri, K., Hattori, M., Minato, N., Irie, S., Takatsu, K., and Kinashi, T. (2000) Rapl is a potent activation signal for leukocyte function-associated antigen 1 distinct from protein kinase C and phosphatidylinositol-3-kinase, *Mol. Cell. Biol.* **20**, 1956–1969.

34. Reedquist, K. A., Ross, E., Koop, E. A., Wolthuis, R. M., Zwartkruis, F. J., van Kooyk, Y., Salmon, M., Buckley, C. D., and Bos, J. L. (2000) The small GTPase, Rapl, mediates CD31-induced integrin adhesion, *J. Cell Biol.* **148**, 1151–1158.

35. Li, Y., Yan, J., De, P., Chang, H. C., Yamauchi, A., Christopherson, K. W., 2nd, Paranavitana, N. C., Peng, X., Kim, C., Munugulavadla, V., Kapur, R., Chen, H., Shou, W., Stone, J. C., Kaplan, M. H., Dinauer, M. C., Durden, D. L., and Quilliam, L. A. (2007) Rapla null mice have altered myeloid cell functions suggesting distinct roles for the closely related Rapla and 1b proteins, *J. Immunol.* **179**, 8322–8331.

36. Chrzanowska-Wodnicka, M., Smyth, S. S., Schoenwaelder, S. M., Fischer, T. H., and White, G. C., 2nd. (2005) Raplb is required

for normal platelet function and hemostasis in mice, *J. Clin. Invest.* **115**, 680–687.

37. Duchniewicz, M., Zemojtel, T., Kolanczyk, M., Grossmann, S., Scheele, J. S., and Zwartkruis, F. J. (2006) Rap1A-deficient T and B cells show impaired integrin-mediated cell adhesion, *Mol. Cell. Biol.* **26**, 643–653.
38. Chu, H., Awasthi, A., White, G. C., 2nd, Chrzanowska-Wodnicka, M., and Malarkannan, S. (2008) Rap1b regulates B cell development, homing, and T cell-dependent humoral immunity, *J. Immunol.* **181**, 3373–3383.
39. Regelmann, A. G., Danzl, N. M., Wanjalla, C., and Alexandropoulos, K. (2006) The hematopoietic isoform of Cas-Hef1-associated signal transducer regulates chemokine-induced inside-out signaling and T cell trafficking, *Immunity* **25**, 907–918.
40. Katagiri, K., Maeda, A., Shimonaka, M., and Kinashi, T. (2003) RAPL, a Rap1-binding molecule that mediates Rap1-induced adhesion through spatial regulation of LFA-1, *Nat. Immunol.* **4**, 741–748.
41. Katagiri, K., Ohnishi, N., Kabashima, K., Iyoda, T., Takeda, N., Shinkai, Y., Inaba, K., and Kinashi, T. (2004) Crucial functions of the Rap1 effector molecule RAPL in lymphocyte and dendritic cell trafficking, *Nat. Immunol.* **5**, 1045–1051.
42. Lafuente, E. M., van Puijenbroek, A. A., Krause, M., Carman, C. V., Freeman, G. J., Berezovskaya, A., Constantine, E., Springer, T. A., Gertler, F. B., and Boussiotis, V. A. (2004) RIAM, an Ena/VASP and Profilin ligand, interacts with Rap1-GTP and mediates Rap1-induced adhesion, *Dev. Cell* **7**, 585–595.
43. Han, J., Lim, C. J., Watanabe, N., Soriani, A., Ratnikov, B., Calderwood, D. A., Puzon-McLaughlin, W., Lafuente, E. M., Boussiotis, V. A., Shattil, S. J., and Ginsberg, M. H. (2006) Reconstructing and deconstructing agonist-induced activation of integrin alphaIIbbeta3, *Curr. Biol.* **16**, 1796–1806.
44. Parmo-Cabanas, M., Garcia-Bernal, D., Garcia-Verdugo, R., Kremer, L., Marquez, G., and Teixido, J. (2007) Intracellular signaling required for CCL25-stimulated T cell adhesion mediated by the integrin alpha-4beta1, *J. Leukoc. Biol.* **82**, 380–391.
45. Katagiri, K., Imamura, M., and Kinashi, T. (2006) Spatiotemporal regulation of the kinase Mst1 by binding protein RAPL is critical for lymphocyte polarity and adhesion, *Nat. Immunol.* **7**, 919–928.
46. Zhou, D., Medoff, B. D., Chen, L., Li, L., Zhang, X. F., Praskova, M., Liu, M., Landry, A., Blumberg, R. S., Boussiotis, V. A., Xavier, R., and Avruch, J. (2008) The Nore1B/Mst1 complex restrains antigen receptor-induced proliferation of naive T cells, *Proc. Natl. Acad. Sci. USA.* **105**, 20321–2326.
47. Dong, Y., Du, X., Ye, J., Han, M., Xu, T., Zhuang, Y., and Tao, W. (2009) A cell-intrinsic role for Mst1 in regulating thymocyte egress, *J. Immunol.* **183**, 3865–3872.
48. Laudanna, C., Kim, J. Y., Constantin, G., and Butcher, E. (2002) Rapid leukocyte integrin activation by chemokines, *Immunol. Rev.* **186**, 37–46.
49. Giagulli, C., Scarpini, E., Ottoboni, L., Narumiya, S., Butcher, E. C., Constantin, G., and Laudanna, C. (2004) RhoA and zeta PKC control distinct modalities of LFA-1 activation by chemokines: critical role of LFA-1 affinity triggering in lymphocyte in vivo homing, *Immunity* **20**, 25–35.
50. Bolomini-Vittori, M., Montresor, A., Giagulli, C., Staunton, D., Rossi, B., Martinello, M., Constantin, G., and Laudanna, C. (2009) Regulation of conformer-specific activation of the integrin LFA-1 by a chemokine-triggered Rho signaling module, *Nat. Immunol.* **10**, 185–194.
51. Sakata, D., Taniguchi, H., Yasuda, S., Adachi-Morishima, A., Hamazaki, Y., Nakayama, R., Miki, T., Minato, N., and Narumiya, S. (2007) Impaired T lymphocyte trafficking in mice deficient in an actin-nucleating protein, mDia1, *J. Exp. Med.* **204**, 2031–2038.
52. Krawczyk, C., Oliveira-dos-Santos, A., Sasaki, T., Griffiths, E., Ohashi, P. S., Snapper, S., Alt, F., and Penninger, J. M. (2002) Vav1 controls integrin clustering and MHC/peptide-specific cell adhesion to antigen-presenting cells, *Immunity* **16**, 331–343.
53. Fukui, Y., Hashimoto, O., Sanui, T., Oono, T., Koga, H., Abe, M., Inayoshi, A., Noda, M., Oike, M., Shirai, T., and Sasazuki, T. (2001) Haematopoietic cell-specific CDM family protein DOCK2 is essential for lymphocyte migration, *Nature* **412**, 826–831.
54. Nombela-Arrieta, C., Lacalle, R. A., Montoya, M. C., Kunisaki, Y., Megias, D., Marques, M., Carrera, A. C., Manes, S., Fukui, Y., Martinez, A. C., and Stein, J. V. (2004) Differential requirements for DOCK2 and phosphoinositide-3-kinase gamma during T and B lymphocyte homing, *Immunity* **21**, 429–441.
55. Nishikimi, A., Fukuhara, H., Su, W., Hongu, T., Takasuga, S., Mihara, H., Cao, Q., Sanematsu, F., Kanai, M., Hasegawa, H., Tanaka, Y., Shibasaki, M., Kanaho, Y., Sasaki, T., Frohman, M. A., and Fukui, Y. (2009)

Sequential regulation of DOCK2 dynamics by two phospholipids during neutrophil chemotaxis, *Science* **324**, 384–387.

56. Mor, A., Wynne, J. P., Ahearn, I. M., Dustin, M. L., Du, G., and Philips, M. R. (2009) Phospholipase D1 regulates lymphocyte adhesion via upregulation of Rap1 at the plasma membrane, *Mol. Cell. Biol.* **29**, 3297–3306.
57. Monks, C. R. F., A., F. B., Kupfer, H., Sciaky, N., and Kufper, A. (1998) Three-dimensional segregation of supramolecular activation clusters in T cells, *Nature* **395**, 82–86.
58. Grakoui, A., Bromley, S. K., Sumen, C., Davis, M. M., Shaw, A. S., Allen, P. M., and Dustin, M. L. (1999) The immunological synapse:A molecular Machine controlling T cell activation, *Science* **285**, 221–226.
59. Dustin, M. L., and Colman, D. R. (2002) Neural and immunological synaptic relations, *Science* **298**, 785–789.
60. Mempel, T. R., Henrickson, S. E., and Von Andrian, U. H. (2004) T-cell priming by dendritic cells in lymph nodes occurs in three distinct phases, *Nature* **427**, 154–159.
61. Miller, M. J., Safrina, O., Parker, I., and Cahalan, M. D. (2004) Imaging the single cell dynamics of CD4+ T cell activation by dendritic cells in lymph nodes, *J. Exp. Med.* **200**, 847–856.
62. Dustin, M. L. (2009) The cellular context of T cell signaling, *Immunity* **30**, 482–492.
63. Lucas, J. A., Miller, A. T., Atherly, L. O., and Berg, L. J. (2003) The role of Tec family kinases in T cell development and function, *Immunol. Rev.* **191**, 119–138.
64. Woods, M. L., Kivens, W. J., Adelsman, M. A., Qiu, Y., August, A., and Shimizu, Y. (2001) A novel function for the Tec family tyrosine kinase Itk in activation of beta 1 integrins by the T-cell receptor, *EMBO J.* **20**, 1232–1244.
65. Labno, C. M., Lewis, C. M., You, D., Leung, D. W., Takesono, A., Kamberos, N., Seth, A., Finkelstein, L. D., Rosen, M. K., Schwartzberg, P. L., and Burkhardt, J. K. (2003) Itk functions to control actin polymerization at the immune synapse through localized activation of Cdc42 and WASP, *Curr. Biol.* **13**, 1619–1624.
66. Fagerholm, S., Hilden, T. J., and Gahmberg, C. G. (2002) Lck tyrosine kinase is important for activation of the CD11a/CD18-integrins in human T lymphocytes, *Eur. J. Immunol.* **32**, 1670–1678.
67. Goda, S., Quale, A. C., Woods, M. L., Felthauser, A., and Shimizu, Y. (2004) Control of TCR-mediated activation of beta 1 integrins by the ZAP-70 tyrosine kinase interdomain B region and the linker for activation of T cells adapter protein, *J. Immunol.* **172**, 5379–5387.
68. Jordan, M. S., Maltzman, J. S., Kliche, S., Shabason, J., Smith, J. E., Obstfeld, A., Schraven, B., and Koretzky, G. A. (2007) In vivo disruption of T cell development by expression of a dominant-negative polypeptide designed to abolish the SLP-76/Gads interaction, *Eur. J. Immunol.* **37**, 2961–2972.
69. Katagiri, K., Shimonaka, M., and Kinashi, T. (2004) Rap1-mediated LFA-1 activation by the T cell antigen receptor is dependent on PLC-gamma1, *J. Biol. Chem.* **279**, 11875–11881.
70. Peterson, E. J., Woods, M. L., Dmowski, S. A., Derimanov, G., Jordan, M. S., Wu, J. N., Myung, P. S., Liu, Q. H., Pribila, J. T., Freedman, B. D., Shimizu, Y., and Koretzky, G. A. (2001) Coupling of the TCR to integrin activation by Slap-130/Fyb, *Science* **293**, 2263–2265.
71. Griffiths, E. K., Krawczyk, C., Kong, Y. Y., Raab, M., Hyduk, S. J., Bouchard, D., Chan, V. S., Kozieradzki, I., Oliveira-Dos-Santos, A. J., Wakeham, A., Ohashi, P. S., Cybulsky, M. I., Rudd, C. E., and Penninger, J. M. (2001) Positive regulation of T cell activation and integrin adhesion by the adapter Fyb/Slap, *Science* **293**, 2260–2263.
72. Wang, H., Moon, E. Y., Azouz, A., Wu, X., Smith, A., Schneider, H., Hogg, N., and Rudd, C. E. (2003) SKAP-55 regulates integrin adhesion and formation of T cell-APC conjugates, *Nat. Immunol.* **4**, 366–374.
73. Wang, H., Liu, H., Lu, Y., Lovatt, M., Wei, B., and Rudd, C. E. (2007) Functional defects of SKAP-55-deficient T cells identify a regulatory role for the adaptor in LFA-1 adhesion, *Mol. Cell. Biol.* **27**, 6863–6875.
74. Kliche, S., Breitling, D., Togni, M., Pusch, R., Heuer, K., Wang, X., Freund, C., Kasirer-Friede, A., Menasche, G., Koretzky, G. A., and Schraven, B. (2006) The ADAP/SKAP55 signaling module regulates T-cell receptor-mediated integrin activation through plasma membrane targeting of Rap1, *Mol. Cell. Biol.* **26**, 7130–7144.
75. Burbach, B. J., Srivastava, R., Medeiros, R. B., O'Gorman, W. E., Peterson, E. J., and Shimizu, Y. (2008) Distinct regulation of integrin-dependent T cell conjugate formation and NF-kappa B activation by the adapter protein ADAP, *J. Immunol.* **181**, 4840–4851.

76. Menasche, G., Kliche, S., Chen, E. J., Stradal, T. E., Schraven, B., and Koretzky, G. (2007) RIAM links the ADAP/SKAP-55 signaling module to Rap1, facilitating T-cell-receptor-mediated integrin activation, *Mol. Cell. Biol.* **27**, 4070–4081.
77. Mueller, K. L., Thomas, M. S., Burbach, B. J., Peterson, E. J., and Shimizu, Y. (2007) Adhesion and degranulation-promoting adapter protein (ADAP) positively regulates T cell sensitivity to antigen and T cell survival, *J. Immunol.* **179**, 3559–3569.
78. Medeiros, R. B., Dickey, D. M., Chung, H., Quale, A. C., Nagarajan, L. R., Billadeau, D. D., and Shimizu, Y. (2005) Protein kinase D1 and the beta 1 integrin cytoplasmic domain control beta 1 integrin function via regulation of Rap1 activation, *Immunity* **23**, 213–226.
79. Batista, F. D., Iber, D., and Neuberger, M. S. (2001) B cells acquire antigen from target cells after synapse formation, *Nature* **411**, 489–494.
80. Carrasco, Y. R., and Batista, F. D. (2006) B cell recognition of membrane-bound antigen: an exquisite way of sensing ligands, *Curr. Opin. Immunol.* **18**, 286–291.
81. Spaargaren, M., Beuling, E. A., Rurup, M. L., Meijer, H. P., Klok, M. D., Middendorp, S., Hendriks, R. W., and Pals, S. T. (2003) The B cell antigen receptor controls integrin activity through Btk and PLCgamma2, *J. Exp. Med.* **198**, 1539–1550.
82. Lin, K. B., Freeman, S. A., Zabetian, S., Brugger, H., Weber, M., Lei, V., Dang-Lawson, M., Tse, K. W., Santamaria, R., Batista, F. D., and Gold, M. R. (2008) The rap GTPases regulate B cell morphology, immune-synapse formation, and signaling by particulate B cell receptor ligands, *Immunity* **28**, 75–87.
83. Arana, E., Vehlow, A., Harwood, N. E., Vigorito, E., Henderson, R., Turner, M., Tybulewicz, V. L., and Batista, F. D. (2008) Activation of the small GTPase Rac2 via the B cell receptor regulates B cell adhesion and immunological-synapse formation, *Immunity* **28**, 88–99.
84. Togni, M., Swanson, K. D., Reimann, S., Kliche, S., Pearce, A. C., Simeoni, L., Reinhold, D., Wienands, J., Neel, B. G., Schraven, B., and Gerber, A. (2005) Regulation of in vitro and in vivo immune functions by the cytosolic adaptor protein SKAP-HOM, *Mol. Cell. Biol.* **25**, 8052–8063.
85. Hynes, R. O. (2002) Integrins: Bidirectional, allosteric signalling machines, *Cell* **110**, 673–687.
86. Zhu, J., Carman, C. V., Kim, M., Shimaoka, M., Springer, T. A., and Luo, B. H. (2007) Requirement of alpha and beta subunit transmembrane helix separation for integrin outside-in signaling, *Blood* **110**, 2475–2483.
87. Kim, M., Carman, C. V., and Springer, T. A. (2003) Bidirectional transmembrane signaling by cytoplasmic domain separation in integrins, *Science* **301**, 1720–1725.
88. Giagulli, C., Ottoboni, L., Caveggion, E., Rossi, B., Lowell, C., Constantin, G., Laudanna, C., and Berton, G. (2006) The Src family kinases Hck and Fgr are dispensable for inside-out, chemoattractant-induced signaling regulating beta 2 integrin affinity and valency in neutrophils, but are required for beta 2 integrin-mediated outside-in signaling involved in sustained adhesion, *J. Immunol.* **177**, 604–611.
89. Totani, L., Piccoli, A., Manarini, S., Federico, L., Pecce, R., Martelli, N., Cerletti, C., Piccardoni, P., Lowell, C. A., Smyth, S. S., Berton, G., and Evangelista, V. (2006) Src-family kinases mediate an outside-in signal necessary for beta2 integrins to achieve full activation and sustain firm adhesion of polymorphonuclear leucocytes tethered on E-selectin, *Biochem. J.* **396**, 89–98.
90. Abram, C. L., and Lowell, C. A. (2009) The ins and outs of leukocyte integrin signaling, *Annu. Rev. Immunol.* **27**, 339–362.
91. Kasirer-Friede, A., Kahn, M. L., and Shattil, S. J. (2007) Platelet integrins and immunoreceptors, *Immunol. Rev.* **218**, 247–264.
92. Mocsai, A., Zhou, M., Meng, F., Tybulewicz, V. L., and Lowell, C. A. (2002) Syk is required for integrin signaling in neutrophils, *Immunity* **16**, 547–558.
93. de Virgilio, M., Kiosses, W. B., and Shattil, S. J. (2004) Proximal, selective, and dynamic interactions between integrin alphaIIbbeta3 and protein tyrosine kinases in living cells, *J. Cell Biol.* **165**, 305–311.
94. Abtahian, F., Bezman, N., Clemens, R., Sebzda, E., Cheng, L., Shattil, S. J., Kahn, M. L., and Koretzky, G. A. (2006) Evidence for the requirement of ITAM domains but not SLP-76/Gads interaction for integrin signaling in hematopoietic cells, *Mol. Cell. Biol.* **26**, 6936–6949.
95. Mocsai, A., Abram, C. L., Jakus, Z., Hu, Y., Lanier, L. L., and Lowell, C. A. (2006) Integrin signaling in neutrophils and macrophages uses adaptors containing immunoreceptor tyrosine-based activation motifs, *Nat. Immunol.* 7, 1326–1333.
96. Zou, W., Kitaura, H., Reeve, J., Long, F., Tybulewicz, V. L., Shattil, S. J., Ginsberg, M.

H., Ross, F. P., and Teitelbaum, S. L. (2007) Syk, c-Src, the alphavbeta3 integrin, and ITAM immunoreceptors, in concert, regulate osteoclastic bone resorption, *J. Cell Biol.* **176**, 877–888.

97. Bezman, N., and Koretzky, G. A. (2007) Compartmentalization of ITAM and integrin signaling by adapter molecules, *Immunol. Rev.* **218**, 9–28.
98. Graham, D. B., Stephenson, L. M., Lam, S. K., Brim, K., Lee, H. M., Bautista, J., Gilfillan, S., Akilesh, S., Fujikawa, K., and Swat, W. (2007) An ITAM-signaling pathway controls cross-presentation of particulate but not soluble antigens in dendritic cells, *J. Exp. Med.* **204**, 2889–2897.
99. Clemens, R. A., Lenox, L. E., Kambayashi, T., Bezman, N., Maltzman, J. S., Nichols, K. E., and Koretzky, G. A. (2007) Loss of SLP-76 expression within myeloid cells confers resistance to neutrophil-mediated tissue damage while maintaining effective bacterial killing, *J. Immunol.* **178**, 4606–4614.
100. Moser, M., Legate, K. R., Zent, R., and Fassler, R. (2009) The tail of integrins, talin, and kindlins, *Science* **324**, 895–899.
101. Calderwood, D. A., and Ginsberg, M. H. (2003) Talin forges the links between integrins and actin, *Nat. Cell Biol.* **5**, 694–697.
102. Larjava, H., Plow, E. F., and Wu, C. (2008) Kindlins: essential regulators of integrin signalling and cell-matrix adhesion, *EMBO Rep.* 9, 1203–1208.
103. Plow, E. F., Qin, J., and Byzova, T. (2009) Kindling the flame of integrin activation and function with kindlins, *Curr. Opin. Hematol.* **16**, 323–328.
104. Malinin, N. L., Zhang, L., Choi, J., Ciocea, A., Razorenova, O., Ma, Y. Q., Podrez, E. A., Tosi, M., Lennon, D. P., Caplan, A. I., Shurin, S. B., Plow, E. F., and Byzova, T. V. (2009) A point mutation in KINDLIN3 ablates activation of three integrin subfamilies in humans, *Nat. Med.* **15**, 313–318.
105. Moser, M., Bauer, M., Schmid, S., Ruppert, R., Schmidt, S., Sixt, M., Wang, H. V., Sperandio, M., and Fassler, R. (2009) Kindlin-3 is required for beta2 integrin-mediated leukocyte adhesion to endothelial cells, *Nat. Med.* **15**, 300–305.
106. Svensson, L., Howarth, K., McDowall, A., Patzak, I., Evans, R., Ussar, S., Moser, M., Metin, A., Fried, M., Tomlinson, I., and Hogg, N. (2009) Leukocyte adhesion deficiency-III is caused by mutations in KINDLIN3 affecting integrin activation, *Nat. Med.* **15**, 306–312.
107. Kuijpers, T. W., van de Vijver, E., Weterman, M. A., de Boer, M., Tool, A. T., van den Berg, T. K., Moser, M., Jakobs, M. E., Seeger, K., Sanal, O., Unal, S., Cetin, M., Roos, D., Verhoeven, A. J., and Baas, F. (2009) LAD-1/variant syndrome is caused by mutations in FERMT3, *Blood* **113**, 4740–4746.
108. Moser, M., Nieswandt, B., Ussar, S., Pozgajova, M., and Fassler, R. (2008) Kindlin-3 is essential for integrin activation and platelet aggregation, *Nat. Med.* **14**, 325–330.
109. Manevich-Mendelson, E., Feigelson, S. W., Pasvolsky, R., Aker, M., Grabovsky, V., Shulman, Z., Kilic, S. S., Rosenthal-Allieri, M. A., Ben-Dor, S., Mory, A., Bernard, A., Moser, M., Etzioni, A., and Alon, R. (2009) Loss of Kindlin-3 in LAD-III eliminates LFA-1 but not VLA-4 adhesiveness developed under shear flow conditions, *Blood* **114**, 2344–2353.

Chapter 18

Rap1 and Integrin Inside-Out Signaling

Koko Katagiri and Tatsuo Kinashi

Abstract

In leukocytes, integrins play important roles in adhesive interactions with endothelium, antigen-presenting cells, and effector functions such as cytotoxicity. This chapter describes methods to study Ras proximity 1 (Rap1), a signaling molecule that has been increasingly recognized as an important regulator of integrin-mediated cell adhesion in the immune system as well as hemostasis. Rap1 is activated by a wide variety of external stimuli including chemokines and antigens. Signaling via Rap1 transmits an inside-out signal to the integrins, thereby increasing adhesiveness to ligands such as immunoglobulin superfamily proteins as well as extracellular matrix proteins and plasma proteins. This process induces leukocyte cell adhesion to the endothelium and antigen-presenting cells. In addition to integrin regulation, activated Rap1 induces cell polarity of lymphocytes, which is coordinated with LFA-1 redistribution to the leading edge.

Key words: Rap1, LFA-1, ICAM-1, Chemokine, TCR, RAPL, Mst1

1. Introduction

Ras proximity 1 (Rap1) belongs to the Ras superfamily of GTPase that cycles between GDP-bound inactive and GTP-bound active forms. Conversion from the GDP-bound form to the GTP-bound form is mediated by guanine nucleotide exchange factors (GEFs) that facilitates the release of GDP and binding to GTP. The inactivation of Rap1 is induced through GTPase-activating protein (GAP) by accelerating hydrolysis of GTP (1). Rap1 has two isoforms, Rap1a and Rap1b, which share approximately 95% amino acids. The discovery of the function of Rap1 (also known as Krev-1) as a suppressor of mutated K-Ras-transformed phenotypes in fibroblasts (2) prompted numerous groups to initiate extensive studies of Rap1 as a Ras antagonist in the 1990s. RalGDS-RBD (Ras-binding domain) was shown to specifically bind to Rap1-GTP

Motomu Shimaoka (ed.), *Integrin and Cell Adhesion Molecules: Methods and Protocols*,
Methods in Molecular Biology, vol. 757, DOI 10.1007/978-1-61779-166-6_18, © Springer Science+Business Media, LLC 2011

in a Pull-down assay. In 2000, two separate groups published the observation that an active Rap1 mutant can dramatically increase adhesion through β1 and β2 integrins without changes to integrin surface expression in lymphoid cell lines (3, 4). The significance of this finding was confirmed using a transgenic mouse model which further demonstrated that activated Rap1 did not inhibit MAP kinases (5). Rap1 activation was detected in agonist-stimulated platelets (6) and in B cells through the B-cell receptor (BCR) ligation (7). Active Rap1 increased the affinity of LFA-1 to soluble ICAM-1-Fc chimera and extension-dependent LFA-1 epitope in lymphocytes (4, 8).

The physiological relevance of Rap1 function on integrins was first reported in lymphocytes and T-cell lines, in which Rap1 was rapidly activated by TCR ligation and chemokines. Inhibition of Rap1 activation by overexpression of a Rap1GAP abrogated the adhesion mediated by LFA-1 and VLA-4 (9, 10). In addition to the stimulatory effect on integrins, Rap1 activation was necessary and sufficient for the development of polarized cell shapes of lymphocytes upon stimulation with chemokines, which was independent of cell attachment (10). In resting lymphocytes, LFA-1 uniformly exists on the plasma membrane. Upon stimulation with chemokines or ectopic expression of active Rap1, LFA-1 clustering occurred on the leading edge of the plasma membrane (11). Thus, Rap1 can regulate affinity and valency of LFA-1 in concert with cell polarization. This modality of integrin regulation stimulates cell migration on ICAM-1-coated surface (12).

Rap1 activation occurs by second messengers such as diacylglycerol (DAG), calcium, or cAMP (1), which implies the involvement of Rap1-GEFs. Although the DAG analog phorbol 12-myristate 13-acetate (PMA) is a potent stimulator of integrins, cAMP is rather inhibitory (13), suggesting that DAG-dependent Rap1-GEFs could be responsible for Rap1 activation. In support of this notion, DAG-producing PLC activity was required for chemokine- or TCR-triggered LFA-1 activation (14). Deficiency of calcium and the DAG-dependent Rap1-specific GEF, CalDAG-GEFI, caused defective β1, β2, and β3 integrin activation in platelets and leukocytes in mice (8, 15–17). In patients with a rare adhesion deficiency syndrome termed LAD-III, the CalDAG-GEFI gene has a splice junction mutation (17). Targeted deletion of *rap1a* or *rap1b* resulted in impaired activation of lymphocyte (α4β1, β2) and platelet integrins (αIIbβ3), respectively (18–21). To date, double deficiency of Rap1a and Rap1b has not yet been reported. Overall, these studies establish that Rap1 plays a crucial role in integrin activation.

There are several effector mechanisms of Rap1 signaling, as described in Chapter 23. The RAPL-Mst1 pathway regulates Rap1-dependent integrin function in lymphocytes stimulated with chemokines and via the TCR. Upon stimulation with chemokines or following TCR ligation, RAPL binds to Rap1-GTP.

RAPL may associate with LFA-1 depending on the αL cytoplasmic region downstream of the GFFKR motif (11). RAPL further associated with and activated Mst1 kinase, resulting in increased adhesion by LFA-1 upon stimulation. Gene targeting of RAPL and Mst1 demonstrated defective lymphocyte trafficking due to impaired adhesion through LFA-1 and α4 integrins (22, 23). In addition, Mst1$^{-/-}$ lymphocytes exhibited poor cell polarity and reduced cell migration within lymphoid tissues (23).

Following TCR ligation, the SLP76-ADAP-SKAP55 pathway could lead to regulated T-cell adhesion. Previous studies have shown that ADAP$^{-/-}$ or SKAP55$^{-/-}$ T cells displayed defective LFA-1 clustering and adhesion (24–26). The putative mechanism by which ADAP regulates T-cell adhesion involves an association with Rap1-GTP-interacting adaptor molecule (RIAM) (27). RIAM was shown to require membrane localization of activated Rap1 via the binding of the RA and PH domains, which leads to Rap1-induced adhesion (28). This proposed model was supported by evidence from Jurkat T cells, such that HA-tagged RIAM associated with Rap1 after TCR ligation. Furthermore, knockdown of RIAM by shRNA decreased active Rap1-induced adhesion to ICAM-1 and fibronectin (28). Although RIAM constitutively interacted with actin regulators Ena/VASP, this was shown to be unnecessary for Rap1 localization to the membrane and adhesion. In a heterologous cell system, RIAM may form a complex with integrins and talin (29), and as such, there are at least three proposed models. In one model, SKAP-55 binds to RIAM, which leads to the formation of a complex with integrins and talin (27). In another model, SKAP55 binds to RAPL and forms a complex with LFA-1, resulting in adhesion to ICAM-1 following TCR ligation (Rudd, C.E. personal communication). The third model is the Rap1-PKD1-β1 integrin pathway (30). Upon TCR ligation, the PH domain of PKD1 binds to Rap1 and further associates with the carboxy-terminal end of β1 integrins, leading to integrin clustering and adhesion. The kinase activity of PKD1 is not required for complex formation. It is unknown whether this pathway also activates β2 integrins.

To clarify the roles of Rap1 on integrins, it is crucial to monitor the spatiotemporal regulation of Rap1 activation and integrins. Below we will describe the basic protocols for Rap1 activation and adhesion assays.

2. Materials

2.1. Pull-Down Assay for Detection of Rap1 Activation

1. The ras-binding domain (RBD) of RalGDS (97 amino acids) cloned in pGEX vector (6).
2. Competent bacterial strains for transformation: Use JM109 and BL21. DH5α (Invitrogen) is also good.

3. LB medium with 100 μg/ml ampicillin.
4. Isopropyl-β-thiogalactopyranoside (IPTG): Dissolve IPTG with water to make 100 mM stock solution. Store at −20°C.
5. Bacterial lysis buffer: 1% Triton X-100, 100 mM NaCl, 50 mM Tris–HCl, pH 7.5, 2 mM $MgCl_2$, 1 mM phenylmethylsulfonyl fluoride (PMSF), and 2 μg/ml aprotinin.
6. Glutathione *S*-transferase (GST)-Sepharose 4B (GE Healthcare BioScience).
7. Cell culture medium: RPMI1640 and Hanks balanced salt solutions (HBSS).
8. Chemokines: CXCL12 (rhSDF-1) and mouse CCL21 (6Ckine) (R&D Systems).
9. Antibodies: Hamster anti-mouse CD3 antibody 2C11 (American Type Culture Collection, ATCC), mouse anti-Rap1 (Transduction lab), and goat anti-mouse antibody conjugated to horse radish peroxidase (HRP) (Santa Cruz Biotechnology).
10. Cell lysis buffer: 1% Triton X-100, 100 mM NaCl, 50 mM Tris–HCl, pH 7.5, 5 mM $MgCl_2$, 1 mM PMSF, and 2 μg/ml aprotinin (see Note 1).
11. Western blotting wash solution: 0.05% Tween20 in PBS.
12. Bovine serum albumin (BSA) (Fraction V, γ-globulin free).
13. Polyvinylidene difluoride (PVDF) (Immobilon-P, Millipore).
14. Enhanced chemiluminescense (Western Lightning-ECL, PerkinElmer Inc.).
15. Image analyzer (LAS 3000, Fuji film).

2.2. LFA-1 Clustering

1. Antibodies: Phycoerythrin (PE)-conjugated antibody to LFA-1 (M17/4, PharMingen) and fluorescent isothiocyanate (FITC)-conjugated rat monoclonal antibody to mouse CD44 (IM7, PharMingen).
2. Poly-l-lysine (PLL): Dilute PLL solution with water at 1:10. Add diluted PLL to an eight-chamber slide (BD Falcon) and incubate for 5 min at room temperature. Aspirate and dry at 50°C for 2 h or at room temperature overnight.
3. T-cell and B-cell isolation kits (MidiMACS, Mitenyl Biotech).
4. Paraformaldehyde (PFA): Prepare freshly 3.3% PFA solution in phosphate-buffered saline (PBS) solution. Dissolve 1 g in 30 ml of PBS and add 90 μl of 6N NaOH to facilitate solubilization. Stir the solution with heat (approximately 70–80°C) for 5–10 min until PFA is dissolved completely.

Add 90 μl of 6N HCl. Cool PFA solution to room temperature before use.

5. Anti-fade mounting medium (Prolong Anti-fade kit, Molecular Probes).
6. Confocal laser microscope (LSM510, Zeiss).

2.3. Static Adhesion Assay of T Cells

1. Recombinant mouse ICAM-1-Fc (R&D Systems, Inc.).
2. Anti-CD3 antibodies: OKT3 (human) and 2C11 (mouse).
3. Bovine serum albumin (BSA): Prepare 1% (w/v) BSA in PBS and store at −20°C. Use at the final concentration of 1–10 μg/ml.
4. Linbro-Titertek 96-well plate (or equivalent 96-well culture plate with high absorbance to protein).
5. Adhesion medium: RPMI1640 supplemented with 5% fetal calf serum (FCS) and 10 mM HEPES, pH 7.4.
6. 2′, 7′-bis(2-carboxyethyl)-5 (and -6) caboxyfluorescein (BCECF, Invitrogen): Dissolve BCECF at 1 mg/ml in dimethyl sulfoxide (DMSO) and store aliquots at −20°C. Use once after thawing.
7. Fluorescence multi-well plate reader (Cytofluor4000, Applied Biosystems).

2.4. T-APC Conjugate Assay

1. Antigen-specific T-cell lines: Hen egg lysozyme (HEL)-specific $I\text{-}A^k$-restricted T cells (3A9 T-cell hybridoma) (31) and CH27 B lymphoma cells (32).
2. 5,6-carboxyfluorescein diacetate (CFSE): Dissolve CFSE at 5 mg/ml in DMSO and store aliquots at −20°C. Use the PKH-26 staining kit (Sigma–Aldrich) according to the manufacturer's instruction.
3. HEL: Dissolve HEL at 10 mg/ml in PBS and store aliquots at −20°C.
4. Flow cytometry (BD FACSCalibur, BD).

2.5. Detachment Assay

1. A parallel plate flow chamber (FCS2 system, Bioptechs).
2. ICAM-1-Fc and capture antibody as described in Subheading 2.3.
3. Capture antibody: rabbit antihuman IgG-Fc.
4. A syringe pump (pump11, Harvard Apparatus).
5. T-cell and B-cell isolation kit (MidiMACS, Mitenyl Biotech).
6. A CCD camera (C2741, Hamamatsu Photonics) connected to a monitor (PVM-96, SONY) and VCR (VZ-S50, SANYO).

3. Methods

3.1. Pull-Down Assay for Detection of Rap1 Activation

3.1.1. Preparation of GST-RalGDS-RBD Coupled with Glutathione-Sepharose

1. Transform competent *E.coli* (JM109) with pGEX-RalGDS-RBD. Incubate the bacteria in 5 ml of LB medium containing 100 μg/ml ampicillin at 37°C overnight with shaking.
2. Add 1 ml of culture to 200 ml of LB medium and incubate the bacteria at 37°C for 2–3 h with shaking until an OD_{600} of 0.6. Reserve 5 μl of the culture to assess expression of the GST fusion protein. Add IPTG to a final concentration of 0.1 mM and incubate the bacteria for 3 h at 37°C. Keep 5 μl of the culture to assess the expression of the GST fusion protein.
3. Transfer the bacteria into 50-ml conical tubes and centrifuge in a fixed-angle rotor at 5,500 × *g*, 4°C for 10 min. Wash the bacteria with ice-cold PBS once.
4. Resuspend the bacterial pellets in 5 ml of ice-cold bacterial lysis buffer in a 15-ml tube. Keep the tube on ice.
5. On ice, sonicate the bacterial suspension in the tube for 15 s and repeat sonication at a 30-s interval until the suspension becomes somewhat clear (usually four times) (see Note 2).
6. Transfer the sonicated bacteria into 1.5-ml tubes and centrifuge at 20,000 × *g* at 4°C for 10 min on a microcentrifuge. Keep 5 μl of the clear lysate for a later check (see Note 3).
7. Collect the supernatant and store 500 μl aliquot at −80°C (stable for 1 year).
8. Take 1 ml of a slurry of glutathione-Sepharose beads in a 1.5-ml tube and centrifuge at 6,200 × *g*, 4°C for 1 min on a microcentrifuge. Wash the beads twice with ice-cold PBS and centrifuge again. Resuspend the beads in 500 μl of PBS (total 1 ml slurry).
9. Thaw the aliquot quickly in a 37°C water bath and place on ice immediately.
10. Add 200 μl of glutathione beads into the thawed lysate and rotate the tube at 4°C for 1 h for coupling to the GST fusion protein.
11. Spin down at 6,200 × *g* at 4°C for 1 min on a microcentrifuge. Wash the beads three times with 1 ml of ice-cold bacterial lysis buffer and once with ice-cold PBS. Resuspend the beads with 200 μl of ice-cold PBS. The beads can be kept at 4°C for 3–4 days.

3.1.2. Detection of Rap1-GTP by Pull-Down Assay with the RalGDS-RBD Beads

1. Wash cells twice with warmed (37°C) HBSS and resuspend the cells in a 1.5-ml tube with 0.5 ml of warmed RPMI1640. Rest the cells by incubating at 37°C for 10 min. Prepare at least two samples before and after stimulation (see Note 4).

2. For stimulation, add 100 nM CCL21 or 10 μg/ml anti-CD3 antibody to the cell suspension, mix briefly, and incubate at 37°C. For Rapl activation in T-antigen-presenting cell (APC) conjugates, use culture conditions described in Subheading 3.3. Take the cells from the culture and proceed to the next step.
3. After incubation, flush spin down in a microfuge (0°C). Resuspend cell pellets with 0.5 ml of ice-cold cell lysis buffer. Keep them on ice for 10 min.
4. Centrifuge cell lysates at 20,000 × *g* at 0°C for 10 min and transfer into 1.5-ml tubes (see Note 5). The cleared cell lysates can be stored at −80°C.
5. Save 10 μl of cell lysates from the samples to estimate the total amount of Rapl in the cell lysate. Add 20 μl of GST-RalGDS-RBD Sepharose beads to the samples and incubate on ice for 45 min. Mix the tube every 5 min. Alternatively, the tube can be rotated at 4°C. Avoid using a cold room with temperature above 4°C for rotation, since this could lead to GTP hydrolysis.
6. Centrifuge the samples at 6,200 × *g* at 0°C for 1 min on a microcentrifuge. Wash the beads three times with ice-cold cell lysis buffer and then resuspend the beads with 20 μl of 2× SDS sample buffer.
7. Boil the samples for 5 min and centrifuge briefly. Separate proteins in a 12% SDS polyacrylamide gel.
8. Transfer the proteins to a PVDF membrane using a semi-dry or wet transfer apparatus.
9. Block the membrane by incubating with 5% BSA in PBS at 4°C overnight.
10. Rinse the membrane briefly with PBS and incubate the membrane with 0.05% Tween20–PBS containing anti-Rapl antibody (1:500 dilution) for 2 h at room temperature.
11. Wash the membrane by shaking three times (once for 15 min, twice for 5 min).
12. Incubate the membrane in 0.05% Tween20–PBS containing goat anti-rabbit antibody conjugated with HRP (1:3,000) for 1 h at room temperature.
13. Wash the membrane with 0.05% Tween20–PBS four times (once for 15 min, three times for 5 min).
14. Detect the bound antibody complex by enhanced chemiluminescence (Fig. 1). Quantify the intensity of the bands with densitometry or an image analyzer, and normalize the amounts of Rapl pull-downed with GST-RalGDS-RBD (Rapl-GTP) to total amounts of Rapl.

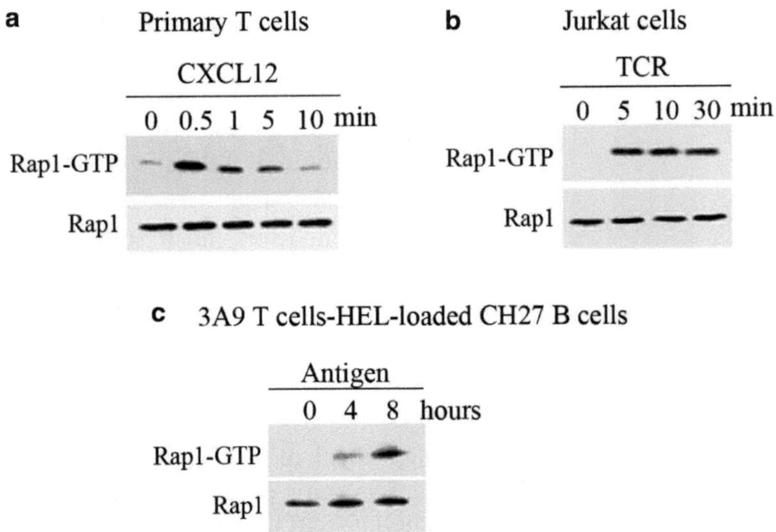

Fig. 1. Rap1 activation triggered by chemokines and TCR ligation. (**a–c**) The pull-down method was employed to detect Rap1 activation in primary T cells (**a**), Jurkat T cells (**b**), after stimulation with chemokines or TCR ligation (anti-CD3 antibody). In (**c**), Rap1 activation in coculture of HEL-specific 3A9 T cells with HEL-loaded CH27 B-cell APC. Time is indicated after stimulation or coculture. Rap1 levels of cell lysates are indicated *lower panels*. This figure shows the time-course experiments of Rap1 activation using the described method. In primary T cells, Rap1-GTP was detected in unstimulated cells and increased over time, with the peak occurring at 30 s and declining thereafter after stimulation with CXCL12. TCR ligation also increased Rap1-GTP in Jurkat T cells. Unlike chemokines, TCR-triggered Rap1 activation was sustained. Activation of T7-tagged Rap1 ectopically expressed in T cells was detected 4 h after coculture with antigen-loaded CH27 B-cell APC upon T-APC cluster formation. Rap1 activation was augmented in parallel with increased T-APC clusters.

3.2. Lymphocyte Polarization with Chemokine Stimulation or by Activated Rap1

Chemokines induce polarized cell shapes with a leading edge membrane and uropod. LFA-1 clustering occurred in the leading edge and CD44 is relocalized to the uropod (33). This process occurs independently of adhesion and requires the Rap1-RAPL-Mst1 signaling pathway (10, 11, 23). The method below can be used to detect LFA-1 and CD44 localization in lymphocytes and mouse pro-B cells (BAF cells) expressing human LFA-1 and constitutively active Rap1 (Rap1V12) (BAF/LFA-1/Rap1V12).

1. Isolate T cells and B cells by magnetic cell sorting according to the manufacturer's instruction. Wash isolated cells with HBSS and suspend at 1×10^6/ml in warmed RPMI1640. For BAF cells expressing Rap1V12, wash the cells with HBSS twice and suspend 2×10^5/ml in warmed RPMI140 (go to step 3).
2. Stimulate lymphocytes with chemokines (100 nM CXCL12 for B cells or CCL21 for T cells) for 10 min at 37°C.

3. Centrifuge the cells at 900 × *g* for 30 s at room temperature and resuspend the cells gently in 3% PFA.
4. Transfer the cells to the PLL-coated eight-chamber slide (400 μl/well) and incubate for 10 min at room temperature. All procedures hereafter are performed at room temperature.
5. Aspirate the supernatant gently, add 400 μl of PBS containing 10% goat serum, and incubate for 30 min (see Note 6).
6. Incubate chemokine-stimulated lymphocytes and BAF/hLFA-1/Rap1V12 with phycoerythrin-labeled anti-LFA-1 and FITC-labeled anti-CD44 for 1 h in the dark.
7. Wash the cells gently with 0.1% BSA in PBS three times, mount with anti-fade reagent, and place a cover glass, sealing with manicure.
8. Observe the stained samples with a confocal laser microscope.

Chemokine-stimulated T cells and B cells exhibit segregated LFA-1 on the leading edge and CD44 on the uropod accompanied with elongated cell shapes (Fig. 2). Similar phenotypes are also observed in BAF/LFA-1/Rap1V12. To quantify cell polarization,

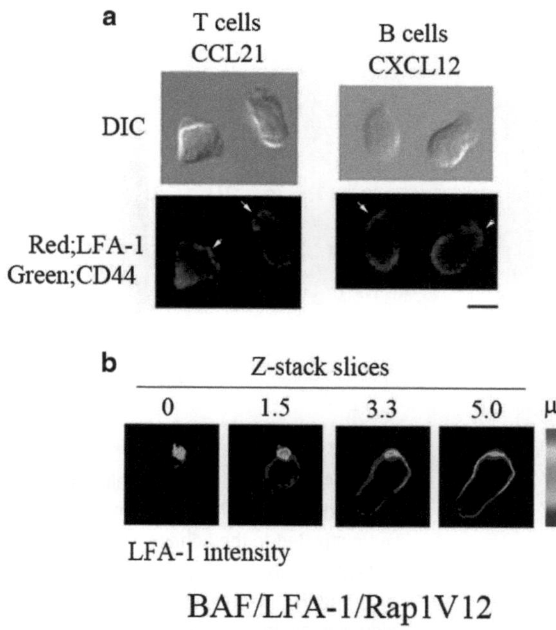

Fig. 2. LFA-1 clustering and cell polarization by chemokine and Rap1. (**a**) T and B cells stimulated with CCL21 and CXCL21, respectively, showed LFA-1 clustering at the ruffling membrane (the leading edge, *arrowed*). In contrast, CD44 is accumulated at uropod. (**b**) Confocal images of LFA-1 clustering of Rap1V12 expressing BAF/hLFA-1, at the different *z* planes relative to the starting image. Note that LFA-1 accumulation occurs at the tip of the leading edge membrane.

measure the length of long axis and short axis, and calculate shape index (long axis/short axis). Cells with a shape index greater than 1.5 with segregation of CD44 and LFA-1 is regarded as polarized cells. Shape index in B cells tends to be lower than that in T cells.

3.3. Static Adhesion of T Cells to ICAM-1 by TCR Ligation

TCR ligation increases adhesiveness of LFA-1 to ICAM-1. The inside-out signaling requires Rap1 activation. The method described below is a simple assay of T-cell adhesion to an ICAM-1-coated surface using Jurkat T cells as an example (Fig. 3). This same method can be applied to primary T cells.

1. Plate coating: The assay is performed in triplicate. Add 100 μl of ICAM-1-Fc (0.2 μg/ml in PBS) to each well in 96-well plates and incubate them overnight at 4°C. BSA (1% in PBS) is used to coat control wells (see Note 7).
2. Wash the wells with PBS three times by repeating addition and aspiration of 100 μl PBS.
3. Block the wells with 1% BSA for 30 min at room temperature.
4. Wash the wells as in step 2 and put 100 μl PBS in the wells until use.

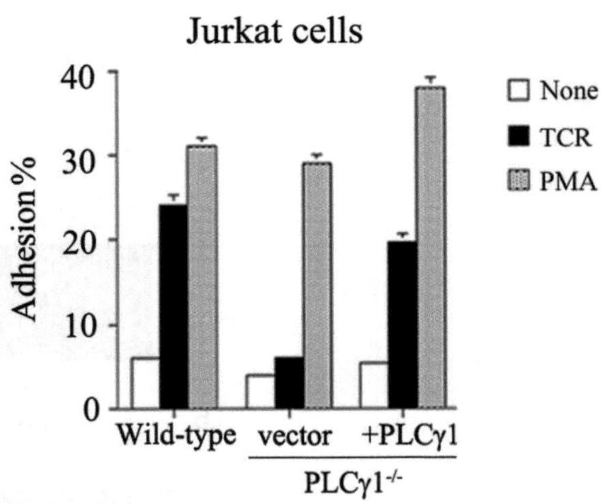

Fig. 3. Adhesion of Jurkat T cells to ICAM-1. Wild-type Jurkat T cells, mutant Jurkat T cells deficient of PLCγ1, and mutant Jurkat T cells re-expressing PLCγ1 by transfection are subjected to the adhesion assay to ICAM-1 in the absence or presence of PMA and anti-CD3 antibody (TCR). This figure shows the result of the adhesion assay using wild-type and PLCγ1-deficient Jurkat T cells with and without transfection with the expression plasmid to re-express PLCγ1. The nonspecific binding to BSA-coated wells is <1%. The basal level of binding to ICAM-1 after four washes is up to 5%. When stimulated with PMA, the binding levels to ICAM-1 are upregulated in all cells indicated. Wild-type Jurkat T cells increased adhesion to ICAM-1 upon TCR ligation. PLCγ1-deficient Jurkat T cells failed to increase the adhesion but restored the adhesive response after re-expression of PLCγ1. The inhibition of TCR-induced adhesion also occurred when Rap1 activation was inhibited by expression of Rap1GAP Spa1 (14).

5. Cell labeling: Centrifuge 1×10^6 cells and suspend the cells in 1 ml of warmed RPMI1640 supplemented 0.1% BSA in a 1.5-ml tube.
6. Add 1 μl of 1 mg/ml BCECF to cells and incubate at 37°C for 20 min.
7. Centrifuge the cells at $900 \times g$ for 30 s on a microcentrifuge. If labeling is efficient, the pellet should turn yellow (see Note 8).
8. Wash the pellet once with 1 ml of RMPI1640/0.1% BSA and resuspend in 500 μl of RPMI1640/5% FCS.
9. Adhesion assay: Add 50 μl of labeled cells into the coated wells.
10. Add 50 μl of RPMI1640/5% FCS containing 1–5 μg/ml anti-CD3 antibody or 20 ng/ml PMA.
11. Incubate the plate at 37°C for 30 min.
12. Measure input fluorescence.
13. Wash the wells by adding 100 μl of warmed RPMI1640/0.1% BSA and remove liquid with a 22-G needle connected with an aspirator. Repeat the wash three times (see Note 9).
14. Measure bound fluorescence after the second and third washes.
15. Calculate adhesion levels by dividing bound fluorescence by input fluorescence.

    For non-adherent cells such as Jurkat T cells, the adhesion level in control wells coated with BSA alone should be <1%.

3.4. T-APC Conjugate Assay

Stable antigen-dependent T-APC conjugate formation is mediated by LFA-1 and ICAM-1. Antigen-specific T cells are either cell lines (3A9 T-cell hybridoma), or derived from TCR transgenic mice. B-cell lines expressing appropriate MHC class II are used as antigen-presenting cells (Fig. 4).

1. Labeling and antigen loading: CH27 B-cell lymphoma cells are used as APCs. Label CH27 B cells with PHK-26 according to the manufacturer's instructions.
2. Incubate labeled CH27 B cells (2×10^5/ml) with 100 μg/ml HEL for 16 h.
3. Labeling: Wash 3A9T cells once with HBSS and resuspend 1×10^6 cells/ml in 1 ml of warmed RPMI1640 medium in a 15-ml tube. Add CFSE to the final concentration of 0.1 μM and incubate for 15 min at 37°C.
4. Wash the cells with RPMI1640 and resuspend at 2×10^5 cells/ml in RPMI1640/10% FCS medium.
5. Conjugation: Mix and incubate 0.5 ml of T cells with 0.5 ml of APCs (10^5 cells of each) for 30 min at 37°C. For the inhibition of antibodies, incubate T cells for 30 min

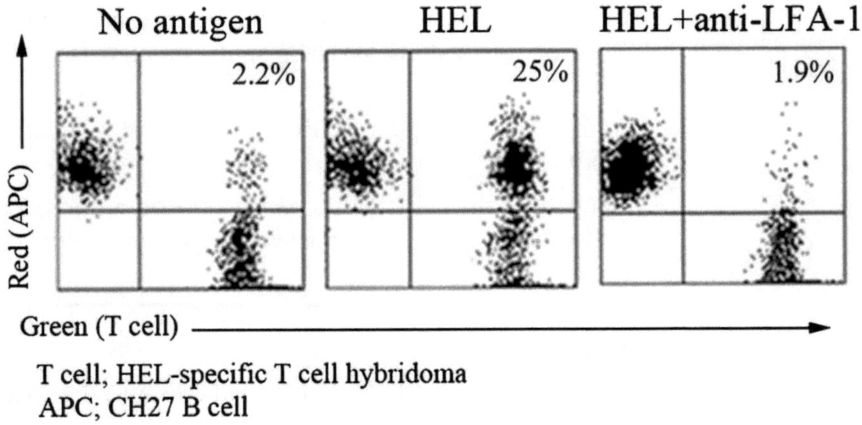

Fig. 4. T cell–APC conjugate assays. T cell–APC conjugates between HEL-specific 3A9 T cells and CH27 B cells without antigen (*left*), with HEL antigen (*middle*), or with HEL antigen in the presence of anti-LFA-1 antibody (*right*).

at 37°C with or without 20 μg/ml anti-LFA-1 antibody before conjugation. Mix T cells with APCs in the presence of the antibody and incubate them for 30 min at 37°C.

6. Vortex briefly to disrupt nonspecific aggregates. Transfer the cells into tubes for FACS analysis.
7. Analyze samples using the FACSCalibur. The percent of conjugates is defined as the number of live-gated, double-positive events in the upper right quadrant divided by the total number of live-gated events.

3.5. Detachment Assay

Compared to TCR-induced activation, chemokine-triggered LFA-1 activation is relatively weak and transient. To detect changes in the adhesiveness after chemokine stimulation, laminar flow is employed to examine the changes in cell adhesiveness accurately. For this purpose, cells are incubated on an ICAM-1-coated surface with chemokines in a parallel flow chamber, and then medium is infused at defined laminar flow rates using a syringe pump to detach non-adherent cells. Adhesive events are monitored in live time and recorded by videomicroscopy for later analysis (see Note 10).

1. Coat the bottom glass disc of the flow chamber overnight at 4°C with 10 μg/ml capture antibody by mounting 100 μl of antibody solution.
2. Wash the disc with PBS three times and mount 200 μl of 0.1 μg/ml ICAM-1Fc and VCAM-1Fc on the same spot coated with the capture antibody for 2 h at room temperature.
3. Wash the disc with PBS three times and incubate with 1% BSA for blocking.
4. Assemble the coated disc in the flow chamber on the microscope with a 20× objective lens, and connect the chamber

through a three-way stop cock and a flexible tube to a 50-ml syringe containing warmed RPMI1640/0.1% BSA of a syringe pump. Set the temperature of the flow chamber at 37°C.

5. Wash isolated T and B cells once with HBSS and suspend at 1×10^6 cells/ml with warm RPMI1640/0.1% BSA.
6. For chemokine stimulation, add 100 nM CCL21 for T cells and CXCL12 for B cells and mix briefly. Infuse the cells with or without chemokines through a three-way stop cock into the flow chamber.
7. Incubate the cells for 10 min (see Note 11).
8. Turn the three-way stop cock to open the connection between the chamber and the syringe pump. Infuse the medium at defined flow rates (1–5 dyn/cm^2) (see Note 12).
9. Record the microscopic images with a VCR. Continue infusion for 1 min.
10. Count the cells before and after washing, and calculate the percentage of bound cells (see Note 13).

Detachment assays using primary T and B cells are shown in Fig. 5. The assay is performed under 2 dyn/cm^2 of shear stress. Data are presented as percentages of attached cells to input cells 1 min

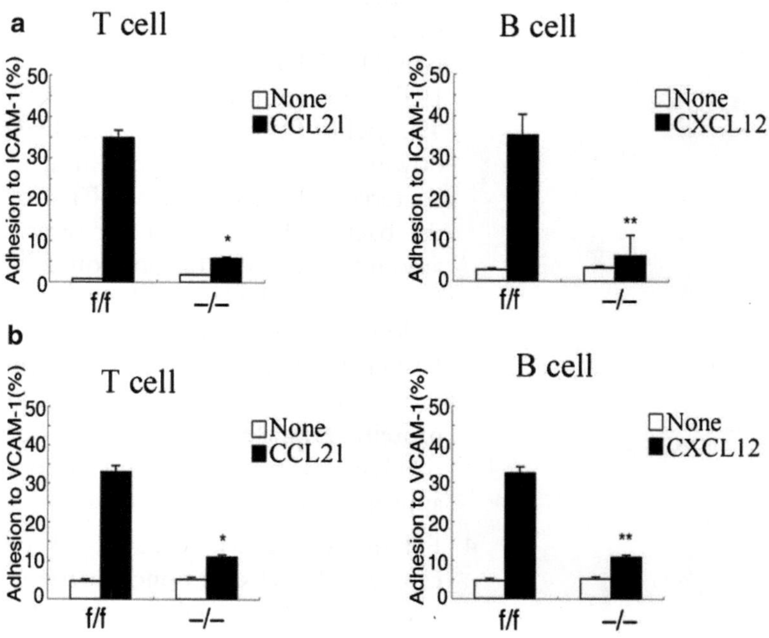

Fig. 5. Chemokine-triggered adhesion of lymphocytes derived from Mst1$^{-/-}$ mice. (**a**, **b**) Detachment assays are performed using T and B cells from control Mst1$^{flox/flox}$ (f/f) and Mst1$^{-/-}$ (–/–) mice. Detachment assays using ICAM-1 (**a**) and VCAM-1 (**b**) are performed with (*filled*) and without (*open*) chemokines. T and B cells are stimulated with CCL21 and CXCL12, respectively. Percentages of bound cells after 1 min under shear stress (2 dyn/cm^2) are indicated.

after application of shear flow. Chemokines (CCL21 and CXCL12) augment the adhesion of control lymphocytes (f/f) to ICAM-1 and VCAM-1. The adhesive responses are severely diminished in both T and B cells from $Mst1^{-/-}$ mice.

4. Notes

1. High Mg^{2+} is required to stabilize GDP- or GTP-bound forms of Rap1. If EDTA is added to the buffer, Mg^{2+} concentration should be increased accordingly.
2. Avoid too much sonication to prevent protein degradation. After sonication of four times or less, the color of the yellowish bacterial suspension turned white and a bit translucent.
3. Induction and purification of GST-RalGDS-RBD can be checked by 10% sodium dodecyl sulfate-polyacrylamide gel elecrophoresis (SDS-PAGE). Add 5 μl of 2× sample buffer (125 mM Tris–HCl, pH 6.8, 4% SDS, 0.1% (w/v) bromophenol blue, 20% (v/v) glycerol, and 2% (v/v) 2-mercaptoethanol) to samples of 5 μl of the bacterial suspension before and after IPTG induction, samples of cleared lysates, and samples after coupling to the beads. Boil for 5 min and cool at room temperature before electrophoresis. After running, the gel is stained with Coomassie brilliant blue. The thick band of GST-RalGDS-RBD (approximately 42 kDa) should be present in IPTG-induced bacterial lysates, cleared lysates, and the beads. The yield of the protein is usually sufficient for 20–30 pull-downs. In case the protein induction is low, check several bacterial colonies, increase IPTG up to 1 mM, or incubate the bacteria lysate at a lower temperature (to 30°C). If the amount of the GST fusion protein is low in the cleared lysate, but higher in the bacterial lysate after IPTG induction, sonication is likely insufficient since solubility of GST-RalGDS-RBD is high. In case of protein degradation, which typically results in a 30-kDa protein instead of the full-length 42-kDa protein in the clear lysate and the beads, check that the bacterial lysis by sonication and coupling steps are performed appropriately.
4. For one pull-down reaction, 10^7 cells for primary lymphocyte, 5×10^6 cells for bone marrow-derived dendritic cells, or $1–5 \times 10^6$ cells for cell lines is a starting point. When Rap1 activation by chemokines or TCR ligation is measured, samples without stimulation and after stimulation should be included. Time points after stimulation should be chosen appropriately (e.g., 10 s, 30 s, 1 min, and 5 min for chemokines, and 5 and 10 min for TCR ligation).

5. Care should be taken to avoid GTP hydrolysis. Ten minutes of incubation is usually enough to prepare cell lysates. Always keep the samples at <4°C.
6. Generally, fixed cells do not attach to PLL-coated surface as firmly as live cells. When washing cells, addition and aspiration should be performed very gently with a pipetman from the corner of the chamber. Blocking PLL with proteins is necessary to avoid nonspecific antibody binding. Serum from the host animal of the antibody is suitable for blocking, but BSA can be also used.
7. The amount of ICAM-1 required for adhesion varies depending on cell types. The described amount is optimal for Jurkat T cells. For primary T cells, increase ICAM-1Fc up to 1 μg/ml. Since TCR is cross-linked with anti-CD3 antibody, coating with the capture antibody for ICAM-1Fc, such as in the detachment assay instead of direct coating, should be avoided.
8. Cells look yellow when labeling with BCECF. BCECF enters into the cytoplasm and is cleaved by endogenous esterase resulting in coloring. The yellow-tainted BCECF (e.g., in case of long-term storage) or poor cell viability results in low efficiency of labeling. It is also important to note cell toxicity of BCECF at high concentrations. The indicated concentration is not harmful to lymphocytes and lymphoid cell lines.
9. For washing, change the aspiration place for uniform washing. For instance, set the aspiration needle at the site corresponding to 12, 3, 6, and 9 o'clock of the well, with the cutting side of the needle outward. If you want to detect weak adhesion, use smaller gauge needles.
10. We use a FCS2 (Bioptechs) for a parallel plate flow chamber, since the flow chamber is directly heated and maintained at 37°C. The method can be modified according to other flow chambers. It is important to control the shear flow and temperature. Live images of the bright microscopic field can be directly recorded at 30 frames/s with an appropriate USB-connected CCD camera and personal computer. We currently use DMK21AU04 (Argo, Japan) for this purpose, which is inexpensive and compatible for standard PC with USB2.0 slots.
11. The incubation time of 10 min is chosen here, since lymphocytes could attach to ICAM-1 and VCAM-1 with a peak of 10 min under this condition (10). Incubation time should be determined empirically for other cells. The endothelial monolayers can be used instead of immobilized ligands. This setting is suitable to observe lymphocyte transmigration under shear flow (10).
12. The formula to convert flow rate to wall shear stress in the parallel plate flow chamber is $Q = 60\, h^2 w\tau / 6\mu$; where Q is flow

rate (ml/min), τ is wall shear stress (dyn/cm^2), μ is viscosity in poise (approximately 0.008 in RPMI1640/0.1% BSA), and h and w is height and width (cm) of the parallel flow chamber gasket, respectively. For typical postcapillary venules, the wall shear stress is ~2 dyn/cm^2. So when a rectangular gasket with 1.2 cm width and 0.01 cm height is used, the syringe pump should be set at 0.3 ml/min to deliver this shear stress.

13. The data are presented at defined time points. Alternatively, you can count the attached cell number at a 5–10-s interval and plot the percentages of attached cells against.

Noted added in proof

After submitting the manuscript, C. Rudd and colleagues have reported that SKAP1 (SKAP55) plays a critical role in TCR-triggered LFA-1 activation by binding directly to RAPL and forming a complex of Rap1-RAPL-LFA-1(Raab, M. et. al., Immunity, (2010) 32:541-556).

References

1. Bos, J. L., de Rooij, J., and Reedquist, K. A. (2001) Rap1 signaling: Adhering to new models, *Nat. Rev. Mol. Cell Biol.* **2**, 369–377.
2. Kitayama, H., Sugimoto, Y., Matsuzaki, T., Ikawa, Y., and Noda, M. (1989) A ras-related gene with transformation suppressor activity, *Cell* **56**, 77–84.
3. Reedquist, K. A., Ross, E., Koop, E. A., Wolthuis, R. M., Zwartkruis, F. J., van Kooyk, Y., Salmon, M., Buckley, C. D., and Bos, J. L. (2000) The small GTPase, Rap1, mediates CD31-induced integrin adhesion, *J. Cell Biol.* **148**, 1151–1158.
4. Katagiri, K., Hattori, M., Minato, N., Irie, S., Takatsu, K., and Kinashi, T. (2000) Rap1 is a potent activation signal for leukocyte function-associated antigen 1 distinct from protein kinase C and phosphatidylinositol-3-kinase, *Mol. Cell. Biol.* **20**, 1956–1969.
5. Sebzda, E., Bracke, M., Tugal, T., Hogg, N., and Cantrell, D. A. (2002) Rap1A positively regulates T cells via integrin activation rather than inhibiting lymphocyte signaling, *Nat. Immunol.* **3**, 251–258.
6. Franke, B., Akkerman, J. W., and Bos, J. L. (1997) Rapid Ca2+–mediated activation of Rap1 in human platelets, *EMBO J.* **16**, 252–259.
7. McLeod, S. J., Ingham, R. J., Bos, J. L., Kurosaki, T., and Gold, M. R. (1998) Activation of the Rap1 GTPase by the B cell antigen receptor, *J. Biol. Chem.* **273**, 29218–29223.
8. Ghandour, H., Cullere, X., Alvarez, A., Luscinskas, F. W., and Mayadas, T. N. (2007) Essential role for Rap1 GTPase and its guanine exchange factor CalDAG-GEFI in LFA-1 but not VLA-4 integrin mediated human T-cell adhesion, *Blood* **110**, 3682–3690.
9. Katagiri, K., Hattori, M., Minato, N., and Kinashi, T. (2002) Rap1 functions as a key regulator of T-cell and antigen-presenting cell interactions and modulates T-cell responses, *Mol. Cell. Biol.* **22**, 1001–1015.
10. Shimonaka, M., Katagiri, K., Nakayama, T., Fujita, N., Tsuruo, T., Yoshie, O., and Kinashi, T. (2003) Rap1 translates chemokine signals to integrin activation, cell polarization, and motility across vascular endothelium under flow, *J. Cell Biol.* **161**, 417–427.
11. Katagiri, K., Maeda, A., Shimonaka, M., and Kinashi, T. (2003) RAPL, a Rap1-binding molecule that mediates Rap1-induced adhesion through spatial regulation of LFA-1, *Nat. Immunol.* **4**, 741–748.
12. Tohyama, Y., Katagiri, K., Pardi, R., Lu, C., Springer, T. A., and Kinashi, T. (2003) The critical cytoplasmic regions of the αL/β2 integrin in Rap1-induced adhesion and migration, *Mol. Cell. Biol.* **14**, 2570–2582.
13. Dustin, M. L., and Springer, T. A. (1989) T-cell receptor cross-linking transiently stimulates adhesiveness through LFA-1, *Nature* **341**, 619–624.

14. Katagiri, K., Shimonaka, M., and Kinashi, T. (2004) Rap1-mediated LFA-1 activation by the T cell antigen receptor is dependent on PLC-gamma1, *J. Biol. Chem.* **279**, 11875–11881.
15. Bergmeier, W., Goerge, T., Wang, H. W., Crittenden, J. R., Baldwin, A. C., Cifuni, S. M., Housman, D. E., Graybiel, A. M., and Wagner, D. D. (2007) Mice lacking the signaling molecule CalDAG-GEFI represent a model for leukocyte adhesion deficiency type III, *J. Clin. Invest.* **117**, 1699–1707.
16. Crittenden, J. R., Bergmeier, W., Zhang, Y., Piffath, C. L., Liang, Y., Wagner, D. D., Housman, D. E., and Graybiel, A. M. (2004) CalDAG-GEFI integrates signaling for platelet aggregation and thrombus formation, *Nat. Med.* **10**, 982–986.
17. Pasvolsky, R., Feigelson, S. W., Kilic, S. S., Simon, A. J., Tal-Lapidot, G., Grabovsky, V., Crittenden, J. R., Amariglio, N., Safran, M., Graybiel, A. M., Rechavi, G., Ben-Dor, S., Etzioni, A., and Alon, R. (2007) A LAD-III syndrome is associated with defective expression of the Rap-1 activator CalDAG-GEFI in lymphocytes, neutrophils, and platelets, *J. Exp. Med.* **204**, 1571–1582.
18. Li, Y., Yan, J., De, P., Chang, H. C., Yamauchi, A., Christopherson, K. W., 2nd, Paranavitana, N. C., Peng, X., Kim, C., Munugulavadla, V., Kapur, R., Chen, H., Shou, W., Stone, J. C., Kaplan, M. H., Dinauer, M. C., Durden, D. L., and Quilliam, L. A. (2007) Rap1a null mice have altered myeloid cell functions suggesting distinct roles for the closely related Rap1a and 1b proteins, *J. Immunol.* **179**, 8322–8331.
19. Chrzanowska-Wodnicka, M., Smyth, S. S., Schoenwaelder, S. M., Fischer, T. H., and White, G. C., 2nd. (2005) Rap1b is required for normal platelet function and hemostasis in mice, *J. Clin. Invest.* **115**, 680–687.
20. Duchniewicz, M., Zemojtel, T., Kolanczyk, M., Grossmann, S., Scheele, J. S., and Zwartkruis, F. J. (2006) Rap1A-deficient T and B cells show impaired integrin-mediated cell adhesion, *Mol. Cell. Biol.* **26**, 643–653.
21. Chu, H., Awasthi, A., White, G. C., 2nd, Chrzanowska-Wodnicka, M., and Malarkannan, S. (2008) Rap1b regulates B cell development, homing, and T cell-dependent humoral immunity, *J. Immunol.* **181**, 3373–3383.
22. Katagiri, K., Ohnishi, N., Kabashima, K., Iyoda, T., Takeda, N., Shinkai, Y., Inaba, K., and Kinashi, T. (2004) Crucial functions of the Rap1 effector molecule RAPL in lymphocyte and dendritic cell trafficking, *Nat Immunol* **5**, 1045–1051.
23. Katagiri, K., Katakai, T., Ebisuno, Y., Ueda, Y., Okada, T., and Kinashi, T. (2009) Mst1 controls lymphocyte trafficking and interstitial motility within lymph nodes, *EMBO J.* **28**, 1319–1331.
24. Peterson, E. J., Woods, M. L., Dmowski, S. A., Derimanov, G., Jordan, M. S., Wu, J. N., Myung, P. S., Liu, Q. H., Pribila, J. T., Freedman, B. D., Shimizu, Y., and Koretzky, G. A. (2001) Coupling of the TCR to integrin activation by Slap-130/Fyb, *Science* **293**, 2263–2265.
25. Griffiths, E. K., Krawczyk, C., Kong, Y. Y., Raab, M., Hyduk, S. J., Bouchard, D., Chan, V. S., Kozieradzki, I., Oliveira-Dos-Santos, A. J., Wakeham, A., Ohashi, P. S., Cybulsky, M. I., Rudd, C. E., and Penninger, J. M. (2001) Positive regulation of T cell activation and integrin adhesion by the adapter Fyb/Slap, *Science* **293**, 2260–2263.
26. Wang, H., Liu, H., Lu, Y., Lovatt, M., Wei, B., and Rudd, C. E. (2007) Functional defects of SKAP-55-deficient T cells identify a regulatory role for the adaptor in LFA-1 adhesion, *Mol. Cell. Biol.* **27**, 6863–6875.
27. Menasche, G., Kliche, S., Chen, E. J., Stradal, T. E., Schraven, B., and Koretzky, G. (2007) RIAM links the ADAP/SKAP-55 signaling module to Rap1, facilitating T-cell-receptor-mediated integrin activation, *Mol. Cell. Biol.* **27**, 4070–4081.
28. Lafuente, E. M., van Puijenbroek, A. A., Krause, M., Carman, C. V., Freeman, G. J., Berezovskaya, A., Constantine, E., Springer, T. A., Gertler, F. B., and Boussiotis, V. A. (2004) RIAM, an Ena/VASP and Profilin ligand, interacts with Rap1-GTP and mediates Rap1-induced adhesion, *Dev. Cell* **7**, 585–595.
29. Han, J., Lim, C. J., Watanabe, N., Soriani, A., Ratnikov, B., Calderwood, D. A., Puzon-McLaughlin, W., Lafuente, E. M., Boussiotis, V. A., Shattil, S. J., and Ginsberg, M. H. (2006) Reconstructing and deconstructing agonist-induced activation of integrin αIIbβ3, *Curr. Biol.* **16**, 1796–1806.
30. Medeiros, R. B., Dickey, D. M., Chung, H., Quale, A. C., Nagarajan, L. R., Billadeau, D. D., and Shimizu, Y. (2005) Protein kinase D1 and the beta 1 integrin cytoplasmic domain control beta 1 integrin function via regulation of Rap1 activation, *Immunity* **23**, 213–226.
31. Allen, P. M., and Unanue, E. R. (1984) Differential requirements for antigen processing by macrophages for lysozyme-specific T cell hybridomas, *J. Immunol.* **132**, 1077–1079.
32. Lombard-Platlet, S., Bertolino, P., Deng, H., Gerlier, D., and Rabourdin-Combe, C.

(1993) Inhibition by chloroquine of the class II major histocompatibility complex-restricted presentation of endogenous antigens varies according to the cellular origin of the antigen-presenting cells, the nature of the T-cell epitope, and the responding T cell, *Immunology* **80**, 566–573.

33. Sanchez-Madrid, F., and del Pozo, M. A. (1999) Leukocyte polarization in cell migration and immune interactions, *EMBO J.* **18**, 501–511.

Chapter 19

Isolation of Focal Adhesion Proteins for Biochemical and Proteomic Analysis

Jean-Cheng Kuo, Xuemei Han, John R. Yates III, and Clare M. Waterman

Abstract

Focal adhesions (FAs) are discrete plasma membrane-associated adhesive organelles that play dual roles in cell force transduction and signaling. FAs consist of clustered transmembrane heterodimeric integrin extracellular matrix (ECM) receptors and a large number of cytoplasmic proteins that collectively form thin plaques linking the ECM to actin filament bundles of the cytoskeleton. FAs are complex organelles that can change their composition in response to biochemical or mechanical cues. These compositional differences may underlie the ability of FAs to mediate an array of important cell functions including adhesion, signaling, force transduction, and regulation of the cytoskeleton. These functions contribute to the physiological processes of the immune response, development, and differentiation. However, linking FA composition to FA function has been difficult since there has been no method to isolate intact FAs reproducibly and determine their composition. We report here a new method for isolating FA structures in cultured cells distinct from cytoplasmic, nuclear, and internal membranous organellar components of the cell. We provide protocols for validation of the fractionation by immunofluorescence and immunoblotting, procedures for preparing the isolated FAs for mass spectrometric proteomic analysis, tips on data interpretation and analysis, and an approach for comparing FA composition in cells in which small GTPase signaling is perturbed.

Key words: Focal adhesion, Immunofluorescence, Western blotting, Mass spectrometry, Integrin

1. Introduction

Focal adhesions (FAs) are specialized adhesion organelles that can mediate various biological functions and responses through engagement of one specific class of surface receptor, integrins (1–3). Integrins are a family of alpha–beta heterodimer transmembrane receptors that specifically bind extracellular matrix or cell surface-bound ligands. Following integrin activation and binding to ligands, FA-associated proteins are recruited from the cytosol

Motomu Shimaoka (ed.), *Integrin and Cell Adhesion Molecules: Methods and Protocols*,
Methods in Molecular Biology, vol. 757, DOI 10.1007/978-1-61779-166-6_19, © Springer Science+Business Media, LLC 2011

to bind to integrin cytoplasmic tails and one another. Together, these proteins form thin (~200 nm), discrete, (~0.1–10 μm^2) plasma membrane-associated FAs plaques linking the ECM to actin filament bundles of the cytoskeleton. Integrins together with FA-associated proteins perform four basic cellular functions that mediate their physiological roles in development, immune response, cardiovascular function, and maintaining tissue integrity. First, integrins in FAs bind extracellular ligands to mediate cell adhesion to ECM or other cells. Second, FAs mediate signaling from integrin engagement to induce dramatically different cellular responses, including proliferation, differentiation, or death. Third, FAs proteins link the cytoskeleton to integrins, allowing cells to transduce force generated in the actomyosin system to the ECM or other cells to drive tissue morphogenesis and cell movement. And finally, FAs act as regulators of the cytoskeleton, modulating its organization through physical interactions with actin filaments and regulating its dynamics via signaling to control the activity of actin-binding proteins and myosin motors (1–3). In addition to mediating these important cell functions, FAs are dynamic, mechanosensitive structures that change composition, grow or disintegrate in response to physical as well as biochemical cues (1, 4). For example, during cell migration, FAs undergo spatiotemporally coordinated cycles of formation and disassembly that enable directional whole cell motion (5). Thus, FAs are complex, dynamic organelles that integrate several distinct cell responses to mediate a diverse array of physiological functions.

One mechanism by which FAs could mediate differential responses through the same class of receptor is by variation of FA protein composition (3, 6). However, a method for systematic analysis of FA composition under different biological conditions has been lacking. Over several decades of research, it has become apparent that FA composition is exceedingly complex. By a variety of methods including protein localization and analysis of FA–protein interactions, more than 155 components (150 proteins plus 4 lipids and calcium ions) have been identified as associating with FAs which, again based on the literature, may be capable of >750 possible protein–protein interactions (7). The FA proteome includes ECM, integrins, cytoskeletal proteins, adaptor proteins, and enzymes, including but not limited to kinases, phosphatases, nucleotide triphosphatases, and their exchange factors. To gain insight into how FAs mediate the downstream responses of integrins through compositional variation of the FA proteome, we present a method for the systematic biochemical isolation of FAs from fibroblasts in tissue culture, and their full compositional analysis by tandem mass spectrometry to ascertain the constituents of complex mixtures of proteins. The protocol includes procedures for validation of the FA isolation by immunofluorescence and immunoblotting, preparation of isolated FAs for proteomic analysis, and tips on data interpretation and analysis. In addition, we present an example application for comparative analysis of FA proteome compositional change in response to perturbation of small GTPase signaling.

2. Materials

2.1. Biochemical FA Isolation Method

1. Human foreskin fibroblasts (HFF1, ATCC): Cultured in 100-mm tissue culture dishes in Dulbecco's modified Eagle's medium (DMEM) containing 15% fetal bovine serum (FBS), 2 mM l-glutamine, and penicillin/streptomycin.
2. Trypsin/EDTA contains trypsin (0.5 g/L) and EDTA-4Na (0.2 g/L) in Hanks' balanced salt solution without $CaCl_2$, $MgCl_2$-$6H_2O$, and $MgSO_4$-$7H_2O$.
3. Human plasma fibronectin purified protein: Diluted to 15 μg/ml in Dulbecco's phosphate-buffered saline, pH 7.4.
4. TEA-containing low ionic strength buffer: 2.5 mM triethanolamine (TEA), pH 7.0. Store at room temperature.
5. Protease inhibitor tablets (complete mini EDTA-free, Roche) containing serine and cysteine protease inhibitors: Dissolved in 10 ml solution just prior to use.
6. Cell scrapers (L-18, LabScientific, inc.).
7. Waterpik™ dental water jet (Interplak dental water jet WJ6RW, Conair): Available from most drug stores.

2.2. Immunofluorescence

1. Cleaned 22×22 mm #1.5 cover slips: The brand, thickness, and cleanliness of the coverslip are very important for even adhesion of ECM proteins and low-background epifluorescence microscopy (see Note 1).
2. HFF cells plated on cleaned 22×22 mm #1.5 cover slips and cultured in 35-mm tissue culture dishes.
3. Cytoskeleton buffer (CB): 10 mM MES, pH 6.1, 138 mM KCl, 3 mM $MgCl_2$, and 2 mM EGTA. Stored at 4°C.
4. Paraformaldehyde: diluted from a freshly opened ampoule to 4% in CB just prior to use.
5. Columbia coverslip staining jars.
6. Blocking reagent: 10% (w/v) fraction V bovine serum albumin and 0.5 mM NH_4Cl in PBS.
7. 0.5% Triton X-100 in CB: prepared immediately before use.
8. 0.1 M glycine in CB.
9. Tris-buffered saline with Tween-20 (TBS-T): Prepare 10× TBS stock with 1.37 M NaCl, 27 mM KCl, and 250 mM Tris–HCl,

pH 7.4. Dilute 100 ml with 900 ml distilled water and add Tween-20 at 0.1% (v/v) just prior to use.

10. Mouse anti-chicken paxillin antibody (see Note 2).
11. AlexaFluor 488 phalloidin (Gibco/Invitrogen).
12. Secondary antibody: Cy3-conjugated donkey anti-mouse IgG (Jackson ImmunoResearch).
13. Fluorescent mounting medium.
14. Nail polish (Electron Microscopy Science).
15. Microscope slides (Gold Seal Micro Slides) (see Note 3).
16. Epifluorescence microscope equipped with a 60× or 100× oil immersion objective lens and a digital camera (Nikon).
17. Digital image analysis software: We use MetaMorph software (MDS Analytical Technologies); however, freeware such as Image J (http://rsbweb.nih.gov/ij/) and MicroManager (http://www.micro-manager.org/) both have capability sufficient for the analysis described below.

2.3. Western Blotting

1. Radioimmuno precipitation assay (RIPA) buffer without SDS: 50 mM Tris–HCl, pH 8.0, 150 mM NaCl, 1.0% NP-40, and 0.5% sodium deoxycholate. Store at 4°C.
2. Primary antibodies: Mouse anti-chicken vinculin antibody (Sigma), mouse anti-human talin antibody (Chemicon), mouse anti-chicken actin antibody (BD), mouse anti-chicken tubulin antibody (Sigma), rabbit anti-human GAPDH antibody (Open Biosystem), and rabbit anti-mouse Akt antibody (Cell Signaling).
3. Secondary antibodies: Horseradish peroxidase (AffiniPure goat anti-mouse IgG, Jackson ImmunoResearch) and horseradish peroxidase (AffiniPure goat anti-rabbit IgG, Jackson ImmunoResearch).
4. PAGE/Western blot apparatus.
5. Immobilon transfer membrane (IPVH304F0, Millipore).
6. Thermo Scientific Pierce BCA protein assay kits (Fisher).

2.4. Proteomic Analysis

1. Agarose-conjugated goat anti-human actin antibody (Santa Cruz).
2. Agarose-conjugated mouse anti-human fibronectin antibody (Santa Cruz).
3. 5-µm Partisphere strong cation exchange resin (Whatman, Clifton, NJ).
4. 5-µm Aqua C18 resin (Phenomenex, Ventura, CA).
5. 3-µm Aqua C18 resin (Phenomenex, Ventura, CA).
6. Urea.

7. Protein reduction and alkylation solution: Tris (2-carboxyethyl) phosphine hydrochloride and iodoacetamide.
8. Calcium chloride.
9. Trypsin, mass spectrometry grade: Store at 4°C.
10. Formic acid.
11. Acetonitrile.
12. Ammonium acetate.
13. Buffers for a modified 12-step MudPIT analysis (REF): Buffer A, 5% acetonitrile and 0.1% formic acid; buffer B, 80% acetonitrile and 0.1% formic acid; and buffer C, 500 mM ammonium acetate, 5% acetonitrile, and 0.1% formic acid.
14. Zero-dead volume union (Upchurch Scientific, Oak Harbor, WA).
15. Agilent 1100 quaternary HPLC (Agilent, Palo Alto, CA).
16. LTQ linear ion trap mass spectrometer (Thermo Fisher Scientific, San Jose, CA).
17. Xcalibur data system (Thermo Fisher Scientific, San Jose, CA).

2.5. Modulation of FA Composition by Manipulation of Cell Signaling

Electroporation is used to transfect plasmid DNA transiently into HFF1.

1. Human Dermal Fibroblast Nucleofector kit (Amaxa biosystem).
2. Nucleofector II (Amaxa Biosystem): Program: U-024 is used.

3. Methods

Here, several protocols are described for isolating FA from tissue culture cells, including methods to validate the fractionation protocol morphologically and biochemically, to identify the protein composition from the fractionated FA structures by proteomic analysis, and to utilize this fractionation protocol for isolating FA structures from cells under different perturbation conditions. The overall procedure is outlined in Fig. 1.

3.1. FA Isolation from Cultured Fibroblasts

The goal of FA isolation from tissue culture cells is to maintain the native FA protein composition and structure bound to the tissue culture substratum for collection, while carefully removing nuclei; internal membrane-bounded organelles; the bulk of the actin, microtubule, and intermediate filament cytoskeletons; most of the cell plasma membrane; and soluble cytoplasmic proteins. Similar to methods introduced by Fujiwara for stress fiber isolation (8), TEA-containing low ionic strength buffer, a pH-balanced

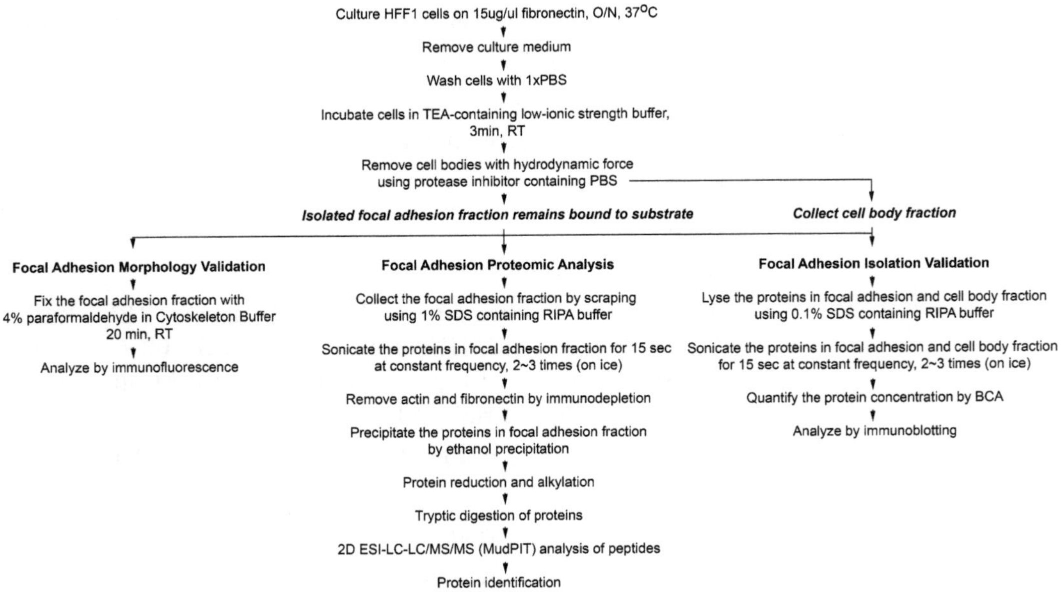

Fig. 1. Diagram of the major steps for isolation of FA from HFF1 cells, followed by validation of FA morphology preservation by immunofluorescence, validation of FA isolation by Western blotting, or determination of FA protein composition by proteomic analysis.

hypotonic solution, is used to create osmotic pressure inside cells, which swells cells and thus weakens the membrane integrity. Once the plasma membrane is weakened, membrane-bound organelles, nuclei, and soluble materials of the cytoplasm can be removed by hydrodynamic forces induced by strong trituration using a protease inhibitor-containing, detergent-free, osmotically balanced buffer (PBS), which is important to avoid protein degradation and maintain the FA proteins in their native state (Fig. 2).

In contrast to the Fujiwara method (8) that was optimized for preservation of actomyosin stress fibers, our method is aimed at isolating native FA. We put in a considerable amount of work in optimizing our protocol to ensure that major FA components were maintained in the FA fraction, as the Fujiwara method alone was unsuccessful for maintaining many FA proteins in the stress fiber preparation. We present methods for validating preservation of FA proteins in the FA fraction in Subheadings 3.2 and 3.3. Although FA fractions isolated by our method will contain a significant amount of actin, this is unavoidable, as the actin cytoskeleton and FA are interdependent structures, and perturbations of the actin cytoskeleton will unavoidably alter FA structure and composition. As noted in Subheading 3.4, following isolation of FA fractions, excess actin can be removed by immunodepletion prior to proteomic or biochemical analysis.

The protocol below shows the details of the FA isolation method optimized for HFF1 cells grown adhered to human

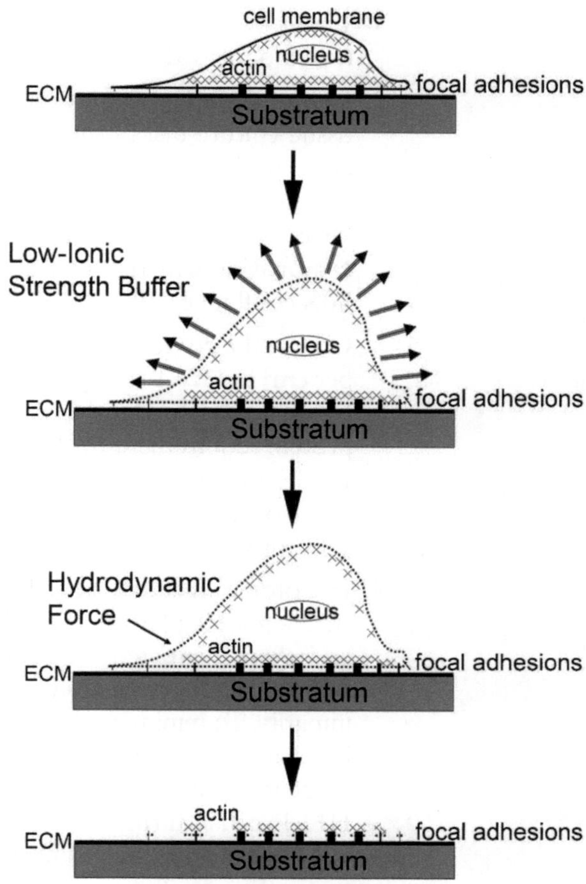

Fig. 2. Schematic diagram of the FA isolation procedure.

fibronectin. This cell type was chosen by us for its human origin, its robust adherent growth in tissue culture, its amenability to transfection, its robust in vitro cell migration behavior, and its well-developed FA morphology. Our protocol could be adapted to other cell types by experimenting with and optimizing the osmotic shock (step 4) and trituration (step 5) steps below. 15 μg/ml human fibronectin was chosen as the growth substrate because this concentration was found to promote optimal migration of HFF1 cells in culture (unpublished observations). Cells could easily be grown adhered to different ECM proteins or a different concentration of human fibronectin simply by modifying step 1.

1. Coat tissue culture dishes or microscope coverslips for immunofluorescence with fibronectin for growth of cells. Tissue culture dishes or coverslips are incubated overnight at 4°C in a solution of 15 μg/ml human fibronectin diluted in PBS. This is followed by rinsing three times with PBS. Subsequently, substrates are incubated in 1% BSA-containing DMEM for

1 h at 37°C for blocking. After blocking, the substrates are rinsed once with PBS prior to plating the cells.

2. Culture HFF1 cells for experiments. HFF1 cells are grown in tissue culture dishes and passaged when approaching confluence with trypsin/EDTA to provide new experimental cultures plated on 15 μg/ml fibronectin-coated 100-mm culture dishes or cover slips. A 1:4 split of near-confluent HFF1 cells can provide experimental cultures that are at approximately 50% confluence after 24 h.

3. Choose the proper amount of cells. The yield of total FA protein per cm^2 of HFF1 cells grown at 50% confluence on 15 μg/ml fibronectin-coated dishes is approximately 0.13 μg. The total protein requirement for Western blot analysis is 4 μg, for 2-D DIGE proteomic analysis 500 μg, and for MudPit Mass spectrometry analysis 60 μg. Plan accordingly when plating cells for your experiments.

4. Hypotonically shock the cells to weaken the plasma membrane. At 24 h after plating, the cells are rinsed once with 1× PBS and then incubated in TEA-containing low ionic strength buffer to induce cell swelling and weaken the cell membrane integrity. To minimize effects on FA composition and structure, incubation time must be as short as possible. To optimize incubation time for HFF1, we analyzed the morphology of HFF1 cells during treatment with TEA-containing low ionic strength buffer using time-lapse phase-contrast microscopy. Figure 3 shows that HFF1 cells start to swell after 3 min of TEA-containing low ionic strength buffer incubation and explode by 50-min treatment. Based on this analysis, we use 3-min hypotonic shock for HFF1 as the minimum time to induce swelling and still allow removal of the cell bodies (Fig. 4).

5. Remove cell bodies with hydrodynamic force. Immediately after the cells are incubated with TEA-containing low ionic strength buffer for 3 min at room temperature, 1× PBS containing protease inhibitor is used in the Waterpik to apply strong, pulsed hydrodynamic forces to the cells in order to remove the

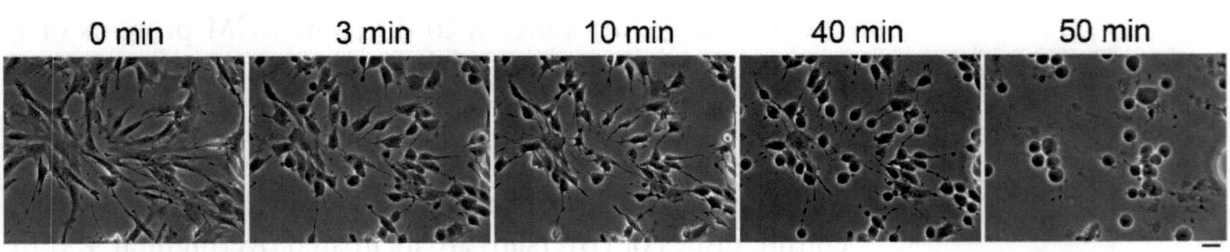

Fig. 3. Time-lapse phase-contrast image series of HFF1 cell morphology after treatment (time in minutes shown) with TEA-containing low ionic strength buffer. As cells appear swollen but not fully rounded up by 3-min treatment, we chose this time point as optimal for our protocol. Scale bar, 20 μm.

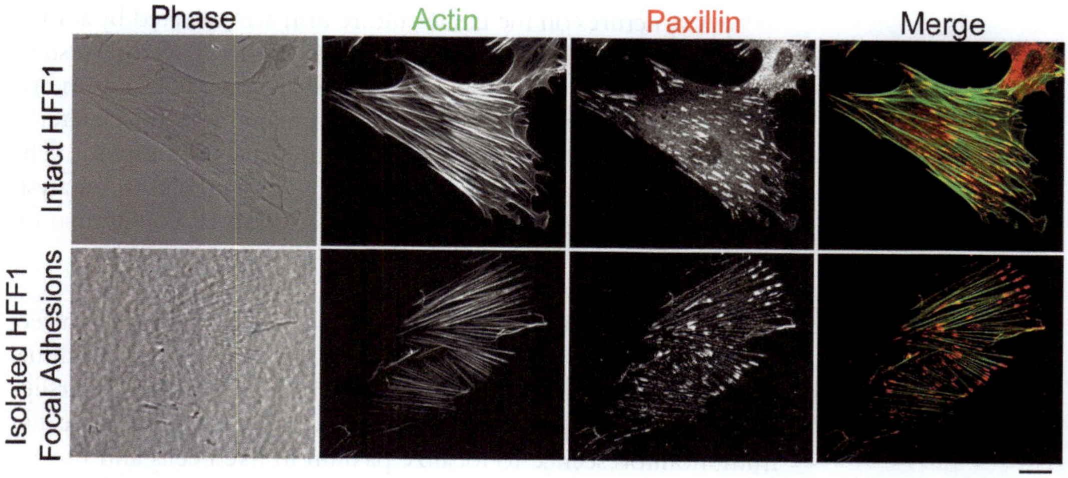

Fig. 4. Comparison of FAs and actin cytoskeleton in HFF1 cells and isolated HFF1 FAs. Intact HFF1 and isolated FAs were fixed and stained with Alexa 488 phalloidin to visualize filamentous actin (*green*) and indirect immuno-localization of paxillin (*red*) to show the distribution of FAs. Scale bar, 20 μm.

cell bodies, membrane-bound organelles, nuclei, cytoskeleton, and soluble materials of the cytoplasm. The power of Waterpik trituration needs to be adjusted carefully, since flushing too strongly will remove some low-abundance FA-associated proteins, and flushing too weakly will fail to remove other cell structures. The Waterpik we use has a 7-setting pressure control, and we adjust it to setting "3" for HFF1 cells. During trituration, we hold the nozzle ~0.5 cm from and at an approximate angle of 90° to the surface of the tissue culture dish, and slowly move the nozzle back and forth across the dish, being careful to spray the whole dish surface evenly, for a total time of about 10 s. These recommendations will have to be adjusted to account for your own dental water jet apparatus, trituration style, and cell type. Validation of the cell body removal procedure is discussed in Subheadings 3.2 and 3.3 below. A total volume of 50 ml buffer is used in the Waterpik reservoir. Following the trituration of one dish of cells, the buffer containing the cell bodies is collected from that dish and recycled into the reservoir for use in the trituration of the next dish of cells. This 50 ml of buffer containing all the cell bodies is collected after trituration of all dishes is complete to serve as a sample for Western blot analysis to validate separation of the FA and cell body fractions (see Subheading 3.3). Subsequently, a total volume of 400 ml fresh buffer is used in the Waterpik reservoir to rinse all the dishes again prior to FA fraction collection.

6. Isolate the FAs that remain attached to the tissue culture dish. Following trituration, excess buffer is removed from the dish using a vacuum manifold, and the remaining buffer and

FA structures on the tissue culture dish are collected by adding 300 μl of buffer (1× RIPA containing 0.1% SDS for Western analysis or 1× RIPA containing 1% SDS for mass spectrometric analysis), thoroughly scraping the remaining cellular material from the dish (Fig. 4), and collecting it with a pipette. The material from all dishes is then pooled and used in the analysis of FA-associated proteins without substantial contamination from proteins in the cell body.

3.2. Validation of FA Isolation by Immunofluorescence

To validate that the FA isolation method does not adversely affect FA morphology prior to the collection step, the size of individual FAs in intact cells and the FAs remaining adhered to coverslips after cell body removal are determined and compared. We use immunofluorescence to localize paxillin in fixed cells and the FA fraction. Paxillin is a well-known FA adaptor protein (9) whose distribution in digital fluorescence images can indicate the size of individual FAs (Fig. 5a). Figure 5b shows that the proportion

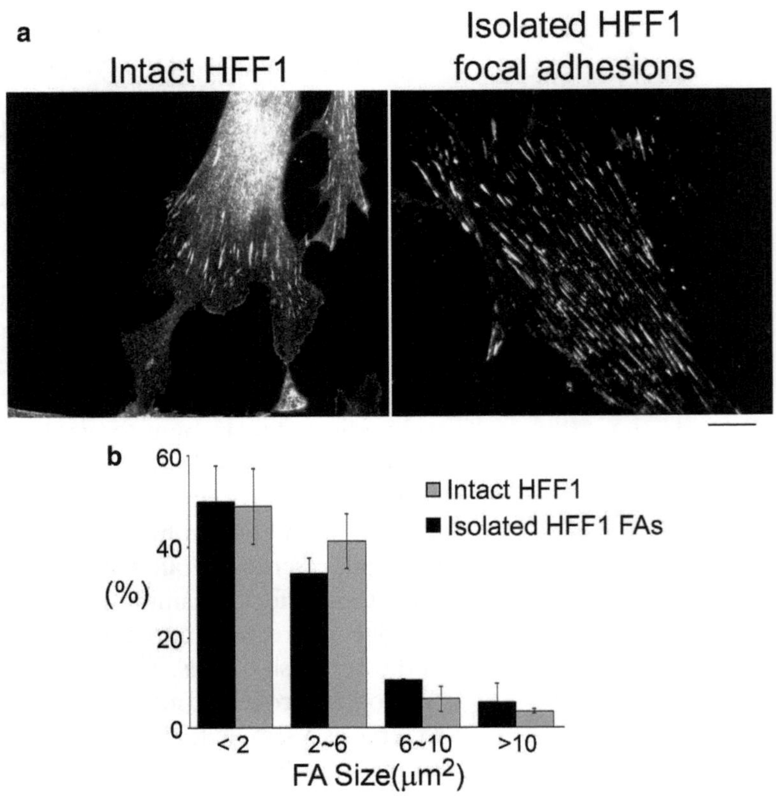

Fig. 5. Immunofluorescence and quantitative analysis for validating the preservation of native FA morphology after FA isolation. (**a**) Intact HFF1 and isolated HFF1 FAs were fixed and processed for indirect immuno-localization of paxillin to show the distribution of FAs. Scale bar, 20 μm. (**b**) Histogram of the size distribution of thresholded paxillin-containing FAs in images of intact HFF1 cells ($n=3$ cells 408 FAs) and the isolated FA fraction ($n=3$ cells, 344 FA). Statistical analysis indicates no significant difference in the percentage of FAs of each size range between FAs from intact cells or isolated FAs.

of different-sized FAs between intact HFF1 and isolated FA fractions does not significantly differ. This indicates that the FA isolation method does not influence the morphology of FA structures. The protocol below shows how to compare the morphology of FAs in cells and isolated FAs by immunofluorescence. It is also a good idea to validate the loss of nuclear, organellar, or cytoskeletal structures from isolated FAs by double labeling with markers for these structures. We recommend staining with DAPI for nuclei, fluorescent phalloidin for actin, and using indirect immunofluorescence with antibodies to tubulin (microtubules) to determine the degree to which these structures are present or absent in the isolated FA fraction.

1. Fix intact HFF1 and the isolated HFF1 FAs adhered to microscope coverslips. Incubate cells/isolated FAs bound to 15 μg/ml fibronectin-coated coverslips with 4% paraformaldehyde in CB for 20 min at room temperature. Use of Columbia coverslip staining jars is helpful for this and remaining steps.
2. Permeablize the cells with 0.5% Triton X-100 in CB for 5 min at room temperature. Since the isolated FA fraction mostly lacks plasma membrane, permeabilization is not necessary.
3. Quench free aldehydes for both the fixed cells and isolated FA fraction with 0.1 M glycine in CB for 10 min at room temperature.
4. Wash the coverslips with TBS-T three times for 5 min each wash.
5. Block the cells and the isolated FA fraction with blocking reagent for 60 min at room temperature or overnight at 4°C.
6. Incubate the permeabilized HFF1 cells and the isolated FA fraction with anti-paxillin antibody, diluted 1:1,000 in blocking reagent, for 60 min at room temperature or overnight at 4°C.
7. Wash the coverslips three times with TBS-T for 5 min each wash.
8. Incubate the coverslips with secondary antibody. We use a Cy3-conjugated donkey anti-mouse IgG diluted 1:300 in blocking reagent for 60 min at room temperature.
9. Wash the coverslips three times with TBS-T for 5 min each wash.
10. Mount the coverslips on microscope slides with fluorescence mounting medium and seal the coverslip to the slide with nail polish.
11. Obtain fluorescence images of FAs in intact cells and isolated FAs. Analysis of FA size distribution requires that images be acquired with a high magnification, high-resolution objective lens and a digital camera. A lens with a magnification >/= 60× should be used (see Note 4) (10). When acquiring images,

utilize the proper filter cube for the fluorophore of your choice, and adjust image brightness by altering the camera exposure time or introducing neutral density filters in the illumination path such that small FAs at the cell periphery are visible, and large FAs do not saturate the camera (Fig. 5a).

12. Measure the segmented area of individual paxillin-marked FA. After acquiring images, open them in your image analysis software and use the image thresholding function to highlight only regions of the cell that are included within FAs. You may need to convolve the image with a smoothing filter to eliminate "hot pixels" that will be counted as bright FAs. Also, an image "flattening" filter can be helpful if cytoplasmic background causes central cell regions that are bright, but contain no FAs, to be included in the threshold. Convert the thresholded regions to areas, and record the area of each FA. The pixel-to-micron conversion factor must be determined using images of a stage micrometer. If the background level in your immunofluorescence staining is high, it will be difficult to segment FAs by thresholding. Experiment with the rinsing and blocking steps, as well as the choice of fluorescent-labeled secondary antibodies to optimize nice, low-background images suitable for efficient FA segmentation (see Note 5). For the FA size comparison, we set the smallest size of FA we can identify in intact HFF1 as the threshold size for FA in the FA fraction to filter out possible background signal due to nonspecific fluorescence that tends to occur on the coverslip containing isolated FAs.
13. To determine if any specific population of FAs is adversely affected by the isolation procedure, bin the individual FA areas from both intact cells and the FA fraction into four groups, including <2, 2–6, 6–10, and >10 μm^2 (Fig. 5b). Other analysis of FA morphometry (ellipticity, length, width, etc.) can also be performed to add support to the conclusions.
14. Reduction in small FA in the FA fraction compared to intact cells is likely the result of either too long osmotic shock or excessive trituration, while the presence of excessive cell structures in the FA fraction would likely be due to the opposite issues.

3.3. Validation of FA Isolation by Western Blotting

To confirm that the FA isolation method is able to separate FA structures from cell body components, the levels of some well-known FA-associated proteins, as well as organellar and cytosolic proteins, are analyzed by Western blot to compare their level in the isolated FA and cell body fractions. We use paxillin, talin, and vinculin as our standard FA-associated proteins, and GAPDH, tubulin, and Akt as the standard cell body proteins. The results in Fig. 6 indicate efficient FA isolation, as the standard FA-associated proteins are concentrated in the isolated FA fraction, but present

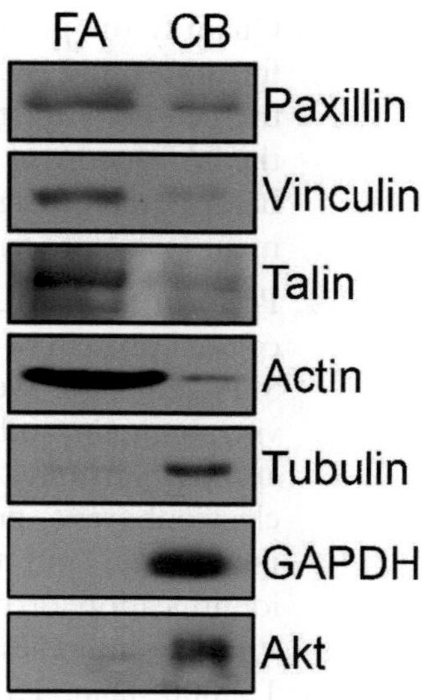

Fig. 6. Western blot comparison of cell body and isolated FA fractions as a validation of the efficiency of FA isolation. Equal total protein of isolated FA (FA) and cell body (CB) fractions were subjected to SDS-PAGE and Western analysis with antibodies to the indicated proteins. Paxillin, talin, and vinculin are known FA-associated proteins, while GAPDH, tubulin, and Akt should not be concentrated in FAs. The results indicate efficient FA isolation, as the FA-associated proteins are concentrated in the isolated FA fraction, but present at much lower levels in the cell body fraction. In contrast, the cell body proteins are nearly absent from the FA fraction and concentrated in the cell body fraction.

at much lower levels in the cell body fraction. In contrast, the cell body proteins are nearly absent from the FA fraction and concentrated in the cell body fraction. It should be noted that most FA proteins are normally cycling between FA-bound and soluble cytoplasmic pools, and that some FA proteins perform functions in the nucleus. These non-FA functions should be taken into consideration in interpreting the distribution of the protein between FA and cell body fractions. In contrast, many cell body and cytosolic proteins have no (known) association with FA, and for a good preparation, these proteins should be close to 100% depleted from the FA fraction. The protocol below shows how to evaluate the efficiency of FA isolation by determining the degree of separation of FA-associated and cell body proteins by Western blotting.

1. Collect the cell body (step 5 of Subheading 3.1) and FA fractions (step 6 of Subheading 3.1) from HFF1 cells plated on 15 μg/ml fibronectin-coated 100-mm tissue culture plates. Remember that ~4 μg total FA protein is needed for Western blot analysis. For HFF1 at 50% confluence, we have found that one dish of cells should suffice. The FA fraction is collected in 1× RIPA buffer containing 0.1% SDS. Add 2× RIPA buffer containing 0.2% SDS to the cell body fraction at a 1:1 volume.
2. Sonicate the denatured fractions on ice at 0°C for 15 s at a constant frequency.

3. Quantify the protein concentration of the cell body and FA fractions using BCA reagent. The concentration of protein in the cell body fraction is generally much higher than that of the FA fraction, so the cell body fraction can be diluted tenfold before quantifying to improve the accuracy of the measurement (see Note 6).
4. Perform SDS-PAGE and Western blot. Load equal amount of total protein of cell body fraction and FA fraction onto an SDS-PAGE for Western blotting. We use an 8% gel to detect vinculin, talin, and Akt; a 12% gel to detect paxillin, GAPDH, and tubulin; and a 15% gel to detect actin (Fig. 6). Following electrophoresis, proteins are electrotransferred to Imobilon membrane, and the membranes processed for immunoidentification of proteins by enhanced chemiluminescence. Primary antibodies are used at the following dilutions: vinculin, 1:2,000; talin, 1:1,000; paxillin, 1:10,000; tubulin, 1:5,000, actin, 1:4,000; GAPDH, 1:200; and Akt, 1:1,000. Horseradish peroxidase-AffiniPure goat anti-mouse or anti-rabbit antibodies are used at 1:5,000 and 1:10,000, respectively.
5. Determine the efficiency of the FA isolation procedure. Western blots are developed with ECL reagent, films exposed, and band intensities quantified. As noted above, FA proteins should be strongly enriched (>70–75%) in the FA fraction compared to the cell body fraction, while cell body proteins should be >95% depleted from the FA fraction.

3.4. Preparation of FA Fractions for Proteomic Analysis

We utilize a "bottom-up" proteomic approach to identify the protein composition of isolated FA fractions. A "bottom-up" approach is a general strategy for analyzing proteomic samples that exploits tandem mass spectrometry to ascertain the constituents of a sample consisting of a complex mixture of fragmented peptides (11). This approach consists of three steps: sample separation, mass analysis, and protein identification. The goal of sample separation is to separate proteins in complex mixtures. Proteins can be separated without enzymatic digestion, by methods such as two-dimensional gel electrophoresis (2D-GE), or after enzymatic digestion into a collection of peptides and are then separated by multidimensional liquid chromatography (LC) approaches. Mass analysis is the driving engine of proteomics that consists of three major elements: ion source, mass analyzer, and detector. Ionization is generally driven by matrix-assisted laser desorption ionization (MALDI) or electrospray ionization (ESI). Mass analysis is generally managed through quadrupole, ion trap (quadrupole ion trap; linear ion trap), time-of flight (TOF), or Fourier-transform ion cyclotron resonance (FTICR) analyzers. Detectors record the ion signals from the analyzer followed by mass spectrometric identification of peptides through sequence database search algorithms.

Our method of choice for protein separation is 2D-LC. Although 2D-GE separates proteins in mixtures by size and by isoelectric focusing, after which spots in the gel are subjected to mass spectrometry, this approach may fail to retain proteins of extreme p*I*, molecular weight, hydrophobicity, or low concentration. Because of these limitations, we perform 2D-LC for the digested peptides separation of the protein mixtures in a technique developed by the Yates lab for comprehensive analysis of complex protein mixtures called multidimensional protein identification technology (MudPIT) (12). In MudPIT, proteins are reduced, alkylated, and digested to peptides and subsequently separated via 2D-LC rather than 2D-GE. The MudPIT separation uses a biphase column to separate the peptide mixture by both their charge and hydrophobicity. The first separation is a strong cation exchange (SCX) column to capture charged peptides. The second separation is a reverse phase chromatography to bind the uncharged peptides. Acetonitrile and ammonium acetate proceed in gradient to elute the peptides and flow them directly into the mass spectrometer. Thus, all the peptides, even from low-abundance proteins in isolated FA fractions, can be analyzed without loss of extreme p*I* and hydrophobicity components.

Here, we describe the protocols including how to prepare the FA fractions for MudPIT analysis and how to identify the protein composition through MudPIT (Fig. 8). Note that we include a critical step (Subheading 3.4.1, step 3) of immunodepleting the two major protein components from the FA fraction, actin and fibronectin. We found that without this important step, many of the low-abundance FA proteins could not be detected.

3.4.1. Preparation of the Isolated FA Proteins for MudPIT Proteomic Analysis

1. Isolate FA fraction from HFF cells at 50% confluence and grown in 15 μg/ml fibronectin-coated 100-mm culture dishes. Once cell bodies are removed, collect the FA fraction in 1× RIPA buffer containing 1% SDS, and incubate for 5 min on ice. Remember that ~60 μg total FA protein is needed for MudPIT analysis. For HFF1 at 50% confluence, six dishes should suffice.
2. Sonicate FA proteins for 15 s on ice at constant frequency.
3. Immunodeplete fibronectin and actin from the denatured FA fraction (Fig. 7). Fibronectin (used to plate the cells on) and actin (which remains adhered to many isolated FAs) are in such excess over most FA proteins that they mask detection of lower abundance FA proteins. To improve the dynamic range of the mass spectrum, the concentration of fibronectin and actin in the samples can be greatly reduced by immunodepletion. For immunodepletion, commercial agarose-conjugated goat antihuman actin and agarose-conjugated mouse-antihuman fibronectin antibodies are used. The linkage between antibody and agarose is covalent and not denatured by 1% SDS. If lots of IgG are detected in mass spectrometry,

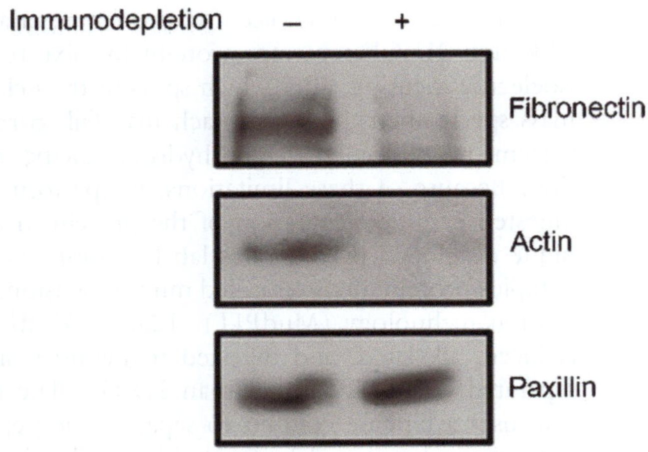

Fig. 7. Western blot analysis of the efficiency of immunodepletion of actin and fibronectin from isolated FA fractions. Equal total protein of isolated HFF1 FA fractions that had been subjected to immunodepletion of actin and fibronectin (+) or not (–) was subjected to SDS-PAGE and Western analysis with antibodies to the indicated proteins. The results indicate substantial depletion of actin and fibronectin, but no effect on the level of paxillin in the samples.

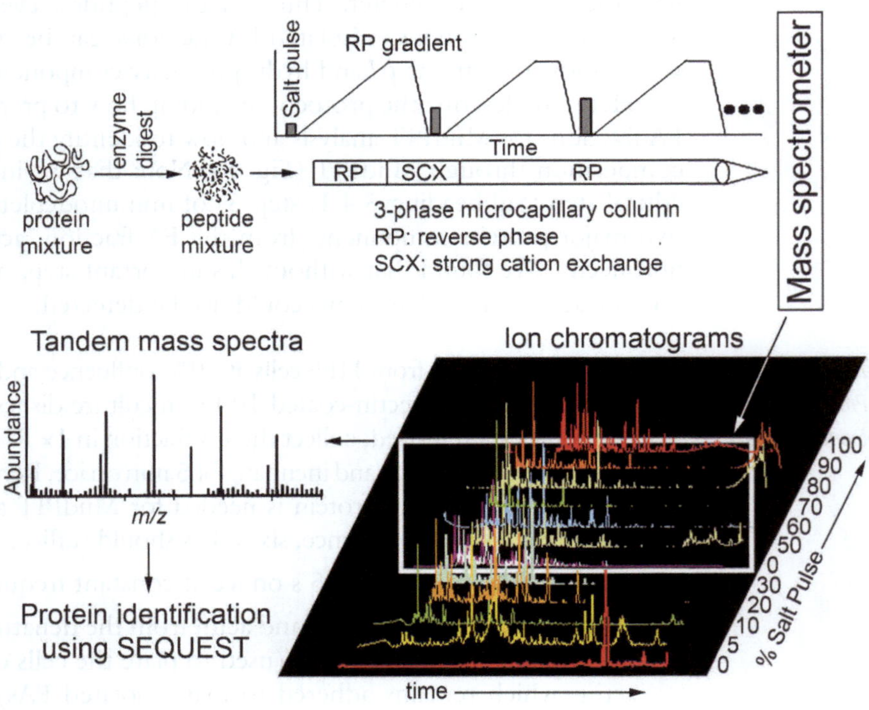

Fig. 8. Flow diagram of the major steps of multidimensional protein identification technology (MudPIT) for protein identification in a complex mixture.

this indicates poor efficiency of cross-linking, and can be overcome by washing with 1× RIPA containing 1% SDS three to five times. Add anti-fibronectin and anti-actin antibody-coupled agarose beads to the FA fraction at a ratio of 1:1:1

(1 μg of fibronectin antibody:1 μg of actin anibody:1 μg FA total protein) and rotate for overnight at 4°C. Subsequently, pellet the agarose-conjugated antibodies and save the supernatant, which consists of the actin/fibronectin-depleted FA fractions. Load equal volume of FA fraction, before and after immunodepletion, onto SDS-PAGE for Western blot analysis of the level of actin, fibronectin, and paxillin in the FA fraction before and after immunodepletion. Figure 7 shows that the majority of actin and fibronectin is depleted, but paxillin still exists in the FA fraction.

4. Concentrate proteins from the actin/fibronectin-depleted FA fractions by ethanol precipitation. 1 volume of the actin/fibronectin-depleted FA fraction is mixed with 9 volumes of ice-cold 100% ethanol. Incubate overnight at −20°C. Collect precipitated proteins by centrifugation for 15 min at 4°C at 15,000 × *g*. After carefully removing the supernatant, the precipitated FA protein pellets can be stored at −80°C without further treatment until use for MudPit analysis.

3.4.2. Protein Identification by Multidimensional Protein Identification Technology

1. Denature proteins and reduce oxidative cross-links. Resuspend the precipitated FA protein pellets (~50 μg) in digestion buffer (8 M urea, 100 mM Tris–HCl, pH 8.5). Proteins are reduced and alkylated to eliminate cysteine cross-links with tris(2-carboxyethyl)phosphine (TCEP) and iodoacetamide. Add 500 mM TCEP stock to bring the solution to 5 mM and incubate for 15 min at room temperature. Subsequently, add 500 mM iodoacetamide stock to 10 mM final concentration and incubate at room temperature for 20 min in the dark. Dilute the mixture to 2 M urea with 100 mM Tris–HCl, pH 8.5, and add $CaCl_2$ to 2 mM.
2. Trypsin digest denatured proteins. Add trypsin at a 1:50 enzyme:protein molar ratio and incubate at 37°C overnight for digestion. Terminate digestion by bringing to 5% formic acid. This mixture of digested protein can then be stored at −80°C without further treatment until mass spectrometry analysis.
3. Load proteins and prepare columns for mass spectrometry. Pressure-load the digested peptide mixtures onto a Kasil-fritted fused silica capillary column (250-μm i.d.) packed with 3 cm of 5-μm Partisphere strong cation exchange resin and 3 cm of 5-μm Aqua C18 resin. Desalt the column with buffer containing 95% water, 5% acetonitrile, and 0.1% formic acid. After desalting, connect the sample-loaded back-end column through a zero-dead-volume union to a 100-μm i.d. capillary column with a 5-μm pulled tip and packed with 10 cm 3-μm Aqua C18 material, and place the entire three-phase column in line with an Agilent 1100 quaternary HPLC.

4. Perform a modified 12-step MudPIT analysis. The first step consists of a 70-min gradient from 0 to 100% buffer B. Steps 2–12 have the following gradient profile: 3 min of 100% buffer A, 5 min of *X*% buffer C, a 10-min gradient from 0 to 10% buffer B, a 70-min gradient from 10 to 45% buffer B, a 10-min gradient from 45 to 100% buffer B, and a 10-min equilibration of 100% buffer A. The 5-min buffer C percentages (*X*) are 5, 10, 20, 30, 40, 50, 60, 70, 80, 90, and 100%, respectively, for the 2–12-step analysis. As peptides are eluted from the microcapillary column, they are electrosprayed directly into an LTQ linear ion trap mass spectrometer with the application of a distal 2.5-kV spray voltage. A cycle of one full-scan mass spectrum (400–1,400 m/z) followed by five data-dependent tandem mass (MS/MS) spectra at a 35% normalized collision energy is repeated continuously throughout each step of the multidimensional separation. Application of mass spectrometer scan functions and HPLC solvent gradients is controlled by the Xcalibur data system, which provides instrument control and data analysis (13).

3.4.3. Data Interpretation

1. Generate tandem mass (MS/MS) spectra via the Xcalibur data system, which consists of an automated spectral quality assessment algorithm (13) to remove poor-quality spectra.
2. Match the filtered MS/MS spectral data of peptides against amino acid sequences in a human protein database (14). The database may be downloaded from the International Protein Index ((IPI), version 3.30 in June 2007) using the SEQUEST™ algorithm (14). The peptide identification is based on the determination of whether detected tandem mass spectra of the target peptide match the theoretical fragmentation pattern generated on the predicted peptides that would be produced from trypsin digestion of proteins in the human proteome. Each protein can thus be represented by multiple mass spectra, each representing different peptide fragments (sequence count), and each individual peptide can be detected multiple times. Thus, the total "spectrum counts" that match to a specific protein roughly represent that protein's abundance in the sample.
3. Assemble the protein "hit list" based on two criteria.

   Criteria 1. There must be a single sequence count (or more) indicating the presence of the protein, and that the chances that it is a false-positive match between the detected peptide and the human protein sequence must be determined to be less than 1%. In the proteomics field, often this first "hit" criterion is two sequence counts and has a false positive rate of 5%. We used a lower cutoff for sequence counts to increase inclusion of important low-abundance proteins, but increase the stringency of the false-positive rate to maintain our confidence in the protein identification. To determine the false-positive protein identification rate, we use the target-decoy approach in which the actual determined peptide sequences are matched to a decoy database made up of the peptides predicted from digestion of the reverse amino acid sequence of the human proteome (15). Basically, SEQUEST converts the character-based representation of amino acid sequences of peptides of similar precursor masses in a protein database to theoretically predicted fragmentation patterns, which are compared against the tandem spectrum generated on the target peptide. The algorithm initially identifies amino acid sequences in the database that match the measured precursor mass of the peptide, compares fragment ions against the MS/MS spectrum, and generates a preliminary score for each amino acid sequence. A cross-correlation analysis is then performed on the top 500 preliminary scoring peptides by correlating theoretical, reconstructed spectra against the experimental spectrum. All of our searches were parallelized and performed on a Beowulf computer cluster consisting of 100 1.2-GHz Athlon CPUs (16). No digestion enzyme specificity is considered for any search. A fixed modification (+57.02146 Da) on cysteines, introduced by reduction and alkylation by our protein preparation protocol, is also considered. SEQUEST results are assembled and filtered using the DTASelect (version 2.0) program (17), a tool for assembling and comparing protein identifications. DTASelect 2.0 uses a linear discriminate analysis to set XCorr and DeltaCN thresholds dynamically for the entire dataset to achieve a user-specified false-positive rate. The false-positive rates are estimated by the program from the number and quality of spectral matches to the decoy database. Finally, the protein hit list contains protein IDs with spectrum counts (18, 19).

Criteria 2. To decrease the chance of variations from sample preparation, comparing protein hits from multiple runs using CONTRAST software (20) is necessary. We include proteins in the hit list that are identified at least twice out of three or four biological replicate runs.

3.4.4. Expectations and Troubleshooting

In our studies, we routinely identify ~1,000 different proteins in our HFF1 FA preps, ~600 of which end up passing the two criteria (Subheading 3.4.3) for inclusion in the final hit list. In examining the results, one should compare what is known to be in FA from the literature. A good resource for this is the recent comprehensive review by Zaidel-Bar and Geiger (7) which contains an assembled list of proteins known to associate with FA based on the literature. You may find that well-known FA proteins that should be present in your FA fraction are absent from your protein hit list. We had this problem with the well-known FA protein, paxillin, which was easily detected in isolated FA fractions by both immunofluorescence and immunoblotting (as described in Subheadings 3.2 and 3.3), but was not detected by MudPIT analysis. Through much effort, we determined that the lack of detection of low-abundance proteins might be due to the presence of too much actin and fibronectin in the FA fractions. This led us to develop the inclusion of the immunoprecipitation step for the removal of actin and fibronectin from the FA fraction to improve the dynamic range of the MudPIT analysis. Other possibilities for loss of low-abundance proteins from the FA fraction include improper hypotonic shock or excess trituration. On the contrary, many proteins are not expected in FA fractions, such as nuclear proteins, soluble proteins, "housekeeping" proteins, or microtubule-associated proteins, but you may find that they pass the two criteria for inclusion in the hit list. Some of them may be novel FA-associated proteins or proteins included by indirect protein–protein interactions as members of multi-protein complexes, only a subset of which may have been shown to associate with FAs. In addition, some proteins in the FA hit list may be from contamination of the sample preparation. Although we set the second criterion of presence in multiple sample replicates to decrease the chance of contamination from sample preparation, it is still possible to get contamination if the skill of FA isolation is not practiced. It is suggested that the proteins that the experimenter finds interesting for their inclusion in FAs be further validated by literature search and assays including localization in cell culture.

3.5. Comparative Analysis of FA Composition Under Different Biological Conditions

The most important use of the FA isolation method is for comparative analysis of FA protein composition under various conditions. The conditions can include pharmacological, genetic, molecular, or physical manipulations of cells, as well as comparison of cells of different tissue origin or from disease models. Results of comparative proteomic analysis of FAs from cells under different conditions can provide insight into molecular mechanisms by which FA mediate responses to specific stimuli, or the role of specific proteins in FA function. In addition, since MudPIT analysis is such a sensitive technique, the presence or absence of specific proteins detected in isolated FA fractions under multiple conditions will add confidence to either including or excluding certain proteins as *bona fide* FA components.

In this section, we describe the approach for comparative analysis of FA proteins under different biological conditions. We describe as a test model the manipulation of Rho family small GTPases by expression of constitutively active mutants of Rac1 and RhoA. It is well known that members of the Rho GTPase family serve as biochemical regulators of FA structure and function

(1, 21). Rac1 GTPase activity initiates formation of small focal complexes beneath the lamellipodium that drive cell protrusion, while RhoA activity promotes FA elongation and maturation that may inhibit motility via activation of myosin II contractility (1, 21). In spite of their known effects on FA morphology, how Rac1 and RhoA affect FA protein composition is not known. Therefore, examples of how the biochemical FA isolation method is applied to analyze the composition of the different types of FA structures induced by constitutive activation of these small GTPases are shown below. Obviously, this procedure could be applied to perturbation of any protein of choice using expression of dominant negative mutants or shRNAi. We include tips on how to obtain homogenous perturbation across a cell population, and how to interpret results and quantitate the changes in FA protein composition under these conditions of genetic perturbation compared to untreated control cells.

3.5.1. Determining the Protein Composition of FAs from Experimental and Control Cells

1. To obtain homogeneous perturbation of Rho family small GTPase function across an entire cell population, isolate HFF1 cells expressing GFP fusions of activated mutants of Rac1 and RhoA. The cDNA constructs driving expression of the mutant proteins should include a selectable marker for isolation of expressing cells by antibiotic selection if desired, or a GFP fusion tag for isolation of expressing cells by flow cytometry. In addition, the promoter should drive high level expression (such as CMV) for relatively homogenous, high level expression. To generate HFF1 cells expressing GFP-tagged versions of the constitutively active Rac1 or RhoA (1, 21), in this example, we utilize transient transfection followed by flow cytometry isolation of expressing cells. Obviously, different methods of protein expression including adeno- or lentivirus-based systems can be used, and stable and/or clonal cell lines can be selected. First, transiently transfect eukaryotic expression vectors coding GFP-tagged constitutively active Rac1 (Rac1Q61L) or RhoA (RhoAV14) into HFF1 cells using your method of choice. GFP-only in a similar vector as the mutant proteins should be transfected into cells as a control. We use a Nucleofector electroporator, which achieves up to 50% transfection efficiency when used according to the manufacturer's instructions (kit: Human Dermal Fibroblast Nucleofector kit, program: U-024). Remember to take the transfection efficiency into account when determining how many cells you will need to transfect to obtain the total amount of protein you will need for FA isolation for immunofluorescence, Western blotting, or proteomic analysis.
2. Following transfection, cells are grown for 24 h and then subjected to flow cytometry to isolate the GFP-positive cells.

3. The isolated GFP protein-expressing HFF1 are plated on 15 μg/ml fibronectin-coated culture dishes or coverslips overnight.
4. The FA isolation method is performed as described above (Subheading 3.1) to isolate the Rac1-mediated small FA structures and the larger, RhoA-induced FA (Fig. 9), as well as FA from cells expressing GFP alone as a control.
5. Perform FA morphology validation. Using the procedure described in Subheading 3.2, validate preservation of FA morphology in isolated FA fractions and intact cells under each experimental and control condition. In the case of Rac1 GTPase activation, the results of this validation reveal that FAs are smaller than those in control and those in cells expressing activated RhoA. This will reduce the total protein isolated in an FA prep from Rac1-expressing cells, and should be taken into consideration when plating cells for FA isolation validation by Western blotting (step 6 below) or FA proteomic analysis (step 7 below).

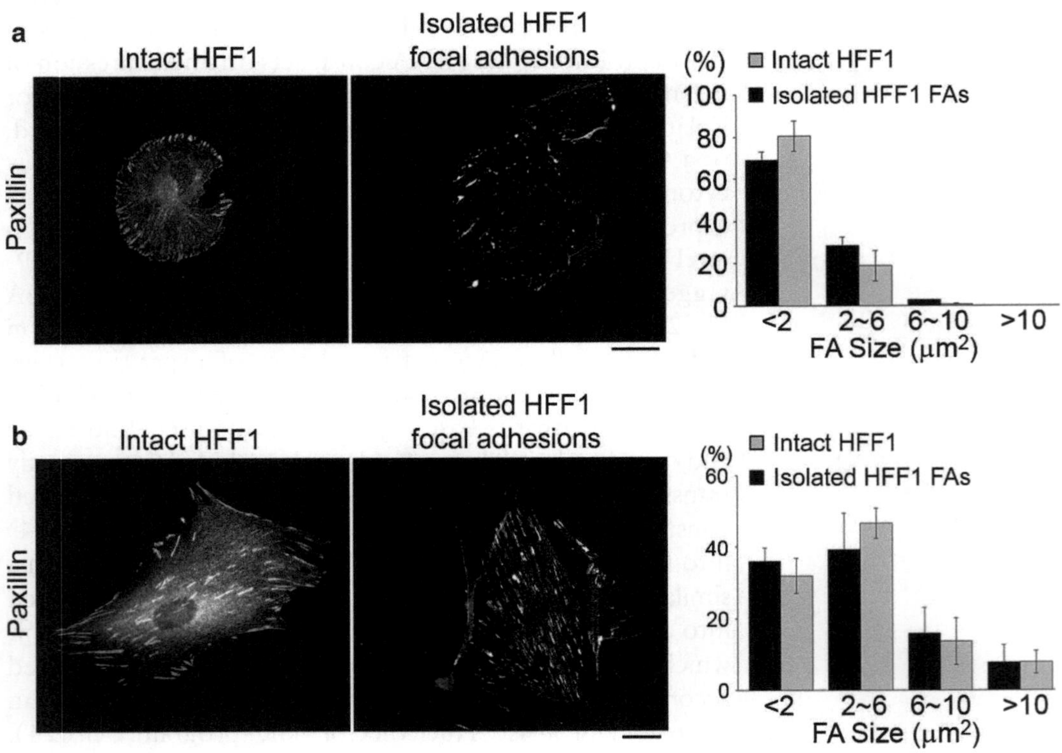

Fig. 9. Immunofluorescence and quantitative analysis for validating the preservation of native FA morphology after FA isolation in HFF1 cells expressing either constitutively activated Rac1 (**a**) or constitutively activated RhoA (**b**). Intact HFF1 and isolated HFF1 FAs were fixed and processed for indirect immuno-localization of paxillin to show the distribution of FAs. Scale bar, 20 μm. *Right* shows histogram of the size distribution of thresholded paxillin-containing FAs in images of intact HFF1 cells ($n=3$ cells, 403 FAs for activated Rac1; $n=3$ cells, 460 FAs for activated RhoA) and the isolated FA fraction ($n=3$ cells, 96 FAs for activated Rac1; $n=4$ cells, 533 FAs for activated RhoA). Statistical analysis indicates no significant difference in the percentage of FAs of each size range between FAs from intact cells or isolated FAs.

6. Perform FA isolation validation. Using the procedure described in Subheading 3.3, validate separation of FA proteins from cell body proteins for each experimental and control condition. It is imperative that the total protein loaded onto the gel be the same for all experimental and control conditions. In examining the results of the validation, in addition to the efficiency of separation of FA and cell body proteins, attention should also be paid to the relative level of FA proteins in control and Rac1- or RhoA-expressing cells. This can serve as validation of the MudPIT-based quantitative comparison of proteins in FAs under the different experimental conditions, and can also provide information for choosing an internal normalization standard for the quantitative comparison between conditions (see Subheading 3.5.2).
7. Perform MudPIT proteomic analysis. It is imperative that the total protein loaded into the 2D-LC system be the same for all experimental conditions. Using the procedure described in Subheading 3.4, identify the protein components in isolated FA fractions for each experimental and control condition that pass both selection criteria, and assemble hit lists of proteins, including spectrum counts, for each experimental condition.

3.5.2. Quantitative Comparison of Relative Protein Abundance in Isolated FA Under Different Experimental Conditions

Changes in FA protein composition of experimental cells versus control fall into two categories. First, there is the binary gain or loss of proteins from experimental FAs compared to control. This occurs when a protein that is present on the hit list of control cells is absent from the hit list of experimental FAs, or vice versa. This tends to occur for lower abundance proteins, and thus its validity can be questionable. Alternatively, the relative abundance of a protein may change in FAs with an experimental perturbation, i.e., its level in FA may increase or decrease relative to control. Here, we discuss the rationale for our quantitation of changes in relative protein abundance in FAs under different experimental conditions.

A semi-quantitative readout of protein abundance in a complex mixture that is obtained from MudPIT analysis is the total spectrum counts for each protein. The number of spectrum counts is related to both the efficiency of the digestion process (dependent on protein sequence and length) and the amount of total protein input (20, 22). Thus, since different proteins are digested with different efficiency, the number of spectrum counts for each protein within a given sample is not a direct measure of the relative abundance of those proteins in the sample. However, since the digestion efficiency of a specific protein should be independent of the origin of the sample, spectrum counts can serve as a quantitative measure of the relative change in abundance of specific proteins in similarly prepared samples from different experimental conditions. Since we are interested in how proteins change relative abundance in FAs under different cellular perturbation conditions, and we can assume that the proteins are digested similarly in the different samples, the quantitative comparison of specific FA proteins can be obtained based on the ratio of spectrum counts for each protein in the hit lists of experimental and control FA. However, since mass spectrometry is much more sensitive that our ability to determine protein concentration accurately by simple assays such as BCA, the spectrum counts should be normalized relative to a medium-abundance protein that can unequivocably shown to remain constant in level between experimental and control samples. To normalize relative to the constant protein, we divide the spectrum counts of this constant protein in the control sample into the spectrum counts of all proteins (including the constant) in the hit list of the experimental sample. Finally, the normalized spectrum counts can be used to obtain the ratio of spectrum counts, and thus the relative protein abundance, between control and experimental FAs.

1. Compare protein hit lists from control and experimental cells to determine proteins with a binary response in terms of their presence in FAs under different experimental perturbations. This will produce a list of proteins that are either gained or lost in FAs of cells expressing activated GTPases compared to FAs from control cells. Since the complete loss or gain of a protein cannot be used to calculate a ratio, the ratio of normalized spectrum counts cannot be used as a measure of the change in relative protein abundance between samples. However, the relative abundance can be estimated from the number of spectrum counts of these binary response proteins in the sample in which they are present. The validity of the binary gain or loss of a protein from FAs should be considered questionable when the protein is present only in very low abundance. If such proteins are of interest, their gain or loss from FAs should be validated by Western blotting of isolated FA under different conditions, as in Subheading 3.3 and by immunofluorescence of intact cells.
2. Determine a protein appropriate for use as a normalization standard whose relative abundance in FAs does not change between control and experimental conditions. Using the same protocol as for Western blot-based validation of the FA isolation procedure (Subheadings 3.3 and 3.5.1), load the same amount of total isolated FA protein (without immunodepletion) from the isolated FA fractions of control and experimental cells into SDS-PAGEs and subsequently blot to Immobilon. Probe the blots with antibodies to known FA proteins that were found from their spectrum counts in MudPIT analysis to be present at similar, intermediate levels in the hit lists from control and experimental FA preps.

Develop the blots and quantitate the band intensities to see which protein(s) keep the same level of relative abundance in FAs across experimental conditions. Once a protein appears to be a good candidate as a constant for normalization, it is a good idea to run a Western analysis of a dilution series of FA proteins from control and experimental cells to provide confidence that the amount of the normalization constant protein is really similar in each sample, independent of signal in the developed blot. One can support the notion that the protein is present in similar levels in FAs of intact cells under the different experimental perturbation conditions by quantifying the average intensity of FAs either in fluorescence images of cells expressing a GFP-tagged version of the protein or in fixed cells processed for immunofluorescence localization of the protein, although this may not be as reliable as Western blotting. It is wise to find a few FA proteins whose relative abundance does not change across experimental conditions to allow comparison of results with different normalization standards, which will add confidence to one's conclusions.

3. Normalize the spectrum counts of proteins in FAs from experimental cells. Using the procedure above, we found that paxillin does not change its relative abundance in FAs under Rac1 or RhoA perturbations compared to in FAs of GFP-expressing control cells, and therefore use the number of spectrum counts of paxillin in control cells as the normalization standard. Divide the spectrum counts for each protein in the hit lists for FA proteins from cells expressing activated Rac1 or activated RhoA by the number of spectrum counts for paxillin in FAs from cells expressing GFP alone.

4. Determine the ratio of the normalized spectra counts for each protein present in both experimental and control conditions. If the ratio is close to 1, it indicates that the experimental condition does not influence the abundance of this protein in FAs. If the ratio is extremely high or low, it indicates the increase or reduction, respectively, in the relative abundance of this protein in FAs dependent on the experimental condition. To categorize proteins according to how strongly their relative abundance in FAs is affected by the perturbation of Rho GTPase signaling, we divide proteins into groups based on the calculated ratio. We consider proteins with less than twofold change in abundance (ratios of >0.5 or <2) to have little change between conditions, those with between two- and five-fold change (ratios of 0.2–0.5 or 2–5) as having intermediate change, and proteins with more than five-fold change (ratios of <0.2 or >5) to be strongly affected by the perturbation in terms of their abundance in FAs. Together, these results

will provide a comprehensive analysis of how Rac1 and RhoA activity affect the protein composition of FAs.

4. Notes

1. Coverslips should be cleaned by sequential 30-min periods of bath sonication immersed in the following solutions: Hot tap water with Versaclean detergent; hot tap water; 1 mM EDTA in distilled, deionized water; distilled, deionized water; 10% ethanol in distilled, deionized water; and 100% ethanol. Each sonication step should be interspersed with 5–10 rinsings in the subsequent (and final) solution. This procedure can be performed in coverslips in bulk, and cleaned coverslips stored in a jar immersed in 100% ethanol until use.
2. This antibody is excellent for both Western blotting and immunofluorescence.
3. The Gold Seal brand is recommended because of its high-quality glass and cleanliness, which reduces background fluorescence in images.
4. The resolution of the lens is defined by the numerical aperture, which is listed below the magnification on the lens body. A lens with a numerical aperture greater than 1, preferably 1.4, should be used. Such high-resolution lenses require immersion oil for full resolution.
5. We always buy Jackson "minimal cross" secondaries as these tend to have the lowest nonspecific background staining. In addition, use of old permeabilization buffer in which the triton was not added immediately prior to use may inhibit the efficiency of permeabilization, which will lead to a high level of cytosolic staining that obscures FA staining in intact cells.
6. Do not use Bradford reagent for protein quantification, as SDS inhibits its accuracy.

References

1. Bershadsky A.D., Balaban N.Q., and Geiger B. (2003). Adhesion-dependent cell mechanosensitivity. *Annu. Rev. Cell Dev. Biol.* **19**, 677–695.
2. Burridge K., Fath K., Kelly T., Nuckolls G., and Turner C. (1988). Focal adhesions: transmembrane junctions between the extracellular matrix and the cytoskeleton. *Annu. Rev. Cell Biol.* **4**, 487–525.
3. Hynes R.O. (2002). Integrins: bidirectional, allosteric signaling machines. *Cell* **110**, 673–687.
4. Discher D.E., Janmey P., and Wang Y.L. (2005). Tissue cells feel and respond to the stiffness of their substrate. *Science* **310**, 1139–1143.
5. Lauffenburger D.A. and Horwitz A.F. (1996). Cell migration: a physically integrated molecular process. *Cell* **84**, 359–369.

6. Zamir E. and Geiger B. (2001). Molecular complexity and dynamics of cell-matrix adhesions. *J. Cell Sci.* **114**, 3583–3590.
7. Zaidel-Bar R., Itzkovitz S., Ma'ayan A., Iyengar R., and Geiger B. (2007). Functional atlas of the integrin adhesome. *Nat. Cell Biol.* **9**, 858–867.
8. Katoh K., Kano Y., Masuda M., Onishi H., and Fujiwara K. (1998). Isolation and contraction of the stress fiber. *Mol. Biol. Cell* **9**, 1919–1938.
9. Brown M.C. and Turner C.E. (2004). Paxillin: adapting to change. *Physiol. Rev.* **84**, 1315–1339.
10. Murphy, D.B. (2001). Fundamentals of light microscopy and electronic imaging. pp. 85–91. Wiley-Liss, New York.
11. Han X., Aslanian A., and Yates J.R., III (2008). Mass spectrometry for proteomics. *Curr. Opin. Chem. Biol.* **12**, 483–490.
12. Washburn M.P., Wolters D., and Yates J.R., III (2001). Large-scale analysis of the yeast proteome by multidimensional protein identification technology. *Nat. Biotechnol.* **19**, 242–247.
13. Bern M., Goldberg D., McDonald W.H., and Yates J.R., III (2004). Automatic quality assessment of peptide tandem mass spectra. *Bioinformatics.* **20**, s49–s54.
14. Eng J.K., McCormack A.L., and Yates J.R., III (94 A.D.). An approach to correlate tandem mass spectral data of peptides with amino acid sequences in a protein database. *J. Am. Soc. Mass Spectrom.* **5**, 976–989.
15. Peng J., Elias J.E., Thoreen C.C., Licklider L.J., and Gygi S.P. (2003). Evaluation of multidimensional chromatography coupled with tandem mass spectrometry (LC/LC-MS/MS) for large-scale protein analysis: the yeast proteome. *J. Proteome. Res.* **2**, 43–50.
16. Peng J., Schwartz D., Elias J.E., Thoreen C.C., Cheng D., Marsischky G., Roelofs J., Finley D., and Gygi S.P. (2003). A proteomics approach to understanding protein ubiquitination. *Nat. Biotechnol.* **21**, 921–926.
17. Sadygov R.G., Eng J., Durr E., Saraf A., McDonald H., MacCoss M.J., and Yates J.R., III (2002). Code developments to improve the efficiency of automated MS/MS spectra interpretation. *J. Proteome. Res.* **1**, 211–215.
18. Chen E.I., Hewel J., Krueger J.S., Tiraby C., Weber M.R., Kralli A., Becker K., Yates J.R., III, and Felding-Habermann B. (2007). Adaptation of energy metabolism in breast cancer brain metastases. *Cancer Res.* **67**, 1472–1486.
19. Tabb D.L., McDonald W.H., and Yates J.R., III (2002). DTASelect and Contrast: tools for assembling and comparing protein identifications from shotgun proteomics. *J. Proteome. Res.* **1**, 21–26.
20. Liu H., Sadygov R.G., and Yates J.R., III (2004). A model for random sampling and estimation of relative protein abundance in shotgun proteomics. *Anal. Chem.* **76**, 4193–4201.
21. Etienne-Manneville S. and Hall A. (2002). Rho GTPases in cell biology. *Nature* **420**, 629–635.
22. Paoletti A.C., Parmely T.J., Tomomori-Sato C., Sato S., Zhu D., Conaway R.C., Conaway J.W., Florens L., and Washburn M.P. (2006). Quantitative proteomic analysis of distinct mammalian Mediator complexes using normalized spectral abundance factors. *Proc. Natl. Acad. Sci. USA* 103, 18928–18933.

Chapter 20

Talin and Signaling Through Integrins

Mohamed Bouaouina, David S. Harburger, and David A. Calderwood

Abstract

Integrin adhesion receptors are essential for the development and functioning of multicellular animals. Integrins mediate cell adhesion to the extracellular matrix and to counter-receptors on adjacent cells, and the ability of integrins to bind extracellular ligands is regulated in response to intracellular signals that act on the short cytoplasmic tails of integrin subunits. Integrin activation, the rapid conversion of integrin receptors from low to high affinity, requires binding of talin to integrin β tails and, once bound, talin provides a connection from activated integrins to the actin cytoskeleton. A wide range of experimental approaches have contributed to the current understanding of the importance of talin in integrin signaling. Here, we describe two methods that have been central to our investigations of talin; a biochemical assay that has allowed characterization of interactions between integrin cytoplasmic tails and talin, and a fluorescent-activated cell-sorting procedure to assess integrin activation in cultured cells expressing talin domains, mutants, dominant negative constructs, or shRNA.

Key words: Integrin, Talin, Activation

1. Introduction

Integrin adhesion receptors are type I transmembrane heterodimeric glycoproteins capable of bidirectionally transmitting signals between the intracellular and extracellular environments. The extracellular domains of the integrins specifically bind in a cation-dependent manner to extracellular matrix (ECM) proteins or to counter-receptors on adjacent cells, and the short cytoplasmic tails link to cytoskeletal elements, adaptors, and signaling proteins inside the cell (1–7). These adhesive interactions permit transmission of mechanical and chemical signals into and out of the cell and are crucial for a great many processes including embryonic development, tissue maintenance and repair, host defense, and hemostasis (8–16). A notable feature of integrins, central to these activities, is

Motomu Shimaoka (ed.), *Integrin and Cell Adhesion Molecules: Methods and Protocols*,
Methods in Molecular Biology, vol. 757, DOI 10.1007/978-1-61779-166-6_20, © Springer Science+Business Media, LLC 2011

their ability to undergo extensive conformational changes that regulate their affinity for extracellular ligands (17). Conformational rearrangement of the extracellular domains occurs in response to signals that impinge on the integrin cytoplasmic tails, a process referred to as "inside-out" signaling or integrin activation (3, 5). It is now appreciated that the large cytoskeletal adaptor protein, talin, plays a central role in integrin function, both as a mechanical link between integrins and the actin cytoskeleton and as an essential regulator of integrin activation (7, 18–22).

Talin is a cytoskeletal actin-binding protein composed of an amino terminal 50-kDa, head followed by a 220-kDa rod domain (18, 19). The talin head (amino acids 1–433) contains a three-lobed FERM (4.1, ezrin, radixin, moesin) domain preceded by an F0 domain and followed by a 33-amino acid stretch (23–25). The third lobe of the talin FERM (F3) is structurally related to a phosphotyrosine-binding (PTB) domain and mediates direct binding to an NP(I/L)Y motif in integrin β tails (23, 25). The talin rod is composed of a series of helical bundles (26, 27) and contains the major binding sites for actin and vinculin, as well as a dimerization site and a second integrin-binding site (28–30). The potential that talin provides a step in an integrin-mediated transmembrane linkage from the ECM to actin was first appreciated more than 20 years ago (31) and more recently, advanced biophysical analysis has demonstrated a role for talin as a molecular mechanosensor and confirmed the importance of talin in providing an initial linkage from integrins to actin that is important for stabilizing cell spreading (21, 22, 32).

In addition to its mechanical role, approximately 10 years ago, talin was first implicated in the regulation of integrin activation (33). Using recombinant mimics of integrin β-subunit cytoplasmic tails, we showed that talin bound directly to integrin β tails, and localized the talin-binding site within the β tail to the first NP(I/L) Y motif and the integrin-binding site to the F3 sub-domain of the talin head (23, 33). We further showed that overexpression of integrin-binding F3-containing fragments of talin triggered activation of αIIbβ3 integrins, but that talin or integrin mutants defective in binding were impaired in activation (33, 34). Based on these data and the finding that knockdown of talin impaired integrin activation, we proposed that talin binding, via its PTB domain-like F3 sub-domain, to the integrin β tail was a final common step in integrin activation (34). This conclusion has extensive in vitro and in vivo support from additional structural and mutational analyses, and from the phenotypes of knockout and knock-in model organisms (35–41).

While a key role for the PTB-like domain–integrin β tail interaction in integrin activation is now well established, additional complexities have recently become apparent. Integrin activation by the F3 sub-domain also requires additional interactions with membrane-proximal integrin residues (36), and in intact talin, this interaction is impaired due to a previously predicted (42), auto-inhibitory talin head–rod interaction (43, 44). Talin binding to integrin β tails is regulated by phospholipid binding and intracellular signaling pathways through control of talin localization and release of the auto-inhibitory activation (43, 45–47). Competition between talin and other integrin β tail-binding proteins provides an additional layer of regulation (36, 48, 49). Surprisingly, while talin can bind most short integrin β tail and talin binding seems to be a general requirement, at least for β1, β2, and β3 integrin activation, differences in the ability of minimal integrin-binding talin domains to activate β1 and β3 integrins have highlighted differences between integrins and shown the importance of domains outside the PTB-like F3 domain for cellular activation of integrins (50). More strikingly, evidence suggests that additional factors that cooperate with talin during integrin activation are important in vivo. The first such factors to be identified are the members of the kindlin family of proteins (20, 51). These proteins are structurally related to talin but bind to a distinct site on integrins and are required for normal integrin activation in vivo. A detailed understanding of how kindlins exert their effects is lacking and whether other proteins play similar roles by cooperating with talin during integrin activation remains to be determined. Thus, in spite of the extensive studies already performed on talin, many questions concerning its role in integrin signaling remain. A wide range of experimental approaches spanning structural biology, biochemistry, biophysics, cell biology, and genetics have shaped our understanding of talin. Amongst them, two general methods have been central to our investigations of talin in integrin signaling: a biochemical assay that allowed characterization of integrin cytoplasmic tail interactions and a robust method to assess integrin activation in cultured cells expressing talin domains, mutants, dominant negative constructs, or shRNA. Here, we will describe the assays we have used to investigate the roles of talin in integrin signaling, techniques that will continue to be useful in the examination of the role of proteins that modulate talin function.

2. Materials

2.1. Generation of Recombinant Talin Head Fragments and Endogenous Talin from Mammalian Cell Lysates

1. Complete Dulbecco's modified Eagle's medium (DMEM): DMEM supplemented with 10% fetal bovine serum, 1% of 100× penicillin (10,000 U/ml)/streptomycin (10,000 μg/ml), 1% of 100× sodium pyruvate (100 mM), and 1% of 100× MEM nonessential amino acids solution (10 mM).
2. Dulbecco's phosphate-buffered saline (DPBS).

3. Lipofectamine Reagent (Invitrogen).
4. Cell lysis buffer: 50 mM NaCl, 10 mM Pipes, 150 mM sucrose, 50 mM NaF, 40 mM $Na_4P_2O_7$, pH 6.8, 0.5% Triton X-100, 0.1% sodium deoxycholate, and EDTA-free protease inhibitor tablet (Roche).
5. BCA protein assay (Pierce).

2.2. Purification of Recombinant Talin Head Fragments from Bacterial Cell Lysates

1. Luria broth base (LB).
2. Isopropyl-beta-D-thiogalactopyranoside (IPTG), dioxane free.
3. GST lysis buffer: phosphate-buffered saline (PBS), pH 7.4, 1% Triton X-100, 1 mM dithiothreitol (DTT), and protease inhibitor tablet (Roche).
4. Glutathione-sepharose beads (GE Healthcare).
5. PBS wash buffer: PBS, 1% Triton X-100, and 1 mM DTT.
6. GST wash buffer: 0.1 M Tris–HCl, pH 7.5, 0.1 M NaCl, 1% Triton X-100, 1 mM DTT, and protease inhibitor tablet.
7. GST elution buffer: GST wash containing 20 mM glutathione.

2.3. Integrin β Tail Binding Assay

1. Ni^{2+}-NTA resin (Novagen).
2. Buffer XT: 50 mM NaCl, 10 mM Pipes, 150 mM sucrose, 50 mM NaF, 40 mM $Na_4P_2O_7$, 1 mM Na_3VO_4, and 0.05% Triton X-100, pH 6.8.

2.4. Generation of GST-Fused Fibronectin (9-11) from Bacteria

1. GST-Fibronectin (9-11) cDNA was engineered as previously described elsewhere (52).
2. BL21 *Escherichia coli* bacterial culture.
3. Luria broth base (LB).
4. Ampicillin sodium salt.
5. IPTG.
6. GST lysis buffer: PBS, pH 7.4, 1% Triton X-100, 1 mM DTT, and protease inhibitor tablet (Roche).
7. Glutathione-sepharose beads.
8. PBS wash buffer: PBS, 1% Triton X-100, and 1 mM DTT.
9. GST wash buffer: 0.1 M Tris–HCl, pH 7.5, 0.1 M NaCl, 1% Triton X-100, 1 mM DTT, and protease inhibitor tablet.
10. GST elution buffer: GST wash buffer with 20 mM glutathione.
11. 5× SDS sample buffer (100 ml): 15.625 ml of 1 M Tris Base, pH 6.8, 10 mg of SDS, 20 ml of glycerol, and 100 mg of 0.1% bromophenol blue. Add 1 ml β-mercaptoethanol to 3 ml of buffer to make 5× working solution.
12. Float-A-Lyser™ (Spectrum Laboratories; 10 mm diameter, 3 ml volume, and 3,500 Da molecular weight cut-off).

2.5. Biotinylation of GST-Fibronectin (9-11)

1. Biotinyl-N-hydroxy-succinimide (EZ-Link™ NHS-Biotin, spacer arm 13.5 Å; Pierce).
2. Dimethyl sulfoxide (DMSO).
3. Fresh $NaHCO_3$ 1 M solution.
4. PD10 desalting columns (Sephadex™ G-25 M; Amersham Bioscience).
5. Bovine serum albumin (BSA).

2.6. Fluorescence-Activated Cell-Sorting Activation Assay

1. 5-ml FACS polystyrene round-bottom tubes (BD Falcon).
2. DPBS.
3. Ethylenediaminetetraacetic acid (EDTA).
4. Nylon screening mesh (80 μm) (Sefar NITEX™).
5. Tyrode's buffer, pH 7.3: 136.9 mM NaCl, 10 mM Hepes, 5.5 mM glucose, 11.9 nM $NaHCO_3$, 2.7 mM KCl, 0.5 mM CaCl, 1.5 mM MgCl, and 0.4 mM NaH_2PO_4.
6. EDTA-0.05% trypsin.
7. Rat Anti-mouse integrin β1 chain antibody (clone 9EG7) (BD Pharmingen).
8. Purified mouse IgM anti-human αIIbβ3 integrin PAC-1 antibody (BD Bioscience).
9. Streptavidin-allophycocyanin (Thermo Scientific).
10. Alexa647-conjugated donkey anti-mouse IgG (H+L) antibody and Alexa 647-conjugated goat anti-mouse IgM antibody.
11. $MnCl_2$.
12. Mouse anti-human αIIbβ3 integrin D57 antibody. D57 is a function-independent antibody used for the detection of αIIbβ3 on the surface of cells (53).

3. Methods

3.1. Biochemical Analysis of Integrin–Talin Interactions

Integrins are large heterodimeric proteins composed of bulky extra-extracellular domains, a transmembrane region, and generally short cytoplasmic tails, and as a consequence, raise challenges for biochemical analysis. In an effort to examine intracellular biochemical interactions with the integrin tail, we utilize a system of bacterial, recombinantly expressed proteins containing the integrin cytoplasmic tails. This system has provided a framework for reliably identifying the molecular requirements for integrin-binding proteins, correlates with findings from genetic and structural experiments, and has been useful in providing extensive information on talin–integrin β tail interactions (23, 33, 34, 48, 50, 54). The integrin β tail pull-down assay has been described in detail in a previous review, with a focus on the role of the integrin tail (55). The assay relies on expression and purification of recombinant integrin cytoplasmic tails consisting of an N-terminal His-tag followed by a thrombin cleavage site, a cysteine-residue linker, a coiled-coil sequence, a glycine spacer, and the integrin cytoplasmic domain (55). The cysteine linker and coiled-coil regions allow parallel dimerization of integrin tails in aqueous solution, act as a spacer between the integrin tail and affinity matrix, and mimic the helical structure of integrin transmembrane domains. The His-tag allows purification of the recombinant proteins by metal ion-affinity chromatography and immobilization of the purified integrin tail on His-bind resin for use in binding assays. In the following sections, we briefly discuss preparation of cell lysates for talin pull-down assays, the expression and purification of recombinant talin fragments in bacteria, and the pull-down assays themselves.

3.1.1. Preparation of Mammalian Cell Lysates Containing Recombinant Talin Head Fragments and Endogenous Talin

Integrin β tail pull-down assays can be performed using a variety of cell lysates (18, 55, 56). However, we generally use Chinese hamster ovary (CHO) cells because they can be efficiently transfected using standard protocols, have consistently provided successful expression for many of our proteins of interest, and are the cell type in which we have optimized the integrin activation assays discussed later in this review.

1. Culture CHO cells in complete DMEM until confluent on a 10-cm tissue culture dish. If using untransfected cells, proceed to step 3.
2. Wash the cells with DPBS and transiently transfect cDNA-encoding fragments of talin fused to epitope tags such as GFP, FLAG, HA, etc., utilizing Invitrogen's recommendations for working with Lipofectamine reagent. Typical talin fragments include F3, F2F3, F1F2F3, and entire head, encoding the amino acids 309–405, 206–405, 86–405, and 1–433, respectively (see Note 1).
3. At 24 h post-transfection, wash the cells with cold DPBS, lyse by scraping in 0.5 ml of cell lysis buffer, collect into a 1.5-ml microcentrifuge tube, and incubate on ice for 20 min.
4. Separate the insoluble fraction from the lysate by centrifuging at 20,000 × *g* for 10 min at 4°C.
5. Determine the protein concentration of the supernatant using the BCA protein assay and either use lysates immediately or store at –20°C until required.

3.1.2. Expression and Purification of Recombinant Talin Fragments from Bacterial Cell Lysates

Based on sequence analysis and insights from structural findings, we have been able to express recombinant fragments of talin fused to a tag useful for protein purification (23, 33, 50). Here, we will describe purification of talin fragment fused to GST; however, other tags, such as His-tag, may also be used (57).

1. Inoculate 100 ml of LB containing the appropriate antibiotics with BL21 *E. coli* culture containing GST-talin fusion protein expression construct. Grow overnight at 37°C in a rotary shaker incubator (radius of gyration: 1 in.) with agitation at 225 rpm.
2. Use the overnight culture to inoculate 2 L of LB with appropriate antibiotic and continue to grow the culture to an approximate OD_{600} of 0.4 (see Note 2).
3. Induce the bacteria with IPTG (0.2 mM final concentration) and grow for 3 h (see Note 2).
4. Harvest the bacteria by centrifugation and resuspend in 60 ml of GST lysis buffer on ice. Lyse by sonication and repeat until the bacterial extract is no longer viscous.
5. Centrifuge at 30,000 × *g* for 25 min at 4°C and collect the supernatant.
6. During this final centrifugation step, prepare 0.5 ml of glutathione-sepharose beads. Wash with 10 volumes of PBS and then mix with supernatant from step 5.
7. Incubate the glutathione-sepharose beads with the bacterial lysate overnight at 4°C with slight agitation to avoid bead sedimentation.
8. Load the beads onto a column. Wash once with 20 volumes of PBS wash buffer followed by two washes with 20 volumes of GST wash buffer.
9. Elute GST fusion proteins with 20 volumes of GST elution buffer and collect eluate in 1-ml fractions.
10. Analyze the elution fractions on a 10% Tris–glycine SDS polyacrylamide gel to identify fractions containing the GST fusion protein. Determine the concentration of fusion protein in positive fractions and keep at 4°C for immediate use or store at –20°C until needed.

3.1.3. Preparation of Recombinant Integrin β Tails and Subsequent Binding Assays

The details for these steps have been described by Lad et al. (55). Briefly, this assay entails the following:

1. Bacterial expression of integrin cytoplasm tails containing an N-terminal His-tag.
2. Purification of integrin cytoplasmic tails from the bacterial lysates through two rounds of sonication and centrifugation, batch absorption to Ni^{2+}NTA resin, and elution from the resin followed by purification via reversed-phase high-pressure liquid chromatography.
3. Preparation of the affinity matrix through the steps of mixing purified integrin tails with Ni^{2+}-charged beads in the presence of binding buffer, followed by washing steps to remove urea and exchange integrin tail-coated resin into buffer XT.

4. Mixing either cell lysates or bacterially purified proteins with the integrin tails for 24 or 2 h, respectively, at 4°C, followed by several washing steps, resuspending beads in SDS sample buffer, boiling samples, and analyzing on SDS polyacrylamide gel either through immunoblotting or using Coomassie stain (see Note 3).

3.2. Integrin Activation Assays

Integrin activation assays provide functional readouts that have complemented our biochemical and structural studies to reveal the effect of different talin fragments on the activation of surface-expressed integrins. These assays rely on fluorescence-activated ell sorting (FACS), to measure protein expression and integrin activation state in individual cells. Here, we will outline the basis for these assays that make powerful tools in screening for players in integrin signaling (58) and discovering new regulators of talin-mediated integrin activation (54, 59–61).

Integrin activation involves conformational changes in the ectodomains that result in an increase in affinity for ligand. As the dynamically regulated integrin domains are exposed on the surface of the cell, and so accessible to reporter ligand or antibody binding, we can use FACS analysis as a means to measure integrin activation in live cells. Our FACS assays rely on the selective binding of integrin ligands or ligand-mimetic antibodies only to activated integrins. When combined with transiently expressed fluorescently tagged talin fragments, this assay permits us to assess the effect of talin on the activation of endogenous or exogenously expressed integrin using multicolor FACS analysis. This technique, coupled to proper data analysis, offers an excellent tool to study the intricate control of integrin activation, and a variety of FACS-based integrin activation assays have been developed (62–65). We will discuss the details of those in use in our laboratory which are based on an assay originally developed by the Shattil and Ginsberg laboratories (63, 66). This sensitive assay, initially developed for αIIbβ3 integrins, has been amended to also allow measurements on fibronectin-binding β1 integrins and has become a powerful tool for understanding modes of regulation of integrin activity.

3.2.1. Probes Used to Assess Integrin Activation by Flow Cytometry

Antibodies

Monoclonal antibodies that recognize neo-epitopes on the extracellular domain of the activated integrin are commonly used to report integrin activation state; examples include Ab24, a β2 integrin-specific mouse antibody (67), and HUTS-21 antibody, a β1-specific mouse antibody (68). The anti-αIIbβ3 mouse monoclonal IgM antibody PAC1, which is the basis of our αIIbβ3-activation assay, is notable as it is a ligand-mimetic antibody that binds to the extracellular ligand-binding site in the integrin in a way that is very similar to that of the specific ligand (69, 70). PAC1 is capable of competing with fibrinogen for binding to active αIIbβ3 integrins (71), and sequencing has revealed that the complementarity-determining region three within PAC1 heavy chain contains an RYD sequence forming the integrin-binding site and behaving like the RGD sequence in fibrinogen (69). Mutagenesis around this site has permitted generation of activation-specific antibodies against human and mouse αIIbβ3 and αVβ3 (72, 73) and the Drosophila αPS2βPS (74).

Ligands

Soluble ligands can also be used to probe integrin activation in flow cytometry because integrins require activation to bind to soluble ligands. Natural ligands are often large, multivalent proteins that bind more than one integrin, making them unsuitable for studying integrin activation; however, intact fluorescently labeled fibrinogen has been successfully used in integrin activation assays (34, 73, 75). More commonly, soluble recombinant fragments from extracellular integrin ligands, engineered to ensure specificity toward the integrin of interest and to lower multivalency, are used in activation assays (50, 76–80). We use a well-characterized GST-fused fragment of fibronectin to probe α5β1 integrin activation. The GST-fibronectin (9-11) (GST-FN9-11) fragment contains the 9th, 10th, and 11th type III repeats of Xenopus fibronectin (52). This fragment carries the α5β1 integrin-binding RGD site (repeat 10) along with the synergy motif "PPSRN" (repeat 9) known to support the spreading of many cell types (81). Fibronectin is well conserved and the Xenopus protein also binds human and mouse integrins.

Specificity controls

Specific binding of ligand-mimetic anti-integrin antibodies or soluble ligands can be inhibited by RGD peptides, or small molecule antagonists such as Ro43-5054 or 3F (69, 82–84). Integrin–ligand binding is also divalent cation dependent and, therefore, can be inhibited by EDTA (85). Antibody or soluble ligand binding in the presence of integrin antagonists or EDTA can, therefore, be used to assess background nonspecific binding in our FACS assays and provides a control for the specificity and selectivity of the activation reporters being used. In addition, binding should be enhanced in the presence of integrin activators such as the divalent cation Mn^{2+} or activating antibodies.

3.2.2. Expression and Purification of GST-FN9-11

1. Inoculate 1 L of LB and 50 μg/ml ampicillin, with 25 ml of overnight culture of BL21 *E. coli* carrying the GST-tagged fibronectin (9-11) construct. Incubate at 37°C in a rotary shaker (radius of gyration: 1 in.) with agitation at 225 rpm until it reaches an OD_{600} of 0.4.

2. Induce the bacterial culture with IPTG (final concentration 0.2 mM) and continue culture for 3 h at 37°C with shaking.
3. Harvest the bacteria by centrifugation for 20 min at 5,500 × *g* at 4°C. Gently resuspend the pellet in 30 ml of GST lysis buffer and avoid making bubbles, then transfer the mixture into a beaker and maintain at 4°C.
4. Lyse the bacteria with 10-s bursts of sonication followed by 50 s on ice to avoid overheating. Repeat until the extract is no longer viscous.
5. Centrifuge at 30,000 × *g* for 25 min at 4°C and collect the supernatant.
6. During the last centrifugation step, prepare 250 μl of glutathione-sepharose beads. Wash with 10 ml of PBS and then mix with the supernatant from step 5.
7. Incubate the glutathione-sepharose beads with the bacterial lysate overnight at 4°C with slight agitation to avoid bead sedimentation.
8. Load the beads onto a column and wash with 10 ml of PBS-Triton X-100 at 4°C, followed by two 10-ml washes with wash buffer.
9. Elute the GST-FN9-11 fragment with 1.5 ml of elution buffer at 4°C. Incubate for 5 min at room temperature and collect the eluted fraction in a 1.5-ml centrifuge tube and store on ice. Repeat to collect 6–8 fractions.
10. Check the levels and purity of the eluted protein in each fraction by SDS-PAGE followed by protein staining.
11. Pool the two fractions containing most GST-FN9-11 protein, load into a Float-A-Lyser™ unit, and dialyze against 4 L of PBS at room temperature overnight to remove the free glutathione. Replace dialysis fluid with fresh PBS and continue dialysis for 2 h.
12. Collect the dialyzed GST-FN9-11 and store on ice.

3.2.3. Biotinylation of GST-FN9-11

To enable detection of the FN9-11 with flurophore-conjugated streptavidin, we covalently attach NHS-biotin (short chain) to the purified, dialyzed protein. For optimal biotinylation, the FN9-11 concentration should be between 0.5 and 2.0 mg/ml.

1. Dissolve 1 mg of biotinyl-*N*-hydroxy-succinimide in 100 μl DMSO.
2. Mix 800 μl of dialyzed FN9-11 with 100 μl of freshly prepared 1 M $NaHCO_3$ and 100 μl of the biotin solution, and incubate for 1 h at room temperature with gentle agitation.
3. During the incubation step, pre-equilibrate a PD10 desalting column. Add 1 ml of 1% BSA to the column to block nonspecific binding sites and wash extensively with 20–40 ml of PBS.

4. Load the biotinylated FN9-11 sample onto the pre-equilibrated PD10 column and elute the FN9-11 in 1–1.5-ml fractions by adding PBS.

5. Identify protein-containing fractions based on absorbance at 280 nm and store at 4°C.

3.2.4. Preparation of Cells for Use in Integrin Activation Assays

The integrin activation assays described below can be used on a wide range of cell types. Platelets and polymorphonuclear neutrophils present a good cell model for studying activation of endogenous integrins because they are abundant and can be readily harvested from blood, and because integrin activation is tightly regulated in these cells (66, 86). These primary cells are, however, not readily transfectable and so to investigate the effects of specific proteins or pathways on integrin activation, a variety of adherent, immortalized, and well-established cell lines have been used, including CHO, NIH-3T3, and Jurkat (34, 48, 87), as well as ES cell-derived megakaryocytes (88) and murine bone marrow-derived megakaryocytes (89, 90).

We make extensive use of CHO cells as lines in which to assess the effect of knocking down integrin regulatory molecules or overexpressing talin, talin fragments, or molecules that cooperate with talin (23, 33, 34, 59). CHO cells have been widely used to investigate β1 integrins (91, 92) and a variety of CHO cell lines that express different wild-type and chimeric integrins has permitted investigation of other integrin families not normally present in CHO cells, most notably the platelet integrin αIIbβ3 (53, 83, 91, 93) CHO cells are cultured as previously described (94) in DMEM supplemented with 10% fetal calf serum (FCS), 100 U/ml of penicillin, 0.1 mg/ml of streptomycin, 2 mM of L-glutamine, and 1% nonessential amino acids at 37°C in 5% CO_2.

To assess the effect of specific integrin regulators, we transiently overexpress fluorescent, GFP- or DsRed-tagged recombinant proteins in the chosen cell line using pcDNA3 (Invitrogen), pEGFP (BD Biosciences Clontech), or pDsRed (BD Biosciences Clontech) vectors as follows:

1. Seed 1×10^6 CHO cells per 10-cm plate for transfection the next day (see Note 4).

2. Mix 250 μl of DMEM with 2–4 μg of DNA and add 4 μl of Lipofectamine/μg of DNA. Incubate the mixture at room temperature for 15 min.

3. Wash the cells once with 5 ml of room temperature PBS to remove the serum and any dead cells, then add 5 ml of DMEM and the DNA–lipofectamine mixture from step 2 to the plate, and incubate for 5 h at 37°C in an incubator (5% CO_2).

4. Wash the plate once with 5 ml of room temperature PBS, then add 10 ml of complete DMEM, and incubate at 37°C, 5% CO_2, overnight.

3.2.5. FACS Integrin Activation Assay

This assay assesses integrin activation by measuring the binding of PAC1 antibody or biotinylated-FN9-11 or to cells expressing constructs of interest. Cells are transfected as described above, and 16–18 h after transfection, integrin activation is assessed by FACS. Bound FN9-11 or PAC1 is detected using streptavidin or anti-mouse IgM antibodies conjugated to APC or Alexa 647 fluorophores, whose emitted fluorescence is measured in the FL4 channel of the FACS machine. The optimum concentrations of PAC1, FN9-11, and fluorophore-conjugated secondary antibody or streptavidin for the cell line under study are determined by titration. Expression levels of the expressed GFP- or Ds-Red conjugated proteins are assessed in the FL1 or FL2 channel, respectively. The separation between the emission optima for the GFP or DsRed fluorophores and the APC or Alexa 647 fluorophores avoids cross talk between the FL1 or FL2 channels and the FL4 channel, so simplifying analysis of the results.

Cells from each plate, overexpressing one fluorescent protein of interest or control, are routinely divided into four tubes to undergo different treatments prior to FACS analysis. In three tubes, integrin activation state is assessed in native conditions, in the presence of an inhibitor of ligand binding such as EDTA or small molecule inhibitor, or in the presence of an exogenous integrin activator such as 1 mM Mn^{2+} or activating antibody (9EG7 for α5β1 integrins and anti-LIBS6 for αIIbβ3 integrins). As described below, these conditions allow for the assessment of the nonspecific background binding and the maximum fully stimulated binding, and the native binding can then be related to these minima and maxima. In the fourth tube, integrin surface expression is assessed with an antibody that binds integrin in an activation-independent manner, we use PB1 (95) to measure hamster α5β1 levels in CHO cells and D57 (53) to assess αIIbβ3 expression levels. These measures can be used to normalize integrin activation results to the total integrin expression level.

The method outlined below describes a basic FACS integrin activation assay. Similar methods are used for α5β1 and αIIbβ3, the major difference being that FN9-11 is used to report α5β1 activation, while PAC1 is used for αIIbβ3.

1. Wash transfected plates once with 5 ml of room temperature PBS to remove dead cells and serum contained in the medium. Add 0.7 ml of 0.5 mM EDTA and 0.05% trypsin, and incubate for 5–10 min at 37°C.
2. Neutralize trypsin with 0.7 ml of complete DMEM, collect the detached cells in a tube, and spin for 5 min at 200 × *g* at room temperature.
3. Wash the cells by re-suspending them in 1.5 ml of room temperature PBS. Filter the cell suspension through an 80-μm

nylon screening mesh (SEFAR NITEX™; see Note 5) and then remove 10 μl for cell counting.

4. Pellet the cells by spinning for 5 min at 200 × *g* at room temperature and re-suspend gently at 6–8 × 10^6/ml in Tyrode's buffer.
5. Transfer 45 μl of cell suspension to each of four FACS tubes; four FACS tubes for each transfected plate allows measurement of reporter binding in native, activated, and inhibited condition as well as one tube to measure integrin expression levels by standard antibody FACS assay performed in parallel using anti-α5 or anti- αIIbβ3 antibodies PB1 or D57, respectively.
6. Add 2.5 μl of 200 mM EDTA to the inhibited tubes and incubate for 10 min at room temperature.
7. Add 2.5 μl of Tyrode's buffer to each native tube.
8. For activated tubes, add 5 μl of 20 mM $MnCl_2$, or if using activating antibodies, 5 μl of 100 μg/ml 9EG7 for α5β1 assays, or anti-LIBS6 for αIIbβ3 assays.
9. Add 2.5 μl of the appropriate dilution of activation reporter (determined by titration) to each tube; biotinylated GST-FN9-11 for α5β1 assays and PAC1 for αIIbβ3 assays.
10. Incubate FACS tubes at 30°C for 30 min.
11. Wash off unbound ligand/antibody by adding 250 μl of Tyrode's buffer to each tube and then pellet the cells for 5 min at 200 × *g* at room temperature. Carefully pour off the supernatant.
12. For α5β1 assays, re-suspend the cells in 50 μl of 2% APC-conjugated streptavidin in Tyrode's buffer. For αIIbβ3 assays, re-suspend the cells in 50 μl of 2.5 μg/ml Alexa 647-conjugated anti-mouse IgM in Tyrode's buffer.
13. Incubate FACS tubes for 30 min on ice in the dark.
14. Wash off unbound antibodies by adding 250 μl of cold Tyrode's buffer to each tube, pellet the cells for 5 min at 200×*g* at 4°C, and then carefully pour off the supernatant.
15. Re-suspend cells in 400 μl of cold PBS.
16. Analyze stained cells using FACS Calibur or LSRII FACS machines and collect 10,000–30,000 cells per condition.

3.2.6. Data Analysis

The FACS analysis described above provides data on FN9-11 or PAC1 binding to transfected and untransfected cells under a variety of conditions. These data can be processed using Flowjo analysis software. Analysis first requires identification of the live cells based on their forward and side scatter (Fig. 1a). Cells with the appropriate forward and side scatter are gated and further analysis is performed on this population. The GFP or DsRed signal (FL1 or

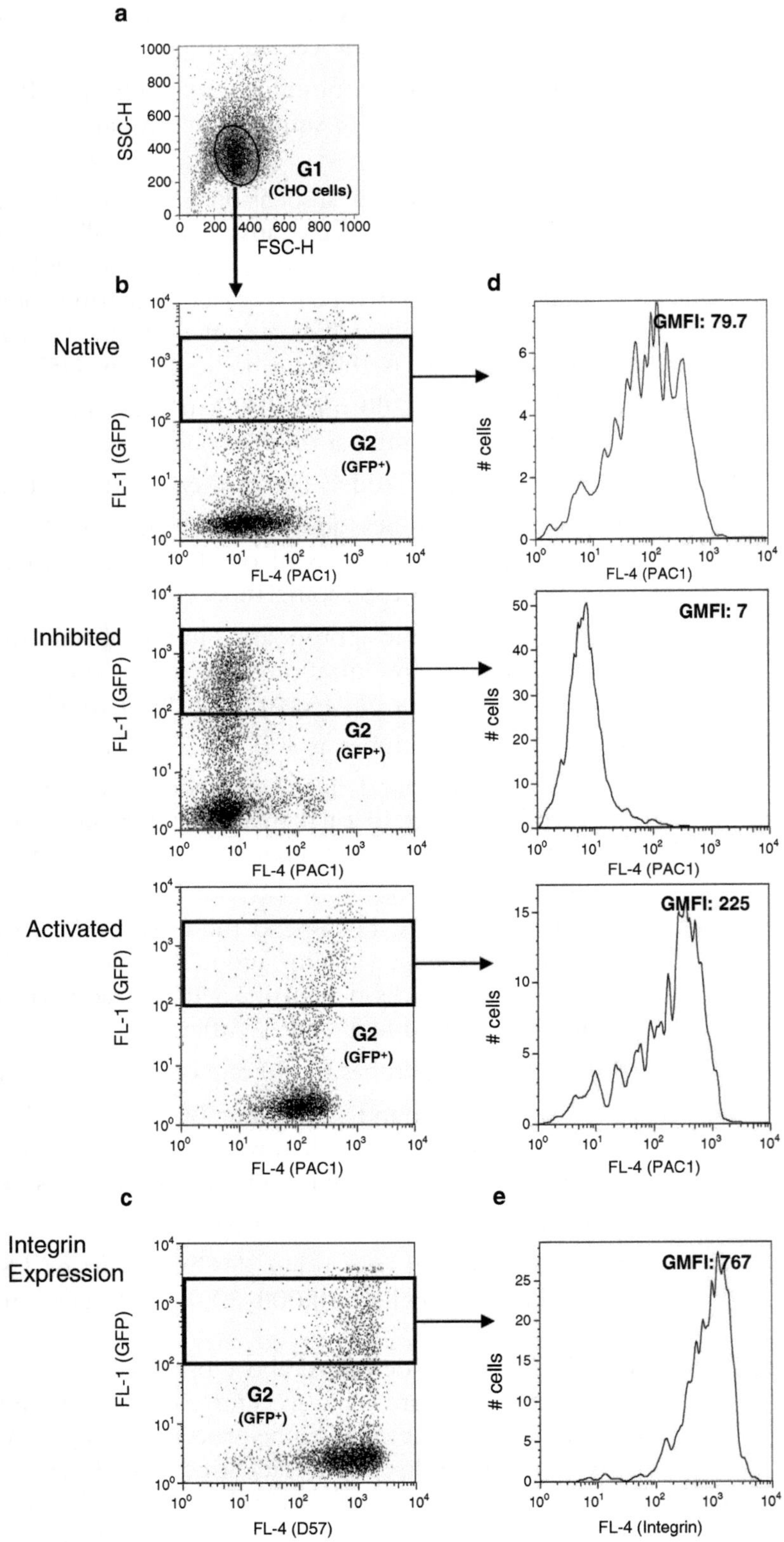
a
G1
(CHO cells)
SSC-H
FSC-H
b
Native
Inhibited
Activated
G2
(GFP+)
FL-1 (GFP)
FL-4 (PAC1)
c
Integrin
Expression
FL-4 (D57)
d
GMFI: 79.7
GMFI: 7
GMFI: 225
cells
e
GMFI: 767
FL-4 (Integrin)

FL2, respectively) is plotted against the PAC1- or FN9-11-binding signal (FL4), providing a visual representation of how the ability of cells to bind soluble integrin ligand changes in response to expression of the fluorophore-tagged recombinant protein (Fig. 1b). Gating of cells based on GFP or DsRed fluorescence permits segregation into expressing and non-expressing populations or further separation into high- and low-expressing cells. The PAC1 or FN9-11 binding to cells in each of these populations can then be expressed as a frequency histogram and geometric mean fluorescence intensity (GMFI) of cells in the population calculated (Fig. 1c). Applying the same multi-gating parameters to all tubes from a given sample allows calculation of GMFI values for the native, activated, and inhibited conditions, and a similar approach allows calculation of integrin levels on the cells based on anti-integrin antibody staining (Fig. 1c). As discussed below, these data permit calculation of the activation state of integrins on the transfected cells. Applying the same gating parameters to cells transfected with different constructs allows comparison of the effect of different talin domains on integrin activation. Furthermore, since all data are expressed based on populations gated for comparable GFP or DsRed expression levels, any differences in expression levels between constructs are evident and can be controlled for.

We routinely use activation index as a measure of integrin activation. This is expressed as $AI = (F - F_0)/(F_{max} - F_0)$, where F is the GMFI of FN9-11 binding, F_0 is the GMFI of FN9-11 binding in the presence of binding inhibitor (peptide, antibody, or EDTA), and F_{max} is the GMFI of FN9-11 binding in the presence of binding activator (antibody or Mn^{2+}). Alternatively, since we can measure integrin expression levels on the gated population, we can express activation as the specific reporter ligand binding normalized for integrin expression using the equation $AI = (F - F_0)/(F_{integrin})$, where $F_{integrin}$ is the normalized GMFI of a function-independent integrin-specific antibody binding to transfected cells (50).

The assay described above readily permits assessment of the effect of expression of fluorophore-tagged proteins on integrin activation state. It is straightforward to modify this assay to assess

Fig. 1. Analysis of αIIbβ3 activation in CHO cells expressing GFP-talin head. CHO cells stably expressing αIIbβ3 were transfected with DNA-encoding GFP-talin head, the following day αIIbβ3 activation and expression were assessed as described in the text. Binding of the ligand-mimetic monoclonal antibody PAC1 was measured under native, activating, and inhibited conditions, and integrin expression was determined using the D57 monoclonal antibody. Data were analyzed using Flowjo software. (**a**) A plot of forward scatter (FCA-A) vs. side scatter (SCA-A) was prepared and the gate G1 was drawn around the CHO cell population based on its homogenous size and granularity; debris and dead cells that scatter differently were excluded. For each condition, using data from cells falling within the G1 gate, GFP signal in the FL1 channel was plotted against the PAC1 (**b**) or the D57 (**c**) signal in the FL4 channel. A rectangular gate, G2, was then drawn to identify cells expressing GFP at high levels. Frequency histograms were plotted as a means of expressing PAC1 (**d**) or D57 (**e**) binding to cells within the G2 gate, and the geometric mean fluorescence intensity was calculated.

the effect of non-tagged proteins that are co-transfected with an expression construct for a fluorescent marker protein (e.g., talin constructs co-transfected with GFP (50, 96)). Alternatively, another nonfluorescent marker protein, such as a transmembrane protein to which an antibody to the extracellular domain is available, can be used, allowing transfection to be assessed by antibody staining in parallel with assessment of integrin activation (33, 97). The disadvantage of such assays is that they do not allow for direct measurement of the levels of the protein of interest and rely instead on the assumption that expression levels of the protein of interest and the marker protein are directly proportional.

The recent discovery of new integrin co-activators, such as kindlin (59–61), has highlighted the importance of assessing integrin activation using more than one protein at a time. The basic integrin activation assay can, therefore, be modified to assess levels of two differently tagged co-expressed proteins simultaneous with measuring integrin activation state. In these assays, we use cells co-expressing GFP- and DsRed-tagged proteins (e.g., DsRed-talin and GFP-kindlin or GFP-talin and DsRed-kindlin (3, 98)). By monitoring the expression of each protein and the integrin activation reporter in different channels, we are able to investigate more accurately the effect of expressing an additional factor, such as kindlin, on talin-mediated integrin activation (see Note 6).

4. Notes

1. When expressing recombinant fragments of talin head in mammalian cells, we find that expression levels and transfection efficiency depend on cell type and on the boundaries of the construct used. Therefore, we vary the amount of cDNA used during the transfection step to optimize protein production.
2. When working with bacterial cultures, recombinant talin expression levels vary depending on the boundaries of the construct and the epitope/affinity tag used. In an effort to maximize protein expression, we sometimes induce at a higher OD_{600}, as high as 0.8, vary the amount of IPTG used for induction, and increase the total volume of LB used to increase the overall yield. To improve purification of talin fragments from bacteria, we also vary the lysis buffer or induce protein expression at lower temperatures, e.g., at 12–16°C overnight.
3. Common areas for optimization in the integrin tail-binding assays include varying the stringency of the binding buffer, varying the amount of cell lysate or purified recombinant

protein used in the binding assay, and lastly decreasing the amount of integrin tails coated on the matrix in an effort to reduce nonspecific binding.

4. Seeding 1×10^6 CHO cells in a 10-cm plate ensures that the cells will be adherent and sufficiently confluent when transfected the following day. We find cell confluency is important for good transfection and efficient expression of recombinant proteins. Under our experimental conditions, 10^6 CHO cells will have 60–70% confluency on the day of transfection, generally yielding 40–50% transfection efficiency. If different cell types are used, the seeding density will need to be optimized for each cell type.

5. Filtration of cell suspensions prior to analysis by FACS is important to remove large cell aggregates that might clog the FACS machine and interfere with data acquisition.

6. Advanced applications of the FACS integrin activation assay using two co-expressed activators are performed as described for the basic integrin activation FACS, but some modifications are required to (1) ensure efficient expression of both proteins, (2) allow accurate measurement of both GFP and DsRed fluorophores in the same cell without extensive cross talk between the FL1 and FL2 channels, and (3) ensure that comparable populations of double transfected cells are used for data analysis. These areas are highlighted below.

   Preparing doubly transfected cell: A key step in assays involving two activator proteins is efficient co-expression of both proteins as in many cases, expression of one protein can impair the expression of the second. The transfection protocol for expressing two proteins is essentially the same as that described above for expressing a single protein, but in a co-transfection setting, the expression of each fluorescent protein should initially be assessed by varying the amounts and ratios of the two DNAs to optimize their co-expression in the double transfected cells. This process needs to be repeated for each pair of proteins and fluorophores under test, and for each cell line being used.

   Preparation of compensation controls: Monitoring both green and red signals in adjacent channels (FL1 and FL2) requires careful setup of the FACS machine using the proper compensation controls prior to acquisition of the double transfected cells. Controls should include untransfected cells, cells expressing only GFP, and cells expressing only DsRed. These control cells are processed along with the experimental samples but do not receive PAC1 or FN9-11 and, therefore, should only exhibit background levels of fluorescence in the FL4 channel. Prior to acquisition of the experimental samples, the acquisition gate and channel voltages are set using untransfected cells. This provides a background signal in each channel. GFP-expressing cells are then run and setting adjusted to compensate for any green signal that bleeds through into the red channel. Likewise, DsRed-expressing cells are then used to compensate for any red signal that bleeds into the green channel.

Data analysis: Data analysis proceeds much as described in Fig. 1, but requires an additional step to gate on cells expressing both green and red fluorescent proteins (Fig. 2). After gating on the live cell population based on forward and side scatter, a population that is positive for both DsRed and GFP

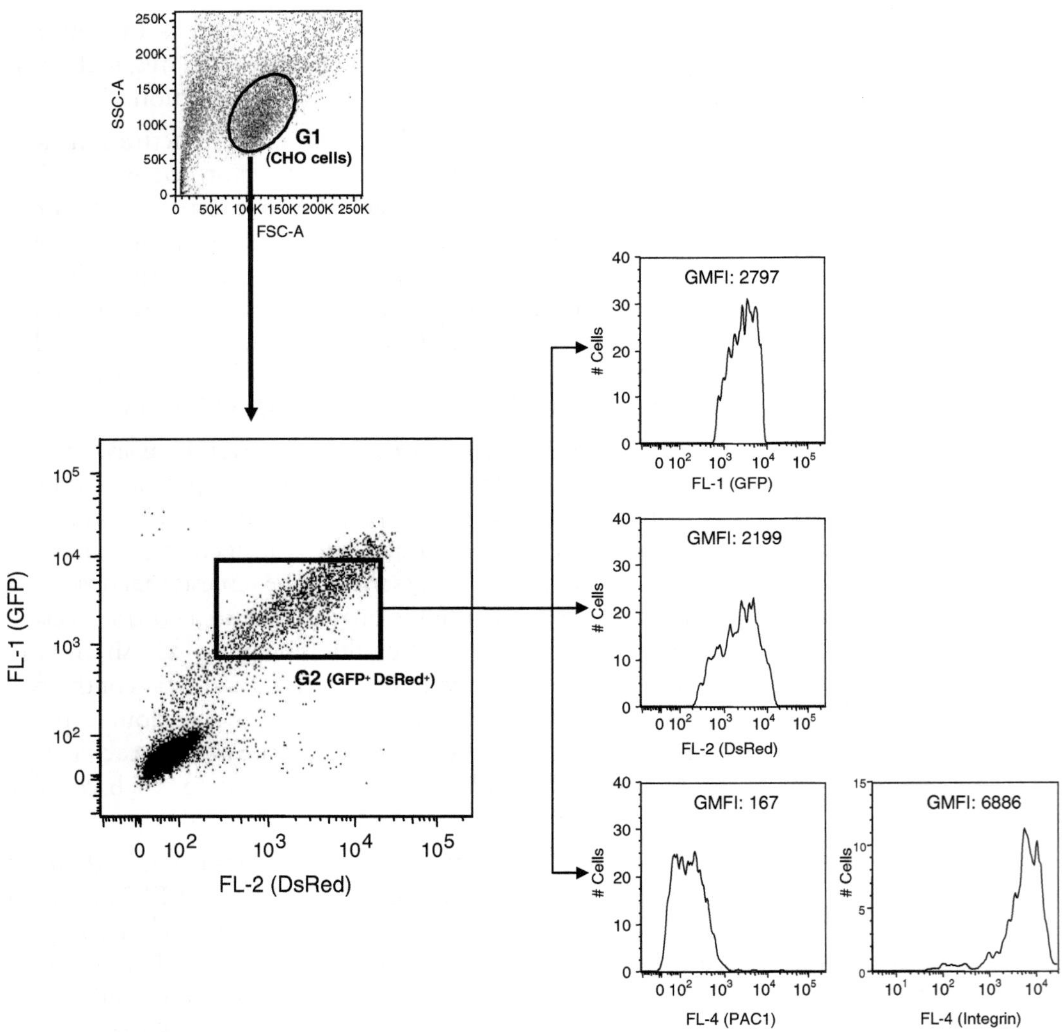

Fig. 2. Analysis of αIIbβ3 activation in CHO cells expressing GFP-talin head and DsRed-kindlin1. CHO cells stably expressing αIIbβ3 were transfected with DNA-encoding GFP-talin head, and DsRed-kindlin1 and αIIbβ3 activation and expression were assessed as described in the text. Results were analyzed as described in Fig. 1, but the G2 gate was drawn to define a double positive (GFP and DsRed) CHO cell population. Histogram plots from cells in this gate were generated to measure the GMFI of GFP, DsRed, and PAC1 or D57.

signals is then selected. The GMFI signal for GFP and DsRed in this gate is then calculated and a similar gate is applied to other experimental samples. This gate is then adjusted to ensure that the gated population in each sample has comparable GMFI for GFP and DsRed. The GMFI of PAC1, FN9-11, or anti-integrin antibody binding within this gate is then calculated (Fig. 2) for each condition and activation indices calculated as described above.

References

1. Luo, B. H., Carman, C. V., and Springer, T. A. (2007) Structural basis of integrin regulation and signaling. *Annu. Rev. Immunol.* **25**, 619–647.
2. Arnaout, M. A., Goodman, S. L., and Xiong, J. P. (2007) Structure and mechanics of integrin-based cell adhesion. *Curr. Opin. Cell Biol.* **19**, 495–507.
3. Harburger, D., and Calderwood, D. A. (2009) Integrin signalling at a glance. *J. Cell Sci.* **122**, 159–163.
4. Hynes, R. O. (2002) Integrins: bidirectional, allosteric signaling machines. *Cell* **110**, 673–687.
5. Askari, J., Buckley, P., Mould, A., and Humphries, M. (2009) Linking integrin conformation to function. *J Cell. Sci.* **122**, 165–170.
6. Chen, J., Salas, A., and Springer, T. A. (2003) Bistable regulation of integrin adhesiveness by a bipolar metal ion cluster. *Nat. Struct. Biol.* **10**, 995–1001.
7. Ginsberg, M. H., Partridge, A., and Shattil, S. J. (2005) Integrin regulation. *Curr. Opin. Cell Biol.* **17**, 509–516.
8. Humphries, J., Byron, A., and Humphries, M. (2006) Integrin ligands at a glance. *J. Cell Sci.* **119**, 3901–3903.
9. Evans, E. A., and Calderwood, D. A. (2007) Forces and bond dynamics in cell adhesion. *Science* **316**, 1148–1153.
10. Legate, K. R., and Fassler, R. (2009) Mechanisms that regulate adaptor binding to β-integrin cytoplasmic tails. *J. Cell Sci.* **122**, 187–198.
11. Hynes, R. O. (2007) Cell-matrix adhesion in vascular development. *J. Thromb. Haemost.* **5 Suppl 1**, 32–40.
12. Brakebusch, C., and Fassler, R. (2005) β1 integrin function in vivo: adhesion, migration and more. *Cancer Metastasis. Rev.* **24**, 403–411.
13. Dupuy, A. G., and Caron, E. (2008) Integrin-dependent phagocytosis: spreading from microadhesion to new concepts. *J. Cell Sci.* **121**, 1773–1783.
14. Aplin, J. D. (1997) Adhesion molecules in implantation. *Rev. Reprod.* **2**, 84–93.
15. Petri, B., Phillipson, M., and Kubes, P. (2008) The physiology of leukocyte recruitment: an in vivo perspective. *J. Immunol.* **180**, 6439–6446.
16. Nieswandt, B., Moser, M., Pleines, I., Varga-Szabo, D., Monkley, S., Critchley, D., and Fassler, R. (2007) Loss of talin1 in platelets abrogates integrin activation, platelet aggregation, and thrombus formation in vitro and in vivo. *J. Exp. Med.* **204**, 3113–3118.
17. Takagi, J., and Springer, T. A. (2002) Integrin activation and structural rearrangement. *Immunol. Rev.* **186**, 141–163.
18. Calderwood, D. A., and Ginsberg, M. H. (2003) Talin forges the links between integrins and actin. *Nat. Cell. Biol.* **5**, 694–697.
19. Critchley, D. R. (2009) Biochemical and structural properties of the integrin-associated cytoskeletal protein talin. *Annu. Rev. Biophys.* **38**, 235–254.
20. Moser, M., Legate, K. R., Zent, R., and Fassler, R. (2009) The tail of integrins, talin, and kindlins. *Science* **324**, 895–899.
21. Jiang, G., Giannone, G., Critchley, D. R., Fukumoto, E., and Sheetz, M. P. (2003) Two-piconewton slip bond between fibronectin and the cytoskeleton depends on talin. *Nature* **424**, 334–337.
22. Zhang, X., Jiang, G., Cai, Y., Monkley, S. J., Critchley, D. R., and Sheetz, M. P. (2008) Talin depletion reveals independence of initial cell spreading from integrin activation and traction. *Nat. Cell Biol.* **10**, 1062–1068.
23. Calderwood, D. A., Yan, B., de Pereda, J. M., Alvarez, B. G., Fujioka, Y., Liddington, R. C., and Ginsberg, M. H. (2002) The phosphotyrosine binding-like domain of talin activates integrins. *J. Biol. Chem.* **277**, 21749–21758.
24. Critchley, D. R. (2004) Cytoskeletal proteins talin and vinculin in integrin-mediated adhesion. *Biochem. Soc. Trans.* **32**, 831–836.
25. Calderwood, D. A., Fujioka, Y., de Pereda, J. M., Garcia-Alvarez, B., Nakamoto, T.,

Margolis, B., McGlade, C. J., Liddington, R. C., and Ginsberg, M. H. (2003) Integrin β cytoplasmic domain interactions with phosphotyrosine-binding domains: a structural prototype for diversity in integrin signaling. *Proc. Natl. Acad. Sci. USA.* **100**, 2272–2277.

26. Papagrigoriou, E., Gingras, A. R., Barsukov, I. L., Bate, N., Fillingham, I. J., Patel, B., Frank, R., Ziegler, W. H., Roberts, G. C., Critchley, D. R., and Emsley, J. (2004) Activation of a vinculin-binding site in the talin rod involves rearrangement of a five-helix bundle. *EMBO J.* **23**, 2942–2951.

27. Gingras, A. R., Ziegler, W. H., Frank, R., Barsukov, I. L., Roberts, G. C., Critchley, D. R., and Emsley, J. (2005) Mapping and consensus sequence identification for multiple vinculin binding sites within the talin rod. *J. Biol. Chem.* **280**, 37217–37224.

28. Lee, H. S., Bellin, R. M., Walker, D. L., Patel, B., Powers, P., Liu, H., Garcia-Alvarez, B., de Pereda, J. M., Liddington, R. C., Volkmann, N., Hanein, D., Critchley, D. R., and Robson, R. M. (2004) Characterization of an actin-binding site within the talin FERM domain. *J. Mol. Biol.* **343**, 771–784.

29. Gingras, A. R., Bate, N., Goult, B. T., Hazelwood, L., Canestrelli, I., Grossmann, J. G., Liu, H., Putz, N. S., Roberts, G. C., Volkmann, N., Hanein, D., Barsukov, I. L., and Critchley, D. R. (2008) The structure of the C-terminal actin-binding domain of talin. *EMBO J.* **27**, 458–469.

30. Burridge, K., and Mangeat, P. (1984) An interaction between vinculin and talin *Nature* **308**, 744–746.

31. Horwitz, A., Duggan, K., Buck, C., Beckerle, M. C., and Burridge, K. (1986) Interaction of plasma membrane fibronectin receptor with talin–a transmembrane linkage. *Nature* **320**, 531–533.

32. del Rio, A., Perez-Jimenez, R., Liu, R., Roca-Cusachs, P., Fernandez, J. M., and Sheetz, M. P. (2009) Stretching single talin rod molecules activates vinculin binding. *Science* **323**, 638–641.

33. Calderwood, D. A., Zent, R., Grant, R., Rees, D. J., Hynes, R. O., and Ginsberg, M. H. (1999) The Talin head domain binds to integrin β subunit cytoplasmic tails and regulates integrin activation. *J. Biol. Chem.* **274**, 28071–28074.

34. Tadokoro, S., Shattil, S. J., Eto, K., Tai, V., Liddington, R. C., de Pereda, J. M., Ginsberg, M. H., and Calderwood, D. A. (2003) Talin binding to integrin β tails: a final common step in integrin activation. *Science* **302**, 103–106.

35. Calderwood, D. A. (2004) Integrin activation. *J. Cell Sci.* **117**, 657–666.

36. Wegener, K. L., Partridge, A. W., Han, J., Pickford, A. R., Liddington, R. C., Ginsberg, M. H., and Campbell, I. D. (2007) Structural basis of integrin activation by talin. *Cell* **128**, 171–182.

37. Czuchra, A., Meyer, H., Legate, K. R., Brakebusch, C., and Fassler, R. (2006) Genetic analysis of beta1 integrin "activation motifs" in mice. *J. Cell Biol.* **174**, 889–899.

38. Tsujioka, M., Yoshida, K., Nagasaki, A., Yonemura, S., Muller-Taubenberger, A., and Uyeda, T. Q. (2008) Overlapping functions of the two talin homologues in Dictyostelium. *Eukaryot. Cell* 7, 906–916.

39. Tanentzapf, G., Martin-Bermudo, M. D., Hicks, M. S., and Brown, N. H. (2006) Multiple factors contribute to integrin-talin interactions in vivo. *J. Cell Sci.* **119**, 1632–1644.

40. Ratnikov, B. I., Partridge, A. W., and Ginsberg, M. H. (2005) Integrin activation by talin. *J. Thromb. Haemost.* **3**, 1783–1790.

41. Wennerberg, K., Fassler, R., Warmegard, B., and Johansson, S. (1998) Mutational analysis of the potential phosphorylation sites in the cytoplasmic domain of integrin β1A. Requirement for threonines 788–789 in receptor activation. *J. Cell. Sci.* **111 (Pt 8)**, 1117–1126.

42. Yan, B., Calderwood, D. A., Yaspan, B., and Ginsberg, M. H. (2001) Calpain cleavage promotes talin binding to the β3 integrin cytoplasmic domain. *J. Biol. Chem.* **276**, 28164–28170.

43. Goksoy, E., Ma, Y. Q., Wang, X., Kong, X., Perera, D., Plow, E. F., and Qin, J. (2008) Structural basis for the autoinhibition of talin in regulating integrin activation. *Mol. Cell* **31**, 124–133.

44. Goult, B. T., Bate, N., Anthis, N. J., Wegener, K. L., Gingras, A. R., Patel, B., Barsukov, I. L., Campbell, I. D., Roberts, G. C., and Critchley, D. R. (2009) The structure of an interdomain complex that regulates talin activity. *J. Biol. Chem.* **284**, 15097–15106.

45. Han, J., Lim, C. J., Watanabe, N., Soriani, A., Ratnikov, B., Calderwood, D. A., Puzon-McLaughlin, W., Lafuente, E. M., Boussiotis, V. A., Shattil, S. J., and Ginsberg, M. H. (2006) Reconstructing and deconstructing agonist-induced activation of integrin αIIbβ3. *Curr. Biol.* **16**, 1796–1806.

46. Lee, H. S., Lim, C. J., Puzon-McLaughlin, W., Shattil, S. J., and Ginsberg, M. H. (2009) RIAM activates integrins by linking talin to ras

GTPase membrane-targeting sequences. *J. Biol. Chem.* **284**, 5119–5127.

47. Martel, V., Racaud-Sultan, C., Dupe, S., Marie, C., Paulhe, F., Galmiche, A., Block, M. R., and Albiges-Rizo, C. (2001) Conformation, localization, and integrin binding of talin depend on its interaction with phosphoinositides. *J. Biol. Chem.* **276**, 21217–21227.
48. Kiema, T., Lad, Y., Jiang, P., Oxley, C. L., Baldassarre, M., Wegener, K. L., Campbell, I. D., Ylanne, J., and Calderwood, D. A. (2006) The molecular basis of filamin binding to integrins and competition with talin. *Mol. Cell* **21**, 337–347.
49. Takala, H., Nurminen, E., Nurmi, S. M., Aatonen, M., Strandin, T., Takatalo, M., Kiema, T., Gahmberg, C. G., Ylanne, J., and Fagerholm, S. C. (2008) β2 integrin phosphorylation on Thr758 acts as a molecular switch to regulate 14-3-3 and filamin binding. *Blood* **112**, 1853–1562.
50. Bouaouina, M., Lad, Y., and Calderwood, D. A. (2008) The N-terminal domains of talin cooperate with the phosphotyrosine binding-like domain to activate beta1 and β3 integrins. *J. Biol. Chem.* **283**, 6118–6125.
51. Larjava, H., Plow, E. F., and Wu, C. (2008) Kindlins: essential regulators of integrin signalling and cell-matrix adhesion. *EMBO Rep.* **9**, 1203–1208.
52. Ramos, J. W., and DeSimone, D. W. (1996) Xenopus embryonic cell adhesion to fibronectin: position-specific activation of RGD/synergy site-dependent migratory behavior at gastrulation. *J. Cell Biol.* **134**, 227–240.
53. O'Toole, T. E., Katagiri, Y., Faull, R. J., Peter, K., Tamura, R., Quaranta, V., Loftus, J. C., Shattil, S. J., and Ginsberg, M. H. (1994) Integrin cytoplasmic domains mediate inside-out signal transduction. *J. Cell Biol.* **124**, 1047–1059.
54. Calderwood, D. A., Tai, V., Di Paolo, G., De Camilli, P., and Ginsberg, M. H. (2004) Competition for talin results in trans-dominant inhibition of integrin activation. *J. Biol. Chem.* **279**, 28889–28895.
55. Lad, Y., Harburger, D. S., and Calderwood, D. A. (2007) Integrin cytoskeletal interactions. *Methods Enzymol.* **426**, 69–84.
56. Pfaff, M., Liu, S., Erle, D. J., and Ginsberg, M. H. (1998) Integrin β cytoplasmic domains differentially bind to cytoskeletal proteins. *J. Biol. Chem.* **273**, 6104–6109.
57. Garcia-Alvarez, B., de Pereda, J. M., Calderwood, D. A., Ulmer, T. S., Critchley, D., Campbell, I. D., Ginsberg, M. H., and Liddington, R. C. (2003) Structural determinants of integrin recognition by talin. *Mol. Cell* **11**, 49–58.
58. Matter, M. L., Ginsberg, M. H., and Ramos, J. W. (2001) Identification of cell signaling molecules by expression cloning. *Sci. STKE* **2001**, PL9.
59. Harburger, D. S., Bouaouina, M., and Calderwood, D. A. (2009) Kindlin-1 and -2 directly bind the C-terminal region of β integrin cytoplasmic tails and exert integrin-specific activation effects. *J. Biol. Chem.* **284**, 11485–11497.
60. Moser, M., Nieswandt, B., Ussar, S., Pozgajova, M., and Fassler, R. (2008) Kindlin-3 is essential for integrin activation and platelet aggregation. *Nat. Med.* **14**, 325–30.
61. Ma, Y. Q., Qin, J., Wu, C., and Plow, E. F. (2008) Kindlin-2 (Mig-2): a co-activator of β3 integrins *J. Cell Biol.* **181**, 439–446.
62. Stewart, M. P., Cabanas, C., and Hogg, N. (1996) T cell adhesion to intercellular adhesion molecule-1 (ICAM-1) is controlled by cell spreading and the activation of integrin LFA-1. *J. Immunol.* **156**, 1810–1817.
63. Ginsberg, M. H., Frelinger, A. L., Lam, S. C., Forsyth, J., McMillan, R., Plow, E. F., and Shattil, S. J. (1990) Analysis of platelet aggregation disorders based on flow cytometric analysis of membrane glycoprotein IIb-IIIa with conformation-specific monoclonal antibodies. *Blood* **76**, 2017–2023.
64. Stallmach, A., Giese, T., Pfister, K., Wittig, B. M., Kunne, S., Humphries, M., Zeitz, M., and Meuer, S. C. (2001) Activation of β1 integrins mediates proliferation and inhibits apoptosis of intestinal CD4-positive lymphocytes. *Eur. J. Immunol.* **31**, 1228–1238.
65. Heilmann, E., Hynes, L. A., Burstein, S. A., George, J. N., and Dale, G. L. (1994) Fluorescein derivatization of fibrinogen for flow cytometric analysis of fibrinogen binding to platelets. *Cytometry* **17**, 287–293.
66. Shattil, S. J., Cunningham, M., and Hoxie, J. A. (1987) Detection of activated platelets in whole blood using activation-dependent monoclonal antibodies and flow cytometry. *Blood* **70**, 307–315.
67. Dransfield, I., and Hogg, N. (1989) Regulated expression of Mg2+ binding epitope on leukocyte integrin β subunits. *EMBO J.* **8**, 3759–3765.
68. Luque, A., Gomez, M., Puzon, W., Takada, Y., Sanchez-Madrid, F., and Cabanas, C. (1996) Activated conformations of very late activation integrins detected by a group of antibodies (HUTS) specific for a novel regulatory region (355–425) of the common β1 chain. *J. Biol. Chem.* **271**, 11067–11075.
69. Taub, R., Gould, R. J., Garsky, V. M., Ciccarone, T. M., Hoxie, J., Friedman, P. A.,

and Shattil, S. J. (1989) A monoclonal antibody against the platelet fibrinogen receptor contains a sequence that mimics a receptor recognition domain in fibrinogen. *J. Biol. Chem.* **264**, 259–265.

70. Shattil, S. J., Hoxie, J. A., Cunningham, M., and Brass, L. F. (1985) Changes in the platelet membrane glycoprotein IIb.IIIa complex during platelet activation. *J. Biol. Chem.* **260**, 11107–11114.
71. Shattil, S. J., Motulsky, H. J., Insel, P. A., Flaherty, L., and Brass, L. F. (1986) Expression of fibrinogen receptors during activation and subsequent desensitization of human platelets by epinephrine. *Blood* **68**, 1224–1231.
72. Pampori, N., Hato, T., Stupack, D. G., Aidoudi, S., Cheresh, D. A., Nemerow, G. R., and Shattil, S. J. (1999) Mechanisms and consequences of affinity modulation of integrin αVβ3 detected with a novel patch-engineered monovalent ligand. *J. Biol. Chem.* **274**, 21609–21616.
73. Bertoni, A., Tadokoro, S., Eto, K., Pampori, N., Parise, L. V., White, G. C., and Shattil, S. J. (2002) Relationships between Rap1b, affinity modulation of integrin αIIbβ3, and the actin cytoskeleton. *J. Biol. Chem.* **277**, 25715–25721.
74. Bunch, T. A., Helsten, T. L., Kendall, T. L., Shirahatti, N., Mahadevan, D., Shattil, S. J., and Brower, D. L. (2006) Amino acid changes in Drosophila αPS2βPS integrins that affect ligand affinity. *J. Biol. Chem.* **281**, 5050–5057.
75. Kashiwagi, H., Tomiyama, Y., Tadokoro, S., Honda, S., Shiraga, M., Mizutani, H., Handa, M., Kurata, Y., Matsuzawa, Y., and Shattil, S. J. (1999) A mutation in the extracellular cysteine-rich repeat region of the β3 subunit activates integrins αIIbβ3 and αVβ3. *Blood* **93**, 2559–2568.
76. Montanez, E., Ussar, S., Schifferer, M., Bosl, M., Zent, R., Moser, M., and Fassler, R. (2008) Kindlin-2 controls bidirectional signaling of integrins. *Genes Dev.* **22**, 1325–1330.
77. Chan, J. R., Hyduk, S. J., and Cybulsky, M. I. (2001) Chemoattractants induce a rapid and transient upregulation of monocyte α4 integrin affinity for vascular cell adhesion molecule 1 which mediates arrest: an early step in the process of emigration. *J. Exp. Med.* **193**, 1149–1158.
78. Chan, J. R., Hyduk, S. J., and Cybulsky, M. I. (2003) Detecting rapid and transient upregulation of leukocyte integrin affinity induced by chemokines and chemoattractants. *J. Immunol. Methods* **273**, 43–52.
79. Weber, K. S., Ostermann, G., Zernecke, A., Schroder, A., Klickstein, L. B., and Weber, C. (2001) Dual role of H-Ras in regulation of lymphocyte function antigen-1 activity by stromal cell-derived factor-1α: implications for leukocyte transmigration. *Mol. Biol. Cell* **12**, 3074–3086.
80. Hughes, P. E., Oertli, B., Hansen, M., Chou, F. L., Willumsen, B. M., and Ginsberg, M. H. (2002) Suppression of integrin activation by activated Ras or Raf does not correlate with bulk activation of ERK MAP kinase. *Mol. Biol. Cell* **13**, 2256–2265.
81. Bowditch, R. D., Hariharan, M., Tominna, E. F., Smith, J. W., Yamada, K. M., Getzoff, E. D., and Ginsberg, M. H. (1994) Identification of a novel integrin binding site in fibronectin. Differential utilization by β3 integrins. *J. Biol. Chem.* **269**, 10856–10863.
82. Heckmann, D., Meyer, A., Marinelli, L., Zahn, G., Stragies, R., and Kessler, H. (2007) Probing integrin selectivity: rational design of highly active and selective ligands for the α5β1 and αvβ3 integrin receptor. *Angew. Chem. Int. Ed. Engl.* **46**, 3571–3574.
83. O'Toole, T. E., Ylanne, J., and Culley, B. M. (1995) Regulation of integrin affinity states through an NPXY motif in the β subunit cytoplasmic domain. *J. Biol. Chem.* **270**, 8553–8558.
84. Kouns, W. C., Kirchhofer, D., Hadvary, P., Edenhofer, A., Weller, T., Pfenninger, G., Baumgartner, H. R., Jennings, L. K., and Steiner, B. (1992) Reversible conformational changes induced in glycoprotein IIb-IIIa by a potent and selective peptidomimetic inhibitor. *Blood* **80**, 2539–2547.
85. Abrams, C., Deng, Y. J., Steiner, B., O'Toole, T., and Shattil, S. J. (1994) Determinants of specificity of a baculovirus-expressed antibody Fab fragment that binds selectively to the activated form of integrin αIIbβ3. *J. Biol. Chem.* **269**, 18781–18788.
86. Bouaouina, M., Blouin, E., Halbwachs-Mecarelli, L., Lesavre, P., and Rieu, P. (2004) TNF-induced β2 integrin activation involves Src kinases and a redox-regulated activation of p38 MAPK. *J. Immunol.* **173**, 1313–1320.
87. Rose, D. M., Liu, S., Woodside, D. G., Han, J., Schlaepfer, D. D., and Ginsberg, M. H. (2003) Paxillin binding to the α4 integrin subunit stimulates LFA-1 (integrin αLβ2)-dependent T cell migration by augmenting the activation of focal adhesion kinase/proline-rich tyrosine kinase-2. *J. Immunol.* **170**, 5912–5918.
88. Eto, K., Leavitt, A. L., Nakano, T., and Shattil, S. J. (2003) Development and analysis of

megakaryocytes from murine embryonic stem cells. *Methods Enzymol.* **365**, 142–158.

89. Shiraga, M., Ritchie, A., Aidoudi, S., Baron, V., Wilcox, D., White, G., Ybarrondo, B., Murphy, G., Leavitt, A., and Shattil, S. (1999) Primary megakaryocytes reveal a role for transcription factor NF-E2 in integrin αIIbβ3 signaling. *J. Cell. Biol.* **147**, 1419–1430.
90. Kashiwagi, H., Shiraga, M., Honda, S., Kosugi, S., Kamae, T., Kato, H., Kurata, Y., and Tomiyama, Y. (2004) Activation of integrin αIIbβ3 in the glycoprotein Ib-high population of a megakaryocytic cell line, CMK, by inside-out signaling. *J. Thromb. Haemost.* **2**, 177–186.
91. O'Toole, T. E., Loftus, J. C., Du, X. P., Glass, A. A., Ruggeri, Z. M., Shattil, S. J., Plow, E. F., and Ginsberg, M. H. (1990) Affinity modulation of the αIIbβ3 integrin (platelet GPIIb-IIIa) is an intrinsic property of the receptor. *Cell. Regul.* **1**, 883–893.
92. Harper, P. A., and Juliano, R. L. (1981) Fibronectin-independent adhesion of fibroblasts to the extracellular matrix: mediation by a high molecular weight membrane glycoprotein. *J. Cell. Biol.* **91**, 647–653.
93. Calderwood, D. A., Huttenlocher, A., Kiosses, W. B., Rose, D. M., Woodside, D. G., Schwartz, M. A., and Ginsberg, M. H. (2001) Increased filamin binding to β-integrin cytoplasmic domains inhibits cell migration. *Nat. Cell Biol.* **3**, 1060–1068.
94. Prevost, N., Kato, H., Bodin, L., and Shattil, S. J. (2007) Platelet integrin adhesive functions and signaling. *Methods Enzymol.* **426**, 103–115.
95. Brown, P. J., and Juliano, R. L. (1985) Selective inhibition of fibronectin-mediated cell adhesion by monoclonal antibodies to a cell-surface glycoprotein. *Science* **228**, 1448–1451.
96. Chou, F. L., Hill, J. M., Hsieh, J. C., Pouyssegur, J., Brunet, A., Glading, A., Uberall, F., Ramos, J. W., Werner, M. H., and Ginsberg, M. H. (2003) PEA-15 binding to ERK1/2 MAPKs is required for its modulation of integrin activation. *J. Biol. Chem.* **278**, 52587–52597.
97. Ramos, J. W., Kojima, T. K., Hughes, P. E., Fenczik, C. A., and Ginsberg, M. H. (1998) The death effector domain of PEA-15 is involved in its regulation of integrin activation. *J. Biol. Chem.* **273**, 33897–33900.
98. Goult, B. T., Bouaouina, M., Harburger, D. S., Bate, N., Patel, B., Anthis, N. J., Campbell, I. D., Calderwood, D. A., Barsukov, I. L., Roberts, G. C., and Critchley, D. R. (2009) The structure of the N-terminus of kindlin-1: A domain important for αIIbβ3 integrin activation. *J. Mol. Biol.* **394**, 944–956.

Chapter 21

Proteomics Method for Identification of Pseudopodium Phosphotyrosine Proteins

Yingchun Wang and Richard L. Klemke

Abstract

Cell migration requires actin/myosin-mediated membrane protrusion of a pseudopodium (or lamellipodium) and its attachment to the substratum. This process guides the direction of cell movement through cytoskeletal remodeling and is regulated by complex signaling networks that act spatially downstream of integrin adhesion receptors. Understanding how these regulatory networks are organized in migratory cells is important for many physiological and pathological processes, including wound healing, immune function, and cancer metastasis. Here, we describe methods for the immunoaffinity purification of phosphotyrosine proteins (pY) from pseudopodia that have been isolated from migratory cells. These methods are compatible with current mass spectrometry-based protein identification technologies and can be utilized for the large-scale identification of the pseudopodium pY proteome in various migratory cell lines, including primary and cancer cells.

Key words: Pseudopodium, Cell body, Immunoprecipitation, Phosphotyrosine, Proteomics, Kinases, Cell migration, Cytoskeleton, Signal transduction

1. Introduction

Reversible protein phosphorylation is a major mechanism for signal propagation in a wide spectrum of signal transduction pathways (1–5). In the advancing pseudopodium, many cytoskeleton and integrin-associated proteins are highly tyrosine phosphorylated as the result of membrane attachment to extracellular matrix proteins (6). We previously described a biochemical approach for the large-scale purification of pseudopodia from migrating cells for proteomic analyses (6, 7). This approach allowed the identification of thousands of proteins and more than 200 phosphoproteins involved in pseudopodium formation and cell migration. However, pY proteins were underrepresented in this study due to

Motomu Shimaoka (ed.), *Integrin and Cell Adhesion Molecules: Methods and Protocols*,
Methods in Molecular Biology, vol. 757, DOI 10.1007/978-1-61779-166-6_21,

their low abundance when compared to the phosphoserine and phosphothreonine proteins (8, 9). Here, we provide detailed methods for the immunoaffinity enrichment and identification of pseudopodial pY proteins using multidimensional protein identification technology (MudPIT) mass spectrometry (10).

2. Materials

2.1. Cell Culture

1. Cell lines: Cos-7 (African green monkey) and NIH 3T3 fibroblasts (ATCC).
2. Growth medium: Dulbecco's modified Eagle's medium supplemented with 10% fetal bovine serum (FBS), 2 mM L-glutamine, 50 μg/ml gentamicin, and 1 mM sodium pyruvate.
3. Starvation medium: same as growth medium, but without FBS.
4. Migration/adhesion medium: same as growth medium, but without FBS and with 0.25% (W/V) radioimmunoassay (RIA) grade fraction V bovine serum albumin (BSA).
5. L-α-Lysophosphatidic acid, oleoyl, sodium (LPA, Sigma): Dissolved in water (1 mg/ml) and stored in aliquots of 10 μl at −80°C for up to 1 year.
6. Human fibronectin (BD Biosciences): Prepared at 1 mg/ml in sterile Milli-Q water and can be kept at 4°C for 2 months (see Note 1).
7. 1× Trypsin–EDTA solution.
8. Sterile phosphate-buffered saline (PBS) 1× solution: Dissolve one package of PBS powder (Gibco, 9.6 g per package) in water to make the final volume 1,000 ml and autoclave. Store at 4°C and warm to 37°C in a water bath before use.

2.2. Pseudopodia Induction and Isolation

1. 24-mm transwell containing polycarbonate membrane with 3.0-μm pores in 6-well plate format (see Note 2).
2. Cotton swab (or cotton-tipped cleaning sticks) (Puritan Medical Products Company LLC, Guilford, Maine).
3. Cell scrapers.
4. 100% Ice-cold methanol.
5. Parafilm.
6. 1% SDS lysis buffer: 1% (W/V) sodium dodecyl sulfate (SDS), 2 mM sodium orthovanadate, 1 mM PMSF, and 1/4 tablet of protease inhibitors (Roche) per 10 ml. Aliquot to 1 ml and store at −80°C.
7. Heat block at 100°C.
8. Micro BCA protein assay kit (Pierce Biotechnology, Rockford, IL).

9. Quick start BSA standard set (Bio-Rad).
10. Bio-Tek µQuant microplate spectrophotometer.

2.3. SDS-Polyacrylamide Gel Electrophoresis (SDS-PAGE)

1. PAGE 4–12% Bis-Tris gel.
2. PAGE 20× MOPS SDS running buffer.
3. 6× Laemmli sample buffer (10 ml): 3.5 ml 1 M Tris–HCl, pH 6.8, 3.6 ml glycerol (36% v/v), 1.0 g SDS (10% w/v), 1.2 mg bromophenol blue, 0.93 g DTT (0.6 M), and 3.5 ml H_2O.
4. Coomassie stain solution.

2.4. Coupling Antibody to Protein A Sepharose Beads

1. PBS-T: 500 ml PBS, 2.5 ml Tween 20, and 2.5 ml 10% sodium azide.
2. Anti-phosphotyrosine antibody, clone 4G10 (Millipore).
3. Protein A Sepharose beads: 1 g beads are finally resuspended in 33.4 ml PBS with 0.25 g NaN_3 to make the final concentration of the beads slurry 30 µg/µl.
4. Albumin from chicken egg white, grade VII.
5. 0.2 M sodium borate: Adjust pH to 9.0 with NaOH.
6. 20 mM dimethylpimelimidate (DMP): 0.0104 g DMP is dissolved in 2 ml 0.2 M sodium borate (pH 9.0).
7. 0.2 M ethanolamine: Add 120 µl 16.6 M ethanolamine to 10 ml water and adjust pH to 8.0 with about 160 µl concentrated HCl.
8. 0.01% Merthiolate: Dissolve 1.0 mg thimerosal with 10 ml PBS.

2.5. Immunoaffinity Purification of pY Proteins

1. Modified RIPA buffer (1% NP-40 buffer): 1% NP-40, 0.1% deoxycholate, 150 mM NaCl, 1 mM EDTA, 50 mM Tris–HCL (pH7.5), 1 mM Na orthovanadate, 10 mM β-glycerol-phosphate, 10 mM sodium fluoride, and 1/4 tablet of protease inhibitors cocktail (Roche) per 10 ml solution.
2. Phenyl phosphate elution buffer: 50 mM Tris–HCl, pH 7, 50 mM NaCl, 1 mM EDTA, 20 mM phenyl phosphate, and 2 mM sodium orthovanadate.

2.6. Desalting and SDS Removal from pY Proteins

1. ReadyPrep 2-D cleanup kit (Bio-Rad).

2.7. Protein Digestion

1. Trypsin.
2. Endoproteinase Lys-C sequencing grade.
3. Iodoacetamide.
4. Tris(2-carboxyethyl)phosphine hydrochloride solution (TCEP).
5. Urea.

3. Methods

The pseudopodia purification system consists of an upper and a lower chamber separated by a microporous filter pre-coated with common ECM proteins such as fibronectin and collagen (Fig. 1). Cells are allowed to attach and spread on the upper surface of the filter. The chemoattractant is placed in the bottom chamber and allowed to diffuse upward through the porous filter, creating a stable gradient. Cells can sense the gradient and protrude their pseudopodia through the 3.0-μm pores to the lower surface of the filter toward the high concentration of chemoattractant. Importantly, the cell body remains on top of the filter since it

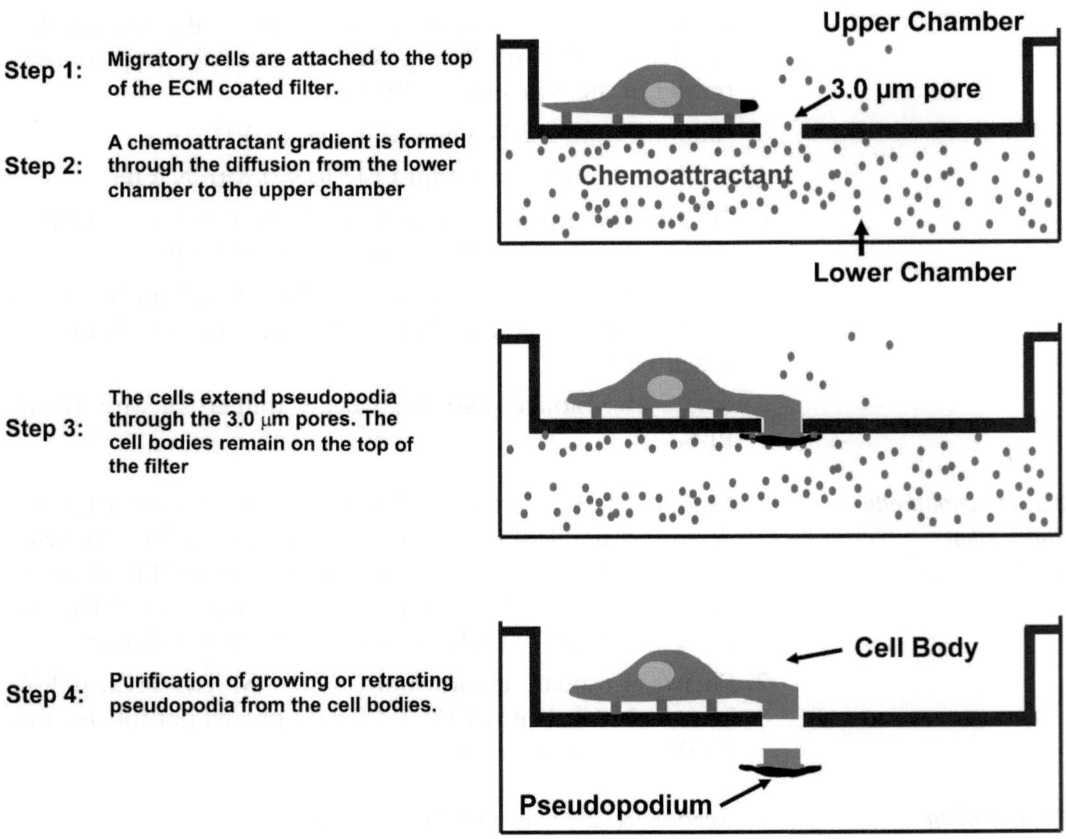

Fig. 1. Schematic illustrating the four basic steps of pseudopodia purification using 3.0-μm porous filters and a chemoattractant gradient. (*Step 1*) Cells are seeded on top of the 3.0-μm porous filters coated with ECM proteins and allowed to attach and spread for 2–3 h. (*Step 2*) The chemoattractant is added to the lower chamber, creating a diffusion gradient through the small holes. The cells sense the gradient and respond with localized amplification of phosphoprotein signals (*black*) on the side facing the gradient. (*Step 3*) The cells become morphologically polarized in the direction of the chemoattractant gradient and protrude pseudopodia through the 3.0-μm pores, which are stabilized by forming new integrin-mediated attachments to the ECM protein coated on the filter. The pseudopodia can also be induced to retract by removing the chemoattractant. (*Step 4*) Growing or retracting pseudopodia can then be harvested by manually shearing off the cell bodies on the top of the filter. Alternatively, the cell bodies can be isolated after shearing off the pseudopodia from the bottom side of the filter.

cannot penetrate through the small opening. Thus, cells achieve morphological polarity with a distinct protruding front and trailing cell body or rear compartment that is characteristic of migrating cells. The polarized cells are then rapidly fixed and their cell bodies and pseudopodia differentially harvested using the methods described below.

The method described here is for the purification of 200–250 μg of pseudopodia proteins using 24-mm size filters and a 6-well format plate. However, it is possible to purify greater than 500–1,000 μg proteins by simply pooling samples from a number of filters or by using larger size filters (100 mm). Also, the methods herein are optimized for purification of pseudopodia from Cos-7 or NIH 3T3 cells induced by a gradient of LPA. While the 3.0-μm porous filters are the optimal size for these cells, in some cases, smaller cells may migrate completely through the 3.0-μm pore and thus it may be necessary to reduce the size of the pore so that the cell body remains on the top of the filter. In addition, it is possible to purify pseudopodia from various cell lines induced by different chemoattractants such as PDGF-BB, EGF, and SDF-1α (6, 7, 11–13).

3.1. Preparation of Cells for Pseudopodia Purification

1. Three days before the pseudopodia preparation assay, split 60–70% confluent low-passage Cos-7 or NIH 3T3 cells from one 100-mm Petri dish into four 100-mm dishes containing 7 ml of growth medium (see Note 3).
2. Allow the cells to grow at 37°C supplemented with 5% CO_2. The cells will reach 70–80% confluence after 3 days (see Note 4).
3. When the cells are about 70% confluent, replace the growth medium with 7 ml of starvation medium to serum-starve the cells overnight.

3.2. Preparation of Microporous Filters for Pseudopodia Purification

1. Before coating the microporous filters with ECM proteins, it is necessary to check whether the filters are completely sealed to the plastic wall of the chamber. To check for leaky filters, add 2 ml of sterile PBS into the lower chambers of the 6-well transwell containing the polycarbonate membrane, and incubate at room temperature for at least 30 min. Verify visually that there is no fluid around the edge of the filter surface where it attaches to the plastic chamber (see Note 5).
2. Remove the PBS from functional chambers and discard the chambers that leak.
3. Add 100 μl of 1 mg/ml human fibronectin solution in PBS to 20 ml of sterile PBS to make the final concentration 5 μg/ml.
4. For chemoattractant gradient-induced pseudopodia purification, add 2 ml of fibronectin solution (5 μg/ml) to the lower chamber and 1.3 ml to the top chamber. Gently rock the plate to allow the fluid to coat both sides of the filter surface evenly (see Note 6).

5. For ECM gradient-induced pseudopodia purification, add 2 ml of fibronectin solution (5 μg/ml) to the bottom chamber only to coat the lower side of the filter. The cells sense the ECM protein gradient and protrude pseudopodia through the pores in the same way as they do in response to a chemoattractant gradient.
6. Incubate the 6-well plate at 37°C for 2 h to allow the filters to be completely coated by fibronectin.

3.3. Detachment of Cells from Culture Dishes

1. Remove the starvation medium and rinse the cells with sterile PBS (see Note 7).
2. Add 2 ml of 1× trypsin–EDTA solution for each 100-mm dish of cells and incubate until the cells are floating or can be gently tapped off the dish.
3. Quench the trypsin activity with 3 ml of warm migration and adhesion medium per dish.
4. Repeat steps 1–3 for all dishes and combine the cells into a 50-ml tube.
5. Spin down the cells at 500 × *g* for 3 min.
6. Carefully remove the supernatant without disturbing the cell pellet.
7. Resuspend the cells with 10 ml of migration and adhesion medium.
8. Count the cells using a microscope and hemocytometer.
9. Dilute the cells with migration and adhesion medium to a final concentration of 1×10^6 cells/ml.

3.4. Attachment of the Cells to the Upper Side of the Microporous Filter

1. Add 2.5 ml of migration and adhesion medium to each well (lower chamber) of a new standard 6-well plastic plate (see Note 8).
2. Remove the microporous filters from the fibronectin solution (see Subheading 3.2) and gently shake off the excess liquid (see Note 9).
3. Seed 1.5×10^6 cells (1.5 ml of cell suspension) to the upper chamber of the microporous filter and immediately place into a well of the 6-well plate to which the migration and adhesion medium was added (Step 1). Repeat for each of the filters (see Note 10).
4. For cells that will be stimulated with a chemoattractant gradient, allow the cells to attach and spread for 2–2.5 h at 37°C in a tissue-culture incubator.
5. For cells responding to an ECM gradient, allow the cells to attach and directly extend pseudopodia through the pores for 2 h at 37°C in a tissue-culture incubator. Now proceed

to Subheading 3.10 for pseudopodia visualization or Subheading 3.6 for pseudopodia or cell bodies purification after the incubation period is complete.

6. Examine cell adhesion and spreading using a standard phase-contrast microscope to ensure proper attachment to the filter before continuing (see Note 11).

3.5. Stimulation of Pseudopodia Growth and Retraction

1. Add 1.5 μl of 1 mg/ml LPA stock to 15 ml of migration and adhesion medium to make the final concentration 100 ng/ml.
2. Add 2.5 ml of LPA-containing medium in step 1 to the lower chamber of each well of a new 6-well plate. As a control, add 2.5 ml of migration and adhesion medium without LPA (see Note 12).
3. Rapidly but gently transfer the filter with the attached cells (see Subheading 3.4) to a well containing LPA or a control well containing medium only. This is now the upper chamber (see Note 13).
4. Gently remove any bubbles that may have been trapped underneath the filter during the transfer by tilting the chamber to one side.
5. Gently transfer the 6-well plate into the 37°C cell-culture incubator and allow pseudopodia to grow for ~30–60 min without disruption (see Note 14). At the end of the incubation period, pseudopodia will be ready for purification as described below.
6. It is also possible to make the pseudopodia retract by removing the chemokine gradient. In this way, retracting or growing pseudopodia can be purified for comparison, as described below. To induce retraction-phase pseudopodia, remove the microporous filters to which the cells are attached after the 60-min LPA stimulation.
7. Carefully aspirate the medium from the upper and the lower surface of the filter without disrupting the cells.
8. Quickly, but gently, place the microporous filter to which the cells are attached into a well of a new 6-well plate containing 2.5 ml of migration and adhesion medium pre-equilibrated to 37°C in the tissue-culture incubator. The medium from the lower chamber will rapidly diffuse to the upper chamber and, therefore, it is not necessary to add more medium to the top chamber.
9. Gently transfer the 6-well plate into the 37°C cell-culture incubator and allow pseudopodia to retract for the appropriate time, which for Cos-7 and NIH 3T3 cells is approximately 30 min (see Note 15).

3.6. Purification of Pseudopodia

1. Thirty minutes before harvesting pseudopodia or cell bodies, add 5 ml of ice-cold methanol to each well of a clean 6-well plate (see Note 16). Incubate the plate on ice.
2. Remove the filters containing pseudopodia and cell bodies from the 6-well plate and quickly, but gently, rinse in 500 ml of ice-cold PBS and shake off the excess liquid.
3. Immediately place the filters into the plate wells from step 1 above containing ice-cold 100% methanol and fix for 30 min on ice.
4. Add 200 μl of 1% SDS lysis buffer to a piece of Parafilm that has been flattened out on top of a glass plate (see Note 17). The lysis buffer will form a small drop on the Parafilm, which allows for the utilization of small fluid volumes.
5. After the fixation is complete, remove the filters from methanol one at a time and gently rinse in a container with 200 ml of 1× PBS solution at room temperature. Shake off excess PBS.
6. Manually scrap off the cell bodies from the top of the filter with a cotton-tipped swab with the tip flattened (see Note 18).
7. Remove cell body debris on the top of the filter by rinsing the filter in 200 ml of 1× PBS.
8. Repeat steps 4–7 for a total of five filters from one 6-well plate, and leave one filter for the purification of cell bodies that is described in the following section (see Note 19).
9. With a sharp razor blade, carefully cut the filters along the outer edge of the chamber to generate a round filter with a smooth edge (see Note 20).
10. Place one of the filters prepared in step 9 onto a 200-μl drop of lysis buffer on the Parafilm with the pseudopodia facing down and allow the proteins to solubilize for at least 30 s (see Note 21).
11. Scrape the pseudopodia from the filter into the lysis buffer with a small plastic cell scraper.
12. Retrieve all residual lysate left on the filter using a pipette and add it back to the drop of lysis buffer on the Parafilm.
13. Repeat steps 9–12 for each of the pseudopodia filters from the 6-well plate. To concentrate pseudopodia proteins, we use 200 μl of lysis buffer per set of five filters. The number of filters can be increased to increase protein yield if desired (see Note 22).
14. Transfer the lysate to a 1.5-ml microfuge tube with a pipette.
15. Immediately freeze the samples at –80°C.

3.7. Purification of Cell Bodies

1. Remove the last methanol-fixed filter described in Subheading 3.6 from the 6-well plate and quickly, but gently, rinse in 200 ml of 1× PBS at room temperature.

2. Shake off excess PBS.
3. Carefully wipe off pseudopodia from the bottom of the filter using a cotton-tipped swab.
4. Rinse the filter in 200 ml of 1× PBS to remove debris (see Note 23).
5. Add 200 μl of the 1% SDS lysis buffer to the top of the filter and scrape off the cell bodies using a small cell scraper.
6. Using a pipette, retrieve the lysis buffer from the filter and transfer the cell body sample into a 1.5-ml microfuge tube.
7. Immediately freeze the samples at −80°C.

3.8. Determining Protein Concentration

1. Centrifuge the collected lysates at 45,000 × *g* for 1 h at 4°C to pellet cell debris.
2. Determine the protein concentration in supernatants from the pseudopodia and cell bodies lysates with the BCA protein assay using a microplate spectrophotometer and a series of pre-diluted BSA protein concentration standards (see Note 24).

3.9. Separating Pseudopodial and Cell Body Proteins by SDS-PAGE and Western Blotting for pY Proteins

1. In separate tubes, mix 20 μg of pseudopodia and cell bodies lysates with 6× Laemmli sample buffer.
2. Boil the sample in a heat block set at 100°C.
3. Separate the proteins with standard SDS-PAGE, for example, with the mini NuPAG 4–12% Bis-Tris gel.
4. To detect phosphotyrosine-containing proteins, perform a Western blot using the 4G10 antibody directed against phosphorylated tyrosine using standard procedures, as previously described (6). PY proteins are highly enriched in the pseudopodia fraction (Fig. 2). This provides an additional marker of purity and validates the enrichment of these proteins in the pseudopodia lysate prior to their affinity purification, which is described below.
5. Alternatively, stain the gel with Bio-Safe Coomassie stain solution and wash the gel with deionized water until the bands are clearly visible (Fig. 2) (see Note 25).

3.10. Visualization and Quantification of Pseudopodia by Brightfield Microscopy

1. Remove the filters with the growth- or retraction-phase pseudopodia from the 6-well plate and quickly but gently rinse in 200 ml of 1× PBS (Recipe 2) at room temperature.
2. Place the filters in 4 ml of crystal violet solution to stain for 15 min.
3. Wash the filter thoroughly with deionized water.
4. Shake off residual water and place in a clean dry well of a 6-well plate.
5. Wipe off the cell bodies from the top of the filter using a cotton swab and then rinse the chamber with water. Repeat wiping

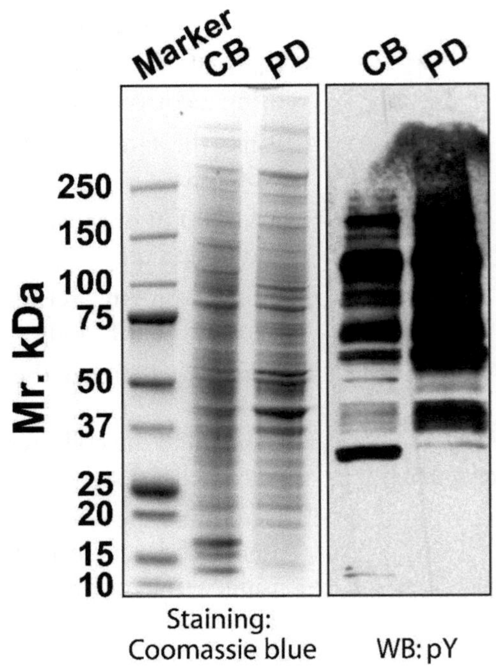

Fig. 2. The total pY protein profile in growing pseudopodia (PD) and cell bodies (CB). The proteins were separated by SDS-PAGE, and then were either Western blotted to detect pY proteins or stained with Coomassie blue to detect total cellular proteins.

and rinsing until the cell bodies have been completely removed from the top of the filter.

6. To visualize stained pseudopodia by standard microscopy, cut the filter along the outer edge of the chamber with a sharp razor blade, then mount the filter on a glass slide, and visualize by brightfield microscopy (see Note 26).
7. Alternatively, to quantify the stained pseudopodia, elute the dye with 10% acetic acid, transfer the solution to a microplate, and measure the absorbance at 590 nm (see Note 27).

3.11. Coupling of anti-pY Antibodies to Protein A Sepharose Beads

1. Pipet 400 µl of the Protein A Sepharose bead slurry into a 2-ml tube (use the same amount of beads for coupling non-pY targeting IgG, e.g., goat anti-mouse secondary antibody for the purpose of pre-clearing samples, see step 5 below) (see Note 28).
2. Spin down the beads at 2,000 × *g* for 2 min (see Note 29).
3. Remove the supernatant. Resuspend the bead pellet with 2 ml of PBS-T and spin down again. Repeat this step 2 times.
4. Resuspend the beads with 2 ml of 1 mg/ml albumin from chicken egg white in PBS-T.

5. Add 5 μg anti-pY Ab (4G10) or 5 μg non-targeting IgG to the beads.
6. Incubate the beads: antibody mixture at room temperature for 1 h with gentle rotation.
7. Spin down the beads. Wash the beads with 1 ml of 0.2 M sodium borate, pH 9.0, for four times.
8. Resuspend the beads in 2 ml of 0.2 M sodium borate, pH 9.0.
9. Add DMP solid to bring the final concentration to 20 mM DMP (pH should be around 8.3 after addition of DMP) and mix for 30 min at room temperature with rotation.
10. Spin down the beads. Stop coupling reaction by washing beads once with 2 ml of 0.2 M ethanolamine, pH 8.0, to block any excess DMP.
11. Spin down the beads and remove the washing ethanolamine. Add 2 ml of fresh 0.2 M ethanolamine, pH 8.0, to beads and incubate for 2 h at room temperature with gentle rotation.
12. Wash the beads 3× with 2 ml of PBS.
13. The antibody is now covalently coupled to the beads and is ready for immunoprecipitation experiments. Alternatively, the beads can be stored at 4°C in PBS with 1% merthiolate and can be re-used up to ten times.

3.12. Pre-clearing of Lysates with IgG-Coupled Beads

1. Transfer 2.5 mg of protein sample to a 50-ml conical centrifuge tube and dilute the sample 10× with non-denaturing modified RIPA buffer (see Note 30).
2. Add the beads that were coupled with non-pY targeting IgG to the sample and incubate at 4°C for 2 h with gentle rotation (see Note 31).
3. Spin down the beads and transfer the supernatant to a fresh 50-ml centrifuge tube for the pY-IP that is described in the following section.

3.13. Immunoprecipitation of pY Proteins

1. Spin down the 4G10 antibody-coupled protein A Sepharose beads prepared as described in Subheading 3.11, and wash the beads 1× with 2 ml of modified RIPA buffer.
2. Transfer the beads to the pre-cleared samples prepared as described above in Subheading 3.12.
3. Incubate the beads: protein sample mixture overnight at 4°C with gentle rotation.
4. Spin down the beads and remove the supernatant. Wash the beads 4 × 5 min with 2 ml of modified RIPA buffer at 4°C with gentle rotation.
5. Add 200 μl of phenyl phosphate elution buffer to the beads and incubate at RT for 20 min with gentle rotation.

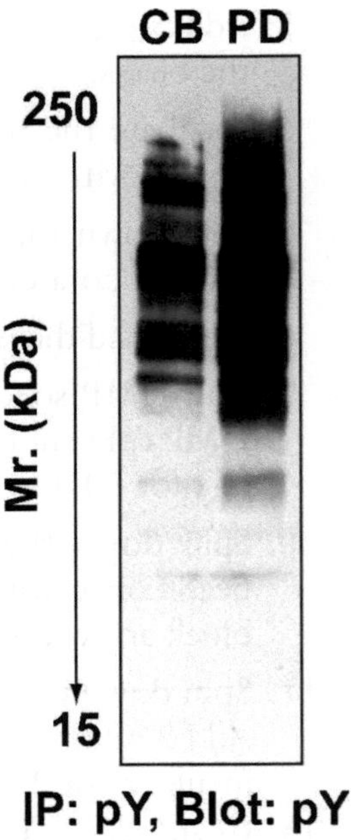

Fig. 3. Western blot showing immunoaffinity-purified pY proteins in PD and CB. The pY proteins were immunoaffinity purified using anti-pY antibody 4G10, and then separated by SDS-PAGE and Western blotted using the anti-pY antibody 4G10.

6. Pellet the beads and collect the supernatant.
7. Repeat steps 5 and 6 and then combine the elutes.
8. Analyze an aliquot (20 μl) of the elute by SDS-PAGE and monitor pY protein capture using anti-pY protein Western blotting and the 4G10 antibody (Fig. 3).

3.14. Sample Reduction and Alkylation for MudPit

1. Purify and precipitate proteins using the Bio-Rad ReadyPrep 2-D cleanup kit according to the manufacturer's instruction (see Note 32).
2. Resolubilize each sample in 20 μl of 8 M urea in 100 mM Tris–HCl, pH 8.5.
3. Add 1 μl of 100 mM TCEP to the sample to make the final concentration 5 mM, and then incubate at room temperature for 30 min.
4. Add 0.4 μl of 500 mM iodoacetamide to the sample to make the final concentration 10 mM, and incubate at room temperature for 30 min in the dark.
5. Add endoproteinase Lys-C with an enzyme:protein ratio of 1:50 (w:w). Incubate the protein/enzyme mixture at 37°C with shaking for 6 h.

6. Dilute the sample 4× by adding 60 μl of 100 mM Tris–HCl, pH 8.5.
7. Add 1.6 μl of 100 mM $CaCl_2$ to make the final concentration 2 mM.
8. Add trypsin at the trypsin: protein ratio of 1:100 (w:w). Incubate the protein/enzyme mixture at 37°C with shaking overnight.
9. Add 90% formic acid to the protein/enzyme mixture with the volume ratio 1:20 to stop the reaction.
10. The peptide digests are now ready for mass spectrometry using MudPIT identification methods as described (10).

4. Notes

1. Fibronectin is used as the ECM protein coating the 3.0-μm porous filters. Prepare the solution in a biological safety cabinet. Filter sterilize Milli-Q water through a 0.22-μm tube-top filter. Add the water to the vial containing the fibronectin and leave at room temperature for 1 h. Avoid vortexing or violent shaking of the vial because this may cause aggregation of the fibronectin.
2. 6.5-mm transwell in 24-well plate format with pore size 3.0 μm can also be used and are available from Corning.
3. Using low-passage (20 or fewer) cells can reduce the frequency of random, background pseudopodia protrusion through the filter pores.
4. This will yield enough cells for ~6, 24-mm chambers and 200–250 μg of purified pseudopodia protein.
5. We typically use a 6-well, 24-mm transwell with a 3.0-μm microporous filter. Before beginning, it is important to make sure that the filter is properly sealed and does not leak around the outer edges. This severely disrupts the normal flow of chemoattractant from the lower chamber to the upper compartment, creating unfavorable gradient conditions and little or no pseudopodia production. If the filter is sealed properly, fluid should not leak through from the lower surface to the upper surface of the filter.

   Cells adhere to different ECM proteins; therefore, different cell types may require different ECM proteins for adhesion to the filter surface. Fibronectin is used here as the ECM protein for NIH 3T3 or Cos-7 cells; however, other ECM proteins such as collagen or vitronectin can be used depending on the cell line.

6. This part of the process is performed while the filters are being coated with ECM proteins (described above) and should be started ~40 min before the coating is complete.
7. Carefully remove bubbles that were introduced during the addition of the medium by aspiration.
8. Always place the filter upside down on the lid of the 6-well plate to avoid disturbing the fibronectin coating on the underside of the filter if the filters must be set down. Sterility is not necessary from this step on because we incubate cells for only 3 h for the pseudopodium protrusion.
9. Bubbles that are trapped between the medium in the lower chamber and the underside of the filter can disrupt the chemoattractant gradient. Remove the bubbles by gently rocking and tapping the upper chamber.
10. Proper cell attachment and spreading on the filter surface are critical for correct gradient sensing and pseudopodium extension.
11. A specific cell line can be induced to extend pseudopodium by more than one type of different chemoattractant. LPA is used here as the chemoattractant for Cos-7 or NIH 3T3 cells. Other chemoattractants such as PDGF or insulin may also be used, but the concentration will have to be determined experimentally for each chemokine and cell line used.
12. It is critical not to shake or bang the dish abruptly because this can disrupt the chemoattractant gradient that is forming between the upper and lower chambers.
13. NIH 3T3 and Cos-7 cells actively extend pseudopodia for 30–60 min, which is optimal for examining factors that mediate membrane protrusion in these cells (6). If other cells or chemokines are used, the time of pseudopodia formation will have to be determined experimentally.
14. NIH 3T3 and Cos-7 cells actively retract pseudopodia from 15 to 60 min (6). The time of retraction will have to be determined experimentally if other cells are used.
15. We typically fix cells with ice-cold methanol to reduce most of the protease and phosphatase activity. This step is necessary for large-scale purification of pseudopodia and cell bodies, which usually takes a relatively long time. For small-scale pseudopodia and cell body purification, the fixation step can be omitted, especially for applications requiring non-denaturing conditions, including kinase activity and immunoprecipitation assays. The old 6-well plates can be re-used for fixing or staining cells with crystal violet.

16. Decreasing the amount of the lysis buffer will increase the final concentration of the purified pseudopodia. To ensure proper lysis of pseudopodia, we usually use 200 μl of lysis buffer to collect pseudopodia from five of the filters from one 6-well plate. Ten filters (two 6-well plates) typically yield about 350 μl of lysate containing 400–500 μg of protein. It is normal to lose 20–30 μl of lysis buffer per five to six filters during manipulation.

    Different buffers can be used to lyse the cellular compartments. For example a buffer containing 8M urea without SDS is compatible for most proteomic analyses. Non-denaturing buffers containing Triton X-100 or NP-40 detergents are compatible for most immunoprecipitation assays. Here, we used 1% SDS lysis buffer to maximize solubilization of proteins, including membrane proteins.

17. Complete removal of the cell bodies and the cell debris from the top is critical for the purity of pseudopodia. It is usually difficult to remove the cell bodies around the edges where the filter attaches to the plastic chamber. Cell bodies in this area can be more efficiently accessed and removed by flattened cotton-tipped swabs.
18. The amount of pseudopodial proteins pooled from five filters is nearly equal to the amount of cell body proteins from one filter.
19. Take care not to break the membrane, because it can fold over on itself and trap lysate or cause the loss of the pseudopodia.
20. The membrane may also be placed into a microfuge tube containing 500 μl of lysis buffer (500 μl per five to six membranes) and the pseudopodia can be directly lysed inside the tube.
21. If necessary, samples from multiple preparations may be pooled to obtain enough quantity for the large-scale proteomic analysis.
22. It is important that all debris be removed from the bottom before protein extraction in lysis buffer.
23. The method used to determine protein concentration of the cell body and pseudopodia lysates will depend on the type of lysis buffer used. For a urea-containing lysis buffer, use a Bradford protein assay. For a SDS-containing lysis buffer, use a standard bicinchoninic acid (BCA) protein assay.
24. The cell body contains the nucleus, whereas the pseudopodium does not. Therefore, the highly abundant nuclear-specific histone proteins are a useful indicator of the purity of the pseudopodia preparation. By SDS-PAGE and Coomassie staining, histones should form several major bands at the

molecular mass of ~15 kD in the cell body fraction, but not in the pseudopodia fraction (Fig. 2). However, most all other housekeeping proteins will be equally distributed between the two fractions. The presence of a significant amount of histones in the pseudopodia fraction is the indicator of incomplete removal of cell bodies during the pseudopodia harvesting steps.

25. Visualization of crystal violet-stained pseudopodia can be used to determine whether pseudopodial extension is random or a specific response to a stimulus and to optimize conditions for pseudopodial stimulation. It is easy to distinguish chemoattractant-stimulated or ECM-stimulated pseudopodia from randomly extended pseudopodia under a microscope by comparing the density (number per field) of the stained pseudopodia.
26. This quantification approach is not used for sample preparation for biochemical or proteomic analysis, but used for evaluating whether the experimental conditions are optimized for pseudopodia growth. If the concentration of the crystal violet in cells exposed to a treatment that should elicit pseudopodia formation (stimulated with a chemoattractant or exposed to an ECM gradient) is the same as that seen under unstimulated control conditions, then either the experimental conditions are not correct (for example, due to leakage of the membrane or disruption of gradient by excessive shaking) or the passage of the cells is too high, resulting in a large number of random protrusions.
27. The amount of protein A Sepharaose beads used here is for immunoprecipitation of pY proteins from 2.5 mg lysate.
28. We use 2,000 × *g* to spin down protein A Sepharose beads in all steps. Higher speed could cause the beads to collapse.
29. The pseudopodia or cell bodies were collected in 1% SDS lysis buffer, which is optimal for solubilizing the methanol fixed cells and to obtain the maximal yield of pseudopodia proteins. However, this lysis buffer is not compatible for experiments requiring non-denaturing condition such as IP. Therefore, we usually dilute the samples in 1% SDS buffer with a non-denaturing modified RIPA buffer to reduce the SDS concentration to 0.1% to make it compatible for IP experiments.
30. The pre-clearing step is necessary to remove some sticky proteins that may nonspecifically bind to the antibody-coupled protein A Sepharose beads.
31. This step is necessary to remove salts and residual SDS from the protein sample.

Acknowledgments

This work was supported by the Susan G. Komen Foundation Grant PDF0503999 (to Y.W.); National Institutes of Health Grants GM068487 (to R.L.K.), CA097022 (to R.L.K.); and Cell Migration Consortium Grant GM064346 (to R.L.K.).

References

1. Hunter, T. (2000) Signaling–2000 and beyond. *Cell* **100**, 113–127.
2. Linding, R., Jensen, L. J., Ostheimer, G. J., van Vugt, M. A., Jorgensen, C., Miron, I. M., et al. (2007) Systematic discovery of in vivo phosphorylation networks. *Cell* **129**, 1415–1426.
3. Miller, M. L., Jensen, L. J., Diella, F., Jorgensen, C., Tinti, M., Li, L., et al. (2008) Linear motif atlas for phosphorylation-dependent signaling. *Sci. Signal.* **1**, ra2.
4. Pawson, T., and Kofler, M. (2009) Kinome signaling through regulated protein-protein interactions in normal and cancer cells. *Curr. Opin. Cell Biol.* **21**, 147–153.
5. Pawson, T., and Nash, P. (2003) Assembly of cell regulatory systems through protein interaction domains. *Science* **300**, 445–452.
6. Cho, S. Y., and Klemke, R. L. (2002) Purification of pseudopodia from polarized cells reveals redistribution and activation of Rac through assembly of a CAS/Crk scaffold. *J Cell. Biol.* **156**, 725–736.
7. Wang, Y., Ding, S. J., Wang, W., Jacobs, J. M., Qian, W. J., Moore, R. J., et al. (2007) Profiling signaling polarity in chemotactic cells. *Proc. Natl. Acad. Sci. USA.* **104**, 8328–8333.
8. Cohen, P. (2000) The regulation of protein function by multisite phosphorylation–a 25 year update. *Trends Biochem. Sci.* **25**, 596–601.
9. Mann, M., Ong, S. E., Gronborg, M., Steen, H., Jensen, O. N., and Pandey, A. (2002) Analysis of protein phosphorylation using mass spectrometry: deciphering the phosphoproteome. *Trends Biotechnol.* **20**, 261–268.
10. Washburn, M. P., Wolters, D., and Yates, J. R., 3rd. (2001) Large-scale analysis of the yeast proteome by multidimensional protein identification technology. *Nat. Biotechnol.* **19**, 242–247.
11. Ge, L., Ly, Y., Hollenberg, M., and DeFea, K. (2003) A beta-arrestin-dependent scaffold is associated with prolonged MAPK activation in pseudopodia during protease-activated receptor-2-induced chemotaxis. *J. Biol. Chem.* **278**, 34418–34426.
12. Howe, A. K., Baldor, L. C., and Hogan, B. P. (2005) Spatial regulation of the cAMP-dependent protein kinase during chemotactic cell migration. *Proc. Natl. Acad. Sci. USA.* **102**, 14320–14325.
13. Jia, Z., Barbier, L., Stuart, H., Amraei, M., Pelech, S., Dennis, J. W., et al. (2005) Tumor cell pseudopodial protrusions. Localized signaling domains coordinating cytoskeleton remodeling, cell adhesion, glycolysis, RNA translocation, and protein translation. *J. Biol. Chem.* **280**, 30564–30573.

Part V

Cell Adhesion and Migration at Organismal Levels in a Physiological Context

Chapter 22

Overview: Studying Integrins In Vivo

Clifford A. Lowell and Tanya N. Mayadas

Abstract

Integrins are adhesive proteins that have evolved to mediate cell–cell and cell–matrix communication that is indispensable for development and postnatal physiology. Despite their widespread expression, the genetic deletion of specific integrin family members in lower organisms as well as mammals leads to relatively distinct abnormalities. Many of the processes in which integrins participate have a requirement for strong adhesion coincident with times of mechanical stress. In Drosophila, the absence of specific integrins leads to detachment of muscle from the gut and body wall and separation of the two epithelial layers in the wing. In mice and humans, a deletion of either subunit of the laminin-binding integrin, α6β4 leads to severe skin blistering and defects in other epithelial layers. In addition, integrins have also evolved to serve more subspecialized roles ranging from the establishment of a stem cell niche in Drosophila and mammals, to the regulation of pathogenic tumor vascularization, platelet adhesion, and leukocyte transmigration in mammalian systems. However, some cells seem to function normally in the absence of all integrins, as revealed by the very surprising finding that deletion of all the major integrin types on dendritic cells of mice has no effect on the ability of these cells to migrate within the interstitium of the skin and enter into lymphatics. In addition to serving as transmembrane mechanical links, integrins in vertebrates synergize with a number of receptors including growth factor receptors, to enhance responses. This leads to the activation of a large signaling network that affects cell proliferation and differentiation, as well as cell shape and migration. In vivo studies, in lower organisms, knockout mouse models as well as in inherited human diseases together have provided important insights into how this major, primordial family of adhesion receptors have remained true to their name "integrins" as their diverse functions have in common the ability to integrate extracellular stimuli into intracellular signals that affect cell behavior.

Key words: Cell adhesion, Extracellular matrix, Signal transduction, Drosophila, Gene knockout, Human integrin deficiency

1. Introduction

Integrins are expressed in all nucleated cells of multicellular animals and are essential for cell–matrix adhesion and, in vertebrates, cell–cell interactions. Since their recognition in 1987, they have been one of the most widely studied family of cell adhesion receptors.

Motomu Shimaoka (ed.), *Integrin and Cell Adhesion Molecules: Methods and Protocols*,
Methods in Molecular Biology, vol. 757, DOI 10.1007/978-1-61779-166-6_22, © Springer Science+Business Media, LLC 2011

Integrins contribute to development, hemostasis, the immune response, as well as diseases such as cancer and autoimmunity. In addition to their role as mechanical links, integrins are important conduits of bi-directional signaling in cells that influence processes from cytoskeletal arrangement and growth factor signaling to gene transcription (1). This overview will highlight a number of the major physiologic functions of integrins in both higher and lower organisms. Use of genetic approaches (i.e., gene knockouts) has provided a wealth of information about integrin function and will be emphasized in this chapter; given the significant published literature in this area we will focus on those examples in which specific functions have been revealed.

2. Integrin Genes in Vertebrates and Invertebrates

Integrins are αβ heterodimeric, transmembrane proteins that are restricted to metazoa. *Caenorhabditis elegans* has one β and two α subunits forming two integrins, Drosophila melanogaster also has orthologs of these primordial integrins referred to as PS1 and PS2. These two specialize in adhesion to basement membrane laminins, and recognition of Arg-Gly-Asp (RGD) peptides present in extracellular molecules such as fibronectin and vitronectin, respectively (1). Mammalian integrin α subunits can be classified into four subfamilies (Fig. 1). One group (α3, α6, and α7), which is related to Drosophila PS1, pairs mainly with the β1 subunit to form the major laminin receptors in mammals (Table 1). Another subgroup (αIIb, αv, α5, and α8), which is related to the Drosophila PS2 proteins, pairs mainly with β1 and β3 subunits to form the primary receptors for RGD-containing extracelluar matrix (ECM) proteins (Table 1). In mammals, this group of integrins plays a particularly important role in early embryonic development. In mammals, the RGD motif is also present in non-ECM proteins (e.g., fibrinogen, latent TGFβ), thus expanding the repertoire of ligands recognized by this integrin subfamily to allow development of novel cell-specific functions. This is evident in the PS2 subgroup whose members perform more specialized functions in bone, kidney, and platelets. For example, αIIbβ3 integrin on platelets recognizes a motif in fibrinogen similar to RGD to promote platelet adhesion. Although the PS1 and PS2 groups of integrins are present in many invertebrates, including *C. elegans*, the PS3-PS5 groups characterized in Drosophila seem to be specific for insects and may align with the laminin group when homology across the whole protein is compared. The third mammalian subgroup, the α4/α9 cluster is present only in chordates. These α subunits pair with β1 or β7 to form integrins that

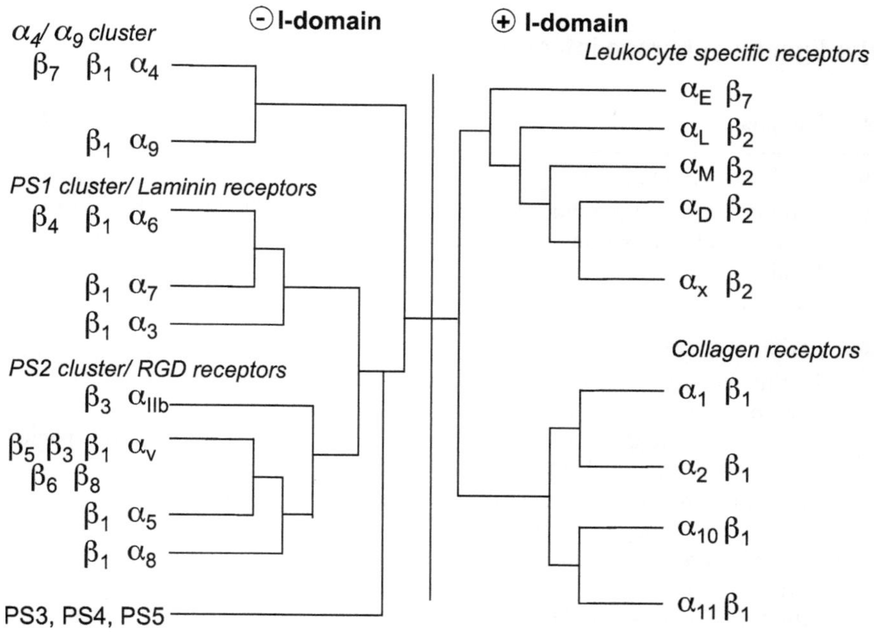

Fig. 1. The integrin family. 18 α and 8 β subunits form 24 distinct, functional integrin pairs in humans and most likely in other vertebrates. Based on their evolutionary lineage, these can be grouped into subfamilies. The α subunits are grouped into those recognizing laminin (PS1 cluster) or the classical RGD sequence (PS2 cluster), and are clearly evolutionarily ancient receptors for ECM proteins. There is also a cluster of α4/α9 integrins, collagen receptors and leukocyte-specific integrins. These subfamilies are further grouped into those containing the "inserted" I domain with structural similarity to the A domain found in von Willebrand factor. I-domain-containing integrins together with the α4/α9 cluster are restricted to chordates. These have evolved to bind counter receptors on vascular cells (e.g., ICAMs, VCAM), soluble molecules found in blood (e.g., fibrinogen, complement components) and pathogens (e.g., viruses and bacteria). While β1 and αv are ubiquitously expressed, β2 and β7 subunits are restricted to leukocytes. Drosophila evolved a small family of its own integrins, PS3-PS5 integrins for which there are no representatives in chordates. (Figure adapted from Johnson et al. 2009. Integrins during evolution: Evolutionary trees and model organisms. *Biochim. Biophys. Acta* **1788**, 779–789).

recognize a host of ECM proteins as well as certain plasma proteins, cell-associated counter receptors of the IgG superfamily such as VCAM-1 and vascular endothelial cell growth factors (2) (Table 1).

A structurally and functionally unique subgroup of integrins arose in chordates from the incorporation of an "inserted" domain or "αI domain," which is not found in invertebrates. This protein motif is also referred to as the "A" domain as it has structural similarity to a domain present in von Willebrand factor. The I (A) domain is present in proteins that are components of multiprotein complexes and promote cell adhesion frequently requiring divalent cations (3). The I domain contains a metal ion-dependent adhesive site (MIDAS) that comprises the ligand-binding domain of the integrin. Four of these integrins (α1, α2, α10, and α11, all of which pair with β1) are collagen receptors while the remaining five (αD, αE, αL, αM, and αX, which pair with β2 and β7) are integrins expressed on leukocytes. Counter receptors for the

Table 1
The major ligands of human integrins

Integrins	Ligands	Primary sites of expression
PS1 cluster –Laminin receptors		
α3β1	Laminin, thrombospondin, uPAR	Skin, kidney, lung
α6β1	Laminin, thrombospondin, ADAM, Cyr61	Macrophages, platelets
α6β4	Laminins	Skin (hemidesmosomes)
α7β1	Laminins	Muscle
PS2 cluster – RGD receptors		
αIIbβ3	Fibrinogen, thromobospondin, fibronectin, vitronectin, vWF, Cyr61, ICAM-4, CD40L	Platelets
α5β1	Fibronectin, osteopontin, fibrillin, thrombospondin, ADAM, COMP	Blood vessels (embryonic)
α8β1	Tenascin, fibronectin, osteoponitin, vitronectin, latent TGFβ, nephronectin	Kidney, inner ear
αvβ1	Latent TGFβ, fibronectin, osteopontin, vitronectin	Poorly defined
αvβ3	Fibrinogen, vitronectin, fibronectin, vWF, thrombospondin, fibrillin, tenascin, PECAM-1, osteopontin, ADAM, Cyr61, MMP, uPAR, uPA, ICAM-4	Osteoclasts, vascular endothelium
αvβ5	Osteopontin, vitronectin, latent TGFβ	Eye, bone
αvβ6	Latent TGFβ, fibronectin, osteopontin, ADAM	Skin, lung
αvβ8	Latent TGFβ, vitronectin	Vascular endothelium
α4/α9 cluster – chordate specific		
α4β1	Thrombospondin, MAdCAM, VCAM-1, ICAM-4, fibronectin	Heart (embryonic)
α4β7	MAdCAM, VCAM-1, fibronectin	Gut homing T cells (Peyers Patch)
α9β1	Tenascin, VCAM-1, osteopontin, ADAM, VEGF-C, VEGF-D	Lymphatic endothelium
I-domain-containing integrins		
α1,α2, α10, α11 cluster – collagen receptors		
α1β1	Collagens, semaphorin 7A	Fibroblasts, mesenchymal tissues
α2β1	Collagens, tenascin, E-cadhrin	platelets, epithelium, fibroblasts, mesenchymal tissues
α10β1	Collagens	Cartilage, chrondocytes
α11β1	Collagens	Peridontal ligaments
αD, αE, αL, αM, αX cluster – leukocyte adhesion receptors		
αDβ2	ICAMs, VCAM-1, fibrinogen, plasminogen	Eosinophils
αEβ7	E-cadherin	Skin/gut homing T and B lymphocytes
αLβ2	ICAMs	Leukocytes (all types)
αMβ2	ICAMs, iC3b, fibrinogen, heparin, factor X	Myeloid leukocytes
αxβ2	iC3b, fibrinogen, heparin, collagen, plasminogen, ICAM-4	Most myeloid leukocytes

leukocyte integrins are present mainly on vascular endothelial cells and include members of the IgG superfamily (ICAMs and VCAM-1), as well as the unrelated protein E-cadherin. The leukocyte integrins also recognize plasma proteins such as complement component iC3b and fibrinogen.

The integrin β subunits can also be classified into three major phylogenetic branches (Fig. 2). Two are represented in vertebrates (group A and B) while the third contains only invertebrate sequences. Group A contains β subunits β1, β2, and β7 that are associated with most of the α-subunits from the PS1, PS2, and I domain subgroups. Of these, β1 is the most widely expressed in many cell types, primarily because it associates with α1–α11 and αv subunits. The β2 and β7 subunits are restricted to hematopoietic cells. The β integrin subunits within Group B (β3, β4, β5, β6, and β8 proteins) tend to have a more restricted tissue distribution and associate with fewer α chains (Table 1). It is important to note that the production of α and β chains within any given cell type may not be balanced; often the widely expressed subunits such as αv or β1, which are promiscuous in their pairing with β and α subunits, respectively, are overproduced relative to other subunits. However, only the intact heterodimeric αβ integrins are found on the cell surface – unpaired α or β chains are retained in the endoplasmic reticulum and degraded. This restricts the combination of α and β subunits and thus the integrin repertoire on the surface (2). The repertoire of integrins can switch at the cell surface as a result of gene transcription that is in some cases regulated by

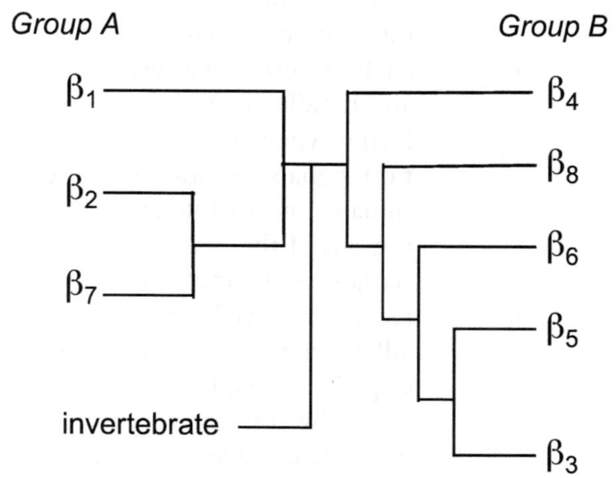

Fig. 2. Generalized organization of the integrin β subunits. The vertebrate β subunits fall into two major groups (A and B) while those present in invertebrates form a clearly distinct lineage. The Group A β subunits form most of the PS1, PS2, and I domain subgroups, integrins with β2 and β7 subunits being restricted to hematopoietic cells. The Group B β subunits have a more restricted tissue distribution and associate with fewer α chains. (Figure from Johnson et al. 2009. Integrins during evolution: Evolutionary trees and model organisms. *Biochim. Biophys. Acta.* **1788**, 779–789).

signaling events elicited by ligated integrins themselves. This in turn has effects on cell fate as distinct ECM components initiate integrin signaling events to promote proliferation vs. differentiation (4). The fact that only specific α and β chains pair in defined patterns suggests that these chains must have co-evolved, even though the genes encoding them are very different and are distributed throughout the genome. This co-evolution must have occurred with the development of novel major functional systems in which the integrins are required; for example, the development of the adaptive immune system necessitated co-evolution of αL, αM, αX, and β2 integrins to form the major receptors needed for guiding immune cells during host defense and inflammatory reactions (2).

3. Integrin Signaling Events: Inside-out and Outside-in

The ability of integrins to bind their ligands (whether ECM proteins or counter receptors on other cell types) is dynamically regulated. This allows cells to carefully control their ability to adhere to matrix or establish connections with other cell types. This property is obviously important during cell migration in development and on circulating immune cells that need to migrate out of the vasculature at sites of infection or injury to mediate host defense. The processes regulating integrin affinity changes are referred to as "inside-out" signaling (Fig. 3). Such events are best recognized for β2 and β3 integrins but are probably common to all integrin types. The leukocyte β2 integrins and platelet integrin αIIbβ3, reside in a resting state (bent/closed conformation) and are rapidly activated (extended/open conformation) to bind ligand when the cells that express these integrins receive activation signals. In addition to conformational changes, inside-out signals control integrin clustering and accumulation at specific regions of the plasma membrane. In leukocytes, these activating signals can be transduced by chemokine and other inflammatory receptors as well as receptors for members of the selectin family of adhesion molecules. In platelets, αIIbβ3 activation is triggered by G-protein-coupled receptors, by the von Willebrand factor receptor or by collagen receptors (such as GPVI) (1). Major insights into integrin activation have come from studies of integrin structure obtained by crystalization of the ligand-binding A domains. Additionally, the identification of various intracellular signaling proteins that associate with the integrin cytoplasmic tails have helped define some of the steps in integrin activation during inside-out signaling responses (1). Although integrin activation is best described in circulating immune cells, as it serves to localize the accumulation of cells to sites of inflammation, it is likely

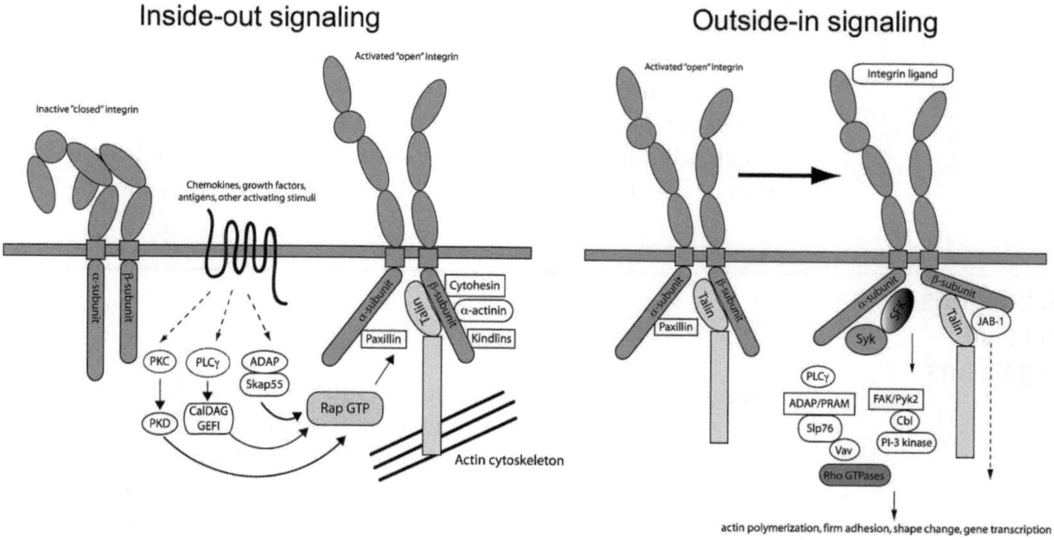

Fig. 3. Integrin signaling. In resting cells, integrins are held in a bent-inactive conformation (*left most diagram*). Following stimulation by any number of agonists (a generalized G-protein-coupled receptor is depicted as a curved line) a number of signaling proteins are engaged which allow the assembly of the signaling complex on the cytoplasmic tails of the both the α and β subunits (*left middle diagram*). This induces an unfolding of the extracellular portion of the integrin to form a high affinity receptor. Shown are a number of proteins involved in these signaling proteins as well as some of the structural proteins that couple the integrin and the actin cytoskeleton. This process is referred to as inside-out signaling events, since signals from inside the cell activate the integrin to a high affinity state. Following binding of the ligand to the integrin (shown in the *right panels*), additional signaling molecules are recruited to the tails of the integrins to lead to activation of downstream functional responses. Many of the molecules that are recruited include tyrosine kinases (Src-family kinases as an example, indicated as SFK) and other signaling adapters that are utilized for a number of different types of signaling pathways. Ligand binding also induces clustering of integrins in the plasma membrane (which is not shown for simplicity), with all these signaling molecules attached to form a macromolecular structure known as the focal adhesion. This entire process is referred to as outside-in signaling since signals from ligand binding to the integrin receptor are communicated into the cell to lead to functional responses, such as actin polymerization and changes in gene transcription. (Figure from Abram and Lowell 2009. The ins and outs of leukocyte integrin signaling. *Annu. Rev. Immunol.* **27**, 339–362).

that dynamic regulation of integrin affinity plays a central role in processes such as vascular development and neurite outgrowth (5).

Integrins are unique among transmembrane receptors in their ability to transmit bidirectional signals. Ligation of integrins by extracellular matrix proteins leads to the assembly of a number of cytoplasmic proteins that link the integrin to the actin and microtubule cytoskeletal network (Fig. 3). Active integrins in turn participate in the regulation of the cytoskeleton and reorganization of the extracellular matrix. Ligation of integrins also leads to the transmission of intracellular signals that modulate many aspects of cell behavior. This includes stimulation of cell cycle via ERK and cyclin D1 and others, inhibition of apoptosis via PI3-kinase, Akt; and NFkB; changes in shape, polarity, and motility via protein tyrosine kinases, phosphatases; and members of the Ras and Rho family of small GTPases. The activation of these signaling pathways, which are referred to as "outside-in" signal transduction events, also leads to the modulation of gene expression (1). Anchorage-dependent signaling through integrins is, in some cases, required for cellular responses to growth factors and other receptors. For example, in immune cells, αLβ2 integrin-mediated T-cell adhesion to antigen-presenting cells is required for induction of T-cell receptor-specific T-cell responses (6).

4. Integrins in Lower Organisms

Integrins are widely expressed in embryonic tissue of Drosophila and *C. elegans* and studies in these organisms have defined the minimal set of integrins required for fundamental processes in development. The two integrins identified in *C. elegans* correspond to the Drosophila PS1 (laminin integrin) and PS2 (RGD integrin). The two alpha subunits, αpat-1 and αpat-2 pair with a single β subunit, pat-3. Drosophila has 5 α-subunits PS1 through PS5, and two β subunits, βPS and Bv. βPS is most similar to both the *C. elegans* β subunit and β subunits found in vertebrates (7). The basement membranes in fruit flies are composed principally of four proteins, type IV collagen, laminin, nidogen/entactin, and the heparan sulfate proteoglycan perlecan. These have homologues in vertebrates suggesting that they form the basis for an early basement membrane. Other well-known vertebrate ECM proteins are absent including fibronectin, vitronectin, osteopontin, fibrinogen, and von Willebrand factor. These ECM proteins, important in vascular development and adhesion functions of blood cells, may have appeared later during the evolution of vertebrates (8).

Loss of function mutations of integrins or their extracellular matrix ligands in fruit flies and nematode worm leads to lethality that is attributed to defects in muscle attachment and contraction, cell migration, gut morphogenesis and adhesion between epithelial cell layers. Distinct roles for the RGD and laminin-binding integrins have been documented in some of these cases. For example, RGD-binding integrins are required for muscle contraction by organizing the actin–myosin contractile structure into sarcomeres in *C. elegans*, and attachment of the ends of the muscle to the ECM in Drosophila. Laminin-binding integrins promote differentiation of endodermal cells required for gut morphogenesis, with integrin deletion leading to misshapen pharynx in the mutant *C. elegans* larvae, and midgut in Drosophila mutant larvae (7). These examples also illustrate the point that despite the large differences between these two simple model organisms, the role of integrin/extracellular matrix interactions for fundamental processes during development are conserved. Roles of both RGD and laminin-binding integrins in post-embryonic development have been identified by generating organisms that are mosaic for homozygous mutant cells. Drosophila PS1 and

PS2 are present in the developing wing which is made of two layers of epidermal cells with PS1 on the dorsal and PS2 on the ventral side. Both integrins are required for providing adhesion between the basal surfaces of the dorsal and ventral epithelia. The complementary expression of integrins may be required to insure asymmetric adhesion and thus prevent wrinkling due to adhesion within the same layer of cells (7). In addition to maintaining structural integrity in animal tissues, laminin integrin anchoring to the ECM in Drosophila is essential for establishing a stem cell niche that in turn is important for the retention of a stable stem cell population. This integrin-mediated function, first established in Drosophila, has been shown to be conserved in vertebrates (9). Finally, studies in both flies and worms suggests that axon guidance is dependent on the laminin-binding integrins (10). The role of the other Drosophilia integrins, PS3-PS5 are not well described. Mutations in PS3 affect short-term olfactory memory (11) while the function of PS4 and PS5 remains unknown.

Focal contacts link integrins to the cytoskeleton in mammalian systems and contain cytoskeletal/adaptor/linker and signal transduction molecules such as talin, α-actinin, vinculin, paxillin, FAK, ILK, p130, and others. These proteins are well conserved in Drosophila and possibly *C. elegans* (8). Genetic deletion of several of these proteins has revealed their relative importance in integrin-dependent functions. For example, talin deficiency mimics integrin deficiency, whereas deletion of ILK results in a much weaker defect (12). The conservation of integrins, their cytoskeletal associated proteins and extracellular matrix ligands in flies, worms and people makes these genetically tractable lower organisms attractive for analysis of ECM–integrin–cytoskeleton connections. Also, the presence of the aforementioned signaling molecules as single genes in flies and worms reduces the possibility of redundancy and thus functional overlap in the system (8). However, it is also evident that vertebrates have evolved proteins with more specialized functions than those found in flies and worms. This is particularly evident in vascular biology, some aspects of neurobiology and likely other uniquely vertebrate functions requiring multicellular adhesive interactions. The advent of genetic engineering in mice has provided important insights into the specialized, nonredundant roles that integrins and their interacting proteins, have developed in higher organisms.

5. Integrins in Mammalian Development and Pathophysiology

All the individual integrin subunits have now been deleted in mice and the reported phenotypes highlight the nonredundant, specialized functions that integrins play in vivo (13). Embryonic lethality upon genetic deletion of some of the integrins may have been predicted from the known expression pattern of the integrin during development. Additional information about the role of these integrins in adult mice will be undoubtedly forthcoming as more conditional knockout animals are generated – we have seen this especially with β1 integrins as outlined below. The phenotypes of knockout mice lacking α or β subunits are grouped according to the organ system most affected by their deletion.

5.1. Vasculature

A number of integrins play central roles in vascular formation during fetal and embryonic development, as well as regulating blood vessel integrity, in adult animals (14, 15). This function is served through both binding of ECM proteins by vascular endothelial precursors (which provides "outside-in" signals to the cells to guide their migration during vascular branching) and through the ability of integrins to specifically bind a number of growth factors (in particular, vascular endothelial growth factors – VEGFs) essential in blood vessel development. Integrin-mediated adhesion events also play a critical function in pathologic angiogenesis, for example, in response to tumor development or chronic inflammation. Most of the integrins involved in vascular development and integrity are members of the PS2 family that recognize RGD peptide motifs in ECM proteins. Vasculogenesis is completely dependent on α5β1 and its ligand fibronectin (16). Mutation of α5 leads to early embryonic lethality characterized by mesodermal defects and poor vascularization of both the yolk sac and the embryo itself (17). This phenotype is similar, but not as severe as that observed in mice lacking fibronectin (18). Mutation of β1 has an even more profound effect, as expected due to the loss of many integrins, which manifests as gastrulation defects and pre-implantation mortality (19). Interestingly, the phenotype of Tie2-Cre; $\beta 1^{flox/flox}$ mutants, in which all β1-integrins are deleted in endothelial cells only, closely resembles the phenotype of α5 mutants, suggesting that endothelial-expressed α5β1 is the dominant integrin that recognizes fibronectin during vascular morphogenesis (20). In contrast, during pathologic vascularization αvβ3 integrin appears to play a central role. This integrin is highly expressed in blood vessels at sites of tumor growth and wound healing (21). Indeed, treatment of animals with anti-αvβ3 mAbs blocks neovascularization in tumor models, (22). In this role it appears that it is the ability of αvβ3 to bind VEGF and even to associate with the VEGF receptor that mediates its role in pathogenic neovascularization. In tumor cells, expression of αvβ3 correlates with overproduction of VEGF (23) while, probably more importantly, in endothelial cells αvβ3 directly associates, and synergizes with VEGF receptor signaling to promote enhanced responses to this growth factor (24, 25).

The physiologic association of integrins with vascular growth factors is also revealed in the phenotype of α9 mutant mice.

Surprisingly, deficiency of α9 produces a lymphatic vascular defect that results in disordered lymphatic development and perinatal lethality due to accumulation of lymph fluid in the thorax (26). Lineage-specific ablation of α9 in blood vessels (using $\alpha9^{flox/flox}$; VE-cadherin-Cre mice) revealed that α9β1 plays a critical role in assembly of fibronectin bundles in the valves of developing lymphatic vessels, which is required to prevent leakage of lymph fluid (27). However, integrin α9β1 is also the major binding protein for VEGF-C and VEGF-D, which together are the two primary growth factors for lymphatic vessels (28). Hence, much of the lymphatic phenotype of α9-deficient mice can be ascribed to loss of VEGF-C and VEGF-D signaling events in the lymphatic precursors.

Deletion of another major group of integrins, those containing the αv subunit, reveals the importance of integrins in organizing the parenchyma surrounding the vasculature. Mice lacking αv are embryonic lethal due to widespread vascular malformations and hemorrhage. However, these arise later in development (after e10.5) and affect primarily the brain in the embryo (29) due to defective interactions of cerebral vessels with surrounding brain parenchyma (30). The specific deletion of αv in endothelium has no effect on vascular development, which is consistent with its relative absence in this microvasculature, whereas its deletion in neural cells, particularly glia, leads to cerebral hemorrhage (31). Of the αv integrins, it appears that αvβ8 is the primary heterodimer involved in this process, since mice lacking β8 also have cranial vascular defects and intracerebral hemorrhage (32). Indeed, targeted deletion of β8 within neuroepithelium ($\beta8^{flox/flox}$; nestin-Cre mice) is sufficient to produce the late developmental cranial hemorrhage phenotype (33). Disruption of α4 also has a later vascular phenotype resulting from failed fusion of the allantois with the chorion during placentation, as well as impairments in cardiac development (34, 35). In this case, the ligand is likely VCAM-1, as mice lacking this member of the IgG superfamily exhibit a similar phenotype to the α4-deficient mice (36).

5.2. Ocular and CNS

A number of studies have demonstrated the critical function of integrins in ocular and CNS development and hemostasis. Within the eye, lens development is best studied, since this tissue consists of relatively simple assemblies of fibronectin, collagen, and laminin (37). Integrins that recognize these ECM proteins have all been implicated. Of them, clearly the β1-containing heterodimers are most important, since mice lacking β1 specifically in the lens ($\beta1^{flox/flox}$; MLR10-Cre animals) are microphthalmic due to apoptosis of the lens epithelium and disintegration of the lens fibers (38). However, whether this major defect is due to loss of fibronectin receptors (such as α5β1) vs. collagen receptors (α1β1) or laminin receptors (α3β1) remains to be seen and will only be sorted out using additional lineage-specific mutagenesis methods.

In lens development, there is also clear evidence of compensation by different integrins within the same class. For example, absence of the α3 or α6 integrins (which when paired with β1 form the major laminin receptors in the developing lens) by themselves have little effect on lens development. In contrast, α3/α6 double-mutant mice show severe dysmorphogenesis of the developing lens (39).

The involvement of integrins in retinal development more closely mirrors their overall contribution to CNS formation, mainly as related to their role in guiding the migration of neuronal precursor cells (40). Though early studies (using blocking mAbs) had suggested members of the β1 integrins play a dominant role in guiding neuronal migration along glial fibers in the developing brain, more modern approaches with conditionally deficient mice have revealed more subtle functions. Specifically, mice lacking the β1 subunit in neurons only ($\beta1^{flox/flox}$; nestin-Cre animals) actually show relatively normal neuronal cell migration but have a completely disorganized cortex as a result of a defect in association between cortical structures and the developing meningeal basement membrane (41). This phenotype resembles the CNS cortical defects seen in mice lacking the α2 chain of laminin (42), suggesting that a laminin β1 integrin receptor (such as those in the PS1 family) is involved. Integrins and other cell adhesive molecules also play a major role in stabilizing neuronal synapses (43). Loss of this stabilizing function, as evidenced primarily by β3 integrin-blocking studies, results in neuronal dysfunction that manifest as defects in long-term potentiation necessary for learning and memory (40). It is likely that other axonal guidance molecules, such as the netrins, semaphorins, and ephrins play the dominant role in guiding neuron migration during development (44).

In contrast to these findings in the CNS, there is better evidence for the involvement of integrins (in particular, β1 family members) in neural crest migration during development of the peripheral nervous system. Mice with conditional deletion of β1 integrins in neural crest cells ($\beta1^{flox/flox}$; Ht-PA-Cre animals) exhibit severe perturbations of the peripheral nervous system, including failure of normal nerve arborization, delay in Schwann cell migration, and defective neuromuscular junction differentiation, all of which can be attributed to defective migration of these cells through the embryonic ECM (45). The β1 integrins also directly affect peripheral nervous system Schwann cells, as conditional deletion of β1 in these cells ($\beta1^{flox/flox}$; P_0-Cre mice) results in reduced survival, proliferation, and differentiation of Schwann cell precursors (46).

The RGD-binding integrin αvβ5 plays a critical role in maintenance of retinal pigment epithelial cells in the eye. Mice lacking the β5 integrin slowly develop age-related blindness due to progressive accumulation of shed pigment from retinal cells. The retinal pigment epithelial cells use the αvβ5 integrin as a phagocytic receptor to mediate uptake of shed photoreceptor particles, which occurs as a natural dinural process in the rods and cones of the retina (47).

Integrins clearly play a role in a number of neuroinflammatory diseases, such as multiple sclerosis. This likely reflects their major function in regulating immune cell trafficking in the body. There is also suggestive evidence that neuronal survival in degenerative diseases such as Parkinson's disease or Alzheimer's disease may be influenced by integrin signaling (44). This has raised interest in integrin-targeted therapeutics as a means to treat neurologic disease.

5.3. Immune

The contribution of integrins to immune system (and general hematologic) function has been extensively studied and in fact serves as the paradigm for our understanding of many facets of integrin biology. The involvement of integrins in regulating neutrophil, monocyte, and lymphocyte trafficking throughout the immune system has been well studied both in vitro and more recently in vivo using a variety of intravital microscopy methods. Furthermore, the overall structural changes that take place during integrin activation have been best characterized in the leukocyte and platelet integrins, allowing a direct, but as yet incomplete understanding of how the protein associations involved in the "inside-out" signaling response mediate integrin subunit unfolding. The primary integrins expressed by immune cells are α4β1, α5β1, all of the β2 integrins, αvβ3, and αEβ7. The primary platelet integrin αIIbβ3 is also found on some immune cells.

The most well-appreciated role for integrins in the immune system is their function in guiding leukocyte exit from the vasculature into the tissues, either within lymph nodes or directly into sites of tissue injury or inflammation (Fig. 4). The process of leukocyte diapedesis from the vasculature is termed the leukocyte adhesion cascade (48). Broadly, this cascade involves three general classes of molecules – the selectins, chemokines, and integrins. The selectins are carbohydrate-recognizing receptors, present both on the leukocyte and vascular endothelium, that allow initial tethering but not full adhesion of the blood cells to endothelial cells. As a result, selectin interactions result in rolling of the leukocyte along the vascular wall. During this process, the leukocytes are exposed to chemokines and other inflammatory mediators, which induces "inside-out" signaling responses leading to integrin affinity modulation and binding to ligands on the surface of activated endothelial cells. This promotes cell arrest. The regional chemokines facilitate the directed, local egress of these cells into tissues. This simple three-step model – rolling mediated by selectins, activation mediated by chemokines, and full arrest mediated

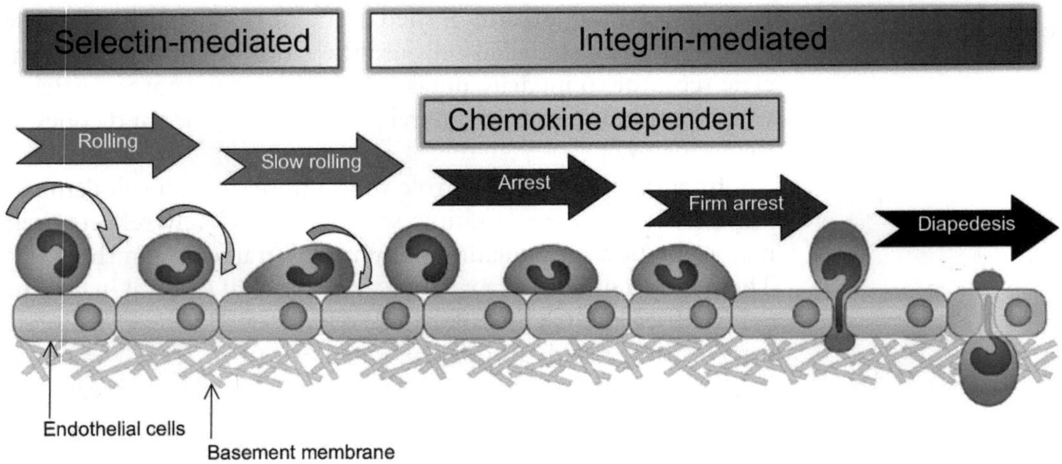

Fig. 4. Integrin-dependent leukocyte recruitment. The steps in leukocyte rolling, arrest, firm adhesion and diapedesis out of the vasculature and into the tissues are schematized. As shown, the initial steps of rolling and slow rolling are largely selectin mediated; however, a number of integrins (such as α4β1) contribute to rolling events. Following cellular activation (which leads to inside-out signaling events and integrin affinity upregulation), leukocytes become firmly adherent to the endothelium and begin the process of crossing through (or around) the vascular endothelial cells. Integrin outside-in signaling events are important in mediating the actin cytoskeletal changes that are needed for cellular shape change during these steps. Direct interactions between the leukocyte and the basement membrane also guide the leukocyte out of the vasculature. (Figure from Ley et al 2007. Getting to the site of inflammation: the leukocyte adhesion cascade updated. *Nat. Rev. Immunol.* **7**, 678–689).

by integrins – is an oversimplification, as there are examples of leukocyte integrins (such as α4β1) that can mediate rolling (49) or which reduce the velocity of selectin-dependent rolling. Similarly, there are many additional steps following firm adhesion, such as adhesion strengthening and signaling events (50) leading to leukocyte shape change that are mediated by integrins. Despite this complexity, the overall role of integrins in guiding leukocyte trafficking out of the vasculature is extremely well supported by the consequences of deficiency of leukocyte integrins. Mice deficient in β2 show profound defects in migration of all leukocyte types, which results in profound T-cell defects, impaired neutrophil recruitment in response to infection and resulting immunodeficiency (51). As expected, these mice show significantly reduced responses to a number of inflammatory stimuli and are protected from immune-mediated tissue damage in a number of disease models, such as cardiac reperfusion injury (52). The susceptibility these mice manifest to infection is similar to that seen in human patients lacking the β2 integrin – a disease referred to as Leukocyte Adhesion Deficiency syndrome (discussed below). Lack of β1 integrin on blood cells affects their migratory capacity, resulting in a failure of fetal hematopoietic stem cells to colonize the fetal liver or spleen during development (53). In mature cells, β1 plays a critical role in guiding lymphocyte migration to the skin, lung, peritoneum, and liver, all primarily through the actions of α4β1 (54). Finally, lack of the β3 integrin results primarily in a loss of platelet function (with subsequent hemorrhage), mainly through absence of the αIIbβ3 family member (55). This phenotype also mirrors what is seen in humans lacking this integrin, which develop a disease referred to as Glanzmann's thrombasthenia (discussed below).

Though there is clear evidence that integrins play a role in tissue-specific recruitment of immune cells during inflammatory or disease processes – for example, the specific involvement of αEβ7 integrin in the migration of T cells to epithelial tissues such as the skin and gut (56) – it is too simplistic to assume that integrins alone play the dominant role in all immune cell trafficking. Indeed, the relative role of integrins in leukocyte migration in some inflammatory responses may be quite minor. This extremely surprising suggestion has emanated from studies of mice lacking ALL the major leukocytes integrins – generated by conditional and inducible deletion of β1, αV combined with complete deletion of β2 and β7 in adult mice (57). Dendritic cells lacking all these adhesion molecules migrate normally into lymph nodes where they present antigens to stimulate immune responses. This indicates that migration through interstitial spaces may be less dependent on integrins than was originally envisioned and instead these molecules may have a dominant role only in allowing leukocyte exit from the vasculature. Whether this general conclusion applies to other immune cells besides dendritic cells remains to be examined. However, it is provocative studies such as this that continue to redefine our understanding of the role of integrins in the immune system.

Beyond their role in guiding cell migration, it is also clear that leukocyte integrins play a central role in activating the effector functions of immune cells through "outside-in" signaling (50). Often this signaling works with other receptors engaged on the cell. For example, co-stimulation of T cells through their antigen receptors (TCR) and integrins (e.g., when the cells are plated on a matrix-coated surface) leads to much more robust proliferative responses than stimulation through the TCR alone (58). Similarly, chemokine activation of neutrophils is dramatically enhanced in cells that are also stimulated through their integrin receptors (59). In some cases, the activation function of the leukocyte integrin is relatively more important than its homing/migration function. For example, in a model of thrombohemorrhagic vasculitis caused by endotoxin and cytokine exposure, deficiency of αMβ2 integrin does not affect the ability of leukocytes to be recruited to the site of vascular inflammation, but once there the cells cease to become activated because they do not recognize complement deposits that form along the vessel walls. As a result, αMβ2-deficient neutrophils fail to release proteases that otherwise would cause vascular damage and hemorrhage (60). This clear role for "outside-in" signaling events is also seen in other physiologic conditions, such as keratinocyte development as outlined below.

5.4. Cutaneous

A number of integrins play a key role in the development and maintenance of epidermal structures. Their primary function, along with other adhesion molecules, is to link keratinocytes to the underlying basement membrane (61). The primary integrins expressed in the epidermis are the laminin receptors (α6β4 and α3β1), the collagen receptor (α2β1) and less abundantly the vitronectin receptor αvβ5 (62). Most of these integrins are localized in focal adhesion contact sites in keratinocytes, where they link the extracellular matrix to the actin cytoskeleton of the keratinocyte to provide firm adhesion. The α6β4 integrin is localized to hemidesmosomes that help link keratinocytes to each other directly.

As in other organ systems, the role these individual integrins play in epidermal biology is best revealed in knockout mouse models. Conditional deletion of all β1 integrins in the epidermis ($\beta 1^{flox/flox}$; Keratin5 or Keratin14-Cre mice) leads to incomplete perinatal mortality due to separation of the epidermis from the underlying dermis (63, 64). Some mice survive up to 6 weeks after birth but show an absence of hair follicles, poor keratinocyte proliferation with fibrous dermal deposition associated with chronic skin inflammation. Interestingly, deletion of the α2β1 integrin or the α3β1 integrin alone (through deletion of the α subunits) has very little effect on skin biology – the loss of α3 produces mild foot blisters – suggesting that the epidermal function of these receptors is largely redundant (65, 66). Most interesting, however, is when the β1 integrin is deleted in adult epidermis, through use of an inducible K14-cre strain. In this case, if the skin is already formed, loss of β1 integrins results in an enhanced proliferative responses of keratinocytes to spontaneous wounding. In addition, the number of melanocytes in the interfollicular epidermis is greatly increased (62). This observation highlights the role of integrin "outside-in" signaling in regulating keratinocyte growth, in this case in a suppressive fashion. Consistent with these observations, knockdown of β1 integrins in mouse keratinocytes in culture leads to increased proliferation and differentiation (67). These results also illuminate the issue of secondary effects of these deletions, such as the cutaneous inflammation that occurs in newborn mice lacking epidermal β1 integrins, which may complicate interpretations of primary effects on the skin. Such is the case in mice lacking the α6β4 integrin, which develop epidermal and dermal separation, similar to that observed in humans born with β4 deficiency (referred to as junctional epidermolysis bullosa syndrome, see below). In this case, alterations in TGFβ responsiveness in the absence of these integrins may contribute to inflammatory responses in the skin that contribute to the epidermal/dermal separation. However, unlike the β1 integrins, it seems that α6β4 does not directly regulate keratinocyte differentiation (68).

5.5. Pulmonary

In terms of development, many of the same integrins that contribute to cutaneous epithelial cell morphogenesis are involved in formation of the pulmonary epithelial layer. This is particularly true for the α3β1 laminin receptor. Integrin α3-deficient mice die at birth due to renal and lung defects associated with disorganized subepithelial basement lamina and poor adhesion of epithelia to the basement lamina (69). This phenotype is, in a sense, a more profound manifestation of epithelial dysfunction than the modest skin blistering in these animals. Combinatorial mutants that more broadly affect laminin binding, such as double-mutant α3/α6 mice have even a more severe epithelial phenotype that results in disordered limb development and severe pulmonary hypoplasia (39).

However, it is loss of the RGD-binding integrin functions, in particular that of αvβ6 and to a lesser extent αvβ5 that result in the progressive pulmonary abnormalities, which have taught us the most about the intersection of integrins and TGFβ regulation. The αvβ6 integrin is expressed primarily in epithelial cells of lung, skin, and kidney where it recognizes both fibronectin, but more importantly the latency-associated peptide (LAP) that is noncovalently associated with newly secreted TGFβ and keeps TGFβ from binding its receptor (70). The LAP of both TGFβ1 and TGFβ3 has an RGD-binding motif, which is recognized by αvβ6 and αvβ8. When these integrins bind the TGFβ LAP, they cause dissociation of LAP/TGFβ complex allowing TGFβ to interact with its receptors present on epithelial cells and resident pulmonary macrophages. Binding of TGFβ serves as a powerful anti-inflammatory stimulus in both cell types – the presence of activated TGFβ6 in the lung helps suppress inappropriate alveolar macrophage activation that would lead to lung injury (71). This amazing biology was revealed by study of the β6 knockout mice. These animals develop progressive pulmonary (and skin) inflammation that is strikingly similar to that seen in TGFβ mutant mice themselves (72). By age 6 months, β6-deficient mice develop emphysema, which is due to overproduction of the alveolar macrophage protease MMP12, as a result of the chronic inflammation present in these mice (73). Re-expression of either β6 or an activated version of TGFβ as a transgene in alveolar and bronchiolar epithelial cells of β6 mutants, rescues the pulmonary inflammation (73). Integrin regulation of TGFβ is also essential in limiting MMP12 release after bleomycin or irradiation-mediated lung damage (74). These phenotypes point to the importance of αvβ6-mediated TGFβ activation in the lung to restrict the inflammatory properties of resident alveolar macrophages.

Interestingly, this modulatory effect of TGFβ on macrophages also affects their phagocytic function. Alveolar macrophages play a major role in recycling pulmonary surfactant; however, this function is lost in both β6 and TGFβ-deficient mice, which is manifested as a progressive accumulation of surfactant material in addition to the inflammation that these mice develop (75). Hence, as seen in other macrophage types, the transition of alveolar macrophages to the inflammatory state (leading to release of MMP12 and other mediators) correlates with the loss of the ability to phagocytose and degrade normal resident proteins.

Pulmonary activation of TGFβ is also mediated by αvβ8 integrin. This may play more of an effect on regulating inflammatory and proliferative responses of airway epithelial cells, vs. macrophages, as determined in cell or organ culture models using functional blocking β8 mAbs (76). Complete loss of β8 results in embryonic lethality due to early vascular malformations; a function that is likely not related to the ability of this integrin to activate TGFβ (32).

Studies of αvβ5 in pulmonary vascular permeability during lung injury have also revealed novel functions for this integrin in the lung. Mice lacking this integrin (or normal animals treated with a β5-blocking mAb) manifest significant reduction in pulmonary edema and vascular permeability following either ischemia-reperfusion or hyperventilation (77). The effect was traced to the ability of αvβ5 to bind and activate VEGF, which in turn directly stimulated the vascular permeability. Indeed, direct treatment of pulmonary epithelial cells with VEGF will induce actin stress fiber formation and cell retraction – a phenotype not observed in αvβ5 mutant cells. Hence, as in vascular development, the ability of integrins to bind and activate vascular growth factors reflects an important function beyond just cell adhesion.

5.6. Musculoskeletal

As in the fruit fly and the worm, a number of RGD, collagen, and laminin-binding integrins serve critical roles in formation of muscle structures. In developing myofibrils, organization of integrin subunits into distinct focal adhesion sites in the sarcolemma allows for nucleation of actin filaments that help organize the actin/myosin structures (78). However, the individual integrins that contribute to this phenotype are less well defined, since loss of specific collagen receptors (α1β1 or α2β1) result in mild musculoskeletal defects. In addition to its role in vascular development, the RGD-binding integrin α5β1 also plays a role in maintenance of muscle structure, as determined by examination of embryonic chimeric mice containing cells derived from both normal and α5-deficient stem cells. These mice develop a progressive skeletal muscular dystrophy phenotype, characterized by diffuse muscle cell degeneration and apoptosis (79). This phenotype seems not to be due to a developmental disorganization of the actin/myosin structure of the muscle cell, but instead to a loss of adhesive signaling between muscle cells and the ECM, which results in poor myoblast survival over time. This illustrates another well-accepted role for integrins in promoting cell survival. Indeed disruption of integrin-mediated anchorage to the ECM leads to apoptosis, a process referred to as "anoikis." Integrin "outside-in" signaling to the PI3K-Akt pathway may be a major regulator of anoikis, with a breakdown of this process in epithelial cells contributing to neoplasia and metastasis (80).

Of the laminin-binding integrins α7β1 clearly has the dominant role in orchestrating normal muscle development. Dystrophin, utrophin, and α7β1 are the major laminin receptors in skeletal muscle. Each of these receptors links laminin in the extracellular matrix and basal lamina to the myocyte actin cytoskeletal network (81). Loss of dystrophin produces human Duchenne's muscular dystrophy, which is characterized by progressive myocyte degeneration that leads eventually to mortality. Mutations in α7 also produce myopathy in humans, though not nearly as severe as Duchenne's dystrophy (82). Mice that lack α7 integrin also develop myopathy, characterized by poor force transmission, compliance, and viscoelasticity in diaphragm muscle (83, 84). In addition, α7β1 plays an important role in maintaining the neuromuscular junctions and muscle–tendon junctions as these structures specifically degenerate in the α7-deficient mice (85). Interestingly, in both Duchenne's patients and the mouse model of this disease (the *mdx* mouse), the levels of integrin α7β1 increase two- to threefold, suggesting that these laminin receptors can functionally compensate for each other. Indeed, in combinatorial mice that lack dystrophin and α7 or utrophin and α7, the myopathy phenotype is dramatically accelerated (86, 87). These observations clearly indicate that these laminin-binding receptors are dispensible for muscle development, but play a critical role in maintaining the structural integrity and functional capacity of skeletal muscle, primarily by linking myocytes to the ECM.

A number of both RGD and collagen-binding integrins play major roles in skeletal development and homeostasis. The most extensively studied is the integrin αvβ3, which is abundantly expressed in osteoclasts and plays a central role in the ability of these cells to adhere to bone matrix and regulate bone morphogenesis. In normal bone, the primary RGD-containing ligands recognized by osteoclast-expressed αvβ3 are osteopontin and bone sialoprotein (88). Adhesion of osteoclasts to these surfaces through αvβ3 causes actin cytoskeletal rearrangement in the osteoclast, spreading, and activates the cell to release proteases into a defined vacuolar space between the osteoclast and the bone, thus causing bone resorption and remodeling. In mice lacking the β3 integrin, the loss of osteoclast function results in bone overgrowth (osteopetrosis); however, in humans lacking β3 (i.e., those with Glanzmann's thrombasthenia) bone density is unaffected.

This may be due to increased expression of compensating β1 integrins in the human patients (89). In the mouse, it is also clear that loss of the "outside-in" signaling events that are elicited by engagement of osteoclast αvβ3 produce the functional equivalent of loss of the receptor. Hence, deficiency of members of the Src-family of tyrosine kinases, Syk kinase, or downstream adapter proteins such as DAP12, or SLP76 or a guanine exchange factor for Rac GTPases, Vav3 all result in varying degrees of osteoclast dysfunction and, in the case of Src kinase mutant mice, severe osteopetrosis resulting in lethality (88, 90, 91).

The collagen-binding integrins α10β1 and α11β1 also play specific, and rather limited, functional roles in musculoskeletal development. Mice lacking α10 develop growth retardation of the long bones in the limbs due to disorganized cartilage formation particularly at the growth plates (92). The chondrocytes in these mice manifest a disordered columnar arrangement, an abnormal shape, and reduced proliferative capacity. As a result, the collagen fibrillar network in the growth plates of α10-deficient animals is disorganized, leading to the growth plate failure and shortened bones. Integrin α11 plays a specific role in dental development through its regulation of periodontal ligament fibroblasts. The α11-deficient mice show severe defects in incisor formation characterized by disorganized periodontal ligaments leading to a block in tooth eruption (93). Embryonic fibroblasts from the α11-deficient mice show reduced cell adhesion and spreading on collagen I as well as reduced cell proliferation. This suggests that α11β1 is specifically required on periodontal ligament fibroblasts for cell migration and collagen reorganization to help generate the forces needed for axial tooth movement.

5.7. Renal

A number of β1 integrins have been implicated as having specific roles in renal development and kidney disease. These functions range from roles in the development of the entire kidney to specific effects on the epithelial structures within the collecting system or in responses of the glomerulus to injury. Studies of α8β1 have revealed its unique role in renal development. Loss of α8 results in a high frequency of complete renal agenesis due to a failure of epithelial–mesenchymal interactions; epithelial-type cells referred to as ureteric bud cells (UB) fail to invade into surrounding mesenchyme to form the metanephric kidney (94). Although α8β1 recognizes a number of RGD-containing proteins, such as fibronectin and tenascin C, in fact the ligand present in the developing kidney which activates α8 function is a novel ECM protein referred to as nephronectin (95). Nephronectin is expressed by the invading UB cells and induces expression of α8β1 in the surrounding mesenchymal cells. Indeed, genetic disruption of nephronectin results in the same type of renal agenesis as seen in α8 mutant animals (96). In this system, it is clear that the nephronectin/α8β1 interactions not only provide adhesive structures to guide invading kidney epithelia, but also are involved in induction of growth factors, specifically glial cell line-derived neurotrophic factor (GDNF), a member of the TGFβ superfamily. GDNF binds to its receptor, the RET tyrosine kinase on the epithelium, to provide growth and proliferative signals to the developing renal epithelial cells (97). Loss of any of these molecules also produces a renal agenesis phenotype similar to α8β1 deficiency. This system of interactions of different cell types in the developing kidney highlights the importance of integrin/ECM interactions in controlling organ morphogenesis by regulating cell growth, migration, and adhesion.

Integrin α3β1 also has a dominant role in renal development. Mice lacking the α3 subunit have a reduced number of collecting ducts in the renal papilla suggesting that this integrin plays a role in the subsequent steps of branching morphogenesis to form the renal tubular collecting system (69). When α3 is specifically deleted in the UB cells ($\alpha3^{flox/flox}$; HoxB7-Cre mice), the kidney papillae are either absent or abnormal (98). As in the earlier steps of renal development, it is clear that α3 interactions in the developing papilla also induce expression of growth factors required for organogenesis. In this case, elaboration of the growth factor Wnt7B following α3 engagement helps regulate epithelial cell survival in the developing tubules to allow formation of the mature renal papilla.

Beyond α3β1 and α8β1 it is likely that other laminin, collagen, and RGD-binding integrins play supportive roles in kidney development. This conclusion derives from deletion of β1 specifically in the UB cells ($\beta1^{flox/flox}$; HoxB7-Cre mice). In this case, a severe defect in branching morphogenesis is observed, worse than just the α8 effect alone (99). Interestingly, when β1 is lost later in renal development ($\beta1^{flox/flox}$; Aqp2-Cre mice) kidney formation is normal; however, the mutant animals develop more severe injury and poor collecting duct healing following ureteric obstruction. In this case, it is likely that the injured renal collecting tubular cells fail to produce GDNF and other growth factors that would support repair of kidney structures following injury (99). Again, this highlights the cooperation of integrins in growth factor responses within the kidney.

It is also clear that β1 integrins play important roles in the development and maintenance of the glomerular structures of the kidney. In the glomerulus, the interactions between mesangial cells and podocytes with the glomerular basement membrane present in the fine vascular capillaries of the glomerulus forms the basic filtration structure of the kidney. These interactions are dependent on β1 integrins, since specific depletion of β1 in podocytes ($\beta1^{flox/flox}$; podocin-Cre mice) leads to severe malformations of the glomerular basement membrane and effacement of podocyte/membrane contacts (100). As a result, the mice are born with massive proteinuria and rapidly progress to death. In the normal glomerulus, continued expression of β1 integrins is important in maintaining the filtration function of the kidney. During glomerular injury, for example, as a result of immune complex deposition, β1 integrins of all types are induced in mesangial cells and podocytes (101). These interactions lead to elaboration of growth factors that can contribute to mesangial cell proliferation during glomerular injury. Indeed, mice lacking α1β1 integrin show protection from glomerular injury and reduced mesangial cell proliferation in a mouse model of Alport's syndrome (102).

6. Human Integrin Deficiencies

Clear support for the importance of integrins and their activation have come from "experiments of nature." Human mutations in integrins within 3 of the 5 mammalian subgroups have been shown to cause disease (103). These include mutations in either the α or β subunits of integrins α6β4 (PS1 cluster, laminin receptors) or αIIbβ3 (PS2 cluster, RGD receptors) or in the β2 integrin subunit (with the I-domain-containing leukocyte receptors). In the case of the β2 and β3 mutations, multiple integrins are affected, which contributes to the variability in the disease phenotype from patient to patient.

Reduced or absent expression of α6β4 leads to the syndrome of junctional Epidermolysis Bullosa (EB) which presents as extreme blistering and fragility of the skin. The skin blistering is caused by the inability of epidermal cells to adhere to basement membranes present in the dermis, leading to epidemeral/dermal separation. The EB syndrome is accompanied by an occlusion of the intestine, termed pyloric atresia, which is often postnatally lethal. The occluded oesophageal, urethral, and pyloric tissues may result from the detachment and subsequent fusion of epithelial linings (103, 104). Mice lacking either α6 or β4 have a very similar epidermal phenotype and suffer perinatal lethality.

Genetic mutations in αIIb or β3 produce a syndrome referred to as Glanzmann's Thrombasthenia. These patients present with a range of problems, the most serious of which is a severe bleeding disorder due to platelet adhesion defects (105). Individuals with β3 mutations will often have bone cysts or other skeletal abnormalities, likely due to loss of αvβ3 function. Mutations in β3 can lead to little synthesis of the protein or allow full expression but no activity of integrin. The mice lacking β3 or αIIb serve as good models of Glanzmann's thrombasthenia (106).

Leukocyte adhesion deficiency I (LADI) syndrome is caused by genetic mutations within the β2 subunit, resulting primarily in deletion or reduction in surface expression of multiple leukocyte integrins or in some cases, leading to near normal levels of integrins but defects in integrin function due to mutations in the ligand-binding domain. These patients present with a variety of immune deficits, ranging from life-threatening soft tissue infections to wound-healing defects and variable forms of infectious gingivitis. The classical clinical hallmark of LAD is the presence of tissue infections without formation of pus. This is due to the fact that impairment of β2 expression leads to reduced leukocyte recruitment and migration to sites of tissue infection because immune cells are unable to adhere to vascular endothelium. All mutations are found in the β2 subunit gene located on chromosome 21. Mutations that effect CD18 expression at the surface can lead to <0.3% to 2.5–31% β2 integrin expression, which correlates with presentation of severe to more moderate disease, respectively. Point mutations within both the ligand-binding domain leading to near normal surface expression but poor ligand recognition have also been described (107). Mice with β2 deficiency also manifest a variety of immune defects due to poor leukocyte adhesion and hence are good models of LAD (52).

Leukocyte adhesion deficiency syndrome can also result from mutations in proteins involved in the inside-out pathway of integrin activation. Recent studies have identified a set of patients with bleeding, immune cell accumulation defects and an osteopetrosis-like bone defect that may result from failure to activate β1, β2, and β3 integrins to high affinity ligand-binding states. This defect in inside-out signaling has been mapped to mutations in the protein Kindlin-3, which binds to specific domains within the cytoplasmic tail of β subunits, leading to separation of the cytoplasmic tails that is required as part of the unfolding of the integrin to form the ligand-binding conformation (108–110). In these patients, their clinical presentation would suggest mainly β2 (immune cell recruitment) and β3 (platelet and osteoclast) functional defects. Indeed, Kindlin-3 expression is restricted to mainly hematopoietic cells. Mutations in ubiquitously expressed Kindlin-1 produce Kindler's Syndrome, which presents as a skin and mucosal blistering disease, due to defects in keritinocyte adhesion probably due to failure to activate β1 integrins in these cells. As in the other integrin deficiencies, mice lacking either Kindlin-1 or Kindlin-3 model the respective human disease, though the *Kindlin-3* knockouts tend to have a less severe bleeding disorder (111). Mice lacking Kindlin-2, which is also widely expressed, die early in gestation due to implantation defects, potentially as a result of poor cell/cell adhesion (112).

7. Concluding Remarks

Some of the challenges in the next decade are to understand how these receptors signal their diverse functions, how the same integrin can specialize to promote specific functions that are dependent on the environment (cell and tissue in which it exists) and how integrins collaborate with other integrins and receptors within a network to deliver cues responsible for cell processes that range from migration to differentiation. This type of knowledge will allow integrin-targeted therapeutic modalities that inhibit pathologic processes in which integrins have been heavily implicated including tumor angiogenesis, inflammation and thrombosis.

References

1. Hynes, R. O. (2002) Integrins: bidirectional, allosteric signaling machines, *Cell* **110**, 673–687.
2. Johnson, M. S., Lu, N., Denessiouk, K., Heino, J., and Gullberg, D. (2009) Integrins during evolution: evolutionary trees and model organisms, *Biochim. Biophys. Acta.* **1788**, 779–789.
3. Whittaker, C. A., and Hynes, R. O. (2002) Distribution and evolution of von Willebrand/integrin A domains: widely dispersed domains with roles in cell adhesion and elsewhere, *Mol. Biol. Cell* **13**, 3369–3387.
4. Meighan, C. M., and Schwarzbauer, J. E. (2008) Temporal and spatial regulation of integrins during development, *Curr. Opin. Cell Biol.* **20**, 520–524.
5. Lauffenburger, D. A., and Horwitz, A. F. (1996) Cell migration: a physically integrated molecular process, *Cell* **84**, 359–369.
6. Dustin, M. L. (2009) The cellular context of T cell signaling, *Immunity* **30**, 482–492.
7. Brown, N. H. (2000) Cell-cell adhesion via the ECM: integrin genetics in fly and worm, *Matrix Biol.* **19**, 191–201.
8. Hynes, R. O., and Zhao, Q. (2000) The evolution of cell adhesion, *J. Cell Biol.* **150**, F89–96.
9. Ellis, S. J., and Tanentzapf, G. (2009) Integrin-mediated adhesion and stem-cell-niche interactions, *Cell Tissue Res.* **339**, 121–30.
10. Montell, D. J. (1999) The genetics of cell migration in Drosophila melanogaster and Caenorhabditis elegans development, *Development* **126**, 3035–3046.
11. Grotewiel, M. S., Beck, C. D., Wu, K. H., Zhu, X. R., and Davis, R. L. (1998) Integrin-mediated short-term memory in Drosophila, *Nature* **391**, 455–460.
12. Delon, I., and Brown, N. H. (2007) Integrins and the actin cytoskeleton, *Curr. Opin. Cell Biol.* **19**, 43–50.
13. Sheppard, D. (2000) In vivo functions of integrins: lessons from null mutations in mice, *Matrix Biol.* **19**, 203–209.
14. Weis, S. M. (2007) Evaluating integrin function in models of angiogenesis and vascular permeability, *Methods Enzymol.* **426**, 505–528.
15. McCarty, J. H. (2009) Integrin-mediated regulation of neurovascular development, physiology and disease, *Cell Adh. Migr.* **3**, 211–215.
16. Astrof, S., and Hynes, R. O. (2009) Fibronectins in vascular morphogenesis, *Angiogenesis* **12**, 165–175.
17. Yang, J. T., Rayburn, H., and Hynes, R. O. (1993) Embryonic mesodermal defects in alpha 5 integrin-deficient mice, *Development* **119**, 1093–1105.
18. Hynes, R. O. (2007) Cell-matrix adhesion in vascular development, *J. Thromb. Haemost.* 5 Suppl **1**, 32–40.
19. Stephens, L. E., Sutherland, A. E., Klimanskaya, I. V., Andrieux, A., Meneses, J., Pedersen, R. A., and Damsky, C. H. (1995) Deletion of beta 1 integrins in mice results in inner cell mass failure and peri-implantation lethality, *Genes. Dev.* **9**, 1883–1895.
20. Carlson, T. R., Hu, H., Braren, R., Kim, Y. H., and Wang, R. A. (2008) Cell-autonomous

requirement for beta1 integrin in endothelial cell adhesion, migration and survival during angiogenesis in mice, *Development* **135**, 2193–2202.

21. Brooks, P. C., Clark, R. A., and Cheresh, D. A. (1994) Requirement of vascular integrin alpha v beta 3 for angiogenesis, *Science* **264**, 569–571.
22. Brooks, P. C., Montgomery, A. M., Rosenfeld, M., Reisfeld, R. A., Hu, T., Klier, G., and Cheresh, D. A. (1994) Integrin alpha v beta 3 antagonists promote tumor regression by inducing apoptosis of angiogenic blood vessels, *Cell* **79**, 1157–1164.
23. De, S., Razorenova, O., McCabe, N. P., O'Toole, T., Qin, J., and Byzova, T. V. (2005) VEGF-integrin interplay controls tumor growth and vascularization, *Proc. Natl. Acad. Sci. USA*. **102**, 7589–7594.
24. Soldi, R., Mitola, S., Strasly, M., Defilippi, P., Tarone, G., and Bussolino, F. (1999) Role of alphavbeta3 integrin in the activation of vascular endothelial growth factor receptor-2, *EMBO J.* **18**, 882–892.
25. Mahabeleshwar, G. H., Feng, W., Phillips, D. R., and Byzova, T. V. (2006) Integrin signaling is critical for pathological angiogenesis, *J. Exp. Med.* **203**, 2495–2507.
26. Huang, X. Z., Wu, J. F., Ferrando, R., Lee, J. H., Wang, Y. L., Farese, R. V., Jr., and Sheppard, D. (2000) Fatal bilateral chylothorax in mice lacking the integrin alpha9beta1, *Mol. Cell. Biol.* **20**, 5208–5215.
27. Bazigou, E., Xie, S., Chen, C., Weston, A., Miura, N., Sorokin, L., Adams, R., Muro, A. F., Sheppard, D., and Makinen, T. (2009) Integrin-alpha9 is required for fibronectin matrix assembly during lymphatic valve morphogenesis, *Dev. Cell* **17**, 175–186.
28. Vlahakis, N. E., Young, B. A., Atakilit, A., and Sheppard, D. (2005) The lymphangiogenic vascular endothelial growth factors VEGF-C and -D are ligands for the integrin alpha9beta1, *J. Biol. Chem.* **280**, 4544–4552.
29. Bader, B. L., Rayburn, H., Crowley, D., and Hynes, R. O. (1998) Extensive vasculogenesis, angiogenesis, and organogenesis precede lethality in mice lacking all alpha v integrins, *Cell* **95**, 507–519.
30. McCarty, J. H., Monahan-Earley, R. A., Brown, L. F., Keller, M., Gerhardt, H., Rubin, K., Shani, M., Dvorak, H. F., Wolburg, H., Bader, B. L., Dvorak, A. M., and Hynes, R. O. (2002) Defective associations between blood vessels and brain parenchyma lead to cerebral hemorrhage in mice lacking alphav integrins, *Mol. Cell. Biol.* **22**, 7667–7677.
31. McCarty, J. H., Lacy-Hulbert, A., Charest, A., Bronson, R. T., Crowley, D., Housman, D., Savill, J., Roes, J., and Hynes, R. O. (2005) Selective ablation of alphav integrins in the central nervous system leads to cerebral hemorrhage, seizures, axonal degeneration and premature death, *Development* **132**, 165–176.
32. Zhu, J., Motejlek, K., Wang, D., Zang, K., Schmidt, A., and Reichardt, L. F. (2002) beta8 integrins are required for vascular morphogenesis in mouse embryos, *Development* **129**, 2891–2903.
33. Proctor, J. M., Zang, K., Wang, D., Wang, R., and Reichardt, L. F. (2005) Vascular development of the brain requires beta8 integrin expression in the neuroepithelium, *J. Neurosci.* **25**, 9940–9948.
34. Grazioli, A., Alves, C. S., Konstantopoulos, K., and Yang, J. T. (2006) Defective blood vessel development and pericyte/pvSMC distribution in alpha 4 integrin-deficient mouse embryos, *Dev. Biol.* **293**, 165–177.
35. Yang, J. T., Rayburn, H., and Hynes, R. O. (1995) Cell adhesion events mediated by alpha 4 integrins are essential in placental and cardiac development, *Development* **121**, 549–560.
36. Kwee, L., Baldwin, H. S., Shen, H. M., Stewart, C. L., Buck, C., Buck, C. A., and Labow, M. A. (1995) Defective development of the embryonic and extraembryonic circulatory systems in vascular cell adhesion molecule (VCAM-1) deficient mice, *Development* **121**, 489–503.
37. Walker, J., and Menko, A. S. (2009) Integrins in lens development and disease, *Exp. Eye. Res.* **88**, 216–225.
38. Simirskii, V. N., Wang, Y., and Duncan, M. K. (2007) Conditional deletion of beta1-integrin from the developing lens leads to loss of the lens epithelial phenotype, *Dev. Biol.* **306**, 658–668.
39. De Arcangelis, A., Mark, M., Kreidberg, J., Sorokin, L., and Georges-Labouesse, E. (1999) Synergistic activities of alpha3 and alpha6 integrins are required during apical ectodermal ridge formation and organogenesis in the mouse, *Development* **126**, 3957–3968.
40. Milner, R., and Campbell, I. L. (2002) The integrin family of cell adhesion molecules has multiple functions within the CNS, *J. Neurosci. Res.* **69**, 286–291.
41. Graus-Porta, D., Blaess, S., Senften, M., Littlewood-Evans, A., Damsky, C., Huang, Z., Orban, P., Klein, R., Schittny, J. C., and Muller, U. (2001) Beta1-class integrins regulate the development of laminae and folia in

the cerebral and cerebellar cortex, *Neuron* **31**, 367–379.

42. Miyagoe-Suzuki, Y., Nakagawa, M., and Takeda, S. (2000) Merosin and congenital muscular dystrophy, *Microsc. Res. Tech.* **48**, 181–191.
43. Benson, D. L., Schnapp, L. M., Shapiro, L., and Huntley, G. W. (2000) Making memories stick: cell-adhesion molecules in synaptic plasticity, *Trends Cell Biol.* **10**, 473–482.
44. Denda, S., and Reichardt, L. F. (2007) Studies on integrins in the nervous system, *Methods Enzymol.* **426**, 203–221.
45. Pietri, T., Eder, O., Breau, M. A., Topilko, P., Blanche, M., Brakebusch, C., Fassler, R., Thiery, J. P., and Dufour, S. (2004) Conditional beta1-integrin gene deletion in neural crest cells causes severe developmental alterations of the peripheral nervous system, *Development* **131**, 3871–3883.
46. Feltri, M. L., Graus Porta, D., Previtali, S. C., Nodari, A., Migliavacca, B., Cassetti, A., Littlewood-Evans, A., Reichardt, L. F., Messing, A., Quattrini, A., Mueller, U., and Wrabetz, L. (2002) Conditional disruption of beta 1 integrin in Schwann cells impedes interactions with axons, *J. Cell Biol.* **156**, 199–209.
47. Nandrot, E. F., Kim, Y., Brodie, S. E., Huang, X., Sheppard, D., and Finnemann, S. C. (2004) Loss of synchronized retinal phagocytosis and age-related blindness in mice lacking alphavbeta5 integrin, *J. Exp. Med.* **200**, 1539–1545.
48. Ley, K., Laudanna, C., Cybulsky, M. I., and Nourshargh, S. (2007) Getting to the site of inflammation: the leukocyte adhesion cascade updated, *Nat. Rev. Immunol.* 7, 678–689.
49. Berlin, C., Bargatze, R. F., Campbell, J. J., von Andrian, U. H., Szabo, M. C., Hasslen, S. R., Nelson, R. D., Berg, E. L., Erlandsen, S. L., and Butcher, E. C. (1995) alpha 4 integrins mediate lymphocyte attachment and rolling under physiologic flow, *Cell* **80**, 413–422.
50. Abram, C. L., and Lowell, C. A. (2009) The ins and outs of leukocyte integrin signaling, *Annu. Rev. Immunol.* **27**, 339–362.
51. Scharffetter-Kochanek, K., Lu, H., Norman, K., van Nood, N., Munoz, F., Grabbe, S., McArthur, M., Lorenzo, I., Kaplan, S., Ley, K., Smith, C. W., Montgomery, C. A., Rich, S., and Beaudet, A. L. (1998) Spontaneous skin ulceration and defective T cell function in CD18 null mice, *J. Exp. Med.* **188**, 119–131.
52. Kakkar, A. K., and Lefer, D. J. (2004) Leukocyte and endothelial adhesion molecule studies in knockout mice, *Curr. Opin. Pharmacol.* **4**, 154–158.
53. Potocnik, A. J., Brakebusch, C., and Fassler, R. (2000) Fetal and adult hematopoietic stem cells require beta1 integrin function for colonizing fetal liver, spleen, and bone marrow, *Immunity* **12**, 653–663.
54. Sixt, M., Bauer, M., Lammermann, T., and Fassler, R. (2006) Beta1 integrins: zip codes and signaling relay for blood cells, *Curr. Opin. Cell Biol.* **18**, 482–490.
55. Hodivala-Dilke, K. M., McHugh, K. P., Tsakiris, D. A., Rayburn, H., Crowley, D., Ullman-Cullere, M., Ross, F. P., Coller, B. S., Teitelbaum, S., and Hynes, R. O. (1999) Beta3-integrin-deficient mice are a model for Glanzmann thrombasthenia showing placental defects and reduced survival, *J. Clin. Invest.* **103**, 229–238.
56. Pauls, K., Schon, M., Kubitza, R. C., Homey, B., Wiesenborn, A., Lehmann, P., Ruzicka, T., Parker, C. M., and Schon, M. P. (2001) Role of integrin alphaE(CD103)beta7 for tissue-specific epidermal localization of CD8+ T lymphocytes, *J. Invest. Dermatol.* **117**, 569–575.
57. Lammermann, T., Bader, B. L., Monkley, S. J., Worbs, T., Wedlich-Soldner, R., Hirsch, K., Keller, M., Forster, R., Critchley, D. R., Fassler, R., and Sixt, M. (2008) Rapid leukocyte migration by integrin-independent flowing and squeezing, *Nature* **453**, 51–55.
58. Pribila, J. T., Quale, A. C., Mueller, K. L., and Shimizu, Y. (2004) Integrins and T cell-mediated immunity, *Annu. Rev. Immunol.* **22**, 157–180.
59. Berton, G., Yan, S. R., Fumagalli, L., and Lowell, C. A. (1996) Neutrophil activation by adhesion: mechanisms and pathophysiological implications, *Int. J. Clin. Lab. Res.* **26**, 160–177.
60. Hirahashi, J., Mekala, D., Van Ziffle, J., Xiao, L., Saffaripour, S., Wagner, D. D., Shapiro, S. D., Lowell, C., and Mayadas, T. N. (2006) Mac-1 signaling via Src-family and Syk kinases results in elastase-dependent thrombohemorrhagic vasculopathy, *Immunity* **25**, 271–283.
61. Watt, F. M. (2002) Role of integrins in regulating epidermal adhesion, growth and differentiation, *EMBO J.* **21**, 3919–3926.
62. Muller, E. J., Williamson, L., Kolly, C., and Suter, M. M. (2008) Outside-in signaling through integrins and cadherins: a central mechanism to control epidermal growth and

differentiation?, *J. Invest. Dermatol.* **128**, 501–516.

63. Brakebusch, C., Grose, R., Quondamatteo, F., Ramirez, A., Jorcano, J. L., Pirro, A., Svensson, M., Herken, R., Sasaki, T., Timpl, R., Werner, S., and Fassler, R. (2000) Skin and hair follicle integrity is crucially dependent on beta 1 integrin expression on keratinocytes, *EMBO J.* **19**, 3990–4003.
64. Raghavan, S., Bauer, C., Mundschau, G., Li, Q., and Fuchs, E. (2000) Conditional ablation of beta1 integrin in skin. Severe defects in epidermal proliferation, basement membrane formation, and hair follicle invagination, *J. Cell Biol.* **150**, 1149–1160.
65. Chen, J., Diacovo, T. G., Grenache, D. G., Santoro, S. A., and Zutter, M. M. (2002) The alpha(2) integrin subunit-deficient mouse: a multifaceted phenotype including defects of branching morphogenesis and hemostasis, *Am. J. Pathol.* **161**, 337–344.
66. DiPersio, C. M., Hodivala-Dilke, K. M., Jaenisch, R., Kreidberg, J. A., and Hynes, R. O. (1997) alpha3beta1 Integrin is required for normal development of the epidermal basement membrane, *J. Cell Biol.* **137**, 729–742.
67. Grose, R., Hutter, C., Bloch, W., Thorey, I., Watt, F. M., Fassler, R., Brakebusch, C., and Werner, S. (2002) A crucial role of beta 1 integrins for keratinocyte migration in vitro and during cutaneous wound repair, *Development* **129**, 2303–2315.
68. DiPersio, C. M., van der Neut, R., Georges-Labouesse, E., Kreidberg, J. A., Sonnenberg, A., and Hynes, R. O. (2000) alpha3beta1 and alpha6beta4 integrin receptors for laminin-5 are not essential for epidermal morphogenesis and homeostasis during skin development, *J. Cell Sci.* **113**, 3051–3062.
69. Kreidberg, J. A., Donovan, M. J., Goldstein, S. L., Rennke, H., Shepherd, K., Jones, R. C., and Jaenisch, R. (1996) Alpha 3 beta 1 integrin has a crucial role in kidney and lung organogenesis, *Development* **122**, 3537–3547.
70. Aluwihare, P., and Munger, J. S. (2008) What the lung has taught us about latent TGF-beta activation, *Am. J. Respir. Cell Mol. Biol.* **39**, 499–502.
71. Takabayshi, K., Corr, M., Hayashi, T., Redecke, V., Beck, L., Guiney, D., Sheppard, D., and Raz, E. (2006) Induction of a homeostatic circuit in lung tissue by microbial compounds, *Immunity* **24**, 475–487.
72. Huang, X. Z., Wu, J. F., Cass, D., Erle, D. J., Corry, D., Young, S. G., Farese, R. V., Jr., and Sheppard, D. (1996) Inactivation of the integrin beta 6 subunit gene reveals a role of epithelial integrins in regulating inflammation in the lung and skin, *J. Cell Biol.* **133**, 921–928.
73. Morris, D. G., Huang, X., Kaminski, N., Wang, Y., Shapiro, S. D., Dolganov, G., Glick, A., and Sheppard, D. (2003) Loss of integrin alpha(v)beta6-mediated TGF-beta activation causes Mmp12-dependent emphysema, *Nature* **422**, 169–173.
74. Pittet, J. F., Griffiths, M. J., Geiser, T., Kaminski, N., Dalton, S. L., Huang, X., Brown, L. A., Gotwals, P. J., Koteliansky, V. E., Matthay, M. A., and Sheppard, D. (2001) TGF-beta is a critical mediator of acute lung injury, *J. Clin. Invest.* **107**, 1537–1544.
75. Koth, L. L., Alex, B., Hawgood, S., Nead, M. A., Sheppard, D., Erle, D. J., and Morris, D. G. (2007) Integrin beta6 mediates phospholipid and collectin homeostasis by activation of latent TGF-beta1, *Am. J. Respir. Cell. Mol. Biol.* **37**, 651–659.
76. Fjellbirkeland, L., Cambier, S., Broaddus, V. C., Hill, A., Brunetta, P., Dolganov, G., Jablons, D., and Nishimura, S. L. (2003) Integrin alphavbeta8-mediated activation of transforming growth factor-beta inhibits human airway epithelial proliferation in intact bronchial tissue, *Am. J. Pathol.* **163**, 533–542.
77. Su, G., Hodnett, M., Wu, N., Atakilit, A., Kosinski, C., Godzich, M., Huang, X. Z., Kim, J. K., Frank, J. A., Matthay, M. A., Sheppard, D., and Pittet, J. F. (2007) Integrin alphavbeta5 regulates lung vascular permeability and pulmonary endothelial barrier function, *Am. J. Respir. Cell. Mol. Biol.* **36**, 377–386.
78. Sparrow, J. C., and Schock, F. (2009) The initial steps of myofibril assembly: integrins pave the way, *Nat. Rev. Mol. Cell. Biol.* **10**, 293–298.
79. Taverna, D., Disatnik, M. H., Rayburn, H., Bronson, R. T., Yang, J., Rando, T. A., and Hynes, R. O. (1998) Dystrophic muscle in mice chimeric for expression of alpha5 integrin, *J. Cell Biol.* **143**, 849–859.
80. Frisch, S. M., and Screaton, R. A. (2001) Anoikis mechanisms, *Curr. Opin. Cell Biol.* *13*, 555–562.
81. Campbell, K. P. (1995) Three muscular dystrophies: loss of cytoskeleton-extracellular matrix linkage, *Cell* **80**, 675–679.
82. Hayashi, Y. K., Chou, F. L., Engvall, E., Ogawa, M., Matsuda, C., Hirabayashi, S., Yokochi, K., Ziober, B. L., Kramer, R. H., Kaufman, S. J., Ozawa, E., Goto, Y., Nonaka, I., Tsukahara, T., Wang, J. Z., Hoffman, E. P., and Arahata, K. (1998) Mutations in the integrin

alpha7 gene cause congenital myopathy, *Nat. Genet.* **19**, 94–97.

83. Lopez, M. A., Mayer, U., Hwang, W., Taylor, T., Hashmi, M. A., Jannapureddy, S. R., and Boriek, A. M. (2005) Force transmission, compliance, and viscoelasticity are altered in the alpha7-integrin-null mouse diaphragm, *Am. J. Physiol. Cell. Physiol.* **288**, C282–289.
84. Mayer, U., Saher, G., Fassler, R., Bornemann, A., Echtermeyer, F., von der Mark, H., Miosge, N., Poschl, E., and von der Mark, K. (1997) Absence of integrin alpha 7 causes a novel form of muscular dystrophy, *Nat. Genet.* **17**, 318–323.
85. Nawrotzki, R., Willem, M., Miosge, N., Brinkmeier, H., and Mayer, U. (2003) Defective integrin switch and matrix composition at alpha 7-deficient myotendinous junctions precede the onset of muscular dystrophy in mice, *Hum. Mol. Genet.* **12**, 483–495.
86. Rooney, J. E., Welser, J. V., Dechert, M. A., Flintoff-Dye, N. L., Kaufman, S. J., and Burkin, D. J. (2006) Severe muscular dystrophy in mice that lack dystrophin and alpha7 integrin, *J. Cell Sci.* **119**, 2185–2195.
87. Welser, J. V., Rooney, J. E., Cohen, N. C., Gurpur, P. B., Singer, C. A., Evans, R. A., Haines, B. A., and Burkin, D. J. (2009) Myotendinous Junction Defects and Reduced Force Transmission in Mice that Lack {alpha}7 Integrin and Utrophin, *Am. J. Pathol.* **175**, 1545–1554.
88. Ross, F. P., and Teitelbaum, S. L. (2005) alphavbeta3 and macrophage colony-stimulating factor: partners in osteoclast biology, *Immunol. Rev.* **208**, 88–105.
89. Horton, M. A., Massey, H. M., Rosenberg, N., Nicholls, B., Seligsohn, U., and Flanagan, A. M. (2003) Upregulation of osteoclast alpha2beta1 integrin compensates for lack of alphavbeta3 vitronectin receptor in Iraqi-Jewish-type Glanzmann thrombasthenia, *Br. J. Haematol.* **122**, 950–957.
90. Mocsai, A., Humphrey, M. B., Van Ziffle, J. A., Hu, Y., Burghardt, A., Spusta, S. C., Majumdar, S., Lanier, L. L., Lowell, C. A., and Nakamura, M. C. (2004) The immunomodulatory adapter proteins DAP12 and Fc receptor gamma-chain (FcRgamma) regulate development of functional osteoclasts through the Syk tyrosine kinase, *Proc. Natl. Acad. Sci. USA.* **101**, 6158–6163.
91. Faccio, R., Teitelbaum, S. L., Fujikawa, K., Chappel, J., Zallone, A., Tybulewicz, V. L., Ross, F. P., and Swat, W. (2005) Vav3 regulates osteoclast function and bone mass, *Nat. Med.* **11**, 284–290.
92. Bengtsson, T., Aszodi, A., Nicolae, C., Hunziker, E. B., Lundgren-Akerlund, E., and Fassler, R. (2005) Loss of alpha10beta1 integrin expression leads to moderate dysfunction of growth plate chondrocytes, *J. Cell Sci.* **118**, 929–936.
93. Popova, S. N., Barczyk, M., Tiger, C. F., Beertsen, W., Zigrino, P., Aszodi, A., Miosge, N., Forsberg, E., and Gullberg, D. (2007) Alpha11 beta1 integrin-dependent regulation of periodontal ligament function in the erupting mouse incisor, *Mol. Cell. Biol.* **27**, 4306–4316.
94. Muller, U., Wang, D., Denda, S., Meneses, J. J., Pedersen, R. A., and Reichardt, L. F. (1997) Integrin alpha8beta1 is critically important for epithelial-mesenchymal interactions during kidney morphogenesis, *Cell* **88**, 603–613.
95. Brandenberger, R., Schmidt, A., Linton, J., Wang, D., Backus, C., Denda, S., Muller, U., and Reichardt, L. F. (2001) Identification and characterization of a novel extracellular matrix protein nephronectin that is associated with integrin alpha8beta1 in the embryonic kidney, *J. Cell Biol.* **154**, 447–458.
96. Linton, J. M., Martin, G. R., and Reichardt, L. F. (2007) The ECM protein nephronectin promotes kidney development via integrin alpha8beta1-mediated stimulation of Gdnf expression, *Development* **134**, 2501–2509.
97. Costantini, F., and Shakya, R. (2006) GDNF/Ret signaling and the development of the kidney, *Bioessays* **28**, 117–127.
98. Liu, Y., Chattopadhyay, N., Qin, S., Szekeres, C., Vasylyeva, T., Mahoney, Z. X., Taglienti, M., Bates, C. M., Chapman, H. A., Miner, J. H., and Kreidberg, J. A. (2009) Coordinate integrin and c-Met signaling regulate Wnt gene expression during epithelial morphogenesis, *Development* **136**, 843–853.
99. Zhang, X., Mernaugh, G., Yang, D. H., Gewin, L., Srichai, M. B., Harris, R. C., Iturregui, J. M., Nelson, R. D., Kohan, D. E., Abrahamson, D., Fassler, R., Yurchenco, P., Pozzi, A., and Zent, R. (2009) beta1 integrin is necessary for ureteric bud branching morphogenesis and maintenance of collecting duct structural integrity, *Development* **136**, 3357–3366.
100. Kanasaki, K., Kanda, Y., Palmsten, K., Tanjore, H., Lee, S. B., Lebleu, V. S., Gattone, V. H., Jr., and Kalluri, R. (2008) Integrin beta1-mediated matrix assembly and signaling are critical for the normal development and function of the kidney glomerulus, *Dev. Biol.* **313**, 584–593.

101. Kagami, S., and Kondo, S. (2004) Beta1-integrins and glomerular injury, *J. Med. Invest.* **51**, 1–13.
102. Cosgrove, D., Rodgers, K., Meehan, D., Miller, C., Bovard, K., Gilroy, A., Gardner, H., Kotelianski, V., Gotwals, P., Amatucci, A., and Kalluri, R. (2000) Integrin alpha1beta1 and transforming growth factor-beta1 play distinct roles in alport glomerular pathogenesis and serve as dual targets for metabolic therapy, *Am. J. Pathol.* **157**, 1649–1659.
103. Hogg, N., and Bates, P. A. (2000) Genetic analysis of integrin function in man: LAD-1 and other syndromes, *Matrix Biol.* **19**, 211–222.
104. Wilhelmsen, K., Litjens, S. H., and Sonnenberg, A. (2006) Multiple functions of the integrin alpha6beta4 in epidermal homeostasis and tumorigenesis, *Mol. Cell. Biol.* **26**, 2877–2886.
105. Kato, A. (1997) The biologic and clinical spectrum of Glanzmann's thrombasthenia: implications of integrin alpha IIb beta 3 for its pathogenesis, *Crit. Rev. Oncol. Hematol.* **26**, 1–23.
106. Hynes, R. O., and Hodivala-Dilke, K. M. (1999) Insights and questions arising from studies of a mouse model of Glanzmann thrombasthenia, *Thromb. Haemost.* **82**, 481–485.
107. Roos, D., and Law, S. K. (2001) Hematologically important mutations: leukocyte adhesion deficiency, *Blood Cells Mol. Dis.* **27**, 1000–1004.
108. Abram, C. L., and Lowell, C. A. (2009) Leukocyte adhesion deficiency syndrome: a controversy solved, *Immunol. Cell. Biol.* **87**, 440–442.
109. Svensson, L., Howarth, K., McDowall, A., Patzak, I., Evans, R., Ussar, S., Moser, M., Metin, A., Fried, M., Tomlinson, I., and Hogg, N. (2009) Leukocyte adhesion deficiency-III is caused by mutations in KINDLIN3 affecting integrin activation, *Nat. Med.* **15**, 306–312.
110. Malinin, N. L., Zhang, L., Choi, J., Ciocea, A., Razorenova, O., Ma, Y. Q., Podrez, E. A., Tosi, M., Lennon, D. P., Caplan, A. I., Shurin, S. B., Plow, E. F., and Byzova, T. V. (2009) A point mutation in KINDLIN3 ablates activation of three integrin subfamilies in humans, *Nat. Med.* **15**, 313–318.
111. Moser, M., Bauer, M., Schmid, S., Ruppert, R., Schmidt, S., Sixt, M., Wang, H. V., Sperandio, M., and Fassler, R. (2009) Kindlin-3 is required for beta2 integrin-mediated leukocyte adhesion to endothelial cells, *Nat. Med.* **15**, 300–305.
112. Montanez, E., Ussar, S., Schifferer, M., Bosl, M., Zent, R., Moser, M., and Fassler, R. (2008) Kindlin-2 controls bidirectional signaling of integrins, *Genes. Dev.* **22**, 1325–1330.

Chapter 23

A Method for the Generation of Conditional Gene-Targeted Mice

Masahiro Yamamoto and Kiyoshi Takeda

Abstract

Integrins and cell adhesion molecules are required for cell adhesion and movement, as well as various biological functions such as development or immune responses. To elucidate their in vivo functions, mice lacking these molecules have been generated and various phenotypes analyzed. In the case of embryonic lethality or a requirement for spatiotemporal ablation of these genes, conditional gene-targeted mice have also been generated. Gene targeting is a powerful tool to study the function of proteins under physiological conditions and is a widely used method. Construction of targeting vectors and selection of homologous recombinant embryonic stem cells require delicate and elaborate technical procedures. In this chapter we describe the methods to generate conventional and conditional knockout mice.

Key words: Embyronic stem cells, Targeting vector, Conditional gene targeting, Cre recombinase, loxP

1. Introduction

Cell adhesion molecules and integrin family proteins play important roles in development, the immune system, and homeostasis (1). Integrins are cell surface molecules that are involved in cell adhesion to the extracellular matrix or intracellular signaling from the matrix. Integrins consist of two subunits: the alpha and beta chains, which form a heterodimer. In addition, combinations of distinct alpha and beta chains are possible. So far, more than 20 subunits of integrin alpha or beta chains have been reported (2). In the cytoplasm, integrins bind to the cytoskeleton such as microfilaments and in cell-to-matrix adhesion. Moreover, recent studies report that some integrins participate in cell-to-cell adhesion. The *in vivo* function of integrins and cell adhesion molecules has been tested using mice lacking single or multiple integrins (3).

Motomu Shimaoka (ed.), *Integrin and Cell Adhesion Molecules: Methods and Protocols*,
Methods in Molecular Biology, vol. 757, DOI 10.1007/978-1-61779-166-6_23, © Springer Science+Business Media, LLC 2011

The knockout technique is an essential tool to test the *in vivo* function of a protein(s) under physiological conditions. The isolation and genetic manipulation of embryonic stem (ES) cells was a milestone for mammalian biology. Based upon basic research regarding pluripotent embryonic carcinoma cells, ES cells were generated from blastocysts in 1981 (4). The ES cells were found to contribute to various organs and tissues, including germ cells, resulting in the generation of chimeric mice (5, 6). Independently, the development of techniques for manipulating ES cells to express exogenous proteins yielded various positive/negative selectable markers such as the neomycin-resistance gene and the thymidine kinase gene, and gave rise to efficient methods for the selection of homologous recombinant clones. Mice lacking the genes required for embryonic development, such as integrin alpha-5 or beta-1, die *in utero*, thereby hampering analysis of the protein function in adult mice (7, 8). To overcome this conundrum, methods for conditional gene targeting have been developed by taking advantage of the bacteriophage Cre-loxP or yeast FRT-Flp recombination systems, permitting the deletion of a target gene in a tissue- and/or time-specific manner (9, 10).

In this protocol, we will describe materials and methods to generate conventional and conditional knockout mice in our laboratory.

2. Materials

2.1. Culturing ES Cells

1. Media for feeder cells (EF media): Dissolve 13.4 g Dulbecco's modified Eagle's medium, 3.7 g $NaHCO_3$, 50 mg Streptomycin, 75 mg Penicillin G in 1,000 ml of Nanopure water. Mix well and bubble with CO_2 until it is colored orange. Filter using bottle top filter. Store at 4°C until use. Add 50 ml of fetal bovine serum (FBS) and 5 ml of 200 mM L-glutamine to a 450 ml of filtered EF medium.
2. Media for ES cells (ES media): Dissolve13.4 g DMEM (powder; Gibco, high glucose), 3.7 g $NaHCO_3$, 20 ml of 100 mM sodium pyruvate in 1,000 ml of Nanopure water. Mix well and bubble with CO_2 until it is colored orange. Filter using bottle top filter (Falcon). Store at 4°C until use. Add 75 ml FBS (final concentration=15%), 3.5 μl 2-mercaptoethanol and 5 ml l-glutamine to a 425 ml of filtered ES medium.
3. G418 (geneticin) solution: Dissolve in Hank's solution at 100 mg/ml. Adjust pH with 1 N NaOH to pH 7–8. Store at 4°C until use.
4. Gancyclovir solution: Dissolve the gancyclovir in Nanopure water at 2 mM. Store at room temperature in dry conditions (see Note 1).

5. Leukemia inhibitory factor (LIF) (ESGRO; Gibco): Add ESGRO to media with a final concentration of 10^3 unit/ml.
6. Gelatin (porcine skin gelatin type A): Dissolve 500 mg in 500 ml Nanopure water (0.1% (w/v)). Autoclave and store at 4°C until use.
7. Mitomycin C (MMC): Prepare at 0.5 mg/ml in phosphate-buffered saline (PBS), and use at a final concentration of 10 μg/ml.

2.2. Extraction of Genomic DNA

1. ES cell lysis buffer: Tris–HCl, pH 8.5, 50 mM KCl, 1.5 mM $MgCl_2$, 10 mM 0.01% Gelatin, 0.45% NP-40, 0.45% Tween-20. Store at 4°C until use. Add Proteinase K at 100 μg/ml just prior to use.
2. Tail lysis buffer: 100 mM Tris–HCl, pH 8.5, 5 mM EDTA, 0.2% SDS, 200 mM NaCl. Store at room temperature until use.

2.3. Deletion of the Floxed Neomycin Gene

1. pMC-Cre/pgk-puro Cre recombinase vector.

3. Methods

3.1. Feeder Cells

3.1.1. Preparation of Feeder Cells

Embryonic fibroblasts for feeder cells must be derived from neomycin-resistant embryos.

1. Dissect embryonic day 13.5–14.5 embryos to remove fetal livers and brains, and trypsinize at 37°C.
2. Add 10 ml of EF medium and mix well.
3. Centrifuge at 180×*g* at 4°C for 5 min.
4. Discard the supernatants and resuspend pellets in 10 ml of EF medium.
5. Seed pellets in new cell-culture grade petri dishes.
6. Passage the cells to confluency in 15-cm dishes.
7. Harvest cells and freeze two vials per confluent 15-cm dish.

3.1.2. Preparation of Mitomycin C-Treated Feeder Cells

1. Thaw one vial of primary feeder cells and resuspend the cells in a 10 ml of ice-cold EF medium.
2. Centrifuge at 180×*g* at 4°C for 5 min.
3. Resuspend the cells in 10 ml of EF medium and seed the suspension onto a 15-cm dish. After 48 h, the 15-cm dish should be confluent.
4. Replace the culture medium with EF medium containing 10 μg/ml MMC.
5. Incubate the cells for 2.5 h at 37°C.

6. Harvest the MMC-treated cells, and freeze two vials per confluent 15 cm dish at −80°C until use (see Note 2).

3.2. Preparation and Passage of ES Cells

We use 129SV-derived E14.1 ES cells (see Note 3).

1. The day before ES cells are thawed, prepare feeder cells (Subheadings 3.1.1 and 3.1.2).
2. Thaw frozen ES cells and resuspend the cells in 10 ml of ice-cold ES medium.
3. Centrifuge at 180 × *g* at 4°C for 5 min.
4. Resuspend the cells in an appropriate volume of ES medium and seed the suspension onto an appropriate dish(es) (see Note 4).

3.3. Construction of the Targeting Vector for Conditional Knockout

We use the pBS-hsvTK-loxP-pgkNEO vector (Fig. 1a) for gene targeting. The vector contains three loxP sites, as shown in Fig. 1a. Before cloning into the targeting vector, four different arms are subcloned.

- Long arm (LA): the longer, the better for homologous recombination, but longer fragments will be more difficult to clone into the targeting vector. We usually use an LA greater than 5 kb.

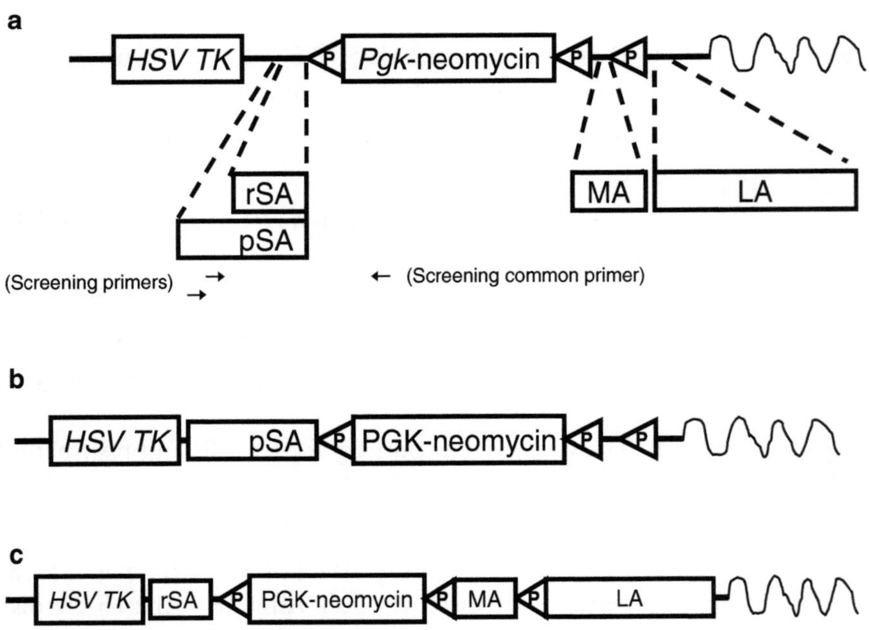

Fig. 1. pBS-hsvTK-loxP-pgkNEO vector, pC, and rC. (**a**) The Back bone of the pBS-hsvTK-loxP-pgkNEO vector is pBluescript. LA or MA is greater or smaller than 5 kb or 3 kb, respectively. Most screening primers are designed within pSA outside rSA. (**b**) Construction of pC. pBS-hsvTK-loxP-pgkNEO vectors are digested to clone pSA. (**c**) Construction of rC. pBS-hsvTK-loxP-pgkNEO vectors are disgested to sequentially clone rSA, LA, and MA.

- Middle arm (MA): the genomic region, including the exon(s) to be deleted. In general, this arm should not be longer than 3 kb, because the deletion efficiency of Cre recombinase decreases in inverse proportion to the length of this arm.
- Short arm for screening homologous recombinants (rSA): this arm should be 0.8–1.2 kb long.
- Short arm for positive control (pSA): this arm contains the rSA plus its flanking genomic regions, with size approximately 0.5 kb, to enable the design of screening primers to detect homologous recombinants by PCR. We usually design two screening primers within this 0.5 kb flanking region to direct to the pgkNEO direction.
- The four arms are amplified by PCR using a high-fidelity DNA polymerase. The amplified fragments are sequenced.
- A positive control vector pC is constructed by cloning the pSA into the pBS-hsvTK-loxP-pgkNEO vector (Fig. 1b).
- A targeting vector for screening of homologous recombinants (rC) is constructed by sequential cloning of the rSA, LA, and MA into the pBS-hsvTK-loxP-pgkNEO vector (Fig. 1c).
- For electroporation, linearize 30 μg of vectors by digestion with an appropriate restriction enzyme. Before electroporation, precipitate linearized pC with EtOH and dissolve in 20 μl sterile PBS.

3.4. Positive Controls for Screening

3.4.1. Preparation of ES Cells

1. Thaw one vial of frozen ES cells (6 cm × 1/2) to a 3.5-cm dish and culture for 48 h until they become subconfluent (NOT confluent).
2. Passage the ES cells on the 3.5-cm dish to a 6-cm dish and culture for 48 h to become confluent.
3. Passage the ES cells on the 6-cm dish to two or three 10-cm dishes to become confluent.

3.4.2. Electroporation of ES Cells for Positive Control

1. Harvest ES cells as prepared in Subheading 3.2 and trypsinize them well.
2. Seed the trypsinized suspension onto three 10-cm dishes, and incubate the dishes at 37°C for 30 min. This step removes feeder cells from the suspension.
3. Collect ES cells by centrifugation at 180 × *g* at 4°C for 5 min.
4. Wash pellets in 10 ml PBS.
5. Collect ES cells by centrifugation at 180 × *g* at 4°C for 5 min.
6. Wash pellets in 10 ml PBS. Use 10 μl of cell suspension for counting cell numbers.
7. Collect ES cells by centrifugation at 180 × *g* at 4°C for 5 min.
8. Discard the supernatants and resuspend the pellets at 1×10^7 cells per 0.8 ml PBS.

9. Add linearized pC in 20 μl PBS to 0.8 ml suspension containing 1×10^7 ES cells.
10. Incubate the mixture on ice for 10 min.
11. Transfer the mixture to an electroporation cuvette (0.4 cm cuvette) and electroporate at 500 μF, 230 V. This setting usually makes a time constant of around nine.
12. Incubate the cuvette on ice for 10 min.
13. Resuspend the electroporated cell suspension in 10 ml of ES medium.
14. Seed the suspension onto a 10-cm feeder cell dish.

3.4.3. Selection of ES Cells by G418

1. Prepare ES media containing 400 μg/ml G418. The ES media is at room temperature just prior to use.
2. Change the culture media promptly to avoid ES cells becoming devoid of LIF-containing ES medium. Culture media are changed on days 1, 2, 4, and 6 for G418 selection. Colonies of selected ES cells are detectable approximately 4–5 days after electroporation.

3.4.4. Picking of ES Cell Colonies

1. Before starting to pick ES cell colonies, prepare ES lysis buffer.
2. Thaw one vial of frozen feeder cell stocks and seed onto one 24-well plate, 24 h before picking ES cell colonies.
3. Change the EF medium to a fresh ES media containing G418.
4. Prepare a 96-well plate containing a 20 μl of 0.25% trypsin per well.
5. Prepare twelve 1.5-ml tubes, each containing 1 ml PBS. The tubes are labeled as screening tubes.
6. Wash the ES cell plate in PBS and pick ES cell colonies under an inverted microscope. The microscope and pipettes are disinfected before use.
7. Transfer picked ES cell colonies to individual wells of the 96-well plate.
8. Resuspend the ES cells in 2×50 μl ES medium from a well of a 24-well feeder cell plate. Return half of the aliquots to the original well and transfer the other half to a screening tube.
9. Incubate the feeder cells at 37°C.
10. Centrifuge the screening tubes at 8,000 rpm, 4°C for 2 min on a microcentrifuge.
11. Discard the supernatants. Resuspend the pellets in 50 μl ES lysis buffer containing 100 μg/ml Proteinase K, and incubate the screening tubes at 55°C overnight.

12. Boil the screening tubes for at 98°C for 15 min. The suspensions are immediately used for the following PCR screening (see Note 5) or stored at –80°C until use.

3.5. Screening of Homologous Recombinant ES Cells

The procedure for screening homologous recombinant ES cells is the same as that for positive controls (Subheadings 3.1–3.4) except for using a linearized 30 μg rC. The frequency for positive clones largely depends on the genomic locus used for the targeting vector (Fig. 2) (see Note 6).

3.6. Expansion of PCR-Positive and Correctly Floxed Homologous Recombinant

1. Mark wells containing PCR-screening-positive ES cells cultured in a 24-well plate.
2. Add 3 drops of trypsin per well and incubate the plate at 37°C for 4 min.
3. Add 1 ml pre-warmed ES medium and collect the suspension in a 15-ml centrifuge tube.
4. Centrifuge at 180 × *g*, at 4°C for 5 min.
5. Discard the supernatants, resuspend the pellets in 1.5 ml ES media and seed them onto feeder cells in a 3.5-cm dish.

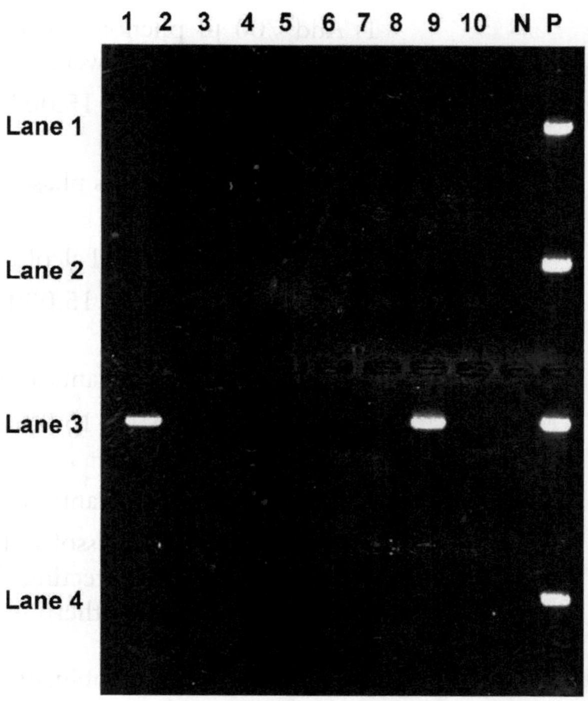

Fig. 2. Screening of homologous recombinant ES cells. A representative result is shown. Total of 40 colonies were picked up and tested to detect homologous recombination by PCR. Colonies corresponding to lanes 3–1 and 3–9 were positive for homologous recombination and expanded to analyze by Southern blotting.

6. Within 48 h, harvest the ES cells and passage on two 6-cm dishes.
7. Within 48 h, harvest and divide the ES cells (9/10 of cells to be frozen, and the rest to passage for Southern blotting).
8. To remove feeder cells, ES cells for Southern blotting are expanded on gelatinized plates. Seed 1/10 ES cells onto a 3.5-cm gelatinized dish and culture until the ES cells become confluent.
9. Harvest the ES cells, seed on two 6-cm dishes and culture until the ES cells become confluent.
10. Harvest the ES cells in two 15-ml centrifuge tubes and wash them once in PBS.
11. After centrifugation, transfer the ES cells to a 1.5-ml tube.
12. Centrifuge the suspension at 4°C, 8,000 rpm, for 2 min on a microcentrifuge.
13. Discard the supernatants, resuspend the pellet in 700 μl Southern lysis buffer with 100 μg/ml Proteinase K, and incubate the suspension at 55°C overnight.

3.7. Extraction of Genomic DNA

Extraction of genomic DNA is performed by a standard phenol–chloroform method.

1. Add 700 μl phenol–chloroform–isoamyl alcohol (25:24:1) solution and vortex well.
2. Centrifuge at 4°C, 15,000 rpm for 10 min on a microcentrifuge.
3. Transfer the aqueous phase (approximately 600 μl) to a new 1.5 ml tube.
4. Add 700 μl isopropyl alcohol and mix well.
5. Centrifuge at 4°C, 15,000 rpm for 10 min on a microcentrifuge.
6. Discard the supernatants and add 1 ml of 70% EtOH.
7. Centrifuge at 4°C, 15,000 rpm for 10 min on a microcentrifuge.
8. Discard the supernatants and dry the pellets quickly.
9. Add 100 μl TE to dissolve the pellets and measure the DNA concentration by detecting the OD at 260 nm. The samples are subjected to Southern blotting (see Note 7).

3.8. Deletion of the Floxed Neomycin Gene

For homologous recombinants possessing a floxed neomycin gene, we transiently transfect the Cre recombinase into these ES cells to delete the neomycin gene. The presence of the neomycin gene has occasionally been shown to affect expression of the targeted gene.

3.8.1. Transient Transfection of Cre Recombinase

1. Transfect 1×10^7 homologous recombinant ES cells with 30 μg of pMC-Cre/pgk-puro Cre recombinase vector.
2. Seed transfected ES cells onto MMC-treated feeder cells at 5×10^6 cells per 10 cm dish and culture them in normal pre-warmed ES media.
3. After 24 h, replace the medium with new pre-warmed ES medium containing 1 μg/ml puromycin to eliminate the growth of recombinant ES cells without Cre expression.
4. After 72 h, harvest the transfected ES cells and seed them onto one or two 10-cm dishes at a density of 1×10^2 or 1×10^3 cells per dish, with normal ES media.
5. Change the medium on days 1, 3, 5, and 7. The colonies of transfected ES cells can be observed around day 5.

3.8.2. Selection of Homologous Recombinant ES Cells with Cre-Mediated Neomycin Deletion

1. Prepare two 24-well plates with normal ES medium or 400 μg/ml G418-containing medium.
2. Pick the transfected ES cell colonies as described in Subheading 3.4.
3. Transfer half of the suspension in each well of the 96-well plate to a well of a 24-well plate with normal ES medium, and transfer the other half to a well with G418-containing medium.
4. Check the G418 sensitivity of each picked clone.
5. Harvest the G418-sensitive recombinant ES clones and expand as described in Subheading 3.6.
6. Extract genomic DNA from the ES cell clones and verify deletion events by Southern blotting (see Note 8).

3.9. Generation of Chimeric Mice and F1 Mice

Verified homologous recombinant ES cells are inject into blastocysts from C57BL/6 mice. As E14.1 ES cells are derived from a male embryo, if the ES cell contribution in chimeric mice is high, the sex distribution will be overwhelmingly biased toward males. Chimeric mice are generated from at least two independent ES cell clones to confirm that any phenotypes observed genuinely result from disruption of the targeted gene (see Note 9).

3.10. Generation of Conditional Gene-Disrupted Mice

The offspring (F1 mice) obtained in Subheading 3.10 possess the flox/+ genotype (i.e., one floxed and one wild-type allele). To induce conditional ablation of the targeted gene, mate the mice with transgenic mice expressing Cre recombinase under the control of a specific promoter. For instance, if the floxed F1 mice are mated with CD19-Cre transgenic mice expressing Cre recombinase under the endogenous CD19 promoter, some of the offspring possess the genotype of CD19-Cre(+) and the targeted gene flox/+. Subsequently, when CD19-Cre(+) and the targeted gene flox/+ mice are mated with flox/+ mice, some of the offspring possess the genotype of CD19-Cre(+) and the targeted gene flox/flox, which lack the targeted locus in cells expressing CD19 (11). Lists of transgenic mice expressing Cre recombinase under various promoters are available at http://www.mshri.on.ca/nagy, http://jaxmice.jax.org/list/ra1511.html or a recent review (12).

4. Notes

1. As gancyclovir becomes inactivated in solution, the aliquots should be desiccated in an evaporator soon after the solution is made. On the day it is to be used, dissolve in 1 ml Hank's solution and filter using 0.22 μm filter. Add the aliquot with a final concentration of 2 μM.
2. The single vial from a confluent 15-cm dish can cover two of 10-cm dishes, four of 6-cm dishes, four of 24-well plates, or eight of 3.5-cm dishes.
3. ES cells proliferate to form colony-like structures and can begin to differentiate if the cell number and density are too high or too low. Therefore, the following three points are essential for the culture of ES cells, culturing ES cells at the appropriate concentration; disrupting colonies thoroughly to single cells when ES cells are passaged; and passaging ES cells within 48 h.
4. In the case of E14.1 ES cells, one confluent 6-cm dish would contain $3–5 \times 10^6$ cells. The cells should be passaged within 48 h in 1/5–1/10 volumes.
5. We use a Perkin Elmer Thermal Cycler 480™. At 85°C, add the primer mix to the PCR reaction solution (total 50 μl). The cycling conditions are: 35 cycles of 94°C for 30 s, 30 s ramp time to 67°C, 67°C for 1 min, 74°C for 1 min, then a final 74°C for 10 min prior to cooling to 4°C. The primer mix contains one forward screening primer, as designed in Subheading 3.3, and the reverse common primer, which is designed within the *Pgk* promoter. Five microliters from each screening tube should be used for PCR. Electrophorese the PCR products on 1% agarose gels to check the amplification. Almost all picked ES cell colonies are usually positive for this PCR screening protocol; however, we have occasionally experienced a smear or no bands at all. In these cases, we re-design the screening primers within the 0.5 kb flanking region of the pSA and re-test for positivity. If no positive bands are obtained using any screening primers, design the pSA within a different genomic region and try the positive control again.

6. The homologous recombinant ES cells must be dealt with very carefully, otherwise the cells will not generate chimeric mice and thereby fail to transmit the mutated genome to the offspring. In addition, PCR is so sensitive that it can unexpectedly detect false-positive clones containing nonhomologous recombinant ES cells. To eliminate such false-positive ES cell clones, we usually expand the cells, extract the DNA, and perform Southern blotting. Moreover, the most important step of this screening protocol is to detect homologous recombinant ES cells with a correctly floxed MA. The distal loxP site is sometimes deleted during homologous recombination, in which the recombination between genomic DNA and the targeting vector occurs within the MA rather than the LA. To avoid this, design PCR primers as shown in Fig. 1a, and perform genomic PCR using these primers.
7. We usually digest 30 μg of genomic DNA with one or two restriction enzymes. The probe to detect homologous recombinants and the presence of the distal loxP site should be designed outside the targeting vector. In most cases, we design the probes within the genomic region just outside the rSA. In addition, to exclude the ES cell clones that possess multiple neomycin cassettes, a neomycin-specific probe should be used on another Southern blot.
8. This step is important to exclude heterologous colonies containing non-deleted and/or G418-sensitive deleted ES cells. In addition, some G418-sensitive clones delete not only floxed neomycin but also floxed MA. Therefore, the presence of floxed MA should also be confirmed by PCR and Southern blotting. In our laboratory, the efficiency of deletion of the floxed neomycin cassette is approximately 25%.
9. We usually try mating chimeric male mice with C57BL/6 female mice, and test for germline transmission by checking for agouti coat color in the offspring. Genomic DNA is extracted from the tails of the offspring with agouti coat and analyzed by Southern blotting using the same probe used to detect homologous ES cell clones.

Acknowledgments

We thank C. Hidaka for excellent secretarial assistance, Y. Magota for technical assistance, and K. Kimura for figure making. This work was supported by grants from the Ministry of Education, Culture, Sports, Science and Technology, the Strategic International Cooperative Program (Research Exchange Type), the Japan Science and Technology Agency (JST), and the Takeda Science

Foundation; KANAE FOUNDATION FOR THE PROMOTION OF MEDICAL SCIENCE; The Cell Science Research Foundation; THE ICHIRO KANEHARA FOUNDATION; Kato Memorial Bioscience Foundation; and THE UEHARA MEMORIAL FOUNDATION. The authors have no conflicting financial interests.

References

1. Hynes, R.O. (1992) Integrins: versatility, modulation, and signaling in cell adhesion. *Cell* **69**, 11–25.
2. Hynes, R.O. (2002) Integrins: bidirectional, allosteric signaling machines. *Cell* **110**, 673–687.
3. Frenette, P.S., and Wagner, D.D. (1997) Insights into selectin function from knockout mice. *Thromb. Haemost.* **78**, 60–64.
4. Evans, M.J., and Kaufman, M.H. (1981) Establishment in culture of pluripotential cells from mouse embryos. *Nature* **292**, 154–156.
5. Evans, M.J. (1989) Potential for genetic manipulation of mammals. *Mol. Biol. Med.* **6**, 557–565.
6. Zijlstra, M., Li, E., Sajjadi, F., Subramani, S., and Jaenisch, R. (1989) Germ-line transmission of a disrupted beta 2-microglobulin gene produced by homologous recombination in embryonic stem cells. *Nature* **342**, 435–438.
7. Yang, J.T., Rayburn, H., and Hynes, R.O. (1993) Embryonic mesodermal defects in alpha 5 integrin-deficient mice. *Development* **119**, 1093–1105.
8. Fässler, R., and Meyer, M. (1995) Consequences of lack of beta 1 integrin gene expression in mice. *Genes Dev.* **9**, 1896–1908.
9. Kühn, R. , Schwenk, F., Aguet, M. , and Rajewsky, K. (1995). Inducible gene targeting in mice. *Science* **269**, 1427–1428.
10. Nagy A. (2000) Cre recombinase: The universal reagent for genome tailoring. *Genesis* **26**, 99–109.
11. Branda, C.S., and Dymecki, S.M. (2004) Talking about a revolution: The impact of site-specific recombinases on genetic analyses in mice. *Dev. Cell* **6**, 7–28.
12. Wang, X. (2009) Cre transgenic mouse lines. *Methods Mol. Biol.* **561**, 265–273.

Chapter 24

T-Cell Homing to the Gut Mucosa: General Concepts and Methodological Considerations

Jaime De Calisto, Eduardo J. Villablanca, Sen Wang, Maria R. Bono, Mario Rosemblatt, and J. Rodrigo Mora

Abstract

Effector/memory T cells can migrate to most extra-lymphoid tissues in the body. However, migration to the intestinal mucosa requires the expression of very specific homing receptors on T cells, integrin α4β7 and chemokine receptor CCR9. These receptors are induced by all-*trans* retinoic acid (RA), a vitamin A metabolite that is specifically synthesized by gut-associated dendritic cells (DC), but not by extra-intestinal DC. Here we summarize some general concepts on T cell homing with an emphasis on the gut mucosa. We also discuss experimental strategies to generate gut-homing T cells in vivo and in vitro and the techniques to track gut-homing T cells.

Key words: Gut homing, α4β7, CCR9, CCL25, Retinoic acid, Small intestine, RALDH, GALT, DC, Homing, Chemotaxis

1. Introduction

Leukocyte adhesion to the endothelium and subsequent migration into different tissues is at the heart of both protective and pathological immune responses. On one hand, leukocyte adhesion is essential for protecting the body against pathogenic organisms, as illustrated by genetic defects in leukocyte adhesion, such as leukocyte adhesion deficiency (LAD)-I (lack of β2 integrins), LAD-II (defective generation of selectin ligands) or LAD-III (defects in integrin activation), which are characterized by variable degrees of immunodeficiency (1). On the other hand, the key role of lymphocyte adhesion in immune-driven pathology is clearly demonstrated by the clinical effectiveness of therapies targeting specific adhesion receptors, as seen in multiple sclerosis,

Motomu Shimaoka (ed.), *Integrin and Cell Adhesion Molecules: Methods and Protocols*,
Methods in Molecular Biology, vol. 757, DOI 10.1007/978-1-61779-166-6_24, © Springer Science+Business Media, LLC 2011

psoriasis, and inflammatory bowel disease (1). Thus, understanding how lymphocytes are targeted to different organs is important not only from a basic immunological standpoint, but also has clinical implications. Lymphocyte adhesion is a multistep process (2–4), involving the "capture" of lymphocytes by the endothelial cells (*tethering*), followed by loose adhesion (*rolling*) and then by an activation step (*activation*) that finally leads to firm arrest (*sticking*) and transmigration of lymphocytes into different tissues. Each of these steps is specifically controlled by distinct sets of adhesion and chemokine receptors expressed on lymphocytes and by their respective ligands, which are displayed on endothelial cells. Upon initial interaction with an endothelial cell a lymphocyte must follow a sequential algorithm with multiple yes/no decision points, all of which must be successful for a lymphocyte to finally adhere and transmigrate to the extravascular space (5). It is important to realize that the multistep nature of lymphocyte adhesion allows for a higher versatility and specificity in lymphocyte migration and that a particular migratory code for a given tissue (molecular "zip code") is determined by the combination of homing receptors involved in each step and not by a single adhesion-receptor pair. This versatility and specificity also makes lymphocyte adhesion readily amenable to manipulation for experimental or therapeutic purposes.

First we will summarize our understanding of lymphocyte migration with a special emphasis in homing to the gastrointestinal mucosa and the skin, which are the two major surfaces exposed to pathogens in the body and are also the tissues with the best characterized homing requirements. Next, we will focus on describing some useful methods and techniques aimed at exploring lymphocyte adhesion and homing to the gut mucosa.

1.1. Lymphocyte Migration to Lymphoid and Nonlymphoid Compartments

Naïve T and B cells constantly transit between the blood and secondary lymphoid organs (SLO) with very high efficiency. Classic experiments using thoracic duct cannulation determined that most lymphocytes recirculate between the blood and the lymph around ten times per day (6, 7). More recent experiments using the immunosuppressant drug FTY720, a sphingosine 1-phosphate analogue that blocks the exit of lymphocytes from secondary lymphoid organs, showed that a high proportion of lymphocytes disappeared from the blood in only a few hours upon administration of the drug due to sequestering of lymphocytes in the lymphoid compartments. These observations illustrate the steady and massive migration of lymphocytes into SLO.

Migration of Naïve T and B cells to SLO, in particular lymph nodes (LN) and Peyer's patches (PP) depends on the expression of L-Selectin (CD62L), integrin LFA-1 ($\alpha L\beta 2$) and chemokine receptor CCR7 on lymphocytes (5). Naïve B cells are less dependent on CCR7 and they can also use CXCR4 and CXCR5 to migrate to LN and PP (1). The integrin $\alpha 4\beta 7$, which is key for lymphocyte homing to the gut mucosa, also plays a relevant role in naïve lymphocyte migration to PP, where it can partially compensate for the lack of L-selectin (1). Lymphocyte adhesion in LN and PP occurs in the high endothelial venules (HEV), specialized postcapillary venules that express the necessary adhesion ligands and chemokines (collectively termed *addressins*) (Table 1) (1). If T cells are activated by their cognate antigen presented by dendritic cells (DC) they proliferate and become effector and/or memory T cells and then leave the lymphoid compartments in order to reach different peripheral/effector sites in the body. Some activated T cells maintain their expression of L-selectin and CCR7 and can migrate to LN. These T cells are referred to as central memory T cells (T_{CM}) (8). On the other hand, a large proportion of T cells loses their expression of L-selectin and/or CCR7 and therefore cannot migrate back to the LN. Those T cells are termed effector (T_{EFF}) or effector memory (T_{EM}) T cells. While initially defined as cells with no effector capacity, it has become clear that T_{CM} can exhibit potent effector activity and can also confer protection against infectious challenges (9–12). Moreover, in addition to L-selectin and CCR7, T_{CM} might also express homing receptors involved in extra-lymphoid migration, similar to T_{EFF}/T_{EM} (12). Therefore, the main functional difference between T_{CM} and T_{EFF}/T_{EM} is that T_{CM} can migrate to LN,

Table 1
T cell homing receptors: inductive and effector sites

	Inductive sites		Effector sites	
Target tissue	**Lymph nodes**	**Peyer's patches**	**Skin**	**Small bowel/LP**
T cell subtype	Naïve and T_{CM}[b]	Naïve and Tcm	T_{EFF}/T_{EM}[c]	T_{EFF}/T_{EM}
Tethering/ rolling	L-selectin/PNAd α4β7/MAdCAM-1- [only MLN]	L-selectin or α4β7/ MAdCAM-1	P- and E-selectin ligands/P- and E-selectin	α4β7^{High}/ MAdCAM-1* α4β1/ VCAM-1**
Integrin activation	CCR7/CCL19, CCL21	CCR7/CCL19, CCL21	CCR4/CCL17, CCL22* CCR10/ CCL27*	CCR9/CCL25 CCR6/CCL20 (Th17)
Firm Adhesion	LFA-1/ICAM-1	LFA-1/ ICAM-1*	LFA-1/ICAM-1* α4β1/VCAM-1**	α4β7^{High}/ MAdCAM-1 α4β1/ VCAM-1**

[a] Pairs of homing receptors expressed on T cells/ligand on endothelial cells
[b] T_{CM}: central memory T cells
[c] T_{EFF}/T_{EM}: effector/memory T cells
*Expressed constitutively and increased during inflammation
**Expressed only during inflammation

whereas T_{EFF}/T_{EM} cannot home to this lymphoid compartment (12).

Recently activated T cells can migrate to multiple extra-lymphoid tissues upon activation, including liver and lungs, apparently without specificity (13, 14). It should be noted, however, that liver, lungs, and the spleen do not have tight postcapillary venules and exhibit a more open circulation due to the presence of sinusoids, which probably explains why T-cell migration to these organs does not seem to be selective (15, 16). However, whether lymphocytes require tissue-specific adhesion or chemokine receptors for their interstitial migration in these organs (e.g., to the alveolar space in the lungs) remains to be determined.

T-cell migration to other tissues in the body, in particular the intestinal mucosa, the skin, and the central nervous system, requires the expression of specific adhesion and chemokine receptors on lymphocytes and their respective ligands to be expressed in the postcapillary venules of these tissues. Homing to the central nervous system requires P-selectin ligands, α4β1 integrin and probably CCR6 on T cells (1, 17). However, the best-characterized tissues in terms of specific homing are the gut mucosa and the skin, which are also the largest surfaces exposed to the environment in the body (the total surface area of the intestinal mucosa in an adult is similar in size to a tennis court). T cell homing to the skin requires P- and E-selectin ligands (P- and E-Lig, respectively) (12, 18), whose synthesis relies on the expression of fucosyltransferase-VII (FucT-VII) (19), among other enzymes (12). In addition, T-cell migration to the skin requires the integrins LFA-1 (and probably α4β1) and the chemokine receptors CCR4 (20) and/or CCR10 (12, 21, 22). T-cell migration to the small intestine requires the integrin α4β7 (23, 24) (and also LFA-1 (25)) and CCR9 (12, 26–28). Interestingly, although recently activated CD8 T cells clearly need CCR9 for migrating to the small bowel, homing of CD4 T cells to this compartment is, at least in part, CCR9-independent (29). Homing to the colon has some very distinctive features, as it requires either α4β7 or α4β1, but not CCR9 (1, 12). Even though IgA-ASC rely on CCR10 for migrating to most mucosal sites, including the colon (1), it is currently unknown which is/are the chemokine receptors involved in T-cell homing to this tissue and why migration to the small bowel and colon is differentially regulated.

1.2. Induction of Gut-Tropic T Cells

How do naïve T and B cells acquire specificity for migrating to the gut mucosa? Whereas systemic immunization induces T cells with multiple homing capacities (13, 14, 17), the site of antigen entry determines, at least in part, the adhesion receptors that T cells acquire (30). Vaccination via the oral route induces preferential expression of α4β7, whereas parenteral immunization does not significantly induce the expression of this gut-homing integrin (1, 12).

Moreover, T cells activated in mesenteric lymph nodes (MLN) upregulate α4β7 and CCR9, whereas those activated in skin-draining peripheral lymph nodes (PLN) acquire the expression of E- and P-Lig, but not gut-homing receptors (30, 31). In fact, in the lymphoid microenvironment DC and stromal cells are sufficient to imprint tissue-specific tropism on T cells upon activation ex vivo (15, 28, 30, 32–35). DC from MLN, PP, and small intestine lamina propria (gut-associated DC), but not DC from extra-intestinal tissues, induce the expression of α4β7 and CCR9 on T and B cells. This gut-specific imprinting property depends on the selective capacity to metabolize vitamin A (retinol) into all-*trans* retinoic acid (RA) (36) and RA is necessary and sufficient to induce gut-tropism in vitro and in vivo in murine (36, 37) and human T and B cells (37, 38) (Fig. 1). RA not only induces gut-tropic T cells, but also prevents the generation of skin-homing

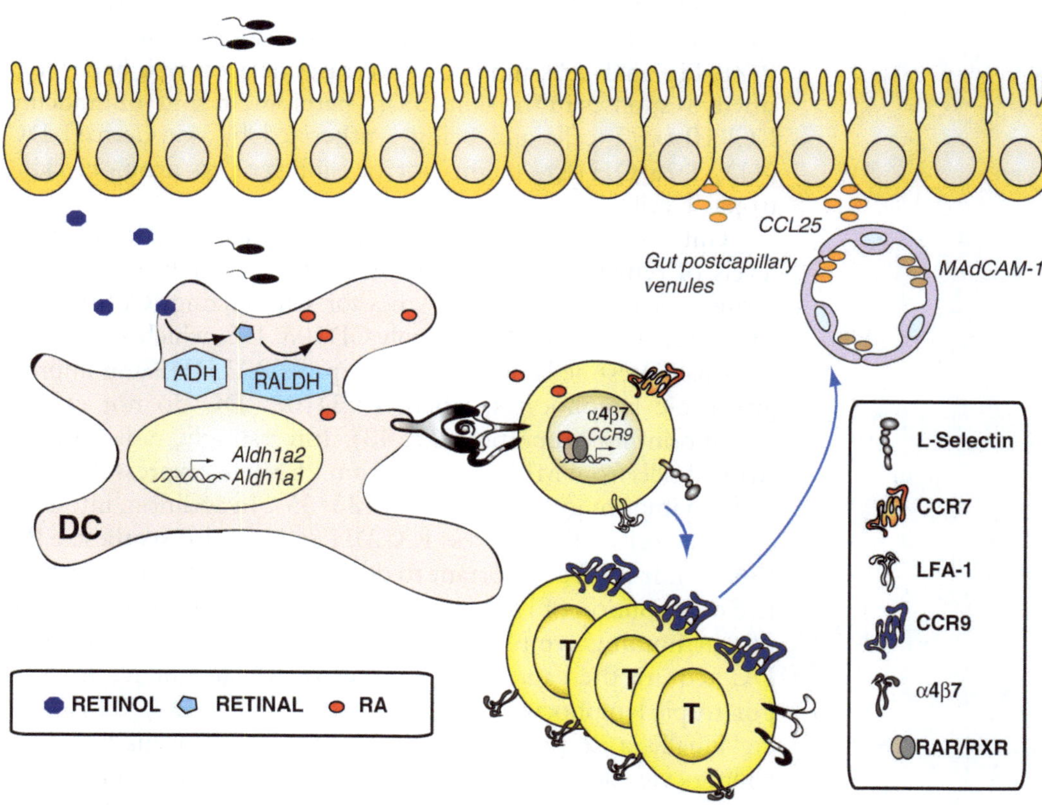

Fig. 1. Tissue-specific homing of antigen-experienced T cells. Naïve T cells in MLN and PP are activated by their cognate antigen presented by gut-associated DC. Gut-associated DC, but not extra-intestinal DC, express retinal dehydrogenases (RALDH1 and RALDH2), which are critical enzymes needed to metabolize vitamin A (retinol) into RA. Alcohol dehydrogenases (ADH, expressed in many tissues) catalyze the first reversible step from retinol to retinal and then RALDH enzymes irreversibly convert retinal into RA. RA induces gut-homing receptors α4β7 and CCR9 on T cells, a process that requires RA signaling through the RAR family of RA receptors. Gut-homing T cells adhere to the endothelial cells in small intestine postcapillary venules, which constitutively express MAdCAM-1 (α4β7 ligand) and CCL25 (CCR9 ligand), and finally transmigrate to the intestinal lamina propria.

receptors by blocking the induction of FucT-VII (39) and P- and E-Lig on T cells (36). In addition, RA decreases the expression of CCR4 mRNA (36, 39). Thus, RA is sufficient to induce gut-tropic T cells, while it reciprocally inhibits the induction of skin-homing T cells.

RA exerts its gut-homing imprinting effect on lymphocytes by acting via RA receptors of the RAR family (36). RAR need to heterodimerize to nuclear receptors of the RXR family, which can also homodimerize and heterodimerize with other nuclear receptors, such as peroxisome proliferator-activated receptor-γ (PPARγ), Vitamin D receptor (VDR) and liver X receptor (LXR) (40). Given this promiscuity in RXR dimerization, it is possible that ligands for other RXR-dependent nuclear receptors might modulate the effect of RA on lymphocytes. In fact, ligation of VDR-RXR by $1,25(OH)_2D_3$ (the physiologically most active form of vitamin D) can inhibit RA-mediated induction of gut-homing T cells (39, 41). Of note, although $1,25(OH)_2D_3$ induces CCR10 (on human T cells) (41), it blocks the expression of FucT-VII and P-/E-Lig on T cells and it also decreases delayed-type hyper sensitivity responses in the skin (39). Therefore, in spite of inducing CCR10 on T cells and being synthesized by skin-derived DC (41), $1,25(OH)_2D_3$ might not be sufficient to promote skin-tropic T cells.

Gut-associated DC, but not extra-intestinal DC, express high levels of retinal dehydrogenases (RALDH1 and RALDH2 isoforms), which are critical enzymes for synthesizing RA (36, 42). Moreover, among MLN-DC, only $CD103^+$ DC, which arrive from the small intestine lamina propria, express RALDH2 and imprint gut-tropic lymphocytes, whereas $CD103^-$ DC do not exhibit gut-imprinting properties (32, 43). Interestingly, radioresistant stromal cells from MLN, but not from PLN, also express RALDH and can imprint gut-tropic T cells (33–35). In addition, intestinal epithelial cells (IEC) express RALDH and can also synthesize RA (36, 44). It will be important to determine the interplay and relative contribution of DC, stromal cells, and IEC in the generation of gut-homing lymphocytes in vivo and how gut-associated DC acquire the capacity to synthesize RA and hence imprint gut-tropic lymphocytes.

Finally, a deeper understating of the mechanisms by which lymphocytes acquire gut-tropism and how DC are educated in the gut could offer simple and straightforward methods for boosting intestinal immune responses for vaccination purposes, such as in infections by *Salmonella*, rotavirus, and HIV, where the small intestine has been identified as a major reservoir for the virus (45) and where recent major vaccine trials have been mostly unsuccessful (46). Below we describe some methods to induce and study gut-tropic T cells in vitro and in vivo.

2. Materials

2.1. Animals

1. C57BL/6 mice: commonly used for DC isolation.
2. OT-1/RAG1$^{-/-}$, OT-2/RAG1$^{-/-}$ and P14/TCR$\alpha^{-/-}$ (Taconic Farms, Germantown, NY): used for or T-cell isolation. In addition, CD45.1 and Thy1.1 congenic strains are available through Jackson Laboratories (Bar Harbor, ME).
3. CCR9$^{-/-}$ mice: There are three published mouse strains with deficiencies in gut-homing receptors, but they are not yet commercially available (47, 48). Although these mice have only a mild phenotype in the steady state, recently activated T cells migrate around 10 times less to the small bowel as compared to wild-type T cells. Similarly, homing of IgA-ASC to the small bowel is impaired in these mice (49).
4. β7 integrin$^{-/-}$ mice (Jackson labs): These mice have a dramatic decrease in the number of lymphocytes even in the steady state. In addition, Peyer's patches are very small in this strain (50).

2.2. Culture Media

1. IMDM (Iscove's Modified Dulbecco's Medium + l-Glutamine + Hepes) plus 10% heat-inactivated FBS (Fetal Bovine Serum) supplemented with 100 U/ml penicillin, 100 mg/ml streptomycin, 0.5 mg/ml fungizone/amphotericin B, and 50 μM β-mercaptoethanol.
2. FBS-free (retinol-free) media (X-vivo 15, Lonza, Basel, Switzerland): regular FBS contains variable amounts of retinol. This culture media support mouse and human T-cell activation without needing the addition of FBS.
3. Red Blood Cell Lysis buffer (RBC, 10 mM $KHCO_3$, 150 mM NH_4Cl, 0.1 mM EDTA, pH 8.0), adjust to pH 7.2–7.4 and store at room temperature).
4. PBS (Phosphate-Buffered Saline).
5. Flow cytometry (FACS) media (PBS or IMDM + 2% FBS + 5 mM EDTA): When staining using Selectin-Fc chimeras, media with 2 mM Ca^{++} should be used in all steps (including FACS acquisition). IMDM is recommended in this case.

2.3. T-Cell Labeling and Adoptive Transfer

1. CFSE (carboxyfluorescein diacetate, succinimidyl ester).
2. CMTMR ((5-(and-6)-(((4-chloromethyl)benzoyl)amino) tetramethylrhodamine) (Molecular Probes®, Invitrogen, Carlsbad, CA). 1,000 × stocks should be made in DMSO (5 mM CFSE, 20 mM CMTMR) and stored at −20°C.

2.4. Flow Cytometry (FACS) Staining

1. Polyclonal activation: CD3 (145-2C11), CD28 (37.51).
2. Lineage mAb: CD4 (GK1.5), CD8a (53-6.7), CD11c (N418), CD90.1 (Thy1.1/HIS51), CD90.2 (Thy.12/53-2.1), CD45.1 (A20), CD45.2 (104).
3. Antibodies to gut-homing receptors: purified CCR9 (CD199/eBioCW-1.2, eBioscience, San Diego, CA) plus secondary reagent F(ab')2 fragment of goat anti-mouse (H+L), α4β7 (LPAM-1/DATK3), isotype control (IgG2a, k), α4 (R1-2), β7 (M293), and αE/CD103 (M290) integrin chains.
4. Skin-homing receptors: P-selectin-Fc (Purified Mouse P-Selectin – IgG Fusion Protein, BD Pharmingen™, San Jose, CA), E-selectin-Fc (Recombinant Mouse E-Selectin/Fc Chimera, R&D Systems, Minneapolis, MN) plus corresponding secondary reagent goat F(ab')2 anti-human IgG R-PE.

2.5. Retinoids and RALDH Inhibitors

1. All-*trans* retinoic acid, retinol (both from Sigma, St. Louis, MO): resuspended in absolute ethanol or DMSO, use a yellow bulb or an indirect source of light during the preparation. Store aliquots in glass vials at −80°C and protect from light at all times.
2. Synthetic RAR-agonists: Am80 (Wako Chemicals, Richmond, VA).
3. RAR antagonists: LE540, LE135 (Wako Chemicals, Richmond, VA).
4. RALDH inhibitors: Citral (Sigma, St. Louis, MO), diethylaminobenzaldehyde (DEAB) and bisdiamine (both from Acros Organics, Geel, Belgium).

2.6. Protein and Peptide for Transgenic Activation

1. OT-1 peptide: chicken ovalbumin$_{257–264}$ (SIINFEKL).
2. OT-2 peptide: chicken ovalbumin$_{323–339}$ (ISQAVHAAHAEI NEAGR).
3. P14 peptide: LCMVgp$_{33–41}$ (KAVYNFATC) (New England Peptides, Gardner, MA). Make 1–10 mM stocks in culture grade water (*no additives of any kind*) and store at −80°C.

2.7. Primers Real-Time PCR Using SYBRGreen

1. FucT-VII (*Fuct7*): forward, ACTGATGTTGAAACCAAA GAGC, and reverse, GCCCAGTCTTCTCCTTATATCC
2. CCR4: forward, GGTACCTAGACTACGCCATCC, and reverse, ATGTACTTGCGGAATTTCTCC
3. α4 integrin chain: forward, AAACACTGGGATTAGCATGG, and reverse, ATTGCCCTGTAGTTGTCTGG
4. CCR9: forward, AGGTTAGTCAGCCAATGTACAGC, and reverse, ATCCTTTCCTAGTTTGTGCTTGC
5. RALDH1 (*Aldh1a1*): forward, ACAAGGTGGCCTTCACT GGA and reverse, GCAAACACAATGCAAGGGCT

6. RALDH2 (*Aldh1a2*): forward, TGGGTGAGTTTGG CTTACGG and reverse, AGAAACGTGGCAGTCTTGGC
7. RALDH3 (Aldh1a3): forward, TGTAGAAAGGGACCGAG CGAT, and reverse, CCCGGCAAAATATCTGAAGGT
8. GAPDH: forward, CAACTTTGTCAAGCTCATTTCC and reverse, GGTCCAGGGTTTCTTACTCC. mRNA levels can be expressed relative to GAPDH mRNA for each sample.

2.8. TaqMan qPCR for Aldh1a2 (Codifying for RALDH2)

1. RNeasy (Qiagen).
2. iScript cDNA synthesis kit (Bio-Rad).
3. TaqMan PCR master mix and the following TaqMan kits (Applied Biosystems): *Aldh1a2* (Mm00501306_m1) and *β-actin* (4352933) for normalization.

2.9. Aldefluor Assay for RALDH Activity

1. ALDEFLUOR FACS-based assay kit (StemCell Technologies, Vancouver, BritishColumbia, Canada): used for measuring RALDH activity (42).

3. Methods

3.1. Ex Vivo Induction of Gut-Homing Receptors

Gut-homing receptors can be induced on T and B cells either by coculturing them with gut-associated DC or by activating them in the presence of RA or a synthetic RA receptor (RAR) agonist, such as Am80 (40). Gut-associated DC allow us to study defined DC subsets involved in imprinting gut-homing lymphocytes. However, obtaining high numbers of gut-associated DC is technically challenging and logistically unpractical for studies involving human T cells. On the other hand, RA or RAR-agonists allow the generation of large number of gut-tropic T cells expressing high levels of both α4β7 and CCR9.

3.1.1. Ex Vivo Induction of Gut-Tropic T Cells Using Gut-Associated DC

Gut-associated DC can be isolated from either unmanipulated mice (in which DC have not been expanded) or from mice that have been pre-treated with the cytokine Flt3L (see Note 1).

Dendritic Cell Isolation

1. C57BL/6 mice are injected subcutaneously with B16 melanoma cells secreting Flt3-L to expand DC in vivo as described (15, 37).
2. After 12–17 days, the mice are sacrificed and single-cell suspensions are generated by digesting lymph nodes (PP, MLN, spleen) using 250 μg/ml Liberase CI (Roche, Indianapolis, IN) and 100 μg/ml DNase I (Roche, Indianapolis, IN) 40 min at 37°C with mild agitation.
3. DC are immunomagnetically isolated using MACS CD11c microbeads (Miltenyi Biotec, Auburn, CA) according to the

manufacturer's instructions. DC purity will be around 90%, as determined by CD11c staining. FACS-based sorting can follow magnetic separation in order to improve DC purity or for isolating specific DC subsets, such as $CD103^+$ DC. If unmanipulated mice are used, one should start from at least 3–5 mice. In Flt3L-treated mice (recombinant cytokine or injected with B16-Flt3L melanoma), 1 mouse might be enough to obtain around 1×10^6 $CD11c^+$ cells from PP and 4×10^6 from MLN.

4. Collect SLO (MLN, PP and PLN as controls) and isolate DC by negative or positive immunomagnetic selection. Check purity by FACS (see Note 2).
5. Load DC with peptide or protein (see Subheading 2 for working concentrations) in complete IMDM for 2 h at 37°C. Wash twice and resuspend in complete IMDM (see Note 3).

Labeling with CFSE

Naive $CD4^+$ and/or $CD8^+$ T cells can be obtained from splenocytes after red blood cell lysis using ACK buffer (see Subheading 2), followed by either negative or positive immunomagnetic selection. The proportion of T cells will be much higher when using splenocytes from TCR transgenic mice in a $RAG^{-/-}$ background due to the lack of B cells (see Note 4).

1. Splenocytes from C57BL/6 mice or TCR transgenic mice are used for T-cell isolation. Cells are centrifuged at $300 \times g$ for 5 min and resuspended in 2 mL of RBC lysis buffer for 3 min at room temperature (longer incubation times might cause significant decrease in T-cell viability).
2. Cells are washed with IMDM, centrifuged as before and finally resuspended in 10 ml of complete IMDM. T cells can be isolated by negative selection, as described (15) or by using CD4/CD8 microbeads for positive selection or the Pan T-cell isolation kit for negative selection (both from Miltenyi Biotec). Negative selection might be preferred, as it does not "touch" T cells.
3. For CFSE labeling, lymphocytes are washed twice with warm PBS to remove the serum. The cell concentration is adjusted to 10^7/ml in PBS (see Note 5).
4. Add CFSE to a final concentration of 1–5 μM, vortex and incubate for 10–15 min in a water bath or an incubator at 37°C. The optimal working concentration should be determined depending on the final application. For in vitro studies it is better to use <5 μM, since it makes it easier to adjust the compensations for FACS analysis. On the other hand, 5–10 μM CFSE gives more consistent results for in vivo adoptive transfer experiments.

5. After the incubation period, add 1 volume of FBS to quench the CFSE and incubate for 1–5 min at room temperature. Then, dilute ten times with warm PBS and centrifuge for 5 min at 300 × *g*. After that, perform an additional washing step with PBS and resuspend the cells in complete IMDM. It should be kept in mind that around 20–30% of the cells can be lost during the labeling, so incubation times should be limited to a minimum.

Polyclonal T-Cell Activation

1. 24-well or 96-well plates are incubated for 2 h at 37°C with 300 μL or 50 μl PBS, respectively, containing 10 μg/ml anti-CD3 plus anti-CD28 antibodies. After that, the culture plates are washed two times with PBS and used immediately for T-cell culture. Plates can be stored at 4°C for 24 h, provided that the wells are covered with PBS to avoid drying.
2. Dynabeads coated with anti-CD3/anti-CD28 ("artificial APC," Dynal, Invitrogen, Carlsbad, CA) can also be used for T-cell activation instead of plate-bound antibodies. However, it should be kept in mind that the expression levels of homing receptors might vary depending on the method of T-cell activation. Therefore, the method of choice should be adapted depending on the final aim of the experiment.

DC/T Cell Co-cultures

1. Flat-bottom 96-well or 24-well plates can be used. For polyclonal activation use anti-CD3 plus anti-CD28 pre-treated plates (see Note 6).
2. Mix 1–2 × 10^6 naïve T cells at 1:1 T:DC ratio (antigen pulsed-DC). Coculture in complete IMDM in a final volume of 1.5–2.0 mL for 24-well plates or 150–200 μL for 96-well plates. Culture at 37°C and 5% CO_2 for 5–7 days. At day 3 replace half of the media if necessary. After that, analyze the cells by FACS (see Note 7).

3.1.2. Ex Vivo Induction of Gut-Tropic T Cells with RAR-Agonists

1. Resuspend T cells at 1 × 10^6/ml in complete IMDM and add all-*trans* RA (or RAR-agonists, such as Am80 or Am580) to a final concentration of 10–200 nM. The induction of gut-homing receptors is dose-dependent and reaches a plateau around 100–200 nM RA (see Note 8).
2. Add 1.5–2.0 mL of the cell suspension to each well of a 24-well plate pre-treated with anti-CD3 plus anti-CD28 and culture at 37°C and 5% CO_2.
3. After 72 h, transfer the cells (in the same media) into a new untreated 24-well plate and culture for an additional 2–3 days. Depending on cell density and proliferation the media might become yellow (acidic), in which case it might be necessary to replace half of the media with fresh complete IMDM.
4. Evaluate the expression of homing markers (see Note 9).

3.1.3. Analysis of Gut-Homing Receptors

The induction/expression of homing markers can be analyzed at the mRNA level using quantitative RT-PCR, and/or at the protein level by FACS. The mRNA levels are usually expressed relative to a housekeeping gene such as GAPDH or β-actin (see Note 10). Among the receptors analyzed are the gut-homing receptors α4β7 and CCR9 and skin-homing receptors including E- and P-selectin ligands and some of the enzymes involved in their synthesis (Core-2/C2GlcNAcT-I, FucT-VII, FucT-IV), and chemokine receptors CCR4 and CCR10. Of note, gut- and skin-homing receptors are reciprocally regulated by RA. Whereas RA induces α4β7 and CCR9, it markedly suppresses the expression of P- and E-selectin ligands, CCR4 and FucT-VII (40) (see Note 11).

Different time points can be used for analyzing the expression of homing receptors on T cells, starting as early as day 2 of culture. Gut-homing receptors (α4β7 and CCR9) are clearly upregulated starting at day 3 of culture with gut-associated DC or RAR agonists.

3.1.4. FACS Analysis

1. For each staining collect between $0.2–0.5 \times 10^6$ cells and centrifuge for 5 min at $300 \times g$ at 4°C.
2. Incubate with the primary antibodies anti-α4β7 or anti-CCR9 for 20 min at 4°C in the dark. E-selectin-Fc and P-Selectin-Fc chimeric proteins can be used to detect the skin-homing receptors P- and E-selectin ligands (see Note 12).
3. After the incubation, centrifuge for 5 min at $300 \times g$ 4°C. Remove the media and incubate with the appropriate fluorochrome-conjugated secondary antibody. In the case of P- and E-selectin-Fc chimeras, an anti-human Fc antibody can be used as secondary reagent. Incubate for no more than 30 min at 4°C.
4. Centrifuge for 5 min at $300 \times g$ 4°C, resuspend in staining buffer, and analyze by FACS.

3.2. Functional Assessment of Gut-Homing T Cells

3.2.1. Chemokine Receptor Functionality

Chemotaxis experiments can be performed after 4–5 days of culture. Each assay should be performed at least in duplicate.

1. Make 10 μM chemokine stocks in PBS + 0.1% BSA and store them at −80°C. At the moment of using the chemokines dilute them in complete IMDM to reach a final working concentration of 0.1–500 nM, depending on the chemokine (see Note 13).
2. Add 600 μL complete IMDM (control condition) or complete IMDM + chemokine in the lower chamber of the 24-well transwell (Corning, 5 μm pore size for T cells).
3. Carefully place the transwells into the wells (see Note 14).
4. Add 100 μl of the cell suspension ($2–5 \times 10^6$/ml) in the upper chambers (i.e., $2–5 \times 10^5$ cells/well). See Note 15.

5. Cover the plate and incubate 1 h at 37°C and 5% CO_2 (see Note 16).
6. After the incubation period, carefully remove the transwell inserts from each well and resuspend the media in the lower chamber. Take an aliquot and read by FACS (see Note 17).
7. The *chemotactic index* is calculated as: the number of cells migrating to the media plus chemokine divided by the number of cells migrating to the media alone. The *percentage of migration* is calculated as the number of cells migrating to the media alone, or the media plus chemokine divided by the total number of cells (obtained from the well in to which the total number of cells were added). See Note 18.

3.2.2. In Vivo Induction of Gut-Homing T Cells

In vivo induction of gut-homing T cells is a fast and relatively simple strategy to assess the impact of genetic and/or pharmacological manipulations in the generation of gut-tropic T cells. TCR transgenic T cells are commonly used to evaluate the expression of gut-homing molecules upon in vivo immunization. T cells can be pre-labeled with CFSE in order to discriminate transferred T cells from the endogenous population and to track T-cell activation and proliferation. In addition, CFSE labeling can also be combined with congenic markers (CD45.1/2 or Thy1.1/2), which makes it easier to identify the transferred T cells by FACS.

1. Inject 3–5 × 10^6 $CFSE^+$-labeled TCR transgenic T cells i.v. (via tail vein) into C57BL/6 recipient. OT-1 or OT-2 T cells are commonly used for this purpose and they can be in a $RAG^{-/-}$ background to exclude the possibility of having preformed effector/memory T cells. In order to easily discriminate the transferred from the endogenous T cells, congenic $Thy1.1^+$ recipient mice can be used. Given that only T cells will be activated upon immunization, total splenocytes (without T-cell isolation) can also be used for CFSE labeling and adoptive transfer (see Note 19).
2. Next day, immunize with specific protein or peptide plus adjuvant i.p. or via oral gavage (see Subheading 2). The adjuvant is usually LPS, but other adjuvants, such as alum, can also be used (28) (see Note 20). If OT-1 CD8 or OT-2 CD4 T cells are used, most protocols immunize using 5 mg ovalbumin plus 100 μg LPS i.p in 300 μl PSB (total volume).
3. Mice can be euthanized at different time points, starting on day 2 post-immunization. However, clear induction of gut-homing receptors in MLN (and skin-homing receptors in PLN) is only seen starting at day 3. At that time, T cells can also be found in the small bowel lamina propria and IEL compartment (see Note 21).
4. The expression of α4β7 and CCR9 should be significantly higher when analyzing T cells activated in MLN and PP as

compared to those activated in PLN. Conversely, P- and E-selectin ligands (detected by the corresponding P- and E-selectin-Fc chimeras) are readily induced in PLN, but at a much lower degree in MLN and PP (31).

3.2.3. Competitive Homing Assays

Competitive homing experiments are used to directly compare the migratory capacity of different T-cell populations in the same mouse. This allows us to minimize the effects of variables such as the number of injected T cells, number of isolated/analyzed T cells in each organ, or eventual difference in the animal size. In a competitive homing experiment, two different cell populations (e.g., T cells treated plus/minus RA or T cells from CCR9 knockout vs. wild-type mice) are mixed in a 1:1 ratio and adoptively transferred into a recipient mouse (ideally a congenic Thy1.1^{+} or CD45.1^{+} host). The aim of the experiment is to determine the relative migration of one T-cell population with respect to the other (differentially labeled) T-cell population in a given tissue. Since the results are expressed as the ratio of two T-cell populations, the analyses do not rely on absolute cell counts. If absolute numbers are needed, special care needs to be taken in order to control the total number of T cells injected and recovered from each mouse. Moreover, total T cells need to be carefully counted in every tissue (see Note 22).

1. The different T-cell populations injected can be distinguished by using a congenic marker, e.g., CD45.1/CD45.2 or Thy1.1/Thy1.2. Alternatively, T cells can be differentially labeled, e.g., using CFSE (green) or CMTMR (orange) (see Note 23).
2. After labeling, both cell populations are mixed together (in equal proportions), resuspended in no more than 250 μl PBS (total volume), and adoptively transferred i.v. via tail vein injection (using a tuberculine syringe with a 30G needle). See Note 24. Some cells should be saved after injection for calculating the input ratio (which should be close to 1).
3. The mice can be euthanized at different time points, although significant T-cell migration to the gut is usually achieved only after 12 h post-injection, even when using bona fide gut-homing T cells (see Note 25).

3.2.4. Labeling T Cells with CFSE & CMTMR

1. Wash T cells twice with PBS to remove the serum. Adjust T-cell concentration to $1.0–1.5 \times 10^{7}$/ml (maximum) in PBS. All media/buffers should be at 37°C.
2. Add CFSE or CMTMR to a final concentration of 5 and 10 μM, respectively, vortex, and incubate for 20 min at 37°C.

3. After the incubation, add 1 volume of FBS and incubate for 1–5 min at room temperature. Dilute ten times with warm PBS and centrifuge for 5 min at 300 × *g*. Wash twice with PBS and resuspend in complete IMDM. Count the cells after the labeling prior to injection.
4. Resuspend the cells in up to 250 μL of warm PBS. Inject i.v. via tail vein (see Note 26). Ideally, 10–20 × 10^6 cells of each T population should be injected (1:1 ratio) (see Note 27).
5. To calculate the input ratio leave 5–10 μL of the injected cell suspension and dilute into 300 μL of FACS buffer and analyze by FACS (see Note 28).

3.2.5. Analysis of T-Cell Homing

The right time to analyze T-cell migration varies depending on the aim of the experiment, the T-cell populations injected, and the tissue analyzed. Analysis of naïve T-cell homing to PLN can be done as fast as 1 h post-injection. The advantage of shorter times is that any differences observed in T-cell homing can be more confidently ascribed to T-cell entry rather than to T-cell exit rate (or proliferation/survival). However, shorter times will not allow for observing clear T-cell migration to some other tissues, such as the gut or the inflamed skin, even when using bona fide gut- or skin-homing T cells. Typically, homing to the gut mucosa is analyzed 12–24 h after T-cell injection. Mice are euthanized and the different tissues are collected in order to make single cell suspensions for FACS analysis. Collect lymphoid tissues including the spleen, PLN, MLN, PP, bone marrow, and peripheral tissues including liver, lungs, small, and large bowel. Isolation of gut lamina propria and intraepithelial lymphocytes has been previously described (15) (see Note 29).

1. FACS staining: depending on the number of cells obtained, the samples will need to be diluted at a density no higher than 3.0 × 10^6 cells/ml. Some tissues, including popliteal lymph nodes, gut lamina propria, and IEL, do not give high T-cell yields. In those cases the whole sample should be used for staining. If congenic CD45.1$^+$ or Thy1.1$^+$ mice were used as recipients, label the cells with the congenic marker (e.g., CD45.2 or Thy1.2) combined with some lineage marker (e.g., TCRβ chain, CD4, CD8, CD45.1$^+$/CD45.2$^+$).
2. Analyze by FACS. During FACS analysis, transferred T cells are distinguished from the endogenous T cells by the congenic marker and because they are labeled with CFSE or CMTMR.
3. The data is usually expressed as the Homing Index (HI), which is calculated as the ratio CFSE/CMTMR (or CMTMR/CFSE) in each tissue divided by the input ratio. If the input

ratio is very close to 1, then the tissue ratios will be equivalent to the HI. In addition, when the HI is significantly different than 1 in the blood, tissue HI can be normalized by blood HI (see Notes 29 and 30).

$$HI = [CFSE]_{tissue} / [CMTMR]_{tissue} : [CFSE]_{input} / [CMTMR]_{input}$$

4. Notes

1. Using unmanipulated mice has the advantage of eliminating any potential known or unknown artifacts derived from DC expansion. For instance, even though Flt3L dramatically expands all the classically described DC subsets in most tissues, including CD103^{+} DC, it might change the relative DC subset composition. Therefore, DC expansion should not be used when studying the physiological composition of DC subsets in different tissues. On the other hand, if the aim is to study gut-specific imprinting, including induction of gut-homing lymphocytes and/or IgA-ASC, we and others have demonstrated that gut-associated DC isolated from Flt3L-treated mice are equivalent to their counterparts isolated from unmanipulated mice (28, 36, 42) and to freshly isolated DC from human MLN (37, 38).
2. While the isolation protocols might affect DC's maturation kinetics, they should not significantly affect the gut-homing imprinting capacity of gut-associated DC. Nonetheless, their viability tends to be lower than extra-intestinal DC, so they should be used for co-cultures immediately after isolation.
3. The concentration of antigenic peptide or protein used to pulse DC plays an important role in the final induction of gut-homing receptors (51). For instance, when activating OT-1 TCR transgenic T cells, pulsing DC with peptide concentrations above 200 nM dramatically decreases the induction levels of $\alpha 4\beta 7$ and CCR9. The mechanism for this effect is unclear, but it also holds true when using exogenous RA instead of gut-associated DC to imprint gut-homing T cells.
4. Using T cells from TCR transgenic mice (like OT-1 or P14 mice), where T cells are activated by a specific peptide presented by antigen-presenting cells, is a valuable tool for studying both in vitro and in vivo induction and expression of gut-homing molecules. Nonetheless, polyclonal T cells can also be used to generate gut-tropic T cells by activating them with polyclonal activators (e.g., anti-CD3/CD28-coated plates or Dynabeads coated with anti-CD3/CD28) in the

presence of either RA or gut-associated DC in *trans* (without presenting antigenic peptide).

5. CFSE labeling is an optional step. It has the advantage of allowing the comparison in the expression of gut-homing receptors among cells that have proliferated equivalently.
6. Adding IL-2 improves T-cell viabilities and final yields. However, adding exogenous IL-2 also decreases the induction of gut-homing receptors in a concentration-dependent manner, so it should be avoided when generating gut-tropic T cells (15).
7. For $CD8^+$ T cells the induction of gut-homing receptors can be analyzed as early as day 3 (52). However, the peak in the induction of gut-homing receptors is reached around day 5. Also, CD4 T cells tend to express lower levels of gut-homing receptors as compared to CD8 T cells.
8. RA is extremely sensitive to light, heat, and oxidation. Ideally, the stock solutions should be made at ~10 mM in ethanol (from an unopened bottle containing little or no water) and replaced every 3 months from a new vial. From those stocks, 1 mM working solutions should be prepared at least every month. All the stock solutions should be kept at −80°C in air-tight glass vials filled with N_2 gas and wrapped with aluminum foil. New RA vials should be opened in the dark (using only a small yellow light). Similarly, all the work involving RA (including co-cultures) should be done in the dark (Makoto Iwata, Tokushima Bunri University, Japan, personal communication).
9. If the protocol was successful, 90–100% of T cells should become $CCR9^+$ and $\alpha4\beta7^+$. It is also important to consider the level of gut-homing receptor expression (MFI), as it is strongly correlated with the final T-cell gut-homing capacity (at least for α4β7) (15). The final MFI will depend on several factors, including the concentration of RAR agonist used, antigenic peptide/protein concentration used to load DC (lower concentrations will lead to higher MFI), addition of cytokines (e.g., IL-2 which has a negative effect), time of culture (>3 days for higher MFI), and cell viability.
10. Even though it is normally assumed that "housekeeping" genes do not change with different treatments and therefore can be used to normalize the expression of other genes, there might be some exceptions. In particular, GAPDH might not be a good choice when comparing naïve T cells vs. activated/effector T cells, as they markedly differ in their metabolic rate, which might cause significant variations in the GAPDH levels.
11. When looking at the expression of α4β7 integrin, it should be kept in mind that it is composed of two chains (as are all integrins)

and that therefore its final expression could be determined by the expression and interplay of at least four integrin chains. α4 can pair with either β7 (α4β7) or β1 (α4β1) integrin chains. On the other hand, the β7 chain can pair with either α4 (α4β7) or αE/CD103 integrin (αEβ7) chains. Therefore, isolated analysis of individual α4 or β7 integrin chains might not necessarily reflect the surface level of the α4β7 heterodimer. The same consideration should be kept in mind when analyzing mRNA for individual integrin chains. Of note, the levels of α4 integrin chain (mRNA and surface protein) seem to correlate better with the levels of α4β7 as compared to other integrin chains, at least for CD8 T cells (52). Since CD4 T cells do not normally express αEβ7, the β7 chain can be more reliably used in this case as a surrogate for α4β7. Finally, a mAb is available to specifically label α4β7 heterodimer in the mouse (DATK32 clone) (53). Another clone exists for labeling human α4β7 (Act-1) (54), although it is currently not commercially available.

12. The binding of E- and P-selectin to E- and P-Lig on T cells is calcium dependent. Therefore, media with a physiological concentration of Ca^{++} should be used at all times (including FACS acquisition) when labeling with these selectin chimeras. DMEM is a good choice and can be supplemented with 1% FBS + 20 mM Hepes. Also, since the binding of selectins to their ligands has a much lower affinity as compared to antibodies, all pipetting and manipulations should be done gently. As negative controls, staining with E- and P-selectin chimeras can be done in a buffer with 5 mM EDTA (Ca^{++} chelator).
13. When using a chemokine for the first time in a given cell type it is recommended to perform a dose–response using 4–5 different chemokine concentrations (many chemokines exhibit a bell-shaped dose–response curve).
14. If media diffuses immediately to the upper chamber it could mean that the transwell is damaged and should be replaced.
15. Include additional wells with 500 μl media plus 100 μl of cells (directly added to the lower well, without transwells) in order to have a maximum total number of cells and then calculate percentages of migration with respect to total cells.
16. The optimal final incubation period will depend on the cell type, pore diameter, chemokine, etc. Nonetheless, try to use the same incubation times among different experiments in order to obtain more reproducible results.
17. Usually 30 s of FACS acquisition per sample should be enough. Acquire only the cells that are in a gate of viable cells. An aliquot of fluorescent beads can be added to each well in order to normalize the cell counting (optative). In addition, the migrated T cells can be stained for different markers (optative).

18. Although chemotaxis is a convenient standard readout for chemokine functionality, it does not necessarily assess the actual physiological function of a given chemokine receptor. For instance, very high concentrations of CCL25/TECK (200–300 nM) are needed in order to detect CCR9 functionality with this assay (in comparison, peak responses to CCL21/SLC are obtained at 10 nM). However, the physiological concentrations of CCL25/TECK needed to signal via CCR9 and activate α4β7 in order to trigger T-cell arrest on endothelial cells might be much lower. In fact, clear calcium flux responses can be obtained at 1–10 nM CCL25/TECK (55), which is significantly lower than what is needed to trigger chemotaxis (200–300 nM). In order to test integrin activation by chemokines ex vivo, a better and more physiological assay is the flow chamber (for a detailed description on this technique, see Ref. (56)). Finally, the in vivo physiological roles of a chemokine–chemokine receptor pair (or any given adhesion receptor) should be assessed by performing homing assays and/or intravital microscopy.
19. Ideally T cells and recipient mice should be matched by sex. Nonetheless, cells can be adoptively transferred from male into male or from female into male, but not from male into female (male cells are rejected in female hosts).
20. Oral gastric gavage will mostly target gut-associated DC and will lead to a preferential activation of T cells in PP and draining MLN. By contrast, i.p. injection will induce T-cell activation in all SLO, including PP, MLN, PLN, and spleen.
21. Spleen should be analyzed first, as this is a tissue where T cells can be easily detected by FACS. Analysis of the spleen will also give a clear idea regarding the efficiency of T-cell activation (CFSE dilution) and additionally can be used to set the gating strategy and FACS compensations.
22. Homing experiments provide valuable information about the migration of total cell populations in any given tissue. The main advantages of homing experiments are that they are relatively fast to set up, they can be done competitively by co-injecting 2 differentially labeled T-cell populations, and they allow for the simultaneous examination of multiple tissues/organs and cell populations by FACS. However, it should be kept in mind that homing experiments do not directly assess endothelial adhesion and therefore do not discriminate in which step(s) of the multistep adhesion cascade (tethering/rolling, activation, or sticking) a given homing receptor is acting. The gold standard to define the specific role of a homing receptor in the adhesion cascade is intravital microscopy (IVM), in which individual fluorescently labeled T cells (or other cells, or even fluorescently labeled beads) are directly observed interacting

with endothelial cells in real time. IVM is a powerful technique that has greatly contributed to define the precise step where a homing receptor is acting in the adhesion cascade. Moreover, IVM can additionally provide information about other valuable parameters, such as lymphocyte velocity, vessel diameter, and the particular venules/vessels involved in lymphocyte adhesion in a given tissue. However, IVM is time consuming, involves mouse anesthesia and laborious surgical procedures and requires a large number of pure and homogeneous cell populations to label and inject. Moreover, some tissues, such as the central nervous system and the thymus, are difficult to access for IVM. Finally, IVM preparations require tissue immobilization that is difficult to attain in some organs/ tissues, such as liver, lungs, and intestine. In fact, few studies have been done using IVM in the gut (57). For a detailed discussion on IVM techniques, see ref. (58). In summary, homing experiments and IVM provide unique and often complementary information about the role of specific homing receptors in tissue-specific cell migration in vivo.

23. In order to control for unwanted effects of the fluorescent labeling on lymphocyte migration (this is especially important for B cells (59), the labeling should be swapped among different experiments or among different injected mice (e.g., cells labeled with CFSE should be labeled with CMTMR in the next experiment and vice versa).

24. When labeling the two T-cell populations with CFSE and CMTMR, the differentially labeled cells should be mixed only immediately before i.v. injection (using 200 μl PBS). This will help to prevent label transfer among T cells and T-cell clumping.

25. It should be kept in mind that results obtained after longer time points might be influenced not only by T cell entry, but also by eventual differences in T-cell exit and, sometimes, by differential T-cell proliferation and/or death.

26. The mice can be warmed with an infrared lamp 5–10 min before the adoptive transfer to dilate the tail vein and make i.v. injection easier.

27. Mice should not be injected with more than 50×10^6 total cells (utilize a lower number if T cells are recently activated blasts), as it increases the likelihood of embolism, respiratory distress, and death of the animal.

28. Due to the high fluorescence intensity of CFSE and CMTMR sometimes it is hard to compensate the FL1-H and FL2-H channels properly. However, rather than perfect compensations, one should aim to clearly distinguish both T-cell populations in order to calculate their ratio.

29. Blood should also be obtained in order to determine whether both T-cell populations are present in the circulation or whether one T-cell population is decreased with respect to the other (e.g., preferential trapping of one T-cell population in lungs/liver or decreased viability might affect the homing index [HI] in blood). This might be a problem when comparing naïve or resting memory T cells vs. recently activated T cells, since the latter might be trapped to a greater extent in the lungs. If the HI in blood is significantly different from 1, one can normalize HI in tissues by the HI in blood. Blood can be obtained via cardiac puncture from mice anesthetized (with avertin or isofluorane prior to euthanasia) and should be lysed twice with ACK buffer before using for FACS staining.
30. The main advantage of expressing the data as HI is that it makes the results independent of the numbers of T cells injected/collected from every tissue or from the size of the animals. However, it does not provide information about the actual magnitude of T-cell migration into each tissue. In fact, in tissues with very poor T-cell migration and in which very few events can be detected by FACS (such as in the non-inflamed colon), the HI calculation can be very misleading. As a general rule, HI should be used only when a sufficient number of events can be detected by FACS, so that at least one of the labeled T-cell populations can be clearly indentified in any given tissue. In fact, HI should be ideally combined with absolute counting and calculations of the numbers of T cells migrating to a given tissue with respect to the total number of transferred T cells (e.g., usually expressed as number of T cells in a tissue per 10^6 transferred T cells). It should be kept in mind that determining absolute numbers of T cells in each tissue is time consuming, as it requires meticulous lymphocyte isolation and precise cell counting in each tissue analyzed. It also relies on careful i.v. injection technique to assure that most T cells are actually injected into the mice.

Acknowledgments

We thank Susan Davis and Allison McNulty for editorial assistance. JRM is grateful to Ingrid Ramos for constant support. EJV was supported by a grant from Crohn's & Colitis Foundation of America (CCFA). MR was supported by Fondecyt grant 1100448. MRB was supported by Fondecyt grant 1100557. JRM was supported by grants from CCFA, Cancer Research Institute (CRI), Howard H. Goodman (MGH), Massachusetts Life Science Center (MLSC) and NIH Director's New Innovator Award DP2 2009A054301.

References

1. Mora, J.R., Von Andrian, U.H. (2006). Specificity and plasticity of memory lymphocyte migration. *Curr. Top Microbiol. Immunol.* **308**, 83–116.
2. von Andrian, U.H., Chambers, J.D., McEvoy, L.M., Bargatze, R.F., Arfors, K.E., Butcher. E.C. (1991). Two-step model of leukocyte-endothelial cell interaction in inflammation: Distinct roles for LECAM-1 and the leukocyte β_2 integrins in vivo. *Proc. Natl. Acad. Sci. USA.* **88**, 7538–7542.
3. Springer, T.A. (1990). Adhesion receptors of the immune system. *Nature* **346**, 425–433.
4. Butcher, E.C. (1991). Leukocyte-endothelial cell recognition: Three (or more) steps to specificity and diversity. *Cell* **67**, 1033–1036.
5. von Andrian, U.H., Mackay, C.R. (2000). T-cell function and migration. Two sides of the same coin. *N. Engl. J. Med.* **343**, 1020–1034.
6. Gowans, J.L., Knight, E.J. (1964). The route of re-circulation of lymphocytes in the rat. *Proc. R. Soc. Lond. B* **159**, 257–282.
7. Gowans, J.L. (1996). The lymphocyte–a disgraceful gap in medical knowledge. *Immunol. Today* **17**, 288–291.
8. Sallusto, F., Lenig, D., Forster, R., Lipp, M., Lanzavecchia, A. (1999). Two subsets of memory T lymphocytes with distinct homing potentials and effector functions. *Nature* **401**, 708–712.
9. Unsoeld, H., Krautwald, S., Voehringer, D., Kunzendorf, U., Pircher, H. (2002). Cutting edge: CCR7+ and CCR7- memory T cells do not differ in immediate effector cell function. *J. Immunol.* **169**, 638–641.
10. Wherry, E.J., Teichgraber, V., Becker, T.C., Masopust, D., Kaech, S.M., Antia, R., von Andrian, U.H., Ahmed, R. (2003). Lineage relationship and protective immunity of memory CD8 T cell subsets. *Nat. Immunol.* **4**, 225–234.
11. Debes, G.F., Hopken, U.E., Hamann, A. (2002). In vivo differentiated cytokine-producing CD4(+) T cells express functional CCR7. *J. Immunol.* **168**, 5441–5447.
12. Mora, J.R., von Andrian, U.H. (2006). T-cell homing specificity and plasticity: new concepts and future challenges. *Trends Immunol.* **27**, 235–243.
13. Masopust, D., Vezys, V., Marzo, A.L., Lefrancois, L. (2001). Preferential localization of effector memory cells in nonlymphoid tissue. *Science* **291**, 2413–2417.
14. Reinhardt, R.L., Khoruts, A., Merica, R. , Zell, T. Jenkins, M.K. (2001). Visualizing the generation of memory CD4 T cells in the whole body. *Nature* **410**, 101–105.
15. Mora, J.R., Bono, M.R., Manjunath, N., Weninger, W., Cavanagh, L.L., Rosemblatt, M., von Andrian, U.H. (2003). Selective imprinting of gut-homing T cells by Peyer's patch dendritic cells. *Nature* **424**, 88–93.
16. Klonowski, K.D., Williams, K.J., Marzo, A.L., Blair, D.A., Lingenheld, E.G., Lefrancois, L. (2004). Dynamics of blood-borne CD8 memory T cell migration in vivo. *Immunity* **20**, 551–562.
17. Reboldi, A., Coisne, C., Baumjohann, D., Benvenuto, F., Bottinelli, D., Lira, S., Uccelli, A., Lanzavecchia, A., Engelhardt, B., Sallusto., F. (2009). C-C chemokine receptor 6-regulated entry of TH-17 cells into the CNS through the choroid plexus is required for the initiation of EAE. *Nat. Immunol.* **10**, 514–523.
18. Picker, L.J., Kishimoto, T.K., Smith, C.W., Warnock, R.A., Butcher., E.C. (1991). ELAM-1 is an adhesion molecule for skin-homing T cells. *Nature* **349**, 796–798.
19. Maly, P., Thall, A.D., Petryniak, B., Rogers, C.E., Smith, P.L., Marks, R.M., Kelly, R.J., Gersten, K.M., Cheng, G., Saunders, T.L., Camper, S.A., Camphausen, R.T., Sullivan, F.X., Isogai, Y., Hindsgaul, O., von Andrian, U.H., Lowe., J.B. (1996). The α(1,3) fucosyltransferase Fuc-TVII controls leukocyte trafficking through an essential role in L-, E-, and P-selectin ligand biosynthesis. *Cell* **86, 643–653.**
20. Campbell, J., Haraldsen, G., Pan, J., Rottman, J., Qin, S., Ponath, P., Andrew, D.P., Warnke, R., Ruffing, N., Kassam, N., Wu, L., Butcher, E.C. (1999). The chemokine receptor CCR4 in vascular recognition by cutaneous but not intestinal memory T cells. *Nature* **400**, 776–780.
21. Morales, J., Homey, B., Vicari, A.P., Hudak, S., Oldham, E., Hedrick, J., Orozco, R., Copeland, N.G., Jenkins, N.A., McEvoy, L.M., Zlotnik, A. (1999). CTACK, a skin-associated chemokine that preferentially attracts skin-homing memory T cells. *Proc. Natl. Acad. Sci. USA.* **96**, 14470–14475.
22. Reiss, Y., Proudfoot, A.E., Power, C.A., Campbell, J.J., Butcher, E.C. (2001). CC chemokine receptor (CCR)4 and the CCR10 ligand cutaneous T cell-attracting chemokine (CTACK) in lymphocyte trafficking to inflamed skin. *J. Exp. Med.* **194**, 1541–1547.
23. Berlin, C., Berg, E.L., Briskin, M.J., Andrew, D.P., Kilshaw, P.J., Holzmann, B., Weissman, I.L., Hamann, A., Butcher, E.C. (1993). α4 β7 integrin mediates lymphocyte binding to

the mucosal vascular addressin MAdCAM-1. *Cell* **74**, 185–195.

24. Hamann, A., Andrew, D.P., Jablonski-Westrich, D., Holzmann, B., Butcher, E.C. (1994). Role of α_4-integrins in lymphocyte homing to mucosal tissues in vivo. *J. Immunol.* **152**, 3282–3293.
25. Marski, M., Ye, A.L., Abraham, C. (2007). CD18 is required for intestinal T cell responses at multiple immune checkpoints. *J. Immunol.* **178**, 2104–2112.
26. Zabel, B.A., Agace, W.W., Campbell, J.J., Heath, H.M., Parent, D., Roberts, A.I., Ebert, E.C., Kassam, N., Qin, S., Zovko, M., LaRosa, G.J., Yang, L.L., Soler, D., Butcher, E.C. , Ponath, P.D., Parker, C.M., Andrew, D.P. (1999). Human G protein-coupled receptor GPR-9-6/CC chemokine receptor 9 is selectively expressed on intestinal homing T lymphocytes, mucosal lymphocytes, and thymocytes and is required for thymus-expressed chemokine-mediated chemotaxis. *J. Exp. Med.* **190**, 1241–1256.
27. Svensson, M., Marsal, J., Ericsson, A., Carramolino, L., Broden, T., Marquez, G., Agace, W.W. (2002). CCL25 mediates the localization of recently activated CD8alphabeta(+) lymphocytes to the small-intestinal mucosa. *J. Clin. Invest.* **110**, 1113–1121.
28. Johansson-Lindbom, B., Svensson, M., Wurbel, M.A., Malissen, B., Marquez, G., Agace, W. (2003). Selective generation of gut tropic T cells in gut-associated lymphoid tissue (GALT): requirement for GALT dendritic cells and adjuvant. *J. Exp. Med.* **198**, 963–969.
29. Stenstad, H., A. Ericsson, B. Johansson-Lindbom, M. Svensson, J. Marsal, M. Mack, D. Picarella, D. Soler, G. Marquez, M. Briskin, and W.W. Agace. (2006). Gut associated lymphoid tissue primed CD4+ T cells display CCR9 dependent and independent homing to the small intestine. *Blood* **107**, 3447–3454.
30. Mora, J.R. (2008). Homing imprinting and immunomodulation in the gut: role of dendritic cells and retinoids. *Inflamm. Bowel. Dis.* **14,** 275–289.
31. Campbell, D.J., Butcher, E.C. (2002). Rapid acquisition of tissue-specific homing phenotypes by CD4(+) T cells activated in cutaneous or mucosal lymphoid tissues. *J. Exp. Med.* **195**, 135–141.
32. Jaensson, E., Uronen-Hansson, H., Pabst, O., Eksteen, B., Tian, J., Coombes, J.L., Berg, P.L., Davidsson, T., Powrie, F., Johansson-Lindbom, B., Agace, W.W. (2008). Small intestinal CD103+ dendritic cells display unique functional properties that are conserved between mice and humans. *J. Exp. Med.* **205**, 2139–2149.
33. Ahrendt, M., Hammerschmidt, S.I., Pabst, O., Pabst, R., Bode, U. (2008). Stromal cells confer lymph node-specific properties by shaping a unique microenvironment influencing local immune responses. *J. Immunol.* **181**, 1898–1907.
34. Hammerschmidt, S.I., Ahrendt, M., Bode, U., Wahl, B., Kremmer, E., Forster, R., Pabst, O. (2008). Stromal mesenteric lymph node cells are essential for the generation of gut-homing T cells in vivo. *J. Exp. Med.* **205**, 2483–2490.
35. Molenaar, R., Greuter, M., van der Marel, A.P.. Roozendaal, R., Martin, S.F., Edele, F., Huehn, J., Forster, R., O'Toole, T., Jansen, W., Eestermans, I.L., Kraal, G., Mebius, R.E. (2009). Lymph node stromal cells support dendritic cell-induced gut-homing of T cells. *J. Immunol.* **183**, 6395–6402.
36. Iwata, M., Hirakiyama, A., Eshima, Y., Kagechika, H., Kato, C., Song, S.Y. (2004). Retinoic acid imprints gut-homing specificity on T cells. *Immunity* **21**, 527–538.
37. Mora, J.R., Iwata, M., Eksteen, B., Song, S.Y., Junt, T., Senman, B., Otipoby, K.L., Yokota, A., Takeuchi, H., Ricciardi-Castagnoli, P., Rajewsky, K., Adams, D.H., von Andrian, U.H. (2006). Generation of gut-homing IgA-secreting B cells by intestinal dendritic cells. *Science* **314**, 1157–1160.
38. Eksteen, B., Mora, J.R., Haughton, E.L., Henderson, N.C., Turner, L.L., Villablanca, E.J., Curbishley, S.M., Aspinall, A.I., von Andrian, U.H., Adams, D.H. (2009). Gut Homing Receptors on CD8 T Cells Are Retinoic Acid Dependent and Not Maintained by Liver Dendritic or Stellate Cells. *Gastroenterology* **137**, 320–329.
39. Yamanaka, K.I., Dimitroff, C.J., Fuhlbrigge, R.C., Kakeda, M., Kurokawa, I., Mizutani, H., Kupper, T.S. (2007). Vitamins A and D are potent inhibitors of cutaneous lymphocyte-associated antigen expression. *J. Allergy. Clin. Immunol.* **121**, 148–157.
40. Mora, J.R., Iwata, M., von Andrian, U.H. (2008). Vitamin effects on the immune system: vitamins A and D take centre stage. *Nat. Rev. Immunol.* **8**, 685–698.
41. Sigmundsdottir, H., Pan, J., Debes, G.F., Alt, C., Habtezion, A., Soler, D., Butcher, E.C. (2007). DCs metabolize sunlight-induced vitamin D3 to 'program' T cell attraction to the epidermal chemokine CCL27. *Nat. Immunol.* **8**, 285–293.
42. Yokota, A., Takeuchi, H., Maeda, N., Ohoka, Y., Kato, C., Song, S.Y., Iwata, M. (2009). GM-CSF and IL-4 synergistically trigger dendritic

cells to acquire retinoic acid-producing capacity. *Int. Immunol.* **21**, 361–377.

43. Johansson-Lindbom, B., Svensson, M., Pabst, O., Palmqvist, C., Marquez, G., Forster, R., Agace, W.W. (2005). Functional specialization of gut CD103+ dendritic cells in the regulation of tissue-selective T cell homing. *J. Exp. Med.* **202**, 1063–1073.
44. Lampen, A., Meyer, S., Arnhold, T., Nau, H. (2000). Metabolism of vitamin A and its active metabolite all-trans-retinoic acid in small intestinal enterocytes. *J. Pharmacol. Exp. Ther.* **295**, 979–985.
45. Veazey, R.S., DeMaria, M., Chalifoux, L.V., Shvetz, D.E., Pauley, D.R., Knight, H.L., Rosenzweig, M., Johnson, R.P., Desrosiers, R.C., Lackner, A.A. (1998). Gastrointestinal tract as a major site of CD4+ T cell depletion and viral replication in SIV infection. *Science* **280**, 427–431.
46. Sekaly, R.P. (2008). The failed HIV Merck vaccine study: a step back or a launching point for future vaccine development? *J. Exp. Med.* **205**, 7–12.
47. Wurbel, M.A., Malissen, M., Guy-Grand, D., Meffre, E., Nussenzweig, M.C., Richelme, M., Carrier, A., Malissen, B. (2001). Mice lacking the CCR9 CC-chemokine receptor show a mild impairment of early T- and B-cell development and a reduction in T-cell receptor gammadelta(+) gut intraepithelial lymphocytes. *Blood* **98**, 2626–2632.
48. Uehara, S., Grinberg, A., Farber, J.M., Love, P.E. (2002). A role for CCR9 in T lymphocyte development and migration. *J. Immunol.* **168**, 2811–2819.
49. Pabst, O., Ohl, L., Wendland, M., Wurbel, M.A., Kremmer, E., Malissen, B., Forster, R. (2004). Chemokine receptor CCR9 contributes to the localization of plasma cells to the small intestine. *J. Exp. Med.* **199**, 411–416.
50. Wagner, N., Lohler, J., Kunkel, E.J., Ley, K., Leung, E., Krissansen, G., Rajewsky, K., Muller, W. (1996). Critical role for β7 integrins in formation of the gut-associated lymphoid tissue. *Nature* **382**, 366–370.
51. Svensson, M., Johansson-Lindbom, B., Zapata, F., Jaenssona, E., Austenaa, L.M., Blomhoff, R., Agace, W.W. (2008). Retinoic acid receptor signaling levels and antigen dose regulate gut homing receptor expression on CD8+ T cells *Mucosal Immunol.* **1**, 38–48.
52. Mora, J.R., Cheng, G., Picarella, D., Briskin, M., Buchanan, N., von Andrian, U.H. (2005). Reciprocal and dynamic control of CD8 T cell homing by dendritic cells from skin- and gut-associated lymphoid tissues. *J. Exp. Med.* **201**, 303–316.
53. Andrew, D.P., Berlin, C., Honda, S., Yoshino, T., Hamann, A., Holzmann, B., Kilshaw, P.J., Butcher, E.C. (1994). Distinct but overlapping epitopes are involved in $\alpha_4\beta_7$-mediated adhesion to vascular cell adhesion molecule-1, mucosal addressin-1, fibronectin, and lymphocyte aggregation. *J. Immunol.* **153**, 3847–3861.
54. Farstad, I.N., Halstensen, T.S., Lien, B., Kilshaw, P.J., Lazarovits, A.I., Brandtzaeg, P. (1996). Distribution of beta 7 integrins in human intestinal mucosa and organized gut-associated lymphoid tissue. *Immunology* **89**, 227–237.
55. Norment, A.M., Bogatzki, L.Y., Gantner, B.N., Bevan, M.J. (2000). Murine CCR9, a chemokine receptor for thymus-expressed chemokine that is up-regulated following pre-TCR signaling. *J. Immunol.* **164**, 639–648.
56. Goetz, D.J., Greif, D.M., Shen, J., Luscinskas, F.W. (1999). Cell-cell adhesive interactions in an in vitro flow chamber. *Methods Mol. Biol.* **96**, 137–145.
57. Hosoe, N., Miura, S., Watanabe, C., Tsuzuki, Y., Hokari, R., Oyama, T., Fujiyama, Y., Nagata, H., Ishii, H. (2004). Demonstration of functional role of TECK/CCL25 in T lymphocyte-endothelium interaction in inflamed and uninflamed intestinal mucosa. *Am. J. Physiol. Gastrointest. Liver Physiol.* **286**, G458–G466.
58. Mempel, T.R., Scimone, M.L., Mora, J.R., von Andrian, U.H. (2004). In vivo imaging of leukocyte trafficking in blood vessels and tissues. *Curr. Opin. Immunol.* **16**, 406–417.
59. Nolte, M.A., Kraal, G., Mebius, R.E. (2004). Effects of fluorescent and nonfluorescent tracing methods on lymphocyte migration in vivo. *Cytometry A* **61**, 35–44.

Chapter 25

In Vivo Quantitative Proteomics: The SILAC Mouse

Sara Zanivan, Marcus Krueger, and Matthias Mann

Abstract

Mass spectrometry-based proteomics is a field that has been quickly developing, enabling increasingly giving in-depth characterization of the proteomes of cells and tissues. Current technology allows identifying thousands of proteins in a single experiment. Stable isotope labeling with amino acid in cell culture (SILAC) was originally developed for high accuracy quantitative proteomic studies in cell lines. We have shown that SILAC can be extended to in vivo animal model by fully labeling C57BL/6 mice with $^{13}C_6$-Lysine (Lys6). We used SILAC mice technology to map quantitative proteomic changes in mice lacking the expression of β1 integrin, β-Parvin, or the integrin tail-binding protein Kindlin-3. This approach confirmed the absence of the proteins and revealed a role of Kindlin-3 in red blood cells. Here we describe a practical method to generate and maintain a colony of SILAC mice and optimal strategies to perform in vivo quantitative proteomic experiments.

Key words: SILAC mouse, Mass spectrometry, In vivo quantitative proteomics, Integrins, Cell adhesion

1. Introduction

Cell adhesion and motility are biological processes with a key role in physiological (i.e., embryonic development (1)), and pathological (i.e., tumor metastatization (2)) conditions. Clearly, it is important to characterize the function of the involved molecules in vivo. Prominent examples are integrins, ubiquitously expressed transmembrane proteins which play crucial roles in cell-to-extracellular matrix adhesion (3), and Kindlins, recently identified as a new family of proteins involved in adhesion as well (4).

The generation of knock-out mice is a commonly used technology that enables insights into protein function by the observation of altered phenotype in mice lacking the expression of

Motomu Shimaoka (ed.), *Integrin and Cell Adhesion Molecules: Methods and Protocols*,
Methods in Molecular Biology, vol. 757, DOI 10.1007/978-1-61779-166-6_25, © Springer Science+Business Media, LLC 2011

the protein. Despite the enormous usefulness of the knock-out technology, this approach by itself gives little or no information about the molecular mechanisms in which the protein is involved. Mass spectrometry (MS)-based proteomics applied to knock-out mice can address this shortcoming. Indeed high accuracy MS can now map proteomes or phosphoproteomes in great depth (5, 6). Moreover, in combination with stable isotope labeling technologies such as SILAC (7), the approach becomes quantitative and allows comparing samples in different states (8–10).

A recent study has shown that we can extend the SILAC technology to the in vivo mouse model by fully labeling mice with Lys6, feeding them with Lys-free food supplemented with Lys6. Moreover, we have shown that the combination of the SILAC mouse and knock-out technologies is a powerful strategy to perform quantitative proteomic studies in vivo. Comparing the proteomes of normal mice and mice lacking the expression of β1 integrin, β-Parvin, or Kindlin-3, we first confirmed the absence of the proteins. We then showed that lack of β1 integrin also affects α2, and α6 integrin expression but not of other integrins, and we revealed the role of Kindlin-3 in red blood cells (11).

2. Materials

2.1. Mouse Labeling with 13C6 Lysine

1. Lys6-SILAC Mouse Diet (Silantes, Germany): This SILAC diet contains 1 g of heavy Lys6 per 100 g of food, according to standard mouse nutritional requirements (12). Alternatively, it is possible to prepare the SILAC diet in-house by mixing Lys6 (Silantes) into a lysine-free diet (TD.99386, Harlan-Teklad, Madison, WI). After vigorous mixing with a blender, the powder has to be compressed with a cylinder and pestle with an inner diameter of approximately 1.5 cm and length of 10 cm. After the compression, the pellets can be taken out by removing the base of the cylinder as shown in Fig. 1. The pellets are then dried overnight and stored at room temperature (RT) for few days, or longer at 4°C.
2. Standard mouse strain: C57BL/6 mice, female, adult.

2.2. Sample Preparation for Mass Spectrometric Incorporation Check

1. Urea/thiourea lysis buffer: 6 M urea or 2 M thiourea in 10 mM HEPES, pH 8.0.
2. Digestion buffer: 50 mM ammonium bicarbonate (ABC) in water.
3. Reduction buffer: 10 mM DTT dissolved in digestion buffer. Prepare fresh buffer from 1 M DTT stock solution stored at –20°C.

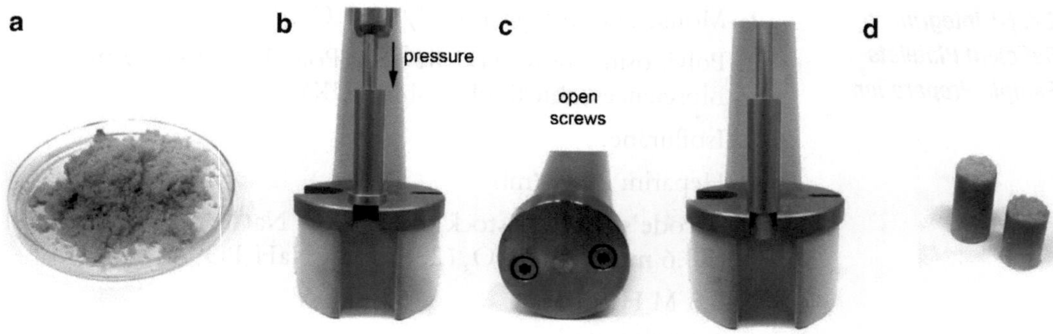

Fig. 1. Food preparation. (**a**) 100 g of powder were mixed with 1 g of Lys6, and mixed with a blender. (**b**) Approximately 10 g of the powder were compressed with a mandrel press by hand. (**c**) Open the base by loosening the two screws and afterwards pushing the compressed food through the cylinder. (**d**) Compressed food.

4. Alkylation buffer: 50 mM iodoacetamide (IAA) dissolved in digestion buffer. Prepare fresh buffer from 550 mM IAA stock solution in digestion buffer stored at –20°C.
5. Endoproteinase Lys-C: 0.5 µg/µl Lys-C dissolved in digestion buffer and stored at –20°C.
6. Ethanol.
7. Trifluoracetic acid (TFA).
8. Sonicator.
9. Table centrifuge for 1.5 ml tubes.
10. Scissor/capillary.

2.3. Sample Preparation for Proteomic Analysis

1. SDS lysis buffer: 100 mM Tris–HCl pH 7.6, 4% SDS, 100 mM dithiothreitol (DTT).
2. 8 M urea buffer: 8 M urea in Tris–HCl pH 8.5.
3. Alkylation buffer.
4. Endoproteinase Lys-C.
5. Sonicator.
6. Blender (Ultra-Turrax T8, IKA Works).
7. Thermo-mixer.
8. Table centrifuge for 1.5-ml tubes.
9. Microcon centrifugal filters (Ultracel YM-30Millipore, Billerica, MA).
10. Ultracentrifuge and ultracentrifuge tubes (optional).
11. Wet chamber.

2.4. β1 Integrin-Deficient Platelets Sample Preparation

1. Mouse line: Integrin $\beta1^{fl/fl}$/Mx1 Cre.
2. Polyinosinic-poly-cytidylic acid, (Poly (I)-Poly (C), Amersham Biosciences, Buckinghamshire, UK).
3. Isoflurane.
4. Heparin: 20 U/ml.
5. Tyrode's buffer stock, pH 7.3: NaCl (2.73 M), KCl (53.6 mM), $NaHCO_3$ (238 mM), NaH_2PO_4 (8.6 mM).
6. 0.5 M HEPES.
7. 0.1 M $MgCl_2$.
8. 0.1 M $CaCl_2$.
9. Tyrode's buffer for 100 ml: Mix 5 ml of Tyrode buffer stock, 2 ml of 0.1 M $CaCl_2$, 1 ml of 0.1 M $MgCl_2$, 1 ml of 0.5 M HEPES, 1 ml of 10% Glucose (add prior to use), 3.6 ml of 10% BSA (add prior to use), and H_2O 86.4 ml, pH 7.35. Before adding the stock solutions, pour 50 ml dd H_2O to the beaker first!
10. PGI_2: Dilute 10 μM prostacyclin stock solution at 1:10 in Tyrode's buffer.
11. Lysis buffer: 1% NP-40, 0.1% sodium deoxycholate, 150 mM NaCl, 1 mM EDTA, 50 mM Tris–HCl, pH 7.5, and protease inhibitors (Complete tablets, Roche, add prior to use).
12. Sample separation: 10–12% precast Bis Tris gels (NuPAGE™, Invitrogen).
13. Colloidal Blue Staining Kit (Invitrogen).
14. Reagents for reduction (DTT), alkylation (IAA), and protein digest (Lys-C): Same as described for the in solution digest (Subheading 2.3).

2.5. Mass Spectrometry Analysis

1. Buffer A: 0.5% acetic acid.
2. Buffer B: 80% acetum nitrile (ACN), 0.5% acetic acid.
3. Buffer A*: 2% ACN, 0.1% TFA.
4. Empore Disk C18 (Varian, Walnut Creek, CA).
5. P200 pipette tips.
6. Syringe.
7. 16- or 20-mm gauge-needle.
8. 96-well plate, PCR plate, and Silicon cover plates (Nerbeplus).
9. Speed-vac with adaptor for 96-well-plate.
10. High resolution mass spectrometer: LTQ-Orbitrap XL, LTQ-Orbitrap; Velos (Thermo Fisher Scientific).
11. Online HPLC system: Easy-nLC (Proxeon Biosystems).
12. Data processing software: MaxQuant (http://www.maxquant.org/).

3. Methods

SILAC is a technique commonly used to perform quantitative proteomic studies. The approach is based on the use of stable isotopes, like ^{13}C and ^{15}N, which are characterized by not being radioactive, having the same chemical and physical properties of the normal ^{12}C and ^{14}N atoms, but with a different mass.

The stable isotopes are used to synthesize arginine and lysine amino acids, which are then termed "heavy" because of their higher mass compare to the normal ones. The choice of arginine and lysine is preferred in cell culture since they are essential (not synthesized at all by the cell) or semi-essential (can be generated by the conversion of proline) amino acids, respectively. Moreover, they are well suited for the standard digestion of proteins into peptides with trypsin, which cuts C-terminally to Arg and Lys or endoprotease Lys-C, which cuts C-terminally to Lys.

In the case of the SILAC mouse we use only Lys, since it is an essential amino acid. In particular, the heavy lysine contains six atoms of ^{13}C, and therefore has a mass of 6 Da more than the natural lysine isotope (see Note 1).

With SILAC applied to mice, it is then possible to perform quantitative MS-based proteomic studies in vivo. In order to obtain precise quantification, it is important to label mice with an incorporation efficiency of the Lys6 of at least 96%.

To check the incorporation and to perform quantitative in-depth analysis of the proteome (see below), it is required to measure the samples with a high resolution and high accuracy mass spectrometer. Our laboratory uses the linear triple quadrupole (LTQ)-Orbitrap. A further advantage of using these instruments is that the data analysis is supported by the software MaxQuant (13). This software allows an automatic processing of the data and support accurate and in-depth quantitative analysis of SILAC-labeled samples.

Here, we describe how to generate SILAC mice and prepare the samples to be used to perform in vivo quantitative proteomic experiments according to the more recent protocols developed in our laboratory. Moreover, we show the application of SILAC mouse in combination with MS to study mice lacking the expression of protein β1 integrin. However, this is only an example of the use of the SILAC-mouse technology. Clearly, it can be used in a large variety of circumstances, since it allows comparing different states of any organ, tissue, fluid, or primary cells that can be dissected or isolated from mice (see Subheading 3.4).

3.1. Mouse Labeling with Lys6

3.1.1. Generation of the Fully Labeled SILAC Mouse

1. A SILAC mouse colony is started from a fertile C57BL/6 female mouse (F0 generation) (see Note 2). Pre-labeling of the F0 mother increases the label efficiency of the following F1 generation. For this purpose the F0 mouse is fed daily with 3–5 g of the SILAC diet, until the incorporation of the Lys6 is approximately 85–90% (which generally takes 8–10 weeks) (Fig. 2). The incorporation is checked every week by sampling blood. While the degree of incorporation increases, the weight of the mouse will not decrease.
2. Following the initial pre-labeling period, a nonlabeled C57BL/6 male mouse is used for breeding. During the breeding 3–5 g of SILAC diet is also to be provided for the male. However, soon after the observation of a vaginal-plug, the male needs to be removed to avoid unnecessary use of SILAC diet.
3. The food amount is increased up to 5–10 g during pregnancy and up to 10–15 g during the lactation and weaning, according to the number of pups (F1 generation).
4. Approximately 3 weeks after delivery, each pup is fed with 1.5 g per day and the mother again with 3 g per day. After 1 week the amount of food per pup will be 2 g per day and after one more week 3 g. In total, we estimate approximately 1 kg of food for one generation.
5. In all the steps involving a change of the amount of food per day, we suggest to check daily the leftover food. If there is some, the amount of food should be slightly reduced.

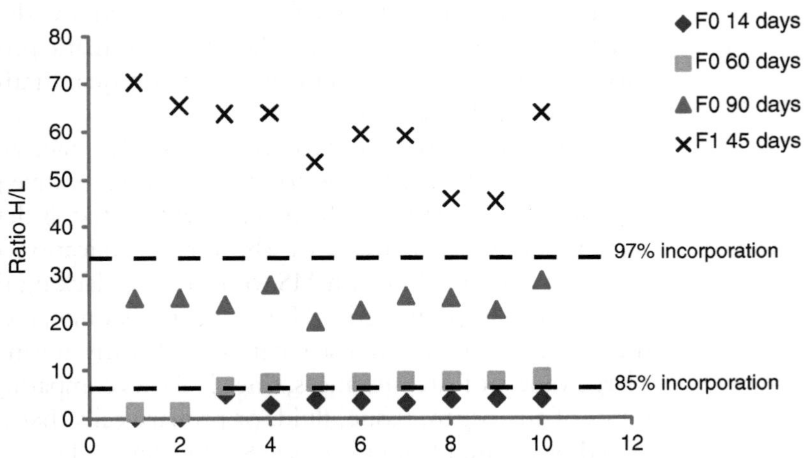

Fig. 2. Lys6 incorporation in blood. Increasing incorporation of Lys6 in blood proteins (representative example). The ratio H/L ("heavy" to "light") indicates the amount of protein with incorporated Lys6 over the amount of protein carrying no Lys6. The incorporation efficiency is calculated as {1−[1/ratio(H/L)] × 100}. Therefore, a ratio of 33.3 indicates that 97% of the Lys0 in the indicated protein are substituted with Lys6.

6. Organs, tissues and fluids from F1 mice usually show an incorporation efficiency of more than 97% (Fig. 2). However, before using them for experiments, always check that all proteins are labeled with the same efficiency. If it is the case, use F1 mice for experiments. Otherwise, continue breeding to generate F2 SILAC mice which will be completely and uniformly labeled (11). Blood is suitable to check incorporation but we strongly recommend verifying the incorporation of the interested organ/tissue/fluid before any use.

3.1.2. Keeping the Colony

Once the F1 generation is established, the feeding of the mice should follow the same schedule as before. Each mouse is fed with 3–5 g of SILAC food per day, again with increased amounts during breeding and pregnancy. It is important never to swap the diet back to food containing normal lysine. If this happens, labeling of mice needs to be started from the beginning.

3.1.3. Blood Preparation for Incorporation Check

1. Prepare the following materials:
   (a) Urea/thiourea lysis buffer.
   (b) Two 1.5-ml tubes per mouse (one to collect the blood and another to transfer the clean lysate afterwards).
   (c) Sterile scissor (to take blood from the tail by snipping the 1 mm distal part of the tail), and capillary (to take blood from the orbital plexus).
2. After sampling 20 μl of blood from the orbital plexus or tail vein under anesthesia, add 80 μl of urea/thiourea lysis buffer immediately to avoid blood coagulation.
3. Sonicate each sample and centrifuge at 14,000 × g for 10 min at RT. Then transfer the supernatant into a new tube for protein concentration measurements by Bradford assay.
4. Use 20 μg of proteins for in solution digest.
5. Add 1 volume of reduction buffer for every 10 volumes of sample and incubate for 1 h at RT.
6. Add the alkylation buffer, 1 volume for every 10 volumes of sample, and incubate for 40 min at RT in the dark.
7. Dilute the sample with 4 volumes of digestion buffer to dilute the concentration of urea.
8. Add the endoprotease Lys-C, 1 μg every 50 μg of lysate and incubate at RT overnight.
9. Acidify the digested peptides with TFA, final concentration 0.1–1%. Samples are now ready to be loaded on C18-StageTips for MS analysis (see below).

3.1.4. Sample Preparation for Proteomics Experiments (see Note 3)

1. Prepare fresh SDS lysis buffer, 8 M urea buffer and a thermo-mixer settled to 95°C. Wash the tip of the blender with PBS and prepare a clean beaker with PBS to clean the tip of the blender between the samples. All the lysis procedures are performed at RT.
2. Take the dissected organ or tissue that you are interested in. If the sample is fresh, start immediately with the homogenization. If the sample is frozen, first reduce it to small pieces or powder with the help of a mortar kept cold with liquid nitrogen. Then transfer the sample into a clean tube.
3. Add SDS lysis buffer, 0.6–1.0 ml for every 100 mg of sample, and immediately homogenize it until it becomes uniformly liquid. A lot of foam will form due to the SDS. If the state of your homogenate is not apparent, wait a few minutes so that the foam reduces and then proceed with further homogenization.
4. Transfer the sample into the thermo-mixer and incubate for 5 min at 95°C.
5. Sonicate it at duty cycle 20% and output control 4, for 1–2 min.
6. Centrifuge the sample, 14,000 × g, for 15 min at RT and transfer the cleared lysate into a fresh tube. Be careful not to take the fatty part that can appear as white and dense foam at the top of the sample.
7. If the collected lysate is not clear and transparent, use the ultracentrifuge: transfer the lysate into an appropriate tube for ultracentrifugation and centrifuge it for 30 min at 435,000 rcf. Collect the clean supernatant into a new tube.
8. Quantify the lysate with the Bradford method according to the manufacturer protocol. Dilute the sample sufficiently so that the SDS and DTT concentration do not affect the protein quantification.
9. Now the sample is ready for checking incorporation (A) or to be used for quantitative proteomic analysis (B). For proteomic study, the SILAC lysate is first mixed in a 1 to 1 ratio with samples of interest (Fig. 3).

   (A) Methanol–Chloroform precipitation and in solution digestion.

   1. Add 4 volumes of methanol and vortex.
   2. Add 1 volume of chloroform and vortex.
   3. Add 3 volumes of deionized water and vortex.
   4. Centrifuge 14,000 × *g* for 1 min.
   5. Discard the aqueous phase.

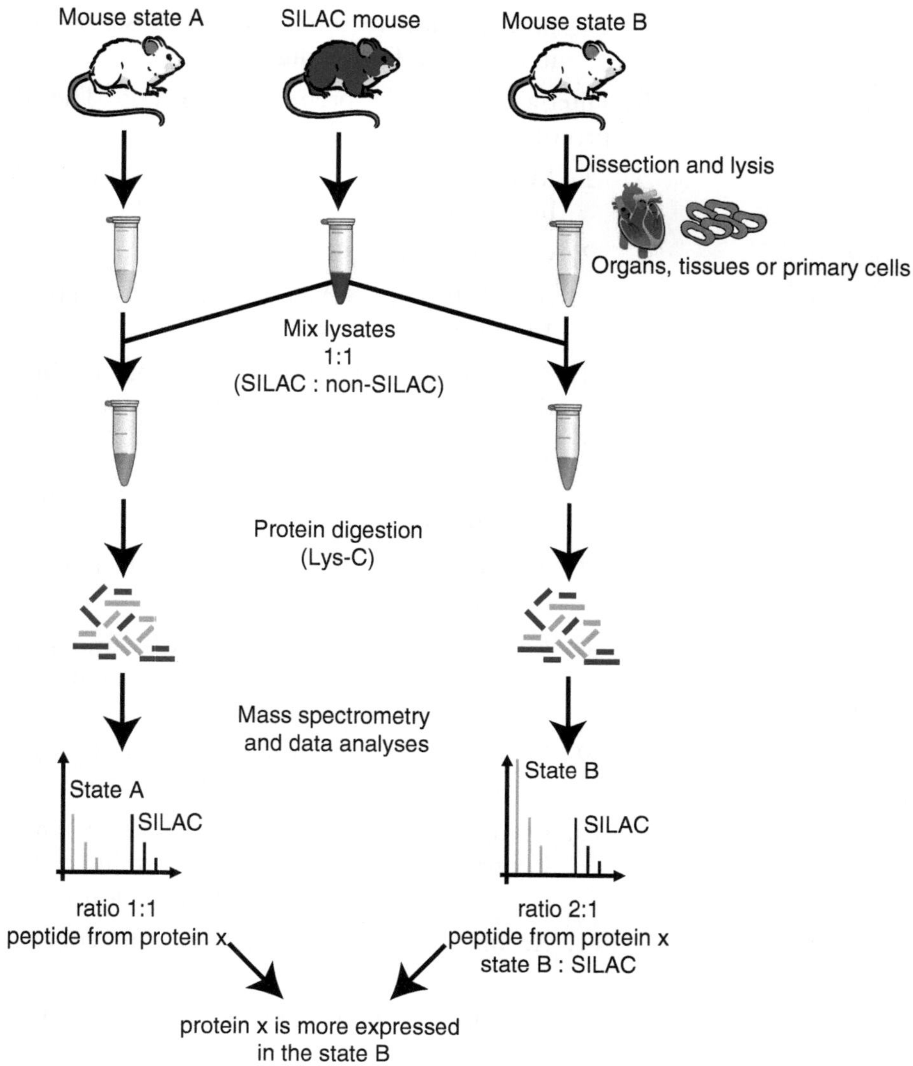

Fig. 3. SILAC mouse as internal standard: experimental workflow. Cells, tissues, or organs are isolated from the mice in the experimental conditions of interest (mouse state A and mouse state B, where A and B can be, as an example, old and young). The same sample is then extracted from a SILAC mouse. After the lysis, the SILAC lysate is mixed into each sample with a ratio 1:1. Then proteins are digested with Lys-C and further analyzed with MS. Bioinformatic analysis by MaxQuant generates, separated for each experiment, a list of proteins with their relative quantification (amount in the sample/amount in the SILAC mouse). In the pictured case, there is no difference in the amount of protein x between SILAC and state A (A/S = Ratio A/SILAC = 1), while in the state B the protein is twice more abundant than in the SILAC skin (B/S = Ratio B/SILAC = 2). The relative abundance of protein x between state B and A can be calculated as follows: B/A = (B/S)/(A/S) = 2. Therefore, protein x in sample B is twice the amount that in sample A, so it is upregulated twofold in sample B.

6. Add 4 volumes of methanol and vortex.
7. Centrifuge 14,000 × *g* for 1 min.
8. Discard the liquid and keep the pellet.
9. Resuspend the precipitated proteins in urea/thiourea lysis buffer (1–10 μl for 1 μg) and follow the same

protocol described to check the Lys6 incorporation in the blood for the digestion.

(B) Filter-aided sample preparation, FASP. The detailed protocol is available in Nature Methods (5).

1. Mix 200–300 μg of lysate with 200 μl of 8 M urea buffer and load it on a Microcon filter.
2. Centrifuge at 14,000 × *g* for 15 min.
3. Add 200 μl of 8 M urea buffer and centrifuge at 14,000 × *g* for 10 min.
4. Add 100 μl of alkylation buffer, mix at 600 rpm in thermo-mixer and incubate for 20 min.
5. Centrifuge at 14,000 × *g* for 15 min.
6. Add 100 μl of 8 M urea buffer and centrifuge at 14,000 × *g* for 15 min (3 times).
7. Add 100 μl of 50 mM ABC and centrifuge at 14,000 × *g* for 10 min (3 times).
8. Add 40 ml of 50 mM ABC with Endoproteinase Lys-C (1 mg every 50 mg of lysate) and incubate O/N at 37°C in a wet chamber.
9. In a new tube, collect the digested peptides by centrifugation at 14,000 × *g* for 10 min.
10. Complete the elution adding 50 μl of 50 mM ABC. Centrifuge at 14,000 × *g* for 10 min (collect peptides in the same tube as the step before).
11. Acidify the peptides with TFA, final concentration 0.1–1%.
12. Peptides are now ready to be prepared for MS analysis.
13. For single analysis at the MS, it is possible to load 10–20 μg of digested peptides on C18-StageTips (see below). To perform more in depth analysis of the proteome, 50–150 μg of the digested peptides can be further fractionated by "OFFGEL" isoelectric focusing (14) or pipette-based Strong Anion Exchange (SAX) (15). This is followed by LC MS/MS of each fraction. Avoiding the step of homogenization, the "sample preparation for experiments" protocol can be used also to prepare cell lysate.

3.2. SILAC Mouse as Internal Standard for Quantitative Proteomics

The SILAC mouse is a source of organs, tissues and primary cells labeled with Lys6, which can be used as internal standard for quantitative proteomics (Fig. 3) (see Note 4).

3.2.1. Quantitative Proteomic Analysis of Integrin β1-Deficient Platelets

As an example, we describe the workflow based on the one that was used for quantitative proteomic of platelets studies in vivo (11).

1. Generation of integrin $\beta1^{fl/fl}$/Mx1 Cre mice was done by intercrossing integrin β1 floxed mice ($\beta1^{fl/fl}$) with a transgenic mouse expressing MX1 promoter-driven, interferon-inducible Cre (MX1-Cre) (16, 17).
2. For extensive downregulation of the β1 integrin receptor in platelets, we injected intraperitoneal dose of 250 μg Poly (I)-Poly (C) diluted in 0.5 ml phosphate-buffered saline (PBS) in a 2-day interval. Mice are sacrificed 10–14 day after injection (18).
3. Mice are bled under anesthesia from the retroorbital plexus and the blood was collected into a tube containing 300 μl of 20 U/ml heparin in TBS, pH 7.3. Blood was centrifuged at 70 × *g* for 6 min at RT to obtain platelet-rich plasma (prp). To wash platelets, prp was centrifuged at 800 × *g* for 5 min in the presence of prostacyclin (PGI_2) (0.1 μg/ml) and the pellet was resuspended in 1 ml of Tyrode's buffer containing PGI_2 (0.1 μg/ml) and apyrase (0.02 U/ml). After a 5 min incubation step at 37°C for 5 min, the platelets were centrifuged at 800 × *g* for 5 min. After a second centrifugation step, platelets were resuspended in the same buffer and incubated at 37°C for 5 min. Platelets were finally centrifuged as above, resuspended in Tyrode's buffer containing apyrase (0.02 U/ml) (500 μl) and left to incubate for at least 30 min at 37°C.
4. For platelet protein isolation prp were centrifuged 5 min at 800 × *g* and the pellet were washed 2×in 1 ml of PBS containing 5 mM EDTA, centrifuged at 2,800 rpm for 5 min. The pellet was resuspended in lysis buffer and incubated 10 min at RT. To pellet cellular debris was centrifuged at full speed for 5 min. A Bradford assay was performed to determine protein concentrations.
5. According to this concentration measurement, samples from labeled wild-type animals were mixed 1:1 with nonlabeled wild-type platelets, and with nonlabeled, induced $\beta1^{(fl/fl)}$× Mx1-cre platelets.
6. After 1D-SDS Page using a NuPAGE 4–12% Bis Tris gel (Invitrogen) and Colloidal Blue Staining (Invitrogen) equally sized gel pieces were excised from the gel and processed for in gel digest as described in (19). We used Lys-C as the protease.
7. As expected, all measured protein ratios between labeled and nonlabeled wild-type platelets were 1:1 (Fig. 4a). The analysis of β1 integrin-deficient platelets revealed a clear downregulation

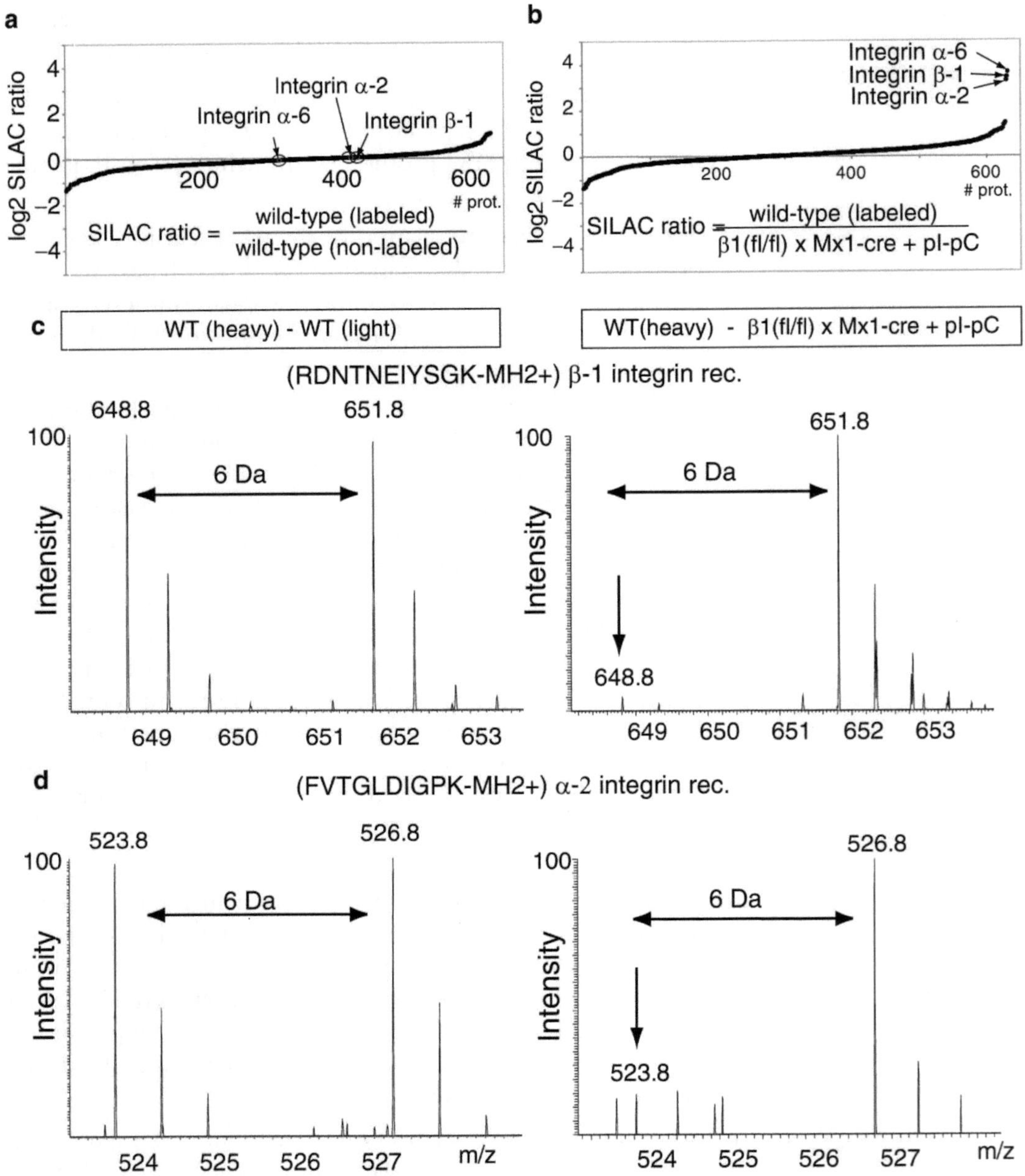

Fig. 4. β1 integrin results. (**a**, **b**) Platelets from SILAC-labeled mice were mixed with either platelets from a nonlabeled wild-type control mouse (**a**) or plpC induced $\beta1^{fl/fl}$; Mx1-Cre mouse (**b**). The β1, α2, and α6 SILAC-ratios are indicated by *circles*. (**c**, **d**) Selected MS-Spectra of β1 (**c**) and α2 (**d**) SILAC-ratios. The *arrows* in **c** and **d** indicate reduced peak intensity of the nonlabeled β1 and α2 peptide peak, respectively. *Horizontal arrows* indicate the 6 Da mass distance of the light and heavy peak.

of the integrin β1 and the direct dimerization partners integrin α2 and integrin α6 (Fig. 4b) (see Note 5).

8. Figure 4c shows the corresponding MS-spectra of selected peptides for integrin β1 and integrin α2. Other integrins, such as the β3 subunit were not affected by the absence of β1.

3.3. Mass Spectrometry and Data Analysis

Since salts can interfere with mass spectrometry (MS) analysis, the digested peptides are desalted making use of the C18-StageTips (20).

3.3.1. Sample Preparation for Mass Spectrometry (Desalting)

1. For each sample, take one p200 pipette tip and pick twice with the gauge-needle the C18 Empore Disk. Insert the picked C18 into the tip.
2. Activate the C18 by adding 50 μl of buffer B to the tip. Let it pass through the filter with the help of a syringe.
3. Wash the C18 material with 50 μl of buffer A.
4. Repeat the wash with buffer A.
5. Load the digested peptides (maximum 10 μg).
6. Wash with 50 μl of buffer A.
7. Now the C18-StageTips with the loaded sample can be store at 4°C until MS analysis.
8. Just before MS analysis, the sample is eluted from the C18-StageTip with 2×20 μl of Buffer B into a clean 96-well plate.
9. To eliminate the ACN, speed-vac the sample down to 5 μl. Then add 5 μl of Buffer A*. Now the sample is ready for MS analysis.

3.3.2. MaxQuant

MaxQuant (13) is a freely available software which supports the analysis of .RAW files generated by LTQ-Orbitrap instruments (see Notes 6 and 7). A detailed protocol for the installation and use of MaxQuant can be found in Nature Protocols (21).

3.4. Use of the SILAC Mouse

Above, we illustrated the usefulness of the SILAC mice for in vivo proteomics studies. However, due to the relatively expensive diet to generate and maintain a SILAC colony, for several applications alternative strategies should be considered. Here we provide some tips to determine whether or not SILAC mice are applicable.

SILAC Mouse are Applicable

1. If you want to analyze a rather big organ (i.e., brain or liver). It is possible to collect a single SILAC organ and use it as an internal standard for hundreds of samples (it is possible to store either the frozen organ or the frozen lysate). In fact, only 50–100 μg of SILAC lysate are needed for a single experiment for in-depth analysis of the proteome.
2. If it is not possible to label cells that can be used as internal standard for the organ, tissue or primary cells of interest may be used (i.e., cells that are not proliferating in culture or cells that lack the expression of important markers needed for the proteomic study).
3. If it is important to analyze an organ or tissue in all its complexity. In the case of tumor studies, the use of SILAC

tissue as internal standard also allows checking the state of vascularization of the tumor. This information may be lost using tumor SILAC-labeled cells as internal standard, since endothelial markers are not expressed.

SILAC Mouse Are NOT Applicable

1. If the interest is in a small organelle like the pancreatic islets or the nephron. Of course, SILAC mice can be used, but just be aware that you need many of them.
2. If there is the possibility to use SILAC-labeled cell lines as internal standard. Indeed, an alternative and cheaper strategy to the SILAC mouse can be the labeling of cell lines similar to the sample of interest. As an example, a mix of white and brown cultured adipocytes SILAC-labeled cell lines was used as internal standard to compare the proteome between white and brawn mouse fat tissues (22). In this regard, it is strongly recommended to perform a preliminary experiment to compare the proteome of the cells and the organ/tissue/primary cells of interest, to verify the similarity between the proteomes.

3.5. The Analysis of Posttranslational Modifications

Phosphoproteomics is an important technology, which in combination with SILAC allows the large-scale quantification of phosphorylated peptides (23). While not described here, the SILAC-mouse can be used equally well to measure the in vivo changes in global cell signaling as reflected in quantitative phosphopeptide changes. This in principle applies to other posttranslational modifications (PTMs) as well.

4. Notes

1. Generally, for quantitative proteomic study, cells are labeled with arginine and lysine heavy amino acids. This allows protein digestion with trypsin, which cuts at the C-terminus of these two amino acids. Since the SILAC mouse is labeled only with heavy Lysine, Lys-C is the endoprotease used to cleave proteins into peptides. If using trypsin, all peptides containing arginine but no lysine, are useless for quantitative purpose. Lys-C instead of trypsin leads to the generation of longer peptides and may therefore result in less identifications in MS/MS (2/3 compared to tryptic digest). We believe that with instrument and software improvements, this will be less of a problem in the future.
2. We show that it is possible to fully label C57BL/6 mice. However, the protocol should work equally well for other mouse strains.

3. We propose SDS lysis buffer and the FASP method for sample preparation and digestion, because this method led to the generation of very clean peptides for MS analysis. However, other lysis protocols can be used according to the requirements. If detergents or other substances that can interfere with the MS analysis are present, proteins must be precipitated and resuspended in suitable buffer for digestion (i.e., urea/thiourea lysis buffer).
4. For large-scale experiments (i.e., many conditions to be compared or different biological replicates), it is better to generate only one batch of SILAC lysate as internal standard; if necessary pooling material from different mice. This makes it straightforward to compare all conditions to each other.
5. Here we showed the use of the SILAC mouse as an internal standard. However an alternative strategy is to generate SILAC mice directly for the condition under study. However, although SILAC mice grow normally, the Lys6 diet used here contains high amount of sucrose. Since this can affect the metabolism of the mice, we discourage the use of cells/tissues/organs from SILAC mice to perform metabolic studies. In principle, this potential problem could be ameliorated by developing a SILAC mouse diet more similar to the ordinary mouse food. Moreover, for direct comparison, we recommend to always first check the similarity in protein expression between SILAC mouse and the mouse of interest for the organ/tissue/cells of interest.
6. Other analysis software can in principle be used for MS data analysis, i.e., MSQuant.
7. Other MS instruments can be used. But at the moment only LTQ-Orbitrap and FT are supported by MaxQuant.

Acknowledgments

The authors would like to thank SILANTES for the development of Lys6-containing mouse diet and Marcus Moser for contributing to the application of the SILAC mouse technology.

References

1. Gumbiner, B. M. (1996) Cell adhesion: the molecular basis of tissue architecture and morphogenesis. *Cell* **84**, 345–357.
2. Hood, J. D., and Cheresh, D. A. (2002) Role of integrins in cell invasion and migration. *Nat. Rev. Cancer.* **2**, 91–100.
3. Hynes, R. O. (2002) Integrins: bidirectional, allosteric signaling machines. *Cell* **110**, 673–687.
4. Meves, A., Stremmel, C., Gottschalk, K., and Fassler, R. (2009) The Kindlin protein family: new members to the club of focal adhesion proteins. *Trends. Cell Biol.* **19**, 504–513.

5. Wisniewski, J. R., Zougman, A., Nagaraj, N., and Mann, M. (2009) Universal sample preparation method for proteome analysis. *Nat. Methods* **6**, 359–362.
6. Zanivan, S., Gnad, F., Wickstrom, S. A., Geiger, T., Macek, B., Cox, J., Fassler, R., and Mann, M. (2008) Solid tumor proteome and phosphoproteome analysis by high resolution mass spectrometry. *J. Proteome Res.* 7, 5314–5326.
7. Ong, S. E., and Mann, M. (2007) Stable isotope labeling by amino acids in cell culture for quantitative proteomics. *Methods Mol. Biol.* **359**, 37–52.
8. de Godoy, L. M., Olsen, J. V., Cox, J., Nielsen, M. L., Hubner, N. C., Frohlich, F., Walther, T. C., and Mann, M. (2008) Comprehensive mass-spectrometry-based proteome quantification of haploid versus diploid yeast. *Nature* **455**, 1251–1254.
9. Graumann, J., Hubner, N. C., Kim, J. B., Ko, K., Moser, M., Kumar, C., Cox, J., Scholer, H., and Mann, M. (2008) Stable isotope labeling by amino acids in cell culture (SILAC) and proteome quantitation of mouse embryonic stem cells to a depth of 5,111 proteins. *Mol. Cell. Proteomics* 7, 672–683.
10. Olsen, J. V., Blagoev, B., Gnad, F., Macek, B., Kumar, C., Mortensen, P., and Mann, M. (2006) Global, in vivo, and site-specific phosphorylation dynamics in signaling networks. *Cell* **127**, 635–648.
11. Kruger, M., Moser, M., Ussar, S., Thievessen, I., Luber, C. A., Forner, F., Schmidt, S., Zanivan, S., Fassler, R., and Mann, M. (2008) SILAC mouse for quantitative proteomics uncovers kindlin-3 as an essential factor for red blood cell function. *Cell* **134**, 353–364.
12. Benevenga, N. J., Calvert, C., Eckhert, C. D., Fahey, G. C., Greger, C. L., Keen, C. L., Knapka, J. J., Magalhaes, H., and Oftedal, O. T. (1995) Nutrient requirements of the mouse, National Academy Press, Washington, D.C.
13. Cox, J., and Mann, M. (2008) MaxQuant enables high peptide identification rates, individualized p.p.b.-range mass accuracies and proteome-wide protein quantification. *Nat. Biotechnol.* **26**, 1367–1372.
14. Hubner, N. C., Ren, S., and Mann, M. (2008) Peptide separation with immobilized pI strips is an attractive alternative to in-gel protein digestion for proteome analysis. *Proteomics* **8**, 4862–4872.
15. Wisniewski, J. R., Zougman, A., and Mann, M. (2009) Combination of FASP and StageTip-based fractionation allows in-depth analysis of the hippocampal membrane proteome. *J Proteome Res.* **8**, 5674–5678.
16. Brakebusch, C., Grose, R., Quondamatteo, F., Ramirez, A., Jorcano, J. L., Pirro, A., Svensson, M., Herken, R., Sasaki, T., Timpl, R., Werner, S., and Fassler, R. (2000) Skin and hair follicle integrity is crucially dependent on beta 1 integrin expression on keratinocytes. *EMBO J.* **19**, 3990–4003.
17. Kuhn, R., Schwenk, F., Aguet, M., and Rajewsky, K. (1995) Inducible gene targeting in mice. *Science* **269**, 1427–1429.
18. Nieswandt, B., Brakebusch, C., Bergmeier, W., Schulte, V., Bouvard, D., Mokhtari-Nejad, R., Lindhout, T., Heemskerk, J. W., Zirngibl, H., and Fassler, R. (2001) Glycoprotein VI but not alpha2beta1 integrin is essential for platelet interaction with collagen. *EMBO J.* **20**, 2120–2130.
19. Shevchenko, A., Tomas, H., Havlis, J., Olsen, J. V., and Mann, M. (2006) In-gel digestion for mass spectrometric characterization of proteins and proteomes. *Nat. Protoc.* **1**, 2856–2860.
20. Rappsilber, J., Mann, M., and Ishihama, Y. (2007) Protocol for micro-purification, enrichment, pre-fractionation and storage of peptides for proteomics using StageTips. *Nat. Protoc.* **2**, 1896–1906.
21. Cox, J., Matic, I., Hilger, M., Nagaraj, N., Selbach, M., Olsen, J. V., and Mann, M. (2009) A practical guide to the MaxQuant computational platform for SILAC-based quantitative proteomics. *Nat. Protoc.* **4**, 698–705.
22. Forner, F., Kumar, C., Luber, C. A., Fromme, T., Klingenspor, M., and Mann, M. (2009) Proteome differences between brown and white fat mitochondria reveal specialized metabolic functions. *Cell Metab.* **10**, 324–335.
23. Macek, B., Mann, M., and Olsen, J. V. (2009) Global and site-specific quantitative phosphoproteomics: principles and applications. *Annu. Rev. Pharmacol. Toxicol.* **49**, 199–221.

Chapter 26

Analysis of Chemotaxis in *Dictyostelium*

Huaqing Cai, Chuan-Hsiang Huang, Peter N. Devreotes, and Miho Iijima

Abstract

Dictyostelium discoideum is an excellent model organism for the study of directed cell migration, since *Dictyostelium* cells show robust chemotactic responses to the chemoattractant cAMP. Many powerful experimental tools are applicable, including forward and reverse genetics, biochemistry, microscopy, and proteomics. Recent studies have demonstrated that many components involved in chemotaxis are functionally conserved between human neutrophils and *Dictyostelium* amoebae. In this chapter, we describe how to define the functions of proteins that mediate and regulate cell motility, cell polarity, and directional sensing during chemotaxis in *Dictyostelium*.

Key words: *Dictyostelium*, Chemotaxis, Intracellular signaling, Cell migration

1. Introduction

Chemotaxis is a dynamic process that involves directional sensing, cell polarity, and cell motility. Cells continuously rearrange their cytoskeleton and plasma membrane, resulting in an asymmetric cell shape, and periodically extend pseudopods. Actin polymerization in pseudopods at the leading edge of the cell is synchronized with contractile forces generated by myosin motor proteins at the rear. A directional sensing system biases pseudopodia formation toward the source of the chemoattractant and thus orients cell movement along the extracellular chemical gradient. To understand the molecular mechanisms of chemotaxis, a variety of assays have been used to dissect the individual subreactions involved in *Dictyostelium* chemotaxis.

This model organism is an excellent system for the study of cell migration. The molecular mechanisms underlying chemotaxis, such as actin polymerization, intracellular signaling, and cell

Motomu Shimaoka (ed.), *Integrin and Cell Adhesion Molecules: Methods and Protocols*,
Methods in Molecular Biology, vol. 757, DOI 10.1007/978-1-61779-166-6_26, © Springer Science+Business Media, LLC 2011

migration, are highly conserved among eukaryotes. Therefore, many powerful experimental tools used to dissect these processes in other organisms are applicable to *Dictyostelium*. When there are sufficient nutrients *Dictyostelium* cells proliferate as haploid single amoebae; however, when nutrients are depleted, starvation immediately triggers the developmental program for surviving harsh conditions through spore formation. Chemotaxis plays an essential role in *Dictyostelium* development. During differentiation, ~100,000 cells migrate toward aggregation centers that release the chemoattractant cAMP and form multicellular structures. Differentiating cells secrete cAMP every 6 min and waves of extracellular cAMP reinforce the expression of the cAMP receptors and other signaling molecules that are required to respond to cAMP. Differentiation normally takes several hours and the chemotactic ability peaks at 5–6 h after starvation. Around this time, cells establish an increased cell polarity due to downregulation of basal cytoskeletal activity and become highly sensitive to chemoattractant stimulation. Thus, the developing amoebae display robust and rapid chemotactic responses.

In this chapter, we will describe how *Dictyostelium* mutants can be analyzed to determine whether and how they are defective in chemotaxis. Cell movement toward chemoattractants can be examined by direct observation using time-lapse microscopy. Quantification of cell movement and shape provides information on cell polarity, directionality, and rate of cell migration. In addition, the many biochemical reactions involved in chemotaxis can be examined in mutant and wild-type cells. Assays for chemoattractant–receptor interactions, G-protein activation, phosphatidylinositol (3,4,5)-triphosphate (PIP_3) production, and activation of Target of rapamycin complex 2 (TORC2) and Ras signaling are described in this chapter. It should be noted that these biochemical reactions are often localized and therefore it is critical to determine where the reactions occur using microscopic approaches in addition to biochemical measurements. These assays will reveal the molecular mechanisms underlying chemotaxis and the function of proteins involved in this process. Identification of genes that are mutated in chemotaxis-defective mutants will help us understand the function of proteins involved in chemotaxis.

2. Materials

2.1. Cell Culture and Development

1. HL5 medium, pH 6.5: 10 g/L Dextrose, 10 g/L Proteose peptone, 5 g/L Yeast extract, 0.67 g/L $Na_2HPO_4 \cdot 7H_2O$, 0.34 g/L KH_2PO_4. Autoclave and store at room temperature. 10 mg/mL Streptomycin sulfate.

2. DB (Development Buffer), pH 6.5: 5 mM Na_2HPO_4, 5 mM KH_2PO_4, 2 mM $MgSO_4$, 0.2 mM $CaCl_2$.
3. DB agar: 1% agar in DB.
4. PM (Phosphate magnesium buffer), pH 6.5: 5 mM Na_2HPO_4, 5 mM KH_2PO_4, 2 mM $MgSO_4$.
5. cAMP: 10 mM stock solution. Store at −20°C.
6. Caffeine: 100 mM stock solution. Store at −20°C.
7. Hydrophobic agar: 1% agar in dH_2O.

2.2. Time-Lapse Imaging

1. Inverted fluorescence microscope.
2. Lab-Tek II chambered coverglasses (Nalge Nunc).
3. FemtoJet Microinjector (Eppendorf).
4. Femtotips microcapillary pipettes (Eppendorf).
5. Micromanipulator (Eppendorf, Narushige).
6. The image processing software NIH ImageJ.

2.3. G-Protein Activation

1. Latrunculin A: 1 mM stock solution in DMSO. Store at −20°C.
2. Alexa Fluor 594 (Molecular Probe): 10 mM stock solution. Store at −20°C.
3. Inverted fluorescence microscope: Required capabilities includes (1) excitation at 457 nm and simultaneous recording for emissions at 480 and 535 nm (for FRET) (e.g., Olympus IX 71 inverted microscope with a 60×, 1.45 NA objective, a Kr/Ar laser line, and the Dual-View system (Optical Insights, LLC) mounted to a Photometrics Cascade 512B CCD for simultaneous image acquisition); and (2) excitation at 550–600 nm and emission around 600–650 nm (for Alexa 594) (e.g., a DsRed fluorescence filter set is used for Alexa 594 fluorescence).

2.4. PIP_3 Production

1. DB-MES, pH 6.5: 20 mM MES, 2 mM $MgSO_4$, 0.2 mM $CaCl_2$.
2. [^{32}P]Phosphorus (5 mCi/mL, NEX).
3. TLC plate (EM Science).
4. Nucleopore filter membrane.
5. Plastic cup.

2.5. TORC2-PKB Activity

1. Mouse anti-phospho PDK docking motif monoclonal antibody (Cell Signaling): Use 1:2,000 dilution in TBST containing 5% (w/v) BSA; detect with anti-mouse IgG-HRP.
2. Rabbit anti-phospho PKC (pan) monoclonal antibody (Cell Signaling): Use 1:2,500 dilution in TBST containing 5% (w/v) BSA; detect with anti-rabbit IgG-HRP.

3. Rabbit anti-phospho PKB substrate monoclonal antibody (Cell Signaling): Use 1:2,500 dilution in TBST containing 5% (w/v) BSA; detect with anti-rabbit IgG-HRP.

2.6. GST-RBD-Byr2 Pull Down Assay

1. GST-RBD-Byr2 expression construct (1).
2. BL21 (DE3) Escherichia coli competent cells. Store at –80°C.
3. LB (Luria broth): 10 g/L NaCl, 10 g/L tryptone, 5 g/L yeast extract. Adjust pH to 7.0 with NaOH. Autoclave and store at room temperature.
4. Ampicillin: 100 mg/mL, filter-sterilize. Store at –20°C.
5. IPTG (isopropyl-β-D-thiogalactopyranoside): 1 M stock solution. Store at –20°C.
6. DTT (dithiothreitol): 1 M stock solution. Store at –20°C.
7. PMSF (phenylmethylsulfonyl fluoride): 100 mM stock solution. Dissolve in isopropanol. Store at –20°C.
8. Lysozyme: 10 mg/mL. Store at –20°C.
9. Glutathione sepharose 4B: Store at 4°C.
10. 10× PBS (Phosphate-buffered saline), pH 7.5: 1.37 M NaCl, 27 mM KCl, 18 mM KH_2PO_4, 100 mM Na_2HPO_4. Store at 4°C.
11. Suspension buffer: 1× PBS, 1 mM DTT, 1 mM PMSF. Add DTT and PMSF before use.
12. 2× Lysis buffer: 20 mM sodium phosphate (pH 7.2), 300 mM NaCl, 1% NP-40, 20% glycerol, 1 mg/mL BSA, 20 mM $MgCl_2$, 2 mM EDTA, 2 mM Na_3VO_4, 10 mM NaF, with one tablet of protease inhibitor (Roche complete) per 25 mL.
13. Wash buffer: 10 mM sodium phosphate (pH 7.2), 150 mM NaCl, 0.5% NP-40, 10% glycerol, 10 mM $MgCl_2$, 1 mM EDTA.
14. Antibodies: Mouse monoclonal anti-pan Ras antibody Ras10 (Calbiochem) and anti-mouse IgG-conjugated HRP. Store at 4°C.

3. Methods

3.1. Assays for Chemotactic Responses

Upon starvation, *Dictyostelium* cells initiate development into fruiting bodies. During development, *Dictyostelium* cells obtain the ability to chemotax toward cAMP in order to form multicellular structures. Therefore, it is important to develop *Dictyostelium* cells to make them competent for robust chemotaxis by depleting nutrients. Methods to induce development and prepare chemotaxis-competent cells are described below. When mutant cells are isolated, it is also important to analyze their developmental phenotypes as chemotaxis-defective mutants often fail to normally develop. To examine chemotaxis, two methods have been extensively used. The first assay observes chemotaxing cells toward cAMP released from a micropipette under a light microscope. The other assay analyzes cells moving toward cAMP on agar plates.

3.1.1. Developing Cells on Agar Plates

1. Grow cells to a density of $2–5 \times 10^6$ cells/mL in HL5 medium at 22°C, shaking at 180 rpm.
2. Centrifuge 5×10^6 cells at $500 \times g$ for 5 min. Remove the supernatant and wash twice with 1 mL DB.
3. Resuspend cells in DB at 5×10^6 cells/mL and spread 200 μL of cell suspension onto a 3.5 cm dish containing 1 mL of 1% DB agar. Remove the DB after cells attach.
4. Incubate cells at 22°C. Cells start aggregating around 6 h after plating. Wild-type cells form mature fruiting bodies within 24 h (Fig. 1a).

3.1.2. Prepare Chemotaxis-Competent Cells

1. Centrifuge 2×10^8 cells at $500 \times g$ for 5 min. Remove the supernatant and wash twice with 40 mL DB.
2. Resuspend cells at 2×10^7 cells/mL in 10 mL DB, transfer cells into a 125 mL flask, and shake for 1 h at 100 rpm.
3. Pulse cells with cAMP at a final concentration of 50–100 nM every 6 min for 4–5 h using a timer-controlled-peristaltic pump. For example, set a pump to drop 50 μL of 20 μM

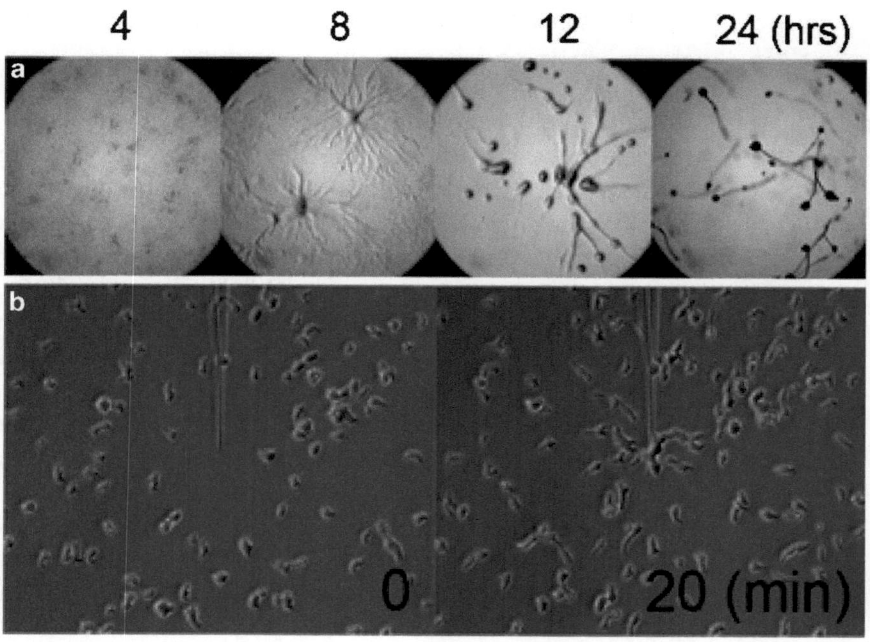

Fig. 1. Development and chemotaxis of *Dictyostelium* cells. (**a**) Development of cells on non-nutrient agar plates. (**b**) Chemotaxis to a pipette filled with 1 μM cAMP.

cAMP every 6 min. Chemotaxis-competent cells start aggregating within 1 h after cells are plated on DB agar as described in the Subheading 3.1.1.

4. To monitor the expression of developmentally regulated genes, take 100 μL of samples every hour and analyze using immunoblotting with antibodies against cAMP receptor 1 (cAR1) and adenylyl cyclase (ACA) (2) (see Note 1).

3.1.3. Needle Assay

1. Dilute developed cells by approximately 50-fold in DB and disrupt aggregates by pipetting gently. Transfer cells (~3 mL) to a one-well Lab-Tek II chambered coverglass, allow them to adhere to the coverglass for 10 min, and ensure that cells are individually separated under a light microscope.
2. Fill a micropipette with 15 μL of 1 μM cAMP. Attach the micropipette to a micromanipulator, which is connected to a microinjector. Use a continuous injection mode with a compensation pressure (Pc) of 100 hPa. Place the micropipette in the middle of the field. Lower the micropipette till it just touches the chambered coverglasss.
3. Observe cells using light microscopy with 10× to 40× objectives and take pictures every 10–30 s for 30 min (Fig. 1b).

3.1.4. Two-Drop Assay

1. Dilute developed cells by approximately fivefold in DB. Spot 5 μL of cell suspension and 5 μL of different concentrations of cAMP (0.1, 1, and 10 μM) onto 1% hydrophobic agar. The distance between spots is 5 mm.
2. Capture images of cells using a 20× objective every 30 s for 1 h.

3.2. Assays for Chemotactic Signalings

3.2.1. Visualization of G-Protein Activation by FRET

During development, *Dictyostelium* cells aggregate by migrating toward high concentrations of cAMP, which binds to the G-protein-coupled receptor cAR1. Upon ligand binding, cAR1 activates the heterotrimeric G-protein composed of Gα2, Gβ, and Gγ, causing Gα2 to dissociate from the Gβ–Gγ complex. This dissociation event can be monitored by the decrease of fluorescence resonance energy transfer (FRET) between Gα2 and Gβ tagged with CFP and YFP, respectively (3). The measurement can be done either at the population level using a fluorometer, or at the single-cell level using fluorescence microscopy (4). By recording the entire spectrum of emission averaged over a large number of cells, the population approach allows more accurate estimation of the absolute FRET change. However, in most instances the kinetics of FRET change is more important than its absolute magnitude, and single-cell measurement with fluorescent microcopy is advantageous due to its ease of applying complex spatial or temporal patterns of stimulation. The following protocol describes the procedure for monitoring G-protein FRET at the single-cell level in real time by fluorescent microscopy, with the use of a micropipette that allows rapid introduction and removal of stimulation.

A. Preparation of Cells

1. *Dictyostelium* cells (Gα2-null or Gβ-null) overexpressing both Gα2-CFP and Gβ-YFP are developed by cAMP pulsing as described in Subheading 3.1.2. Instead of the typical 4–5 h pulsing used for wild-type cells, a longer duration of pulsing (6–7 h) is recommended based on the observation that knock-out cells rescued with G proteins tagged with fluorescent proteins have delayed development.
2. Place cells on a one-well chambered coverglass as described in Subheading 3.1.3.
3. Cells are immobilized by the addition of Latrunculin A to a final concentration of 5 μM. Typically, cells lose their polarized morphology within 10 min of Latrunculin A treatment and round up. This step is to minimize the effect of changes in cell morphology on the analysis of fluorescent signal.

B. Setting Up Micropipettes

1. Fill a Femtotips micropipette with 5 μL of 10 μM cAMP solution containing 50 nM Alexa Fluor 594. Attach the micropipette to a microinjector and micromanipulator. Set a microinjector to continuously inject cAMP with a compensation pressure (Pc) of 100 hPa.
2. Under low magnification, use the micromanipulator to position the micropipette tip at the center of the viewing field. Save the position by holding the "position 1" button till a beeping sound is heard. Lift the micropipette (~0.5 cm) and save the position by holding the "position 2" button till a beeping sound is heard.
3. To obtain the spatial and temporal characteristics of the applied cAMP stimulation, fill a clean Lab-TekII chamber with DB and mount over a 60× objective. Using a dsRed fluorescence filter set, acquire time-lapse movies with a frame rate of one per second while bringing the micropipette tip in (position 1) and out (position 2) of the center of the field. Typically, a stable gradient is established within 10 s of introducing the micropipette.

C. Microscopy and cAMP Stimulation

1. Mount the Lab-Tek II chamber loaded with cells (step A. 3) over a 60× objective. Cells are excited with a 457-nm laser line, and the image of emission at 480 nm (CFP channel) and 535 nm (YFP channel) are acquired simultaneously using the Dual-View system attached to the CCD camera.

2. Look for cells with bright emissions in both channels. Adjust the position of the selected cell(s) to a suitable distance from where the tip of the micropipette will be located when introduced.
3. Take time-lapse movies with a rate of 2 s per frame. Acquire at least ten frames before introducing the micropipette.
4. Use the "position 1" and "position 2" buttons on the micromanipulator to introduce and remove the micropipette while acquiring video.

D. Analysis of G-Protein FRET Response

1. Open the saved video files in ImageJ.
2. For each frame, measure the mean pixel values for cells in both CFP and YFP channels. In ImageJ, the mean pixel value of a selected region of interest can be calculated using the "Measure" function under "Analyze" (see Note 2). Similarly, choose a background region to measure the mean value in both channels. Subtract the background values to obtain the corrected CFP and YFP emissions for cells.
3. Plot the corrected CFP and YFP emissions as well as the YFP/CFP ratio over time. Upon cAMP stimulation, CFP emissions should increase and YFP decrease. The YFP/CFP ratio should decrease and is usually more robust than either CFP or YFP emission because it is corrected for the frame-to-frame noise in excitation intensity (see Note 3). A typical result is shown in Fig. 2.

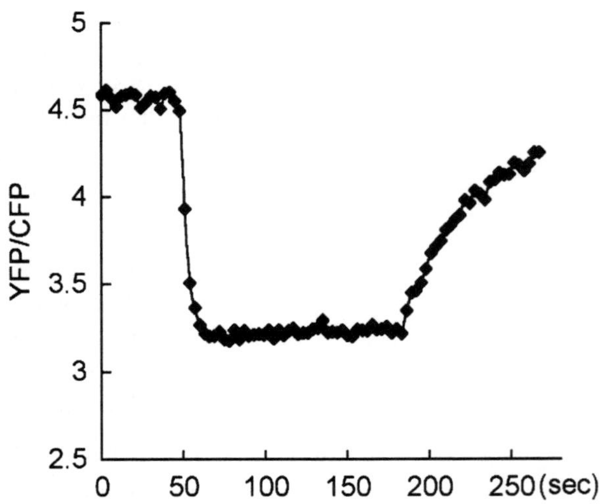

Fig. 2. G-protein activation upon cAMP stimulation monitored by FRET between Gα2-CFP and Gβ-YFP. Loss of FRET (decreased YFP:CFP ratio) was induced when a micropipette containing 10 μM cAMP was introduced at 45 s and persisted until the micropipette was removed at 180 s.

3.2.2. Detection of PIP_3 Production

In chemotaxing cells, PIP_3 is locally produced and activates signaling events that lead to actin polymerization at the leading edge. Pleckstrin homology (PH)-domain-containing proteins that bind to PIP_3 are highly localized at the front of chemotaxing cells and mediate downstream events. The PH-domain-containing proteins include CRAC (5), PKB (6), and PhdA (7). The production of PIP_3 is regulated by PI_3-kinase (PI3K) and PI_3-phosphatase (PTEN). PI3K phosphorylates $PI(4,5)P_2$ to produce PIP_3. On the other hand, PTEN converts PIP_3 to $PI(4,5)P_2$. Both PI3K and PTEN are required for PIP_3 production and chemotaxis toward cAMP in *Dictyostelium* (8–10). Here, we describe methods to measure PIP_3 production and localization upon cAMP stimulation. For in vivo analysis, GFP-fused the PH domain of CRAC is used to visualize PIP_3 by fluorescence microscopy (5–11). Amounts of PIP_3 can be quantitatively analyzed by extracting lipids from cells.

A. Time-Lapse Imaging of PIP_3 Production in Response to Chemoattractant Stimulation

1. Prepare chemotaxis-competent cells expressing GFP-PHcrac as described in the Subheading 3.1.2.
2. Resuspend the cells in DB at $\sim 5 \times 10^5$ cells/mL. Place 360 μL ($\sim 2 \times 10^5$ cells) in a well of eight-well chambered coverglasss. Allow the cells to adhere for 10 min.
3. Place the chamber on an inverted fluorescence microscope.
4. Stimulate cells by adding 40 μL of 10 μM cAMP to the chamber. Carefully squirt cAMP solution to avoid disturbing the cells.
5. Capture images of cells using a 40× objective every 3 s for 1 min. PIP_3 production peaks on the plasma membrane at 5–10 s after stimulation (Fig. 3a).

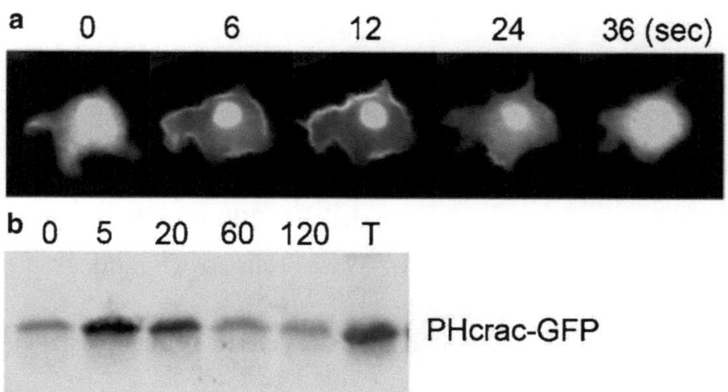

Fig. 3. Visualization of transient PIP_3 production upon uniform cAMP stimulation. Cells expressing PHcrac-GFP were examined by fluorescence microscopy (**a**) and immunoblotting using anti-GFP antibody (**b**) after addition of 1 μM cAMP. T, 20% of input.

B. In Vivo Translocation of PHcrac

1. Prepare chemotaxis-competent cells expressing GFP-PHcrac.
2. Basalate cells by adding caffeine at a final concentration of 2–5 mM and shake for 20 min at 200 rpm.
3. Centrifuge cells at $500 \times g$ for 5 min at 4°C. Remove the supernatant. Wash cells twice with 40 mL ice-cold PM.
4. Resuspend cells in 2.5 mL PM at 8×10^7 cells/mL and keep them on ice until used in the assay.
5. Aliquot cells into small plastic cup and shake at 200 rpm.
6. Add cAMP at a final concentration of 1 μM (e.g., 15 μL of 100 μM cAMP into 1.5 mL of cells).
7. Take samples at 0, 5, 20, 60, 120, and 180 s as described below.
   (a) Lyse 200 μL cells through 5 μm nucleopore filter into 1 mL ice-cold PM.
   (b) Spin at maximum speed for 1 min at RT in microcentrifuge.
   (c) Remove the supernatant.
   (d) Resuspend the pellet with 50 μL of 1× Sample Buffer.
8. Analyze 5 μL samples using SDS-PAGE and immunoblotting with anti-GFP antibody (Fig. 3b).

C. Detection of PIP_3 by Lipid Extractions and TLC

1. Prepare chemotaxis-competent cells in DB-MES.
2. Wash twice with 40 mL ice-cold DB-MES. Resuspend cells at 8×10^7 cells/mL in DB-MES.
3. Aliquot 1.5 mL cells into a small plastic cup and shake at 200 rpm.
4. Add 150 μL of 5 mCi/mL ^{32}Pi (5 mCi Phosphorus P32 radionucleotide: NEX) at a final concentration of 500 μCi/mL and incubate cells for 40 min.
5. Basalate cells by adding 50 μL of 0.1 M caffeine and shake at 200 rpm for 20 min.
6. Wash cells twice and resuspend in 1.5 mL ice-cold DB-MES.
7. Shake cells in a small plastic cup at 200 rpm and stimulate with 1 μM cAMP.
8. Take 150 μL of samples at 0, 5, 20, 60, 120, and 180 s into a glass tube containing 1 mL ice-cold 1N HCl and mix well by vortexing for 5 min.
9. Add 2 mL MeOH/$CHCl_3$ (1:1) and mix well.

10. Spin samples at 1,000 rpm at 4°C for 5 min.
11. Take 1 mL of a lower phase and mix well with 2 mL MeOH/1N HCl (1:1).
12. Spin samples at 1,000 rpm at 4°C for 5 min.
13. Take 800 μL of a lower phase and dry samples in a glass tube under N_2 gas.
14. Pre-run TLC plate in 2% potassium oxalate/dH_2O:MeOH (3:2) over night and dry plates in a chemical hood for 30 min.
15. Activate the TLC plate at 100°C for 3 min.
16. Resuspend dried samples in 30 μL $CHCl_3$:MeOH (2:1).
17. Spot 10 μL of samples on the heat-activated TLC plate.
18. Run the TLC plate with $CHCl_3$:acetone:MeOH:acetic acid:H_2O (30:12:10:9:6) for 200 min at RT.
19. Detect signals by autoradiography or phosphoimaging.

3.2.3. TORC2-PKB Activity

Dictyostelium expresses two Protein Kinase B (PKB) homologs, PKBA and PKBR1. PKBA is recruited to the plasma membrane through a PIP_3-specific Pleckstrin Homology (PH) domain at the N-terminus, whereas PKBR1 is tethered to the plasma membrane via N-terminal myristoylation (6, 12). cAMP activates the two PKBs through phosphorylation within their hydrophobic motifs (HMs) and activation loops (ALs) via Tor complex 2 (TORC2) (13) and Phosphoinositide-Dependent Kinase (PDKs) (14), respectively. Pianissimo A (PiaA), originally isolated in a forward genetic screen for chemotaxis defective mutants, is now recognized as a subunit of TORC2 (15). PKB phosphorylation is significantly reduced in cells lacking PiaA (13). Together, the two PKBs mediate the phosphorylation of several substrates, including Talin B, PI4P 5-kinase, two RasGefs, and a RhoGap (13). PKB phosphorylation occurs at the leading edge of migrating cells and plays important roles in chemotaxis. This section describes methods to assess PKB activity using phospho-specific antibodies.

cAMP Stimulation and Sample Preparation

1. Prepare chemotaxis-competent cells and basalate them as described in Subheading 3.2.2 (see Note 4).
2. Resuspend cells in 5 mL PM at 2×10^7 cells/mL and keep on ice until used in the assay (see Note 5).
3. Transfer ice-cold chemotaxis-competent cells to a 5 mL disposable plastic cup. Immediately add 100 μM of cAMP at a final concentration of 1 μM and start shaking at 200 rpm (see Note 6).

4. At time points of 0, 10, 20, 30, 60, 90, 120, and 180 s, transfer 100 μL cells into a microcentrifuge tube containing 50 μL 3× SDS sample buffer, and boil the sample at 95°C for 5 min.
5. Analyze samples using SDS-PAGE and immunoblotting with antibodies against phospho PDK docking motif, pan phosphor PKC, and phospho PKB substrate (see Notes 7–9) (Fig. 4).

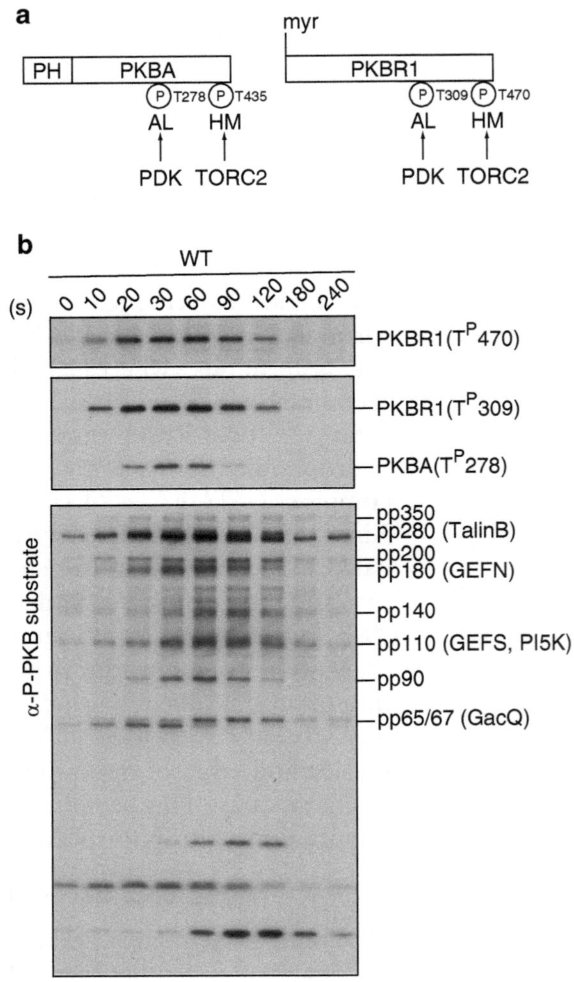

Fig. 4. PKB activation in response to chemoattractant stimulation. (**a**) Schematic representation of the activation of PKBR1 and PKBA. The N-terminus of PKBR1 is myristoylated while the N-terminus of PKBA has a PIP_3-specific PH domain. Upon cAMP stimulation, the two PKBs are activated through phosphorylation of their HMs by TORC2 and ALs by PDK. The phosphorylation sites (P) can be detected by phospho-specific antibodies. (**b**) Wild-type cells were stimulated with cAMP, lysed at the indicated time points, and subjected to immunoblotting using antibodies against phospho-HM (*top panel*), phospho-AL (*middle panel*), and phospho-substrate (*bottom panel*).

3.2.4. Ras Activity

Dictyostelium Ras proteins are important regulators of chemotaxis. Several Ras proteins, including RasB, RasC, RasD, and RasG, are activated upon cAMP stimulation (1, 16, 17). Activated Ras can be detected at the leading edge of cells undergoing chemotaxis (1). Two proteins, PI3K and TORC2, have been suggested to function downstream of Ras in regulating chemotaxis (8, 13, 18). Ras-binding domain (RBD) from yeast Byr2 or human Raf-1, which specifically interacts with the GTP-bound form of Ras, has been used to monitor Ras activation in *Dictyostelium*. The RBD of Byr2 is capable of binding to all four Ras GTPases that are activated by cAMP, whereas the RBD of Rafl is capable of binding to RasB, RasD, and RasG, but not to RasC (1, 17). This section describes biochemical and real-time imaging methods to assess Ras activation. In the biochemical assay, activated Ras proteins in cell lysates are pulled down by GST-fused RBD of Byr2 and glutathione-sepharose beads. In the cell imaging assay, GFP-fused RBD from Rafl is used to detect activated Ras in vivo.

A. Preparation of GST-RBD-Byr2 Attached Beads

1. Transform *Escherichia coli* with GST-RBD-Byr2 expression construct according to the manufacturer's instruction (see Note 10).
2. Pick up a single colony, inoculate into 50 mL LB with 50 μg/mL ampicillin, and grow overnight at 37°C with shaking at 250 rpm.
3. Measure OD600 of the overnight culture. Transfer into 1 L LB containing 50 μg/mL ampicillin at an OD600 of 0.05. Grow cells at 37°C with shaking at 250 rpm until OD600 reaches 0.4–0.6 (see Note 11).
4. Turn the temperature of the air shaker down to 20°C (see Note 12).
5. Add IPTG at a final concentration of 0.5 mM to induce protein expression. Shake the culture at 20°C for 18–20 h (see Note 13).
6. Harvest the cells by centrifugation at 4,000 × *g* for 10 min at 4°C (see Note 14).
7. Remove the supernatant. Resuspend cells in 50 mL suspension buffer (see Note 15).
8. Add 100 μg/mL lysozyme to cell suspension and incubate on ice for 20 min.
9. Sonicate cells. (see Note 16).
10. Add Triton X-100 at a final concentration of 2% and incubate on ice for 10 min.
11. Centrifuge at 10,000 × *g* for 15 min at 4°C (see Note 17).
12. Transfer the supernatant to a 50-mL tube and add 1 mL glutathione-sepharose 4B (see Note 18).

13. Rotate at 4°C for 1 h.
14. Spin down the sepharose by centrifuging at 500 g for 2 min at 4°C.
15. Remove the supernatant and wash five times with 40 mL ice-cold 1× PBS.
16. Remove the supernatant after the final wash and add equal volume of ice-cold PBS to make a 50% slurry. Store at 4°C (see Note 19).
17. Check GST-RBD-Byr2 production by SDS-PAGE and Coomassie Brilliant Blue staining (see Note 20).

B. GST-RBD-Byr2 Pull Down

1. Prepare chemotaxis-competent cells and basalate them as described in Subheading 3.2.2.
2. Wash glutathione-sepharose beads carrying GST-RBD-Byr2 three times with lysis buffer. Aliquot the beads carrying 30 μg GST-RBD-Byr2 into microcentrifuge tubes. Keep on ice until used in the assay (see Note 21).
3. Resuspend basalated cells at 4×10^7 cells/mL and keep on ice.
4. Transfer 3 mL cells to a 10 mL disposable plastic cup. Immediately add 100 μM of cAMP to a final concentration of 1 μM and shake at 200 rpm.
5. At time points of 10, 20, 40, 60 s, transfer 350 μL cells into a microcentrifuge tube containing 350 μL ice-cold 2× lysis buffer. Quickly mix by inverting the tube a few times and put the tube back on ice.
6. For the time point of 0 s, mix 350 μL unstimulated cells with 350 μL 2× lysis buffer.
7. Clarify the lysate by centrifugation at 15,000 ×*g* for 10 min at 4°C.
8. As a total cell lysate control, take 50 μL of the supernatant and mix with 25 μL 3× SDS sample buffer. Heat the sample at 95°C for 5 min.
9. Transfer the remaining supernatant (~600 μL) to the microcentrifuge tube containing GST-RBD-Byr2 beads.
10. Rotate for 45 min at 4°C.
11. Centrifuge at 2,000 ×*g* for 30 s at 4°C.
12. Remove the supernatant carefully. Wash beads four times with 1 mL wash buffer.
13. After the final wash, carefully aspirate the supernatant and add equal bead volume of 2× SDS sample buffer. Quickly move the tubes to a heat-block that is set to 95°C, and boil the sample for 5 min.

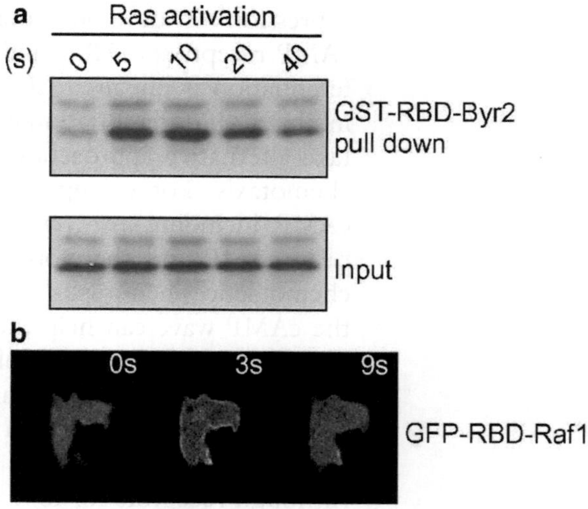

Fig. 5. Ras activation in response to chemoattractant. (**a**) cAMP stimulation triggers rapid and transient activation of endogenous Ras proteins, which can be pulled down by GST-RBD-Byr2. (**b**) Fluorescent images of GFP-RBD-Raf1 in wild-type cells before and after cAMP stimulation.

14. Analyze the pull-down product by immunoblotting with anti-pan-Ras antibody (see Note 22) (Fig. 5a).

C. Time-lapse Imaging of Ras Activation in Response to Global Chemoattractant Stimulation

1. Prepare chemotaxis-competent cells expressing GFP-RBD-Rafl (see Note 23).
2. Resuspend cells with DB at 5×10^5 cells/mL. Place 360 μL (~2×10^5 cells) into one well of a eight-well chambered coverglass. Allow the cells to adhere for 10 min.
3. Place the chamber on a fluorescence microscope.
4. Capture images using a 40× objectives every 3 s (see Note 24).
5. Stimulate Ras activation with 40 μL of 10 μM cAMP (see Note 25) (Fig. 5b).

4. Notes

1. A potential complication, which one might encounter during analysis of chemotaxis mutants, is that mutant cells show defects in differentiation. Failure in formation of multicellular structure is an indication for chemotaxis defects since chemotaxis is required for aggregation. However, this phenotype may simply result from defects in differentiation itself. One way to test if differentiation is affected is to examine the

expression of developmentally regulated genes including cAMP receptors (cAR1) and adenylyl cyclase (ACA). If differentiation is impaired, their expression will be delayed or inhibited. If differentiation is found to be affected, one could take alternative approaches to bypass this defect and study chemotaxis. For example, mutants defective in producing cAMP, including *aca-*, *crac-*, and *piaA-* cells, are unable to relay cAMP and therefore show developmental as well as chemotactic phenotypes. Artificial cAMP pulse mimicking the cAMP wave can help these cells to undergo differentiation and bypass the defects, allowing one to specifically examine chemotaxis in these mutants. It is also possible to analyze chemotaxis toward folate in vegetative cells since folic acid activates essentially the same downstream reactions as cAMP. Although receptors for folic acid have not yet identified and chemotactic responses to folate are relatively weaker compared to cAMP, this experiment will enable one to study chemotaxis independent of differentiation.

2. Instead of measuring a selected region frame by frame manually, the process can be performed with a macro. Under Plugins, click New to open a Macro editing window and type in the following script: for (i = 1; i < 1000; i++) {setSlice(i); run("Measure");} Click Ctrl + R to run the Macro.
3. Since YFP can be directly excited to a smaller extent by 457 nm, YFP emission is not entirely the result of FRET. Due to variations in expression levels of Gα2-CFP and Gβ-YFP from cell to cell, however, it is not possible to determine the absolute magnitude of FRET at the single-cell level only from emissions at 480 nm and 535 nm. Instead, the changes in the intensity of CFP and YFP emissions as well as the YFP/CFP ratio are used as surrogates for FRET changes.
4. Caffeine inhibits the TOR kinase activity and therefore needs to be removed for measuring PKB activation.
5. If cells are not maintained on ice, they will spontaneously secrete cAMP and respond in ~7 min.
6. A styrofoam rack for 50-mL conical tubes makes a convenient holder for multiple cups.
7. The primary antibody can be kept at 4°C in TBST containing 5% (w/v) BSA and reused for additional one to two times.
8. To compare sample loading, stain the membrane with Coomassie Brilliant Blue solution for a few minutes, destain for suitable amounts of time, and finally rinse with H_2O briefly. Let the membrane dry at room temperature.
9. As shown in Fig. 4, the addition of cAMP triggers rapid phosphorylations of the HM of PKBR1 and the ALs of PKBR1

and PKBA, which peak at 30–60 s and decline to pre-stimulus level by 2–3 min. The anti-phospho PKB substrate antibody detects a few bands before stimulation. Following cAMP stimulation, about ten new bands appear and display similar kinetics as that of PKB phosphorylation. The signals from phospho-proteins (pp)-350, 280, 200, 180, 140, 110, 90, and 65/67, are significantly reduced in cells lacking PKB activity, such as *pkbR1*- and *piaA*-, and are therefore considered as PKB substrates. Five of the putative PKB substrates (TalinB, GefN, GefS, PI5K, GacQ) have been identified by immunoprecipitation followed by mass spectrometry. A few of these bands (pp250, pp30, and pp23) are not PKB substrates as their phosphorylations are independent of PKB activity. They presumably contain the consensus motif RXRXXS/T that is recognized by other kinases.

10. It is important to use *E. coli* strains that have low protease activity, such as BL21 or BL21 (DE3).
11. It is important to use exponentially growing cells for protein induction.
12. It usually takes 20–30 min for the air shaker to cool down. The cells continue growing during this period of time to an OD600 of 0.6–0.8.
13. GST-RBD-Byr2 has low solubility when expressed in *E. coli*. Inducing protein production at 20°C increases the solubility.
14. Protein induction can be checked by SDS-PAGE and Coomassie Brilliant Blue staining. Mix cells (0.5 of OD600 unit) before and after IPTG induction with 50 μL 1× SDS sample buffer. Load 10 μL on SDS polyacrylamide gel.
15. Cell suspension can be frozen at −80°C at this stage. When used later, thaw the cell suspension and add fresh 1 mM PMSF.
16. Place the tube containing the cell suspension in an ice-filled plastic cup to keep the sample cool. Sonicate for 30 s each time and six to eight times with 30 s intervals until cells are lysed.
17. It is optional to filtrate the supernatant with 0.45 μm filter attached to a 60 mL syringe. Two or three filters may be required because the filters tend to get clogged.
18. Glutathione-sepharose 4B is stored in 20% ethanol. Before use, wash three times with 10 mL PBS.
19. GST-RBD-Byr2 sepharose beads can also be stored in PBS/50% glycerol at −20°C and are usually stable for a few months.
20. Use BSA or any quantified protein as a control.
21. The volume of the beads containing 30 μg GST-Byr2-RBD should be between 10 and 20 μL.

22. 12–15% SDS polyacrylamide gel is ideal for separating Ras proteins. The Ras10 antibody detects several *Dictyostelium* Ras isoforms.
23. For stable expression of GFP-RBD-Raf1, 20 μg/mL G418 is used.
24. Adjust exposure time and frequency to minimize photobleaching.
25. Add cAMP carefully into the chamber to avoid disturbing the attached cells. Ras activation peaks at about 3–5 s after stimulation.

References

1. Kae, H., Lim, C.J., Spiegelman, G.B., Weeks, G. (2004) Chemoattractant-induced Ras activation during Dictyostelium aggregation. *EMBO Rep.* **5**, 602–606.
2. Parent, C.A., Devreotes, P.N. (1996) Molecular genetics of signal transduction in Dictyostelium. *Annu. Rev. Biochem.* **65**, 411–440.
3. Janetopoulos, C., Jin, T., Devreotes, P. (2001) Receptor-mediated activation of heterotrimeric G-proteins in living cells. *Science* **291**, 2408–2411.
4. Xu, X., Meier-Schellersheim, M., Jiao, X., Nelson, L.E., Jin, T. (2005) Quantitative imaging of single live cells reveals spatiotemporal dynamics of multistep signaling events of chemoattractant gradient sensing in Dictyostelium. *Mol. Biol. Cell* **16**, 676–688.
5. Parent, C.A., Blacklock, B.J., Froehlich, W.M., Murphy, D.B., Devreotes, P.N. (1998) G protein signaling events are activated at the leading edge of chemotactic cells. *Cell* **95**, 81–91.
6. Meili, R., Ellsworth, C., Lee, S., Reddy, T.B., Ma, H., Firtel, R.A. (1999) Chemoattractant-mediated transient activation and membrane localization of Akt/PKB is required for efficient chemotaxis to cAMP in Dictyostelium. *EMBO J.* **18**, 2092–2105.
7. Funamoto, S., Milan, K,. Meili, R., Firtel, R.A. (2001) Role of phosphatidylinositol 3′ kinase and a downstream pleckstrin homology domain-containing protein in controlling chemotaxis in dictyostelium. *J. Cell Biol.* **153**, 795–810.
8. Funamoto, S., Meili, R., Lee, S., Parry, L., Firtel, R.A. (2002) Spatial and temporal regulation of 3-phosphoinositides by PI 3-kinase and PTEN mediates chemotaxis. *Cell* **109**, 611–623.
9. Iijima, M., Devreotes, P. (2002) Tumor suppressor PTEN mediates sensing of chemoattractant gradients. *Cell* **109**, 599–610.
10. Iijima, M., Huang, Y.E., Luo, H.R., Vazquez, F., Devreotes, P.N. (2004) Novel mechanism of PTEN regulation by its phosphatidylinositol 4,5-bisphosphate binding motif is critical for chemotaxis. *J. Biol. Chem.* **279**, 16606–16613.
11. Huang, Y.E., Iijima, M., Parent, C.A., Funamoto, S., Firtel, R.A., Devreotes, P. (2003) Receptor-mediated regulation of PI3Ks confines PI(3,4,5)P3 to the leading edge of chemotaxing cells. *Mol. Biol. Cell.* **14**, 1913–1922.
12. Meili, R., Ellsworth, C., Firtel, R.A. (2000) A novel Akt/PKB-related kinase is essential for morphogenesis in Dictyostelium. *Curr. Biol.* **10**, 708–717.
13. Kamimura, Y., Xiong, Y., Iglesias, P.A., Hoeller, O., Bolourani, P., Devreotes, P.N. (2008) PIP3-independent activation of TorC2 and PKB at the cell's leading edge mediates chemotaxis. *Curr. Biol.* **18**, 1034–1043.
14. Kamimura, Y., Devreotes, P.N. (2010) Phosphoinositide dependent protein kinase (PDK) activity regulates PIP3-dependent and -independent protein kinase B activation and chemotaxis. *J. Biol. Chem.* **285**, 7938–7946.
15. Chen, M.Y., Long, Y., Devreotes, P.N. (1997) A novel cytosolic regulator, Pianissimo, is required for chemoattractant receptor and G protein-mediated activation of the 12 transmembrane domain adenylyl cyclase in Dictyostelium. *Genes Dev.* **11**, 3218–3231.
16. Mondal, S., Bakthavatsalam, D., Steimle, P., Gassen, B., Rivero, F., Noegel, A.A. (2008) Linking Ras to myosin function: RasGEF Q, a Dictyostelium exchange factor for RasB, affects myosin II functions. *J. Cell Biol.* **181**, 747–760.
17. Zhang, S., Charest, P.G., Firtel, R.A. (2008) Spatiotemporal regulation of Ras activity provides directional sensing. *Curr. Biol.* **18**, 1587–1593.
18. Lee, S., Comer, F.I., Sasaki, A., et al. (2005) TOR complex 2 integrates cell movement during chemotaxis and signal relay in Dictyostelium. *Mol. Biol. Cell* **16**, 4572–4583.

Part VI

Methods and Technologies Towards Novel Therapeutics and Diagnostics Targeting Integrins

Chapter 27

Integrins in Tumor Angiogenesis and Lymphangiogenesis

Philippe Foubert and Judith A. Varner

Abstract

Angiogenesis, the formation of new blood vessel, plays an important role for the growth and metastasis of malignant tumors. The recent identification of specific growth factors for lymphatic vessels and of new lymphatic-specific markers provided evidence for an active role of the lymphatic system during the tumor growth and metastasis processes. Tumor lymphangiogenesis has been shown to play a role in promoting tumor growth and metastasis of tumor cells to distant sites. Integrins play keys roles in the regulation of angiogenesis and lymphangiogenesis during normal development and several diseases. Indeed, integrins control vascular and lymphatic endothelial cell adhesion, migration, and survival. Importantly, integrin inhibitors can block angiogenesis and lymphangiogenesis. In this chapter, we will highlight the role of integrins during angiogenesis and lymphangiogenesis as well as the function of individual integrins during vascular development, postnatal angiogenesis, and lymphangiogenesis. We discuss the role of integrins as potential therapeutic targets for the control of tumor angiogenesis, lymphangiogenesis, and metastatic spread in the treatment of cancer. We also describe methods to analyze expression and function of integrins during angiogenesis and lymphangiogenesis.

Key words: Cancer, Integrins, Angiogenesis, Lymphangiogenesis, Migration, Adhesion, Immunohistochemistry

1. Introduction

Angiogenesis and lymphangiogenesis (the development of new blood vessels and lymphatic) play critical roles during embryonic development, physiological processes such as wound healing and reproduction and numerous diseases, including inflammation, tumor progression, and metastasis.

1.1. Angiogenesis

Angiogenesis is the process by which new blood vessels develop from the existing vasculature (1). Angiogenesis is not only a

Motomu Shimaoka (ed.), *Integrin and Cell Adhesion Molecules: Methods and Protocols*,
Methods in Molecular Biology, vol. 757, DOI 10.1007/978-1-61779-166-6_27, © Springer Science+Business Media, LLC 2011

critical physiological mechanism for embryonic development and tissue repair, but it also promotes diseases such as tumor growth, diabetic retinopathy, and arthritis.

The principal cells promoting angiogenesis are endothelial cells, which line all blood vessels. To achieve new blood vessel formation, endothelial cells need to escape from their quiescent and stable location by degrading the basement membrane. Then, endothelial cells migrate toward a gradient of angiogenic factor such as VEGF-A (*Vascular Endothelial Growth Factor-A*) or bFGF (*basic Fibroblast Growth Factor*), released by activated cells. These cells may include platelets, tumor cells, tumor-associated macrophages and fibroblasts. Furthermore, endothelial cells also proliferate, thereby providing enough new cells, which will be organized into the tubular structures that form blood vessels.

All of these steps (basement membrane disruption, cell migration, proliferation, and tube formation) are regulated by members of the integrin family and which can consequently serve as targets to control the development of new vessels (2).

1.2. Lymphangiogenesis

The formation of new lymphatic vessels, or lymphangiogenesis, provides one of the main routes for tumor metastasis, especially for tumors of the breast, lung, and gastrointestinal tract, which frequently colonize draining regional lymph nodes. Compared to the blood vasculature, little is known about the biology of the lymphatic vessels in tumors, the regulation of tumor lymphangiogenesis or the mechanisms that determine the interactions of tumor cells with the lymphatic vessels (3). Recently, specific growth factors inducing the development of lymphatic endothelial cells have been characterized. These factors, VEGF-C and VEGF-D, bind the endothelial cell-specific tyrosine kinase receptors VEGF-R2 and VEGF-R3 (3). VEGF-R2 is a crucial mediator of angiogenesis, whereas VEGF-R3 regulates growth of lymphatic vessels.

Many human tumors express VEGF-C, and increased VEGF-C expression correlates with lymph node metastasis in, for example, thyroid, prostate, gastric, colorectal, and lung cancer. In breast cancer, VEGF-C expression correlates with lymph node positive tumors, whereas VEGF-D showed expression predominantly in inflammatory breast carcinoma (3).

Studies using various rodent models have provided evidence that tumor lymphangiogenesis facilitates lymphatic metastasis. In a transgenic mouse model, overexpression of VEGF-C in the β-cells of the pancreatic islets increased lymphangiogenesis around the primary tumor and enhanced tumor cell spread to the draining lymph nodes (4). More importantly, tumor growth, lymphangiogenesis and lymph node metastasis is inhibited by a blocking antibody VEGFR-3 (5). Finally, new findings indicate that select integrins can modulate lymphangiogenesis and consequently may affect tumor metastasis (6).

1.3. The Integrin Family of Adhesion Receptors

Integrins are a family of heterodimeric transmembrane glycoproteins mediating cell–cell and cell–Extra Cellular Matrix (ECM) interactions. The integrin family consists of 8α and 18β subunits that can associate to form 24 unique integrin heterodimers (7). Each integrin receptor heterodimer binds a specific set of endogenous ligands, which may include ligands in the ECM (collagen, fibronectin, vitronectin, for example), soluble ligands and ligands on other cells surfaces such as VCAM-1 (*Vascular Cell Adhesion Molecule-1*) or ICAM-1 (*InterCellular Adhesion Molecule-1*) (Fig. 1).

Each integrin subunit consists of an extracellular domain, a single transmembrane region and a short cytoplasmic region. Upon ligand binding, as series of intracellular signaling events is initiated. Integrin ligation to its ligand promotes integrin clustering and subsequent integrin-mediated intracellular signal transduction. This integrin signaling promotes endothelial and lymphatic endothelial cell migration, proliferation and survival (8).

1.4. Role of Integrins in Angiogenesis

Angiogenesis depends on a timely and spatially interaction between vascular cells, ECM, growth factors, and proteases. Integrins can mediate cell adhesion to the components of the extracellular matrix and to other cells, as well as make trans-membrane connections to the cytoskeleton and activate many intracellular signaling pathways (9). Endothelial cells are anchorage-dependent cells. Integrins facilitate endothelial cell binding to the ECM. Thus, the upregulation of endothelial cell integrins

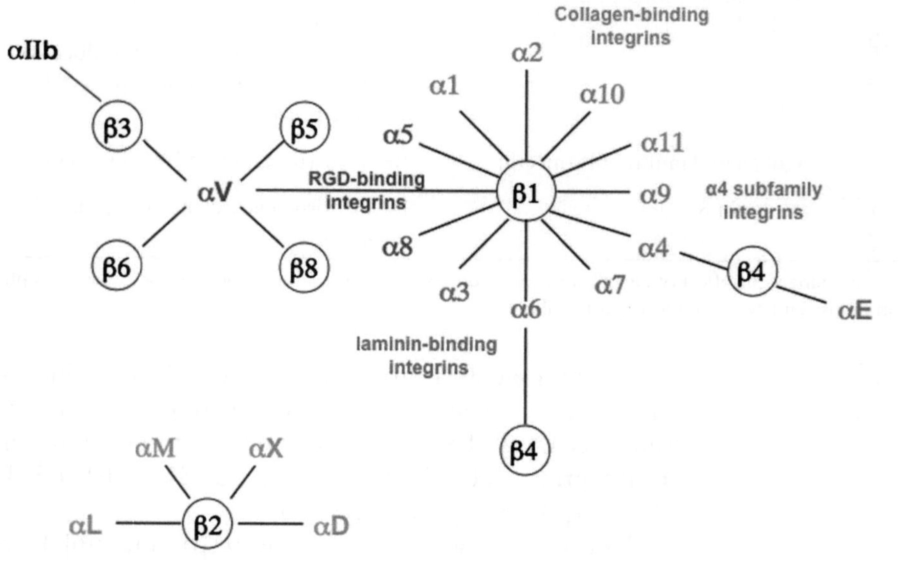

Fig. 1. The integrin family of adhesion of adhesion receptors and their ligands. There are 18α and 8β subunits which assemble to form 24 different heterodimers. Heterodimer composition confers ligand specificity. The main ligands for integrins in the extracellular space are extracellular matrix proteins, such as laminin, collagen, vitronectin, and fibronectin. Moreover, integrins can also bind cellular counter-receptors (VCAM-1 or ICAM-1) and soluble molecules (fibrinogen).

Table 1
Role of integrins in angiogenesis and lymphangiogenesis

Integrin	Major ligands	Mouse phenotype
α1β1	Collagen, laminin	α1$^{-/-}$: normal vascular development; reduced adult angiogenesis
α2β1	Collagen, laminin	α2$^{-/-}$: normal vascular development; enhanced tumor angiogenesis
α4β1	CS1 fibronectin, VCAM-1	α4$^{-/-}$: embryonic lethal; 50% die at E9.5-10.5, failure of chorion-allantois fusion; 50% die at E11.5 owing to cardiovascular defects
α5β1	Fibronectin, L1-CAM	α5$^{-/-}$: embryonic lethal E10-11; yolk sac and embryonic vessel defects
α6β1	Laminin	α6$^{-/-}$: embryonic lethal; lethal skin defects; no vascular defect
α9β1	Tensacin, fibronectin, thrombospondin, VCAM-1, collagen, laminin	α9$^{-/-}$: postnatal lethality P8-P12; chylothrorax (accumulation of lymph in the pleural cavity)
αMβ2	ICAM-1, fibrinogen	αM$^{-/-}$: normal development
αvβ3	Fibronectin, vitronectin, von Willebrand factor, tensacin, DEL-1, osteopontin	αv$^{-/-}$: 80% embryonic lethality E9.5; 20% die P20 with brain hemorrhage β3$^{-/-}$: 50% embryonic and early postnatal lethality; enhanced angiogenesis in surviving adults
αvβ5	Vitronectin, osteopontin, DEL-1	αv$^{-/-}$: 80% embryonic lethality E9.5; 20% die P20 with brain hemorrhage β5$^{-/-}$: normal development; reduced adult angiogenesis in response to certain angiogenic factors
αvβ8	Collagen, laminin, fibronectin	β8$^{-/-}$: disrupted brain blood vessel formation
α6β4	Laminin 5	β4$^{-/-}$: normal vascular development but lethal skin defects

This table summarizes the effect of genetic ablation of different integrin subunits on the vascular and lymphatic development in mouse embryo and during postnatal life

by pro-angiogenic factors sustains cell viability, increases cell sensitivity to growth factors and is required for migration. Endothelial cells have been reported to express up to ten different integrins (Table 1); α1β1, α2β1, α3β1, α4β1, α5v1, α6β1, α6v4, αvβ3, αvβ5, and αvβ8 (10).

During angiogenesis, integrins α1β1 and α6β4 are often downregulated, and integrins α5β1 and αvβ3 are upregulated or expressed de novo (8). Pathological angiogenesis is often associated with upregulation of the expression of certain integrins, including integrin α5β1 and αvβ3 (11). Integrins α2β1 and α1β1 are known to promote cell migration, proliferation, and matrix reorganization, and thus they are important in nonquiescent cells during dynamic situations, such as angiogenesis. VEGF significantly induces their expression on the endothelial cell surface. Inhibiting the function of integrins α1β1 and α2β1 by antibodies leads to selective inhibition of VEGF-driven angiogenesis in vivo without any effects on the pre-existing vasculature. Therefore, it has been suggested that integrins α1β1 and α2β1 play roles in pathological angiogenesis (12). Interestingly, it has been shown that tumor angiogenesis is markedly reduced in α1-null mice. This reduction may be caused by overexpression of metalloproteases (MMPs) and consequent generation of angiostatin, an inhibitor of angiogenesis proteolytically derived from plasminogen (13).

The αv integrins play important roles in angiogenesis. Integrin αvβ3 is selectively expressed on growing blood vessels. Importantly, in vivo angiogenesis in corneal or chorioallantoic membrane model induced by bFGF depends on αvβ3, whereas angiogenesis initiated by VEGF-A depends on αvβ5 (14). While results from studies of integrin antagonists indicate that αv integrins promote angiogenesis, genetic deletion studies indicate that αv integrins are not required for angiogenesis. Integrin αvβ3-deficient mice show normal developmental angiogenesis, but increased pathological angiogenesis (15). In contrast with these genetic studies, blockade of integrin αvβ3 as well as αvβ5 function using integrin antagonists disrupts tumor angiogenesis (16). Both integrins αvβ3 and α5β1 mediate pro-apoptotic signals when they are unligated or occupied by a soluble ligand (17). One hypothesis to explain this conflict is that αv integrins act as negative regulators of angiogenesis; once deleted in development, angiogenesis occurs at an accelerated rate. An alternative hypothesis is that animals lacking αv integrins develop compensatory changes in VEGF signaling that permit angiogenesis to occur during embryogenesis. In fact, β3 null mice exhibit enhanced tumor angiogenesis compared with normal mice, with strongly upregulated VEGFR-2 expression and signaling (18). Taken together, these studies suggest that compensatory VEGF-R2 signaling changes may play a role in the survival of β3-deficient animals.

Additional approaches have clarified the much-disputed role of αv integrins in angiogenesis. Animals expressing the point mutations Y747F and Y759F in the integrin β3 cytoplasmic tail develop normally, but exhibit reduced growth factor and tumor-induced angiogenesis in vivo (19). Mutant endothelial cells exhibit impaired adhesion, spreading, migration, and tube formation, as well as impaired complex formation between VEGF receptor-2 and β3 integrin. These results provide genetic evidence that integrin β3 plays an important role in promoting angiogenesis. Together, these diverse results can be interpreted to indicate that integrin αvβ3 plays an important role in angiogenesis and that loss of expression of this integrin in development can be partially compensated for by upregulation of other angiogenesis signaling pathways. Recently, it has been discovered that αvβ3 integrin binds to MMP-2 and thus this co-operation may regulate endothelial cell migration and other functions necessary for angiogenesis (20). Similar to the integrin β3, integrin β4 plays an important role in angiogenesis. The loss of integrin β4 significantly inhibits tumor angiogenesis suggesting a role for integrin α6β4, although its expression is usually downregulated during angiogenesis (21).

The important role of integrins during in tumor angiogenesis has led to the development of antagonists of integrins as a therapeutic tool for controlling tumor progression. Preclinical studies have suggested that antagonists of several integrins might be useful to suppress tumor angiogenesis and growth either alone or in combination with current cancer therapy (22).

1.5. Role of Integrins in Lymphangiogenesis

Although the role of integrins in angiogenesis is well documented, little is known about the expression and functional relevance of integrins during lymphangiogenesis.

The first evidence of the role of integrins in lymphangiogenesis has been provided by Huang et al. Indeed, this study suggests that the α9 integrin is required for the normal development of the lymphatic system, including the thoracic duct, and that α9 deficiency could be one cause of congenital chylothorax (accumulation of lymph in the pleural cavity) (23). Moreover, in murine embryos, expression of VEGF-R3 and integrin α9 is increased in Prox1-expressing lymphatic endothelial cells (LECs). Knockdown of Prox1 expression in human LECs led to decrease in the expression of integrin α9 and VEGF-R3, resulting in the decreased chemotaxis toward VEGF-C, suggesting integrin α9 may function as a key regulator of lymphangiogenesis acting downstream of Prox1 (24). Importantly, α9β1 integrin can bind VEGF-C and VEGF-D then promotes cell adhesion and migration (25).

Moreover, it has been shown that VEGF-A enhances expression of both integrin α1β1 and α2β1 in lymphatic endothelial cells, promoting their capacity to form cords and their migration. Interestingly, systemic blockade of these integrins potently inhibits wound-associated lymphangiogenesis in vivo (26). More recently, the integrin β1 has been found to be expressed in lymphatic endothelial cells isolated from patients with lymphangioma (27). Several lines of evidence are consistent with a role of α5β1 integrin in lymphangiogenesis mediated through VEGF-R3 signaling. It has been shown that integrin α5 is expressed by human lymphatic endothelial cells in culture (28) and by a subpopulation of resting and proliferating lymphatic vessels in mouse cornea (29). Selective inhibition of α5β1 integrin reduces lymphangiogenesis in a mouse model of suture-induced corneal inflammation (29). More recently, it has been shown that α5β1 integrin blockade reduces lymphatic sprouting and growth in airway inflammation after *M. pulmonis* infection but does not reduce blood vessel remodeling or macrophage recruitment (6). Furthermore, endostatin, which can inhibit endothelial cell migration by binding to α5β1 integrin (30) reduces lymphangiogenesis in skin tumors (31). Our studies reported that the integrin β4 is expressed on tumor lymphatic endothelium. Selective blockade of this integrin can block lymphangiogenesis and tumor metastasis (32). However, it seems that αv integrins do not play a role in tumor lymphangiogenesis (32). Therefore, several integrins can regulate lymphangiogenesis in physiological and pathological conditions (Table 1). Importantly, antagonists of these integrins may be useful to prevent tumor lymphangiogenesis and metastasis.

1.6. Integrins as Therapeutic Agents in Oncology

Blockade of integrin/ECM–ligand interactions inhibits tumor metastasis and angiogenesis and can be achieved by function-blocking antibodies, small organic molecules, and peptidomimetics. Antagonists of pro-angiogenic integrins, such as αvβ3, αvβ5, and α5β1, are under clinical evaluation (Table 2).

Integrins αvβ3 and αvβ5 are involved in angiogenesis and expressed in malignancies such as melanoma, gliomas, and cancers of the breast, prostate, and colon.

Abegrin, (Medi-522), a humanized anti-αvβ3 antibody, was the first anti-integrin therapeutic agent to be tested in clinical trials for cancer (33). A recent study in patients with solid tumors

Table 2
Integrin antagonists tested in clinical trials for cancer therapy

Drug name	Target	Drug type	Tumor type (trial phase)
Abegrin (MEDI-522)	αvβ3	Humanized antibody	Colorectal, melanoma, renal (Phase II)
CNTO 95	αvβ3 and αvβ5	Human antibody	Advanced refractory cancers (Phase I)
Cilengitide	αvβ3 and αvβ5	Peptide	Brain, head and neck, glioblastoma, leukemia, melanoma, prostate (Phase II/III)
Volociximab (M200)	α5β1	Chimeric mouse–human antibody	Non-small lung, melanoma, pancreatic, renal (Phase II)
ATN-161	α5β1	peptide	Malignant glioma (Phase I/II)

This table summarizes the effect of different integrin antagonists on angiogenesis and tumor progression in cancer patients showed that Abegrin had functional efficacy by reducing focal adhesion kinase activity in blood vessels (34). Based on these results, phase III cancer clinical trials are being to be evaluated.

On the basis of preclinical studies showing that both integrins αvβ3 and αvβ5 regulate angiogenesis, a human monoclonal antibody directed against both αvβ3 and αvβ5 integrins, CNTO 95 has been developed. CNTO 95 reduced angiogenesis and tumor growth in human melanoma xenografts in nude mice and rats (35). CNTO 95 is now under evaluation in a phase I/II clinical trial for the treatment of patients with melanoma (36). As CNTO 95 inhibits both integrins αvβ3 and αvβ5, two of the integrins that promote tumor angiogenesis, it might have wide-spread clinical utility. Additionally, most carcinoma cells express integrin αvβ5, which has been shown to promote tumor cell invasion (37). Targeting the αv integrins might thus block both tumor cell invasion and metastasis and tumor angiogenesis.

For these reasons, Cilengitide (EMD-121974), a synthetic cyclic penta-peptide small-molecule inhibitor of αvβ3 and αvβ5 integrins has been developed (38). The peptide has demonstrated anti-angiogenic and antitumor activities in vitro and in vivo. In a phase 1 trial in patients with advanced solid tumors, Cilengitide was administered twice weekly every 28 days and was well tolerated with no dose-limiting toxicities observed at the tested dose levels (39). This agent is currently evaluated in phase II and III trials for glioblastoma, non-small-cell lung cancer, melanoma and pancreatic and prostate cancer (40).

Integrin α5β1 is expressed mainly on vascular endothelial cells and upregulated together with fibronectin in tumor neovasculature. Volociximab is a chimeric human IgG4 against α5β1 that inhibits angiogenesis independent of VEGF/VEGF-R and induces apoptosis in proliferating, but not quiescent, endothelial cells in preclinical experiments (41). Volociximab was evaluated in phase II clinical trials for metastatic melanoma, renal cell carcinoma, and non-small-cell lung cancer (42). Volociximab is being tested in advanced ovarian cancer and in combination with gemcitabine in metastatic pancreatic cancer (43). Another inhibitor of integrin α5β1, the peptide ATN-161, is also developed in clinical trials. In animal models of colon cancer, ATN-161 reduced metastases and improved survival when combined with chemotherapy (44). Thus, these two integrins α5β1-inhibiting agents might offer future benefit to cancer patients.

Nevertheless, as many integrins can promote tumor angiogenesis and metastasis, it is not yet clear whether targeting one or more than one will have the most significant effect on tumor progression. Moreover, it is also likely that integrins antagonists may be combined with radio-chemotherapy or with other angiogenesis inhibitors such as VEGF inhibitors (Avastin).

2. Materials

2.1. Cell Culture

1. Endothelial Growth Medium (EGM) (Cambrex) supplemented with 10% fetal bovine serum (FBS), bFGF, and VEGF.
2. Solution of 0.25% trypsin and 1 mM ethylenediamine tetraacetic acid (EDTA).
3. Incubator 37°C, 5% CO_2.

2.2. Flow Cytometry

1. Wash buffer: Phosphate-Buffered Saline pH 7.4 containing 2% FBS.
2. Primary antibodies diluted in PBS 2% FBS: mouse anti-human α4β1 (HP1/2), mouse anti-human α5β1 (JBS5), mouse anti-human αvβ3 (LM609), and mouse anti-human αvβ5 (P1F6).
3. Fluorochrome-conjugated secondary antibody diluted in PBS 2% FBS.
4. Paraformaldehyde 0.5% in PBS.
5. FACS Calibur flow cytometer (BD Biosciences).

2.3. Cell Adhesion

1. Non-tissue culture 48-well plate.
2. Extracellular Matrix Proteins (ECM): vitronectin, fibronectin, CS-1 fibronectin diluted in PBS (5 μg/ml).
3. Blocking solution: PBS containing 2% heat-denatured bovine serum albumin (BSA).
4. Adhesion Buffer: Hanks Balanced Salt Solution, 10 mM Hepes pH 7.4, 2 mM $MgCl_2$, 2 mM $CaCl_2$, 0.2 mM $MnCl_2$, 1% BSA.
5. Paraformaldehyde: 3.7% in PBS.
6. Crystal violet 2% in sodium borate.
7. 10% acid acetic.
8. Antibodies: anti-human α4β1 (HP1/2), anti-human α5β1 (JBS5), anti-human αvβ3 (LM609), and anti-human αvβ5 (P1F6). These blocking antibodies are used as competitive inhibitors of cells adhesion to ECM proteins.
9. Plate reader to measure absorbance at 560 nm.

2.4. Cell Migration

1. 24-well plate.
2. 8-μm Costar Transwells.
3. Blocking solution: PBS containing 3% BSA.
4. Migration Buffer: DMEM, 10 mM HEPES, pH 7.4, 1.8 mM $MgCl_2$, 1.8 mM $CaCl_2$, 0.2 mM $MnCl_2$, 1% BSA.

5. Paraformaldehyde: 3.7% in PBS.
6. Crystal violet: 2% in sodium borate.

2.5. Immunohistochemistry on Frozen Mouse Tissue Sections

1. Humid chamber.
2. Coplin jars.
3. 100% cold acetone.
4. 1× PBS, pH 7.4.
5. Hydrophobic pen.
6. Normal goat serum.
7. Polyclonal rabbit anti-mouse LYVE-1 (RDI).
8. Monoclonal rat anti-mouse CD31 (MEC 13.3, BD Biosciences).
9. DAPI.
10. Fluorescent mounting media.
11. Clear nail varnish.

3. Methods

3.1. Integrin Expression on Human Umbilical Vein Endothelial Cells and Lymphatic Endothelial Cells

1. HUVEC and LEC are grown in endothelial growth medium (EGM-2) containing 10% fetal bovine serum (FBS), bFGF and VEGF.
2. Expression levels of human integrin α4β1, α5β1, αvβ3, and αvβ5 on HUVEC and LEC are determined by flow cytometry.
3. HUVEC and LEC are washed in PBS then detached with by trypsin treatment for 5 min (see Note 1).
4. Cells are resuspended in EGM-2 containing serum and washed twice in cold PBS containing 2% FBS.
5. Cells are then incubated for 1 h on ice with primary antibodies (1–10 μg antibody diluted in PBS containing 2% FBS). Cells are also incubated with isotype control antibodies IgG2a and IgG2b as negative control.
6. Cells are washed twice with PBS containing 2% FBS and incubated for 1 h with fluorochrome-conjugated goat anti-mouse secondary antibody on ice, in the dark.
7. Cells are washed twice by centrifugation with PBS containing 2% FBS, fixed with 0.5% paraformaldehyde and then analyzed by flow cytometry.

3.2. In Vitro Adhesion Assay

To determine whether specific integrins regulate HUVEC and LEC adhesion, 48-well plates are coated with different ECM proteins.

1. Wells are coated overnight at 4°C with PBS containing 5 μg/ml vitronectin, fibronectin or CS-1 fibronectin. Wells that are not coated are used as negative controls. Perform triplicate samples per group.
2. The next day, plates are blocked with PBS 5% BSA for 2 h at 37°C.
3. After detachment, HUVEC and LEC (250,000 cells/well) are resuspended in adhesion buffer in the presence or not of blocking antibodies (25 μg/ml).
4. Cells are incubated at 37°C, 5% CO_2 for 10–30 min.
5. Plates are carefully washed three times with warm adhesion buffer, and nonadherent cells are removed by aspiration.
6. Adherent remaining cells were then fixed by incubation in 3.7% paraformaldehyde for 1 h.
7. Cells are then stained with 1% crystal violet in sodium borate for 1 h.
8. Plates are well washed with distilled water to remove excess crystal violet, air-dried overnight, and extracted by incubation in 200 μl of acid acetic.
9. 100 μl of each of these extracts is measured at 560 nm using a plate reader.

3.3. In Vitro Migration Assay

To determine whether specific integrins regulate HUVEC and LEC migration, cell migration assays are performed using Costar Transwells.

1. Inserts are coated overnight at 4°C with PBS containing 5 μg/ml vitronectin, fibronectin or CS-1 fibronectin. Inserts that are not coated are used as negative controls. Perform triplicate samples per group.
2. Nonspecific binding sites are blocked by incubation with 3% BSA in PBS for 1 h at 37°C.
3. Resuspend cells in migration buffer.
4. Add 50,000 cells in the presence or not of integrins blocking antibodies to the upper chamber and incubate at 37°C, 5% CO_2.
5. Cells are allowed to migrate from the upper to lower chamber for 4 h at 37°C, 5% CO_2.
6. Remove nonmigrating cells from the upper chamber by wiping the upper surface with a cotton swab.
7. Cells that had migrated to the lower surface of the Transwell insert are fixed for 15 min with 3.7% paraformaldehyde and incubated in a 2% crystal violet in sodium borate.

8. Wash extensively with distilled water to remove excess crystal violet.
9. Count the number of cells that had migrated to the bottom of the insert in 5 random 200× fields per replicate.

3.4. Identification of Blood and Lymphatic Vessels in Tumor Tissue

To identify blood and lymphatic vessels in tissue, mouse tumor cryosections (from Lewis Lung Carcinoma, B16 melanoma models, for example) are immunostained to detect CD31 or Platelet-Endothelial Cell Adhesion Molecule (PECAM), a specific marker for vascular endothelial cells and LYVE-1, a marker of the lymphatic endothelium.

1. Allow slides of sections 5 μm thick to equilibrate at room temperature.
2. Label slides with a pencil, noting specimen.
3. Place slides in a glass coplin jar containing 100% cold acetone (pre-cooled at −20°C) for 2 min to fix.
4. Carefully dry the slides using tissue and draw a box around each specimen with a hydrophobic pen to retain the antibody volumes on the section.
5. Wash slides twice for 5 min in PBS (see Note 2).
6. Create a humidified chamber by placing a damp paper towel in the bottom of a plastic box with a sealing lid.
7. Place slides flat on the staining tray and block nonspecific antibody-binding sites by applying approximately 100 μl of 8% Normal Goat Serum (NGS) in PBS on each encircled section.
8. Incubate 2 h at room temperature overnight at 4°C.
9. Apply primary antibody (anti-CD31 or anti-LYVE1) at 5 μg/ml in 2% NGS in PBS. Apply only block buffer to some section that serves as a negative control (see Note 3).
10. Incubate for 2 h at room temperature.
11. Place slides in a coplin jar containing PBS and wash slides three times for 5 min with agitation.
12. Carefully dry the slides and apply 100 μl of secondary antibody (488-conjugated goat anti-rat IgG for CD31 and 488-conjugated goat anti-rabbit for LYVE1) diluted at 1:1,000 in 2% NGS in PBS (see Note 4).
13. Incubate for 1 h at room temperature.
14. Place slides in a coplin jar containing PBS and wash slides three times for 5 min with agitation.
15. Carefully dry slides and apply 100 μl of DAPI to stain the nuclei.
16. Incubate for 5 min.

17. Place slides in a coplin jar containing PBS and wash slides three times for 5 min with agitation.
18. Carefully dry slides and apply fluorescent mounting media.
19. Place a coverslip gently on the section.
20. Seal each slide by painting around the edge of the coverslip with clear nail varnish.
21. Store slides in the refrigerator in the dark.

3.5. Concluding Remarks

Angiogenesis and lymphangiogenesis have crucial roles in promoting tumor growth and metastasis. A substantial body of experimental evidence indicates that integrins regulate endothelial cell migration and survival during tumor angiogenesis and lymphangiogenesis. Indeed, changes in integrin expression and/or function are directly involved in angiogenesis, inflammation, tumor growth, and metastasis. Therefore, the development of integrin antagonists might be useful in blocking tumor metastasis in cancer patients. Preclinical evidence indicates that integrins are valuable targets for the design of novel cancer therapeutics (Fig. 2).

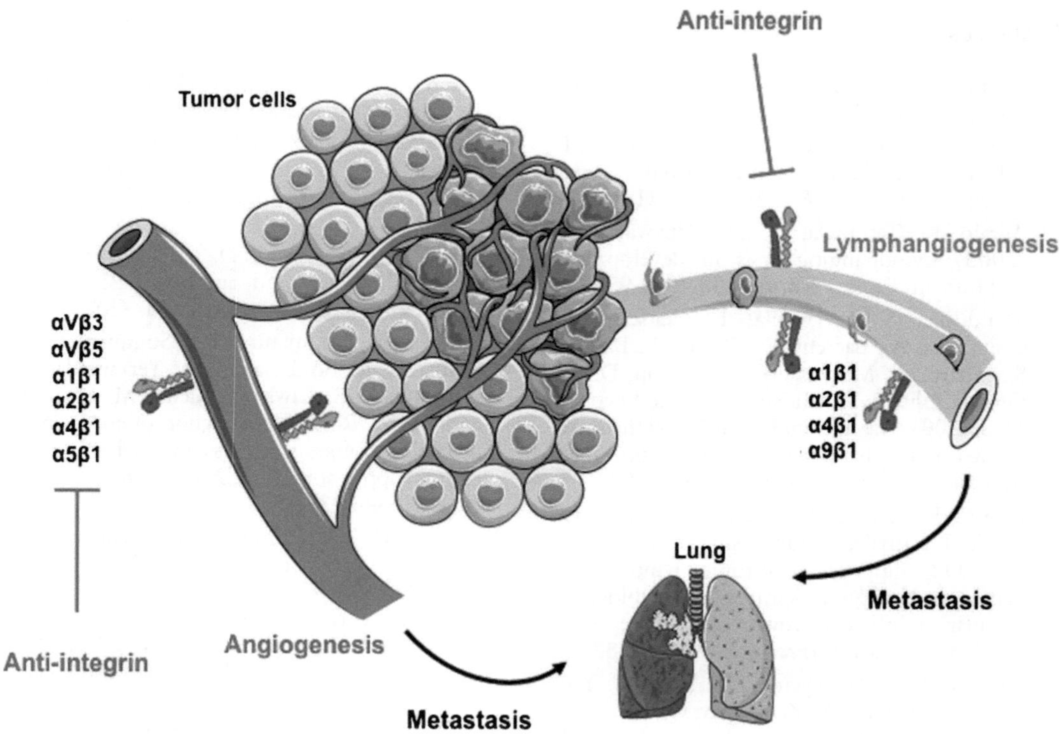

Fig. 2. Role of integrins in tumor angiogenesis and lymphangiogenesis. The tumor microenvironment activates or upregulates expression of integrins such as $\alpha1\beta1$, $\alpha2\beta1$, $\alpha4\beta1$, $\alpha5\beta1$, and $\alpha v\beta3$ on blood vessels and $\alpha1\beta1$, $\alpha2\beta1$, $\alpha4\beta1$, and $\alpha9\beta1$ on lymphatic vessels. Then, these integrins promote endothelial and lymphatic cells migration and survival during invasion of tumor tissue. Angiogenesis and lymphangiogenesis promote metastasis to local and distant organ such as lung.

4. Notes

1. Cell confluency could influence integrin expression. Therefore, always use cells at the same confluence (70–80%) to analyze integrin expression.
2. It is beneficial from this stage onward to ensure that the sections do no dry out. Therefore, the slides should be dried in small batches (2–4 slides) before adding the next solution.
3. The use of a negative control antibody is necessary to confirm the validity of the staining. This should either be an isotype-matched antibody or serum from the relevant species.
4. LYVE1, while a widely used marker for lymphatic endothelium, is expressed by other cell types, including macrophages which are abundant in the tumor microenvironment. Therefore, LYVE1 staining should be identified with care. Moreover, other lymphatic markers such as podoplanin, Prox1 or VEGFR3 should be tested to confirm the presence of lymphatic vessels in tumor.

References

1. Carmeliet, P. (2005) Angiogenesis in life, disease and medicine. *Nature* **438**, 932–936.
2. Conway, E. M., Collen, D., and Carmeliet, P. (2001) Molecular mechanisms of blood vessel growth. *Cardiovasc. Res.* **49**, 507–521.
3. Alitalo, K., Tammela, T., and Petrova, T. V. (2005) Lymphangiogenesis in development and human disease. *Nature* **438**, 946–953.
4. Mandriota, S. J., Jussila, L., Jeltsch, M., Compagni, A., Baetens, D., Prevo, R., Banerji, S., Huarte, J., Montesano, R., Jackson, D. G., Orci, L., Alitalo, K., Christofori, G., and Pepper, M. S. (2001) Vascular endothelial growth factor-C-mediated lymphangiogenesis promotes tumour metastasis. *EMBO J.* **20**, 672–682.
5. He, Y., Kozaki, K., Karpanen, T., Koshikawa, K., Yla-Herttuala, S., Takahashi, T., and Alitalo, K. (2002) Suppression of tumor lymphangiogenesis and lymph node metastasis by blocking vascular endothelial growth factor receptor 3 signaling. *J. Natl. Cancer Inst.* **94**, 819–825.
6. Okazaki, T., Ni, A., Ayeni, O. A., Baluk, P., Yao, L. C., Vossmeyer, D., Zischinsky, G., Zahn, G., Knolle, J., Christner, C., and McDonald, D. M. (2009) alpha5beta1 Integrin blockade inhibits lymphangiogenesis in airway inflammation. *Am. J. Pathol.* **174**, 2378–2387.
7. Hynes, R. O. (2002) Integrins: bidirectional, allosteric signaling machines. *Cell* **110**, 673–687.
8. Cheresh, D. A., and Stupack, D. G. (2008) Regulation of angiogenesis: apoptotic cues from the ECM. *Oncogene* **27**, 6285–6298.
9. Hynes, R. O. (2002) A reevaluation of integrins as regulators of angiogenesis. *Nat. Med.* **8**, 918–921.
10. Serini, G., Valdembri, D., and Bussolino, F. (2006) Integrins and angiogenesis: a sticky business. *Exp. Cell Res.* **312**, 651–658.
11. Sund, M., Hamano, Y., Sugimoto, H., Sudhakar, A., Soubasakos, M., Yerramalla, U., Benjamin, L. E., Lawler, J., Kieran, M., Shah, A., and Kalluri, R. (2005) Function of endogenous inhibitors of angiogenesis as endothelium-specific tumor suppressors. *Proc. Natl. Acad. Sci. USA.* **102**, 2934–2939.
12. Senger, D. R., Claffey, K. P., Benes, J. E., Perruzzi, C. A., Sergiou, A. P., and Detmar, M. (1997) Angiogenesis promoted by vascular endothelial growth factor: regulation through alpha1beta1 and alpha2beta1 integrins. *Proc. Natl. Acad. Sci. USA.* **94**, 13612–13617.
13. Pozzi, A., Moberg, P. E., Miles, L. A., Wagner, S., Soloway, P., and Gardner, H. A. (2000) Elevated matrix metalloprotease and angiostatin levels in integrin alpha 1 knockout mice cause reduced tumor vascularization. *Proc. Natl. Acad. Sci. USA.* **97**, 2202–2207.
14. Friedlander, M., Brooks, P. C., Shaffer, R. W., Kincaid, C. M., Varner, J. A., and Cheresh,

D. A. (1995) Definition of two angiogenic pathways by distinct alpha v integrins. *Science* **270**, 1500–1502.

15. Reynolds, L. E., Wyder, L., Lively, J. C., Taverna, D., Robinson, S. D., Huang, X., Sheppard, D., Hynes, R. O., and Hodivala-Dilke, K. M. (2002) Enhanced pathological angiogenesis in mice lacking beta3 integrin or beta3 and beta5 integrins. *Nat. Med.* **8**, 27–34.
16. Brooks, P. C., Montgomery, A. M., Rosenfeld, M., Reisfeld, R. A., Hu, T., Klier, G., and Cheresh, D. A. (1994) Integrin alpha v beta 3 antagonists promote tumor regression by inducing apoptosis of angiogenic blood vessels. *Cell* **79**, 1157–1164.
17. Stupack, D. G., Puente, X. S., Boutsaboualoy, S., Storgard, C. M., and Cheresh, D. A. (2001) Apoptosis of adherent cells by recruitment of caspase-8 to unligated integrins. *J. Cell Biol.* **155**, 459–470.
18. Reynolds, A. R., Reynolds, L. E., Nagel, T. E., Lively, J. C., Robinson, S. D., Hicklin, D. J., Bodary, S. C., and Hodivala-Dilke, K. M. (2004) Elevated Flk1 (vascular endothelial growth factor receptor 2) signaling mediates enhanced angiogenesis in beta3-integrin-deficient mice. *Cancer Res.* **64**, 8643–8650.
19. Mahabeleshwar, G. H., Feng, W., Phillips, D. R., and Byzova, T. V. (2006) Integrin signaling is critical for pathological angiogenesis. *J. Exp Med.* **203**, 2495–2507.
20. van Hinsbergh, V. W., Engelse, M. A., and Quax, P. H. (2006) Pericellular proteases in angiogenesis and vasculogenesis. *Arterioscler Thromb. Vasc. Biol.* **26**, 716–728.
21. Nikolopoulos, S. N., Blaikie, P., Yoshioka, T., Guo, W., and Giancotti, F. G. (2004) Integrin beta4 signaling promotes tumor angiogenesis. *Cancer Cell* **6**, 471–483.
22. Silva, R., D'Amico, G., Hodivala-Dilke, K. M., and Reynolds, L. E. (2008) Integrins: the keys to unlocking angiogenesis. *Arterioscler. Thromb. Vasc. Biol.* **28**, 1703–1713.
23. Huang, X. Z., Wu, J. F., Ferrando, R., Lee, J. H., Wang, Y. L., Farese, R. V., Jr., and Sheppard, D. (2000) Fatal bilateral chylothorax in mice lacking the integrin alpha9beta1. *Mol. Cell. Biol.* **20**, 5208–5215.
24. Mishima, K., Watabe, T., Saito, A., Yoshimatsu, Y., Imaizumi, N., Masui, S., Hirashima, M., Morisada, T., Oike, Y., Araie, M., Niwa, H., Kubo, H., Suda, T., and Miyazono, K. (2007) Prox1 induces lymphatic endothelial differentiation via integrin alpha9 and other signaling cascades. *Mol. Biol. Cell* **18**, 1421–1429.
25. Vlahakis, N. E., Young, B. A., Atakilit, A., Hawkridge, A. E., Issaka, R. B., Boudreau, N., and Sheppard, D. (2007) Integrin alpha9beta1 directly binds to vascular endothelial growth factor (VEGF)-A and contributes to VEGF-A-induced angiogenesis. *J. Biol. Chem.* **282**, 15187–15196.
26. Hong, Y. K., Lange-Asschenfeldt, B., Velasco, P., Hirakawa, S., Kunstfeld, R., Brown, L. F., Bohlen, P., Senger, D. R., and Detmar, M. (2004) VEGF-A promotes tissue repair-associated lymphatic vessel formation via VEGFR-2 and the alpha1beta1 and alpha2beta1 integrins. *FASEB J.* **18**, 1111–1113.
27. Norgall, S., Papoutsi, M., Rossler, J., Schweigerer, L., Wilting, J., and Weich, H. A. (2007) Elevated expression of VEGFR-3 in lymphatic endothelial cells from lymphangiomas. *BMC Cancer* 7, 105.
28. Zhang, X., Groopman, J. E., and Wang, J. F. (2005) Extracellular matrix regulates endothelial functions through interaction of VEGFR-3 and integrin alpha5beta1. *J. Cell Physiol.* **202**, 205–214.
29. Dietrich, T., Onderka, J., Bock, F., Kruse, F. E., Vossmeyer, D., Stragies, R., Zahn, G., and Cursiefen, C. (2007) Inhibition of inflammatory lymphangiogenesis by integrin alpha5 blockade. *Am. J. Pathol.* **171**, 361–372.
30. Sudhakar, A., Sugimoto, H., Yang, C., Lively, J., Zeisberg, M., and Kalluri, R. (2003) Human tumstatin and human endostatin exhibit distinct antiangiogenic activities mediated by alpha v beta 3 and alpha 5 beta 1 integrins. *Proc. Natl. Acad. Sci. USA.* **100**, 4766–4771.
31. Brideau, G., Makinen, M. J., Elamaa, H., Tu, H., Nilsson, G., Alitalo, K., Pihlajaniemi, T., and Heljasvaara, R. (2007) Endostatin overexpression inhibits lymphangiogenesis and lymph node metastasis in mice. *Cancer Res.* **67**, 11528–11535.
32. Garmy-Susini, B., Makale, M., Fuster, M., and Varner, J. A. (2007) Methods to study lymphatic vessel integrins. *Methods Enzymol* 426, 415–438.
33. Friedlander, M., Theesfeld, C. L., Sugita, M., Fruttiger, M., Thomas, M. A., Chang, S., and Cheresh, D. A. (1996) Involvement of integrins alpha v beta 3 and alpha v beta 5 in ocular neovascular diseases. *Proc. Natl. Acad. Sci. USA.* **93**, 9764–9769.
34. Zhang, D., Pier, T., McNeel, D. G., Wilding, G., and Friedl, A. (2007) Effects of a monoclonal anti-alphavbeta3 integrin antibody on blood vessels – a pharmacodynamic study. *Invest. New Drugs* **25**, 49–55.
35. Trikha, M., Zhou, Z., Nemeth, J. A., Chen, Q., Sharp, C., Emmell, E., Giles-Komar, J., and Nakada, M. T. (2004) CNTO 95, a fully human monoclonal antibody that inhibits alphav

integrins, has antitumor and antiangiogenic activity in vivo. *Int. J. Cancer* **110**, 326–335.

36. Mullamitha, S. A., Ton, N. C., Parker, G. J., Jackson, A., Julyan, P. J., Roberts, C., Buonaccorsi, G. A., Watson, Y., Davies, K., Cheung, S., Hope, L., Valle, J. W., Radford, J. A., Lawrance, J., Saunders, M. P., Munteanu, M. C., Nakada, M. T., Nemeth, J. A., Davis, H. M., Jiao, Q., Prabhakar, U., Lang, Z., Corringham, R. E., Beckman, R. A., and Jayson, G. C. (2007) Phase I evaluation of a fully human anti-alphav integrin monoclonal antibody (CNTO 95) in patients with advanced solid tumors. *Clin. Cancer Res.* **13**, 2128–2135.
37. Brooks, P. C., Klemke, R. L., Schon, S., Lewis, J. M., Schwartz, M. A., and Cheresh, D. A. (1997) Insulin-like growth factor receptor cooperates with integrin alpha v beta 5 to promote tumor cell dissemination in vivo. *J. Clin. Invest.* **99**, 1390–1398.
38. Stupp, R., and Ruegg, C. (2007) Integrin inhibitors reaching the clinic. *J. Clin. Oncol.* **25**, 1637–1638.
39. Hariharan, S., Gustafson, D., Holden, S., McConkey, D., Davis, D., Morrow, M., Basche, M., Gore, L., Zang, C., O'Bryant, C. L., Baron, A., Gallemann, D., Colevas, D., and Eckhardt, S. G. (2007) Assessment of the biological and pharmacological effects of the alpha nu beta3 and alpha nu beta5 integrin receptor antagonist, cilengitide (EMD 121974), in patients with advanced solid tumors. *Ann. Oncol.* **18**, 1400–1407.
40. Friess, H., Langrehr, J. M., Oettle, H., Raedle, J., Niedergethmann, M., Dittrich, C., Hossfeld, D. K., Stoger, H., Neyns, B., Herzog, P., Piedbois, P., Dobrowolski, F., Scheithauer, W., Hawkins, R., Katz, F., Balcke, P., Vermorken, J., van Belle, S., Davidson, N., Esteve, A. A., Castellano, D., Kleeff, J., Tempia-Caliera, A. A., Kovar, A., and Nippgen, J. (2006) A randomized multi-center phase II trial of the angiogenesis inhibitor Cilengitide (EMD 121974) and gemcitabine compared with gemcitabine alone in advanced unresectable pancreatic cancer. *BMC Cancer* **6**, 285.
41. Ramakrishnan, V., Bhaskar, V., Law, D. A., Wong, M. H., DuBridge, R. B., Breinberg, D., O'Hara, C., Powers, D. B., Liu, G., Grove, J., Hevezi, P., Cass, K. M., Watson, S., Evangelista, F., Powers, R. A., Finck, B., Wills, M., Caras, I., Fang, Y., McDonald, D., Johnson, D., Murray, R., and Jeffry, U. (2006) Preclinical evaluation of an anti-alpha5beta1 integrin antibody as a novel anti-angiogenic agent. *J. Exp. Ther. Oncol.* **5**, 273–286.
42. Kuwada, S. K. (2007) Drug evaluation: Volociximab, an angiogenesis-inhibiting chimeric monoclonal antibody. *Curr. Opin. Mol. Ther.* **9**, 92–98.
43. Ma, W. W., and Adjei, A. A. (2009) Novel agents on the horizon for cancer therapy. *CA Cancer J. Clin.* **59**, 111–137.
44. Stoeltzing, O., Liu, W., Reinmuth, N., Fan, F., Parry, G. C., Parikh, A. A., McCarty, M. F., Bucana, C. D., Mazar, A. P., and Ellis, L. M. (2003) Inhibition of integrin alpha5beta1 function with a small peptide (ATN-161) plus continuous 5-FU infusion reduces colorectal liver metastases and improves survival in mice. *Int. J. Cancer.* **104**, 496–503.

Chapter 28

PET-Radioimmunodetection of Integrins: Imaging Acute Colitis Using a ^{64}Cu-Labeled Anti-β_7 Integrin Antibody

Jason L.J. Dearling and Alan B. Packard

Abstract

Integrins are involved in a wide range of cell interactions. Imaging their distribution using high-resolution noninvasive techniques that are directly translatable to the clinic can provide new insights into disease processes and presents the opportunity to directly monitor new therapies. In this chapter, we describe a protocol to image, the in vivo distribution of the integrin β_7, expressed by lymphocytes recruited to and retained by the inflamed gut, using a radiolabeled whole antibody. The antibody is purified, conjugated with a bifunctional chelator for labeling with a radiometal, labeled with the positron-emitting radionuclide ^{64}Cu, and injected into mice for microPET studies. Mice with DSS-induced colitis were found to have higher uptake of the ^{64}Cu-labeled antibody in the gut than control groups.

Key words: Colitis, Inflammatory bowel disease, Radioimmunodetection, Integrin targeting, Copper-64, Monoclonal antibodies, Positron emission tomography imaging

1. Introduction

Inflammatory bowel diseases (IBDs) affect 0.5–2% of the population (1). This is a group of conditions, including Crohn's disease and ulcerative colitis, which can have a severe impact on quality of life. Diagnosis and evaluation of the severity of intestinal inflammation is important but is usually made through invasive techniques, such as colonoscopy, which is operator-dependent, lacks objectivity, and can only assess part of the intestine, giving an incomplete indication of the extent of disease (2). An objective, global method for the detection and description of intestinal inflammation is clearly needed, and medical imaging, such as Positron Emission Tomography (PET), is potentially a way to solve this clinical problem.

Motomu Shimaoka (ed.), *Integrin and Cell Adhesion Molecules: Methods and Protocols*,
Methods in Molecular Biology, vol. 757, DOI 10.1007/978-1-61779-166-6_28, © Springer Science+Business Media, LLC 2011

Here, we describe the steps necessary to use radioimmunodetection to identify the sites and extent of colitis in vivo by targeting an integrin involved in the disease. A major part of gut inflammation involves aberrant behavior of leukocytes, including lymphocytes. These are recruited and retained in the gut through expression of the integrin monomer β_7 (3). The antibody FIB504.64 was originally raised against β_7 in order to investigate its function (4). The antibody was produced from its hybridoma cell line, purified, exchanged into acetate buffer, and then conjugated to a bifunctional chelator. The resulting immunoconjugate was labeled with the positron-emitting radionuclide ^{64}Cu, injected into mice with chemically induced acute colitis, and microPET images were collected over 2 days. These images showed higher uptake of the radiolabeled antibody in the gut of mice with colitis compared to mice in the control groups.

2. Materials

2.1. Conjugation of Bifunctional Chelator with Antibody

1. The anti-β_7 antibody FIB504.64: Produced from its hybridoma (FIB504.64 (4)) and purified using a HiTrap Protein G column (GE Healthcare).
2. Control antibody rat IgG2a.
3. EDC (1-ethyl-3-(3-(dimethylamino)propyl)carbodiimide hydrochloride): Stored over anhydrous calcium sulfate in a −80°C freezer prior to use.
4. *S*-2-(4-aminobenzyl)-1,4,7,10-tetraazacyclododocane tetraacetic acid (*p*-NH_2-Bn-DOTA) (Macrocyclics, Dallas, TX).
5. Metal-naive pipette tips (Rainin, Oakland, CA) (see Note 1).
6. Chelex 100 Resin.
7. 0.1 M Sodium acetate buffer, pH 5.0: Used for both antibody-bifunctional chelator conjugation and antibody radiolabeling.
8. High-Performance Liquid Chromatography (HPLC) equipped with a BioSep SEC-S3000 size-exclusion column (300 × 7.8 mm, Phenomenex, Torrance, CA).
9. 0.1 M NaOAc buffer.
10. Centrifugal filters (0.45 μm pore size).
11. Amicon centrifugal filter devices (30 kDa MWCO): Used for antibody concentration.
12. Chelator solution: Make a stock solution of *p*-NH_2-Bn-DOTA in Me_2SO (e.g., 20 mg in 50 μL = 400 mg/mL), and add 3 volume equivalents of 0.1 M sodium acetate buffer (pH 5.0) to give a final bifunctional chelator concentration of approximately 100 mg/mL. Bring the pH to 5.0 by addition of 1 M or 5 M NaOH. This solution can be aliquoted and stored at −20°C.

2.2. Induction of Acute Colitis in Mice

1. Female C57/BL mice (20–25 g, 2–3 months old): Kept in standard conditions, with food and water available *ad libitum.*
2. Dextran sodium sulfate (DSS, 36–50 kDa MW).

2.3. Imaging of Distribution of Radiolabeled Antibody in Mice

1. Copper-64: Ordered from IsoTrace Technologies (O'Fallon, MO) or MDS Nordion (Ontario, Canada).
2. Ethylenediamine-tetraacetic acid (EDTA): Dissolve Na_4EDTA in phosphate buffer (pH 8) to give a 100 mM solution. Adjust pH to 8.0 by addition of 1 M NaOH and sterile filter.
3. Instant Thin-Layer Chromatography plates (ITLC) (Si Gel impregnated glass fiber sheets; Pall Life Sciences, Ann Arbor, MI): Cut into approximately 6 cm lengths, and dried in an oven at 110°C before use.
4. Saline.
5. Acrodisc 4-mm syringe filters with 0.45-μm nylon membrane.

3. Methods

In this section, we describe the three processes necessary to label an antibody with a radiometal. The antibody is purified using semi-preparative size-exclusion HPLC chromatography and reconcentrated to approximately 10 mg/mL. Then, the antibody is conjugated with the bifunctional chelator. Finally, the immunoconjugate is repurified and frozen in aliquots ready for radiolabeling.

3.1. Conjugation of Bifunctional Chelator with Antibody

3.1.1. Antibody Purification and Re-concentration

1. Equlibrate the Size Exclusion Chromatography column (SEC). High Performance Liquid Chromatography (HPLC) column with degassed ultrapure water, then with degassed NaOAc buffer.
2. The HPLC is run in isocratic mode with 0.1 M NaOAc buffer as mobile phase at 1 mL/min.
3. Filter the antibody through a 0.45 μm centrifugal filter. All proteins should be filtered (0.45 μm) before injection into the HPLC.
4. Clean (i.e., draw up, then expel) a 0.5-mL Hamilton syringe thoroughly with ethanol, water, and buffer (3× each).
5. Draw up the antibody solution (e.g., ~3 mg in 500 μL) into the syringe, and then expel bubbles.
6. Inject the antibody into a 0.5-mL HPLC injection loop, and initiate the run.
7. The antibody (MW ~150 kDa) has a retention time (RT) of approximately 8 min in the system described. A trial run should be carried out with a small aliquot of antibody to establish the RT in each laboratory's system.

8. Collect 1-mL (i.e., 1 min) aliquots of samples into 1.5-mL centrifuge tubes at the time points from 5 min onwards, then 0.5-mL aliquots from 7 min on. Proteins usually elute from the column by 9.5 min. The antibody is usually in approximately 2–3 fractions (2–3 mL).
9. Concentrate the fractions containing antibody using the Centricon filter units. Pipette 400 μL of antibody solution into the top of the filter unit and 360 μL of buffer in the bottom (total 760 μL – the total volume in the bottom of the unit is limited to 660 μL). At least 100 μL will be retained in the top, and the antibody will not dry out. Centrifuge at 5,000 rpm (2.3 rcf), or follow manufacturer's guidelines for different devices and centrifuges. Discard the buffer in the bottom of the cup after each concentration, replace it with 360 μL of fresh buffer, and refill the top to a volume of 400 μL with additional antibody solution. The time taken to reduce the volume in the top will increase as the concentration of antibody in the top of the unit increases.
10. Once all the antibody is concentrated in the top, place 310 μL of fresh buffer in the bottom and add buffer to the retentate to a total volume of 400 μL, then centrifuge. This will concentrate the antibody to a volume of 50 μL in the top. Carefully remove the antibody using a P200 pipette set to 50 μL, adjusting the pipette to remove all liquid, and place in a fresh centrifuge tube. Note total volume of liquid removed from the filter unit. Subtract this volume from 210 μL and divide the result by 2. For example, if the total volume of liquid in the top of the filter was 60 μL, 210 μL − 60 μL = 150 μL. 150 μL/2 = 75 μL. Use this volume of buffer (e.g., 75 μL) to rinse the top of the unit twice, removing as much protein as possible from the unit, and add the washings to the antibody solution in the centrifuge tube. The antibody should now be in a total volume of 210 μL of buffer. Remove bubbles from the solution by briefly centrifuging.
11. Add 10 μL of the antibody solution to 490 μL of buffer and measure the absorbance at 280 nm, using the appropriate molar extinction coefficient and the formula $C = A/E \times l$ (where C = concentration, A = absorbance, E = molar extinction coefficient, and l = path length) to calculate the protein concentration. The antibody concentration should be approximately 10 mg/mL for the conjugation reaction (see Note 2).

3.1.2. Conjugation with the Bifunctional Chelator

1. To conjugate 1 mg of Ab (stock concentration 10 mg/mL), add 100 μL (1 mg, 6.67×10^{-6} mmol) of antibody solution to a 1.5 mL centrifuge tube. Add a 250-fold molar excess of chelator solution to the tube. For the stock chelator solution, this would be 8.5 μL (1.7×10^{-3} mmol). Add buffer to achieve

a final Ab concentration of 5 mg/mL (200 μL total solution volume for 1 mg Ab; e.g., 78.9 μL (200–100 μL Ab solution + 8.5 μL chelator solution + 12.6 μL EDC solution (*vide infra*))). Finally, add 500 molar equivalents of EDC solution, prepared immediately before use in ultrapure water. For an EDC solution of 50 mg/mL (0.261 M), a 500-fold molar excess would be equal to 12.6 μL (3.3×10^{-3} mmol) (See Notes 3 and 4).

2. Gently mix the reaction mixture by pipetting, briefly centrifuge to remove air bubbles, confirm the pH to be 5.0 (see Note 5), and then place the reaction vessel in a 37°C water bath. The final concentration of Me_2SO should always be less than 5% (typically ~1%).

3.1.3. Immunoconjugate Re-purification and Storage

1. After 30 min, separate unbound chelator from the immunoconjugate using size-exclusion HPLC as described in Subheading 3.1.1 (Fig. 1).
2. Pool purified immunoconjugate fractions, concentrate (Subheading 3.1.1, step 9), measure the protein concentration, and store in aliquots at −80°C. Size of aliquots will depend on how much protein is required per radiolabeling. Aliquots of ~50 μg are useful for trial radiolabelings to confirm that the chelator is successfully conjugated to the protein. Repeated freeze–thaw cycles should be avoided, as the protein is denatured (see Note 6).

3.2. Induction of Acute Colitis in Mice

1. Dextran sodium sulfate (DSS), which is known to induce inflammation throughout the bowel (5–8), is included in the

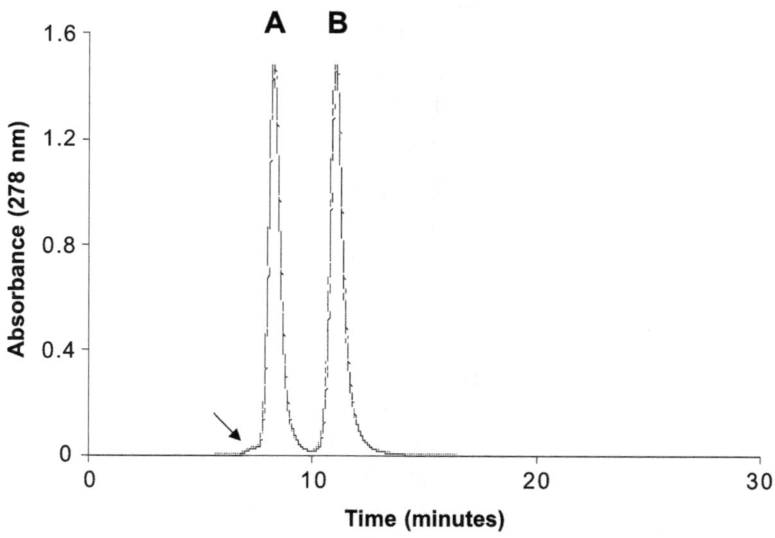

Fig. 1. Immunoconjugate Re-purification by size-exclusion HPLC. A HPLC diagram shows that unbound chelator (**B**) is separated from the immunoconjugate (**A**, and *arrow*).

drinking water for 8 days prior to initiation of the imaging study at 2.0% (W/V).

2. Observe mice and record total body weight for each animal every day.

3.3. Imaging of Distribution of Radiolabeled Antibody in Mice

In this section, we describe the final steps in this process. Initially, the protein is incubated with radionuclide to confirm that it can be radiolabeled. Then it is radiolabeled on the first day of imaging, injected into the mice, and the microPET and microCT images are collected. Finally, tissues are collected to measure the biodistribution at the final time point. This data can be used to calibrate the data obtained from images.

3.3.1. Radioactive Isotopes

The radionuclide used in these studies, ^{64}Cu, emits beta particles, positrons, and gamma rays. Use of any radionuclide presents a safety hazard. Laboratory workers should always follow institutional guidelines, all applicable regulations and minimize exposure to ionizing radiation.

3.3.2. Radiochemical Purity

We use Instant Thin-Layer Chromatography (ITLC) to assay the radiochemical purity of the radiolabeled antibody, i.e. to determine how much "free" radionuclide is present in the sample. As metal ions can be adventitiously attached to antibody molecules, it is necessary to incubate the sample in an EDTA solution before performing the ITLC to remove unchelated radiometal from the antibody.

1. Take an 1-μL aliquot from the radionuclide/immunoconjugate reaction vial.
2. Add this to 9 μL of 100 mM EDTA in pH 8.0 phosphate buffer.
3. Incubate for 5 min, and then place 1 μL of this solution on a TLC strip. Place it at 1 cm above the solvent level in the TLC bath. Allow to air dry.
4. Run the TLC using 100 mM EDTA in pH 8.0 phosphate buffer as the mobile phase.
5. When the solvent front is approximately 0.5 cm short of the top of the strip, remove it from the bath and cut into 4 equal portions.
6. Measure the radioactivity in each portion. Under these conditions, the radionuclide attached to the antibody remains at the baseline, and "free," radionuclide travels with, or close to, the solvent front. Calculate the radiochemical purity (RCP) of the radiolabeled antibody by dividing the activity in the bottom of the strip by the total activity on the strip. The RCP should be >95% for imaging studies.

3.3.3. Test Labeling

1. Thaw one of the vials of the immunoconjugate.
2. Add pH 5.0 acetate buffer to the stock ^{64}Cu solution to neutralize the acid that it is delivered in and bring it to pH 5.0. For example, to 2 mCi ^{64}Cu in 5 μL HCl (0.04 N) add 3 volume equivalents (15 μL) of pH 5, 0.1 M acetate buffer. This should be sufficient to raise the pH. Low pH can damage antibodies and proteins, resulting in loss of function.
3. Incubate immunoconjugate with radionuclide in sodium acetate buffer (0.1 M, pH 5.0) at ^{64}Cu-to-protein ratios of 1:1, 5:1 and 10:1 μCi/μg.
4. Perform TLC measurements of the radiochemical purity of the radiolabeled antibody at the different ^{64}Cu-to-protein ratios. Calculate the radiochemical purity of the radiolabeled antibody for each ^{64}Cu-to-protein ratio.
5. From this data, the amount of radionuclide that can be added to the protein can be measured. For example, if the antibody is 100% labeled at 1:1, 100% labeled at 5:1, and 50% labeled at 10:1, this indicates that the maximum specific activity of the labeled protein is between 5 and 10 μCi/μg. This will vary for different specific activities (i.e., purities) of the radionuclide.

3.3.4. Radiolabeling

For imaging studies, immunoconjugates were radiolabeled with ^{64}Cu as described in the test labeling Subheading 3.3.3.

1. Neutralize the acidic radionuclide solution (see Test Labeling, step 2) and aliquot the amount of radionuclide required into a 1.5-mL microcentrifuge tube.
2. Add the antibody in acetate buffer, and incubate for 30 min at 25°C. The amount of antibody used varies depending on the number of animals to be imaged (50 μg of antibody per mouse).
3. Add 1 μL of the reaction mixture to 9 μL of phosphate buffer (0.1 M, pH 8) containing 100 mM EDTA, and measure radiochemical purity using TLC as described in Subheading 3.3.2 (>95% required for injection).
4. If there is more than 5% free radiometal in the reaction, the antibody needs to be purified using the Centricon filters. The atomic weight of the radionuclide is less than the 30 kDa cut-off for the filters, so it passes through the filter and collects in the bottom of the unit. After three cycles of concentration, re-test the antibody radiochemical purity.
5. Dilute radioimmunoconjugate with saline and dispense to 100-μL aliquots. sterilize samples using a 0.2-μm filter.

3.3.5. Biodistribution Studies

The methods of imaging and biodistribution vary for different antibodies, different radionuclides, different disease models

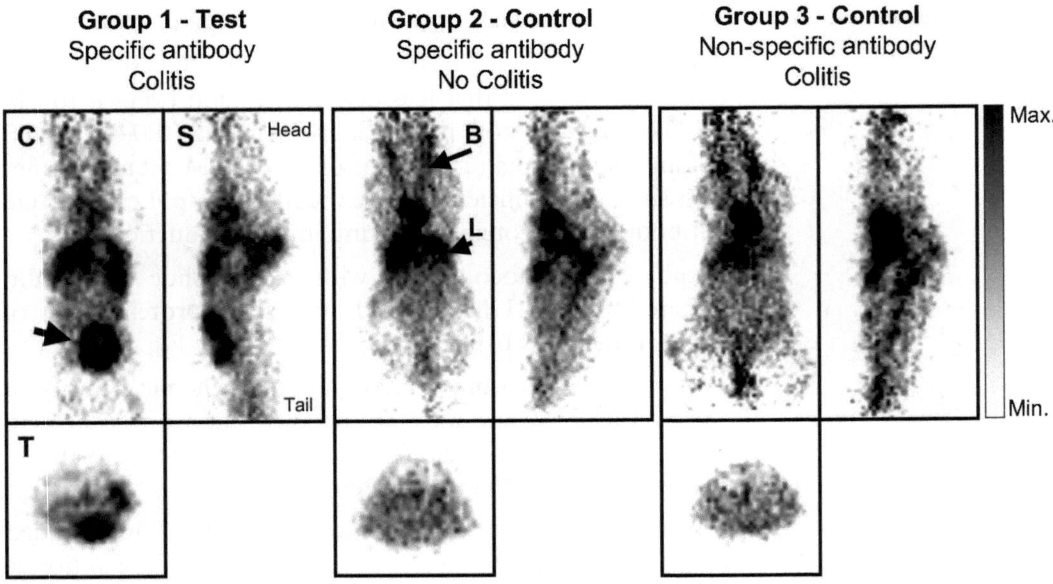

Fig. 2. A representative microPET imaging of ^{64}Cu-labeled anti-β7 integrin antibody in DSS-induced colitis. The data for colitis mice injected with anti-β7 integrin antibody (Group 1), healthy mice injected with anti-β7 integrin antibody (Group 2), and colitis mice injected with isotype control antibody (Group 3) are shown. Coronal (C), sagittal (S), and transaxial (T) views are shown. An arrow in Group 1 indicates an accumulation to the inflamed gut. L, liver; B, blood.

and organs. Figure 2 shows an example of biodistribution data of the antibody radiolabeled using the methods described in this chapter.

4. Notes

1. Throughout this method, the levels of metals in the experimental environment are reduced where possible. This is necessary because the bifunctional chelator will chelate many metals that it encounters. This results in the chelator being occupied and not available for ^{64}Cu chelation, and reduces the specific activity of the radiolabeled antibody (i.e., the ratio of radioactivity to mass of protein).

   Ultrapure water produced by a filtration system with resistivity of greater than 15 M Ohms is used throughout, including in buffer preparation. All glassware should be scrubbed with hot water and a detergent, rinsed thoroughly with distilled water, then with ultrapure water, rinsed with 10% HNO_3, left to soak overnight, then rinsed thoroughly again with ultrapure water. Consider autoclaving glassware before beginning this cleaning process if it is shared, as bioactivity is a major source of metal contamination. Once glassware has been cleaned in this way it should be stored filled with ultrapure water in a cold room (4°C) if it is not going to be used immediately. Plasticware should be washed with 10% HNO_3 if possible, and then rinsed with ultrapure water. Items such as filters, which might be damaged by acidic conditions, should be washed with ultrapure water and then with buffer (e.g., 3× each) before drying in a fume hood.

Use Chelex 100 resin to reduce the metal contaminants in the buffer. This can be accomplished by preparing a column through which the buffer is then passed, or Chelex 100 can be added directly to a volume of buffer, stirred overnight, and then sterile filtered (i.e., 0.2 μm). The resin has to be cleaned before use according to manufacturer's instructions. In addition, the resin must be cleaned to remove organic contaminants, such as iminodiacetate, which as a metal chelator may reduce the specific activity of the final product (9).

2. Retain UV samples by pipetting them out of the cuvette. These can be used as HPLC samples to confirm that the antibody has been purified, which is particularly important after the conjugation reaction. It is very important to confirm that unreacted chelator has been removed from the antibody. The conjugation reaction is very inefficient, with only about 1–2% of chelator reacting with the antibody. Wash the Hamilton syringe carefully and perform an HPLC run with buffer only to see if there is any chelator remaining in the syringe or HPLC injector. Then inject the UV sample and compare the UV absorbance at 278 nm at approximately 8 min (antibody) and 11 min (free chelator) (Fig. 1).

3. EDC is very susceptible to loss of reactivity following interaction with water (10). Upon delivery, the solid should be aliquoted into several glass vials and stored over drying agent in a −80°C freezer. Shortly before the conjugation reaction will be carried out, take the ECD vial out of the freezer, and allow it to come to room temperature before opening in order to avoid condensation collecting in the vial. Weigh out the required amount of powder (~50 mg) in a centrifuge tube, and then immediately reseal the container and close the centrifuge tube.

4. A number of calculations are required for the conjugation reaction, and an Excel spreadsheet is very helpful in carrying these out. The molecular weights, concentrations of stock solutions, and necessary mathematical formulae can be entered to give the volumes of each reactant required for the final reaction. Once the EDC has been weighed, the weight can be entered into the spreadsheet, and the concentration for a given volume of water to be added is calculated (N.B. use water, not buffer, to bring the EDC into solution – it has a longer half-life in pure water). Use the Excel "goal seek" function to alter the volume of buffer to be added to adjust

the concentration of antibody to 5 mg/mL. Add the buffer to the antibody solution and then add the bifunctional chelator solution. Have the necessary pipettes ready for bringing EDC into solution, adding it to the other reactants, and mixing the whole volume. Carry out these three steps as quickly as possible, vortexing the EDC/water solution.

5. Check the pH of the buffers, conjugation reactants, and the conjugation reaction itself with a pH microprobe, *not* pH paper. The reaction is very sensitive to pH variations (11).

6. The BFC used in this method has an $–NH_2$ group which reacts with a –COOH group on the antibody. If there is a –COOH group in the CDR of the antibody then DOTA (1,4,7,10-tetraazacyclododocane-*N, N′, N″, N‴*-tetraacetic acid) can be used (12). DOTA has 4 –COOH groups which could react with $–NH_2$ groups on the antibody. Alternatively, if loss of immunoreactivity is a potential issue, then both methods could be used in a trial study using a small amount of antibody, and antigen binding studies could be carried out to indicate which BFC is most suitable for use with that particular antibody.

References

1. Loftus, E.V. Jr., Sandborn, W.J. (2002) Epidemiology of inflammatory bowel disease. *Gastroenterol. Clin. North Am.* **31**, 1–20.
2. Baron, J.H., Connell, A.M., Lennard-Jones, J.E. (1964) Variation between Observers in Describing Mucosal Appearances in Proctocolitis. *Br. Med. J.* **1**, 89–92.
3. Shaw, S.K., Brenner, M.B. (1995) The beta 7 integrins in mucosal homing and retention. *Semin. Immunol.* 7, 335–342.
4. Andrew, D.P., Berlin, C., Honda, S., Yoshino, T., Hamann, A., Holzmann, B., et al. (1994) Distinct but overlapping epitopes are involved in alpha 4 beta 7-mediated adhesion to vascular cell adhesion molecule-1, mucosal addressin-1, fibronectin, and lymphocyte aggregation. *J. Immunol.* **153**, 3847–3861.
5. Park, E.J., Takahashi, I,. Ikeda, J., Kawahara, K., Okamoto, T., Kweon, M.N., et al. (2003) Clonal expansion of double-positive intraepithelial lymphocytes by MHC class I-related chain A expressed in mouse small intestinal epithelium. *J. Immunol.* **171**, 4131–4139.
6. Kullmann, F., Messmann, H., Alt, M., Gross, V., Bocker, T., Scholmerich, J., et al. (2001) Clinical and histopathological features of dextran sulfate sodium induced acute and chronic colitis associated with dysplasia in rats. *Int. J. Colorectal. Dis.* **16**, 238–246.
7. Cooper, H.S., Murthy, S.N., Shah, R.S., Sedergran. D.J. (1993) Clinicopathologic study of dextran sulfate sodium experimental murine colitis. *Lab. Invest.* 69, 238–249.
8. Okayasu, I., Hatakeyama, S., Yamada, M., Ohkusa, T., Inagaki, Y., Nakaya, R. (1990) A novel method in the induction of reliable experimental acute and chronic ulcerative colitis in mice. *Gastroenterology* **98**, 694–702.
9. Price, N.M., Harrison, G.I., Hering, J.G., Hudson, R.J., Nirel, P.M.V, Palenik, B., et al. (1988/1989) Preparation and chemistry of the artificial algal culture medium *Aquil. Biological Oceanography* **6**, 19.
10. Nakajima, N., Ikada, Y. (1995) Mechanism of amide formation by carbodiimide for bioconjugation in aqueous media. *Bioconjug. Chem.* **6**, 123–130.
11. Di Bartolo, N., Sargeson, A.M., Smith, S.V. (2006) New 64Cu PET imaging agents for personalised medicine and drug development using the hexa-aza cage, SarAr. *Org. Biomol. Chem.* **4**, 3350–3357.
12. Lewis, M.R., Kao, J.Y., Anderson, A.L., Shively, J.E., Raubitschek, A. (2001) An improved method for conjugating monoclonal antibodies with N-hydroxysulfosuccinimidyl DOTA. *Bioconjug. Chem.* **12**, 320–324.

Chapter 29

Integrin-Targeted Nanoparticles for siRNA Delivery

Noa Ben-Arie, Ranit Kedmi, and Dan Peer

Abstract

Integrins are heterodimeric membrane glycoproteins composed of noncovalently associated α and β subunits. Integrins support cell attachment and migration on the surrounding extracellular matrix as well as mediate cell–cell interaction in physiological and pathological settings. Constant recycling of integrins from the plasma membrane to the endosome makes integrins ideal receptors for the delivery of drugs to the cell cytoplasm. RNA interference (RNAi) has evolved not only as a powerful tool for studying gene expression and validating new drug targets, but also as a potential therapeutic intervention. However, the major challenge facing the translation of RNAi into clinical practice is the lack of efficient systemic delivery to specific cell types. Utilizing integrins as delivery target, we have recently devised a strategy to target leukocytes termed Integrin-targeted and stabilized NanoParticles (I-tsNPs) that entrap high RNAi payloads and deliver them in a leukocyte-specific manner to induce robust gene silencing.

Key words: Leukocyte integrins, Gene silencing, Nanoparticles, Hyaluronan, siRNAs, Nanomedicine

1. Introduction

RNA interference is a cellular mechanism that can be induced either by synthetic small-interfering RNAs (siRNAs) or by vectors that express small hairpin RNAs (shRNA) (1). This mechanism mediates sequence-specific gene silencing by cleavage of the targeted messenger RNA or by suppression of the translation machinery (2, 3). To realize the potential of siRNAs for in vivo drug discovery and therapy, there is a need to overcome the considerable hurdles of systemic and intracellular delivery of payloads into the cell (4). siRNAs are subjected to a rapid renal clearance and can be degraded by serum RNases, shortening their half-life in vivo and thus, suitable nanocarriers that protect

Motomu Shimaoka (ed.), *Integrin and Cell Adhesion Molecules: Methods and Protocols*,
Methods in Molecular Biology, vol. 757, DOI 10.1007/978-1-61779-166-6_29, © Springer Science+Business Media, LLC 2011

the RNAi payloads need to be employed (5). Devising such carriers equipped with integrin ligands or their antibodies enable cell-specific delivery and intracellular release of siRNAs to the cell cytoplasm (6, 7). Herein, we describe a method that exploits integrin internalization to introduce siRNAs to the cytoplasm of leukocytes (7), which are known to be difficult to transduce with nucleic acids (8).

2. Materials

2.1. I-tsNPs Production

1. Lipids: L α-Phosphatidylcholine (PC Egg, Chicken), 1,2-Dipalmitoyl-sn-Glycero-3-Phosphoethanolamine (DPPE), cholesterol (Chol) (Avanti polar lipids, Inc., Alabaster, AL). All lipids are stored as powders at −20°C freezer.
2. Rotary evaporator (Buchi Corporation, Switzerland).
3. Thermobarrel Lipex extruder™ (Lipex biomembranes Inc., Vancover, British Columbia, Canada).
4. Nucleopore membranes with 0.1–1 μm pore size.
5. 20 mM HEPES-buffered saline (HBS), pH 7.2.
6. 1 × 0.1 M MES buffer pH 6.0.
7. Hyaluronan (HA, 751 kDa, intrinsic viscosity, 16 dL/g, Genzyme Corp., Cambridge, MA): Stored as a powder at −20°C freezer in a desiccator.
8. 1 M Borate buffer, pH 9.0: Diluted to 0.1 M with H_2O.
9. 400 mmol/L 1-(3-dimethylaminopropyl)-3-ethylcarbodimide hydrochloride.
10. 400 mmol/L Sulfo-*N*-Hydroxysuccinimide.
11. Monoclonal antibody FIB 504.64 Rat anti-mouse IgG2a against β_7 integrin: Adjusted to 10 mg/mL.
12. 1 M ethanolamine hydrochloride, pH 8.5.
13. Size exclusion column with sepharose CL-4B beads.
14. Alpha 1-2 LDplus lyophilizer (Christ, Osterode, Germany).

2.2. Characterization of I-tsNPs

2.2.1. Lipid Mass Assay (7, 9)

1. ^{3}H-hexadecylcholesterol (Perkin Elmer, Boston, MA).
2. 5% (w/v) Ammonium molybdate.
3. Fiske & Subbarow 0.5 g in 3 ml DDW.
4. Perchloric acid (phosphorus free).
5. Phosphate standard.

2.2.2. Particle and Zeta Potential Analysis

1. Malvern Zetasizer nano ZS™ (Malvern Instruments Ltd, Southborough, MA).
2. 1× PBS, pH 7.4, at 20°C.

2.2.3. Binding to Cells and Transfection

1. Flow cytometer: We use FACScan flow cytometer or FACSCalibur (BD Biosciences).
2. FACS buffer: PBS, pH 7.4, 1% FBS, 0.01% sodium azide.
3. TK-1 cells (ATCC): Grown in RPMI1640 supplemented with 1% antibiotics (penicillin and streptomycin), 4 mM L-glutamine, and 10% fetal calf serum (FCS).
4. Primary antibodies: FIB 504.64 (BD Pharmingen) and isotype control rat IgG2a.
5. Secondary antibody: Alexa488-labeled anti-rat IgG2a antibody.
6. Fix-and-Perm Kit (Caltag Laboratories, Burlingame, CA): Used for intracellular staining.
7. Antibody to Ku70 (purified mouse anti-Ku70, Santa Cruz Biotechnology, Santa Cruz, CA): Used for intracellular staining.

2.2.4. siRNA Entrapment Efficiency in I-tsNPs

1. Ku70-siRNAs (Dharmacon Inc., Boulder, CO): The following four sequences are used in equimolar ratios.

   siRNA Sequence #1:

   Sense: 5′-GCUCUGCUCAUCAAGUGUCUGdTdT-3′

   Antisense: 5′-CAGACACUUGAUGAGCAGAGCdTdT-3′

   siRNA Sequence #2:

   Sense: 5′-UCCUUGACUUGAUGCACCUGAdTdT-3′

   Antisense: 5′-UCAGGUGCAUCAAGUCAAGGAdTdT-3′

   siRNA Sequence #3:

   Sense: 5′-ACGGAUCUGACUACUCACUCAdTdT-3′

   Antisense: 5′-UGAGUGAGUAGUCAGAUCCGUdTdT-3′

   siRNA Sequence #4:

   Sense: 5′-ACGAAUUCUAGAGCUUGACCAdTdT-3′

   Antisense: 5′-UGGUCAAGCUCUAGAAUUCGUdTdT-3′

   We recommend to use 2′-*o*-metheyl-modified siRNAs; pre-designed ON-TARGETplus siRNA SMARTpool, Gene ID 14375 for mouse Ku70 (Dharmacon Inc., Boulder, CO); or lock nucleic acid (LNA)-modified siRNAs (Life Technologies, Austin, TX).
2. Nuclease-free water (Ambion Inc., Austin, TX).
3. Human recombinant Protamine (Abnova, TaipeiCity, Taiwan).
4. Quant-iT RiboGreen RNA assay kit for percent entrapment efficiency (Invitrogen, Carlsbad, CA).

2.3. In Vivo siRNA Delivery Using I-tsNPs

1. C57BL/6 mice: Housed in a specific pathogen-free animal facility.
2. 30-gauge needle with a tuberculin syringe: Used to inject to a mouse tail vein.
3. Bath sonicator: Used to briefly sonicate the I-tsNPs suspension prior to intravenous injection to avoid aggregation.

3. Methods

3.1. I-tsNPs Production and Purification

I-tsNPs are nanometer-sized hyaluronan-coated neutral liposomes possessing targeting moieties (antibodies to integrins on leukocytes) on their surface. The preparation involves two critical steps: (1) preparation of stabilized nanoparticles (NPs) by chemical conjugation of hyaluronan that coats the surface of the liposomes and (2) introduction of targeting molecules (mAbs) on the surface of the stabilized NPs (Fig. 1).

1. Prepare multilamellar vesicles (MLV), composed of phosphatidylcholine (PC), dipalmitoylphosphatidylethanolamine (DPPE), and cholesterol (Chol) at molar ratios of 3:1:1 PC:Chol:DPPE using conventional lipid-film hydration method (10, 11). Weight the appropriate amounts of lipids to a final concentration of 40 mg/mL in a round-bottom flask.
2. Dissolve the lipids in the round-bottom flask with 20 mL of 96% ethanol by stirring for 30 min at 65°C. Then, evaporate the ethanol using rotary evaporator until complete dryness (usually takes ~1 h) and look for the appearance of a thin film layer on the round-bottom-flask. Upon completion of the drying process, blow nitrogen to the round bottle flask for 10 min (see Note 1).
3. Hydrate the lipid film with 20mM HEPES pH 7.4 to create MLV (40mg/mL). Thoroughly vortex until a thin milky liposome suspension is formed.

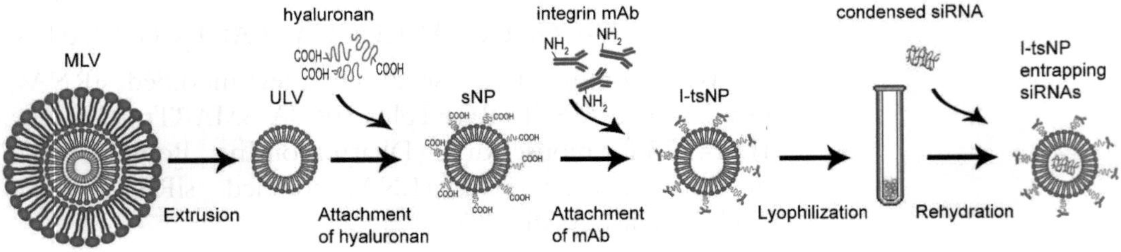

Fig. 1. The steps of producing Integrin-targeted and stabilized nanoparticles (I-tsNP). Multilamellar vesicles (MLV) prepared from neutral lipids are excluded to form unilamellar vesicles (ULV). ULV are surface-modified with a covalent attachment of hyaluronan that stabilized the particles (sNP). An antibody against integrin is then covalently attached to the hyaluronan scaffold, generating I-tsNPs. To encapsulate siRNAs, lyophilized I-tsNPs are rehydrated with a protamine-condensed siRNAs solution.

4. Incubate the MLV suspension in a shaker (~200 rpm) at 37°C for 2 h to ensure complete mixing and homogeneity (see Note 2).
5. Extrude the resulting MLV into small unilamellar vesicles (ULV) with a Thermobarrel Lipex extruder™ with a circulating water bath set at 65°C (above the transition temperature of the lipids) under nitrogen pressures of 300 to 550 psi.
6. Carry out the extrusion in a stepwise manner using progressively decreasing pore-sized membranes (from 1, 0.8, 0.6, 0.4, 0.2 to 0.1 μm), with ten cycles per pore-size.
7. ULV are surface-modified with high molecular weight hyaluronan (HA) (751 kDa, intrinsic viscosity: 16 dL/g) as described below.
8. Dissolve 20 mg HA in 10 mL of double distilled water. Adjust pH 4.5 with HCl. Add 400 mg EDAC and adjust the pH to 4.5. Stir at 37°C for 30 min to fully dissolve HA. Centrifuge the extruded ULV for 1 h in a high speed ultracentrifuge (640,000 × *g*, 4°C) and resuspend the pellet (with the same volume) in 0.1 M borate buffer pH 9.0. Combine the activated HA with the ULV suspension in a 1:1 volume ratio and adjust the pH to 7.4 then incubate for 2 h at incubate 37°C, with gentle stirring. Adjust pH to 8.6 with NaOH and incubate overnight. Separate the resulting HA-ULV (stabilized nanoparticles; sNP) from free HA by washing three times with 0.1 M MES buffer pH 6.0 by ultra-centrifugation (640,000 × *g*, 4°C, for 1 h). Make sure to resuspend to the original volume after the last washing step (see Note 3).
9. Perform the coupling reaction of sNP to mAbs using an amine-coupling method (7). Incubate 50 μL of sNP in 0.1 M MES buffer pH 6.0 with 200 μL of 400 mmol/L EDAC and 200 μL of 400 mmol/L sulfo-NHS (at this order) for 20 min at room temperature (RT) with gentle stirring. Make sure the pH is 6.0 if not adjust accordingly. Separate the sNP from unbound EDC and S-NHS by adding 2.5 ml of 10 mM PBS pH 7.4 and wash with Mini-ultracentrifugation (sorvall) (6.4×10^5 *g*, 40°C, for 1 h) resuspend in PBS to 450 μl. All cross-linkers must be freshly prepared from powder.
10. Mix the EDAC-NHS-activated sNP with 50 μL of mAb (10 mg/mL FIB 504 in HBS or PBS, pH 7.4) and incubate for 3 h at RT with gentle stirring. At the end of the incubation, add 20 μL of 1 M ethanolamine HCl, pH 8.5 to block the reactive residues and stir for 10 min at RT (see Note 4).
11. Purify I-tsNPs to remove uncoupled antibody using a size exclusion column packed with sepharose CL-4B or CL-6B beads equilibrated with HBS, pH 7.4. Test lipid mass and adjust the lipid concentration for in vitro or in vivo use.

12. Prepare the purified I-tsNPs suspensions for lyophilization. Snap freeze 0.2-mL aliquots in a mixture of 96% ethanol and dry ice for ~20–30 min; freeze the aliquots for 2–4 h at −80°C and lyophilize for 48 h using an alpha 1–2 LDplus lyophilizer (see Note 5).
13. Store lyophilized I-tsNPs at −20°C until further use.

3.2. I-tsNP Characterization

3.2.1. Lipid Mass Determination

Determination of phospholipids is preformed either by including a ^{3}H-hexadecylcholesterol trace in the particles membrane (7) or by using a modified Bartlett assay (9). The sensitivity of the assay is in a 0–200 nM phosphate range. The color produced in this assay is proportional to the concentration of phosphorous up to 1.5 μM in each sample.

1. To create a standard curve, add 0, 10, 20, 40, 80, and 100 nM phosphate from phosphate standard (sigma) to six tall glass tubes. Sample volume should not exceed 200 nM phosphate and volume of 200 μL.
2. Add 400 μL of 70% perchloric acid and place the tube in the oven heated to 180°C for 30 min. Use chemical hood to add the perchloric acid and perform all experiments inside a chemical hood.
3. After 30 min, let tubes cool down to RT and add 1.2 mL of H_2O and vortex well.
4. Add 200 μL of 5% (w/v) ammonium molybdate.
5. Add 50 μL of Fiske & Subbarow and vortex well.
6. Place tubes in 100°C water bath and boil for 7 min.
7. Let tubes cool down to RT.
8. Read O.D. values at 830 nm.

3.2.2. Particle and Zeta Potential Analysis

The diameter and surface charge (zeta potential) of nanoparticles are measured using a Malvern Zetasizer nano ZS™ (Table 1).

3.2.3. Binding Efficiency of I-tsNPs

Flow cytometry is used to ensure that conjugated antibodies are functional.

Table 1
Particle size and zeta potential measurements

Particle type	Hydrodynamic diameter (nm)	ζ potential (mV)
IgG sNPs	127 ± 13	−18.5 ± 1.2
β_7 I-tsNPs	139 ± 21	−23.7 ± 2.6

All measurements were done in pH 6.7 (with 10 mM NaCl) at 20°C in a Zetasizer nano ZS, Malvern. Data presented as an average ± SD from $n = 4$ independent experiments

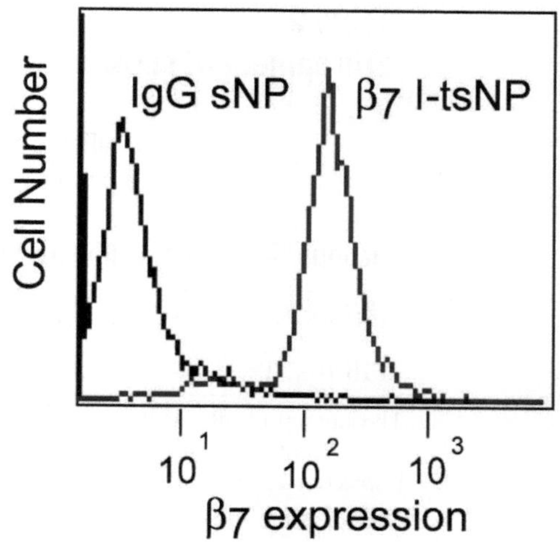

Fig. 2. FACS histograms showing binding of β_7 I-tsNPs to TK-1 cells.

1. Dispense TK-1 cells to FACS tubes at 0.5×10^6 cells per tube.
2. Add to each tube 1 ml of FACS buffer, centrifuge at $300 \times g$, 5 min, 4°C, and aspirate supernatant.
3. Add to each tube 100 μL of FACS buffer containing appropriate antibodies at 10 μg/mL concentrations.
4. Incubate samples on ice for 30 min.
5. Add to each tube 1 mL of FACS buffer, centrifuge at $300 \times g$, 5 min, 4°C, and aspirate supernatant.
6. Add to each tube 100 μL of FACS buffer containing 1 μg/mL secondary antibody Alexa488-Anti-Rat Ab IgG2a.
7. Incubate on ice for 20–30 min.
8. Add to each tube 1 mL of FACS buffer, centrifuge at $300 \times g$, 5 min, 4°C, and aspirate supernatant.
9. Resuspend the cell pellets in appropriate volume of FACS buffer and analyze samples by FACS (Fig. 2).

3.2.4. siRNA Entrapment Efficiency in I-tsNPs

1. Mix siRNAs (e.g., Ku70-siRNAs) with full-length recombinant protamine in a 1:5 (siRNA:protein) mole ratio, in nuclease-free water in a total volume of 200 μL (per each tube) and incubate for 20 min at RT to form a complex.
2. For siRNA entrapment in I-tsNPs, rehydrate the lyophilized nanoparticles (e.g., 0.5–1.5 mg lipids for in vivo experiments and 10–100 μg lipids for in vitro experiments) (see Note 6) by adding 0.2 mL nuclease-free water containing protamine-condensed siRNAs (1–4 nM for in vivo experiments and 50–500 pM for in vitro experiments). Perform the entrapment procedure immediately before use in vitro transfection or intravenous tail-vein injection.

Table 2
Entrapment of siRNAs molecules in I-tsNPs

	siRNAs entrapment (No. of molecules)	Encapsulation efficiency of condensed siRNA
Nanoparticles type	Mean ± SEM	Mean ± SEM
IgG sNPs	3,750 ± 1,300	78 ± 10
FIB I-tsNPs	4,000 ± 1,200	80 ± 12

The amount of siRNAs that was used for encapsulation was known. Upon encapsulation a RiboGreen™ assay (Invitrogen) was preformed according to the manufacturer instructions to assess the amount of siRNAs that was entrapped

3. Determine siRNAs entrapment efficiencies by a Quant-iT™ RiboGreen™ RNA assay (Molecular Probes, Invitrogen) (Table 2).

3.3. Ku70-siRNA Delivery In Vitro in TK-1 Cells

1. Seed TK-1 cells in a six-well microtiter plate at 2.5×10^5 cells in 300 μL per well. Culture cells overnight at 37°C, 5% CO_2.
2. Add to each well 50 μL of β_7 I-tsNP entrapping Ku70-siRNAs. Appropriate controls should be included: cells with no treatment; cells with Ku70 siRNA alone; cells with negative control siRNA (e.g., silencer firefly Luciferase siRNA or scrambled siRNA).
3. Culture cells for 48–72 h at 37°C, 5% CO_2. Perform intracellular staining (as described below) to detect Ku70 protein expression.

3.3.1. Intracellular Staining and Flow Cytometry

1. Transfer TK-1 cells to 96-well V bottom plates.
2. Fix and permeabilize cells with Fix-and-Perm Kit™ (Caltag Laboratories, Burlingame, CA).
3. Add to cells 100 μL-aliquots of FACS buffer containing 10 μg/mL of Ku70 antibody.
4. Incubate samples on ice for 30 min.
5. Counter-stain cells with 1 μg/mL Alexa 488-conjugated goat anti-mouse IgG antibody.
6. Wash cells with FACS buffer and perform FACS analysis to measure Ku70 expression (Fig. 3).

3.4. Delivery of Ku70-siRNA In Vivo

1. Prepare mice for a tail vain injection. Pre-heat mice with a small U.V. lamp for 3–4 min in order to expose the tail veins. Alternatively, expose the mouse's tail to pre-warm water (37°C) until you clearly see the veins.

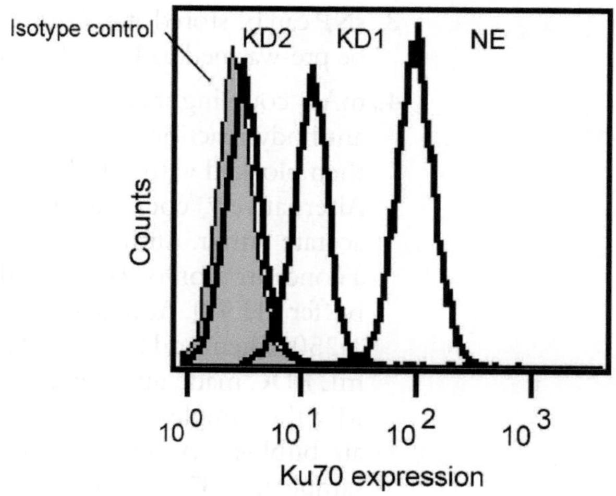

Fig. 3. FACS histograms showing knockdown of the DNA repair protein, Ku70 in mouse T-cell lymphoma (TK-1 cells). Ku70 expression in cells mock-treated (NE), treated with 100 pM Ku70-siRNAs delivered via β_7 I-tsNPs (KD1), and treated with 500 pM Ku70-siRNAs delivered via β_7 I-tsNPs (KD2).

2. Sonicate the I-tsNP suspension for 1.5 min in a bath sonicator to dissolve any potential aggregates.
3. Inject 150 μL of I-tsNP suspension to the tail vein with a 27-gauge needle with a tuberculin syringe (see Note 7). Initially inject 50 μg of siRNA entrapped in 500 μg of I-tsNPs per mouse as a starting dose. After checking the level of silencing, you may modify dosing.
4. At 48 or 72 h after injection, sacrifice the mice and harvest spleens.
5. Prepare a single-cell suspension of splenocytes.
6. Perform an intracellular staining with anti-Ku70 mAb as described in Subheading 3.3.

4. Notes

1. MLV preparation: a water bath of the rotary evaporator is set to 65°C and the chiller is set to be at least in −10°C to make the evaporation process more efficient. After the evaporation of ethanol, it is advisable to pass any inert gas such as argon or nitrogen for 10 min to completely remove traces of ethanol and prevent oxidation of lipids.
2. The MLV suspension can be stored at 4°C until extrusion procedure. However, prior to extrusion, the MLV should be pre-warmed to 37°C to enable easy extrusion process.

3. sNP can be stored at 4°C for 2 weeks. However, the sNP should be pre-warmed to RT before further modification is done.
4. mAb coupling reaction: EDAC/NHS-activated sNP with antibody reaction mixture is incubated overnight at RT and then blocked with 20 μL of 1 M ethanolamine-HCl, pH 8.5. Alternatively, coupling is made at lower pH 5.0 in 0.1 M acetate buffer. Antibody is in 0.1 M sodium acetate pH 5.0 at a concentration of 10 mg/mL. The sNP are in 0.1 M Borate buffer pH 9.0. Added antibody to the sNP at a molar ratio of 1:250. Then, add buffer and 500 molar equivalents of 50 mg/mL EDC made up immediately before use in ultrapure water. Mix the sample gently by pipetting. Centrifuged to remove air bubbles. Adjust pH to 5.0 with 0.1N HCl. Place the sample in a 37°C water bath while gentle shaking for 2 h.
5. Lyophilization of I-tsNPs: Prior to lyophilization of purified liposome fractions, I-tsNP suspensions should be tested for antibody-binding efficiency by flow cytometry. Pool only the fractions that give good binding. Add 0.2 mL aliquots of the I-tsNP suspension into amber glass vials prior to lyophilization.
6. We usually use the following lipid ranges in each vial. For in vitro transfection 10–100 μg lipids per vial. For in vivo delivery, 0.5–1.5 mg lipids per vial.
7. I-tsNPs are in saline containing 5% glucose for in vivo injection.

Acknowledgments

The authors would like to thank Prof. Motomu Shimaoka for his advice and encouragement and Dr. Yoshiyuki Morishita, and Dr. Charudharshini Srinivasan, for technical assistance. This work was supported by the European union – IRG (D.P.), the Alon Foundation (D.P.), and the Lewis Trust for Cancer Research (D.P.).

References

1. de Fougerolles, A., Vornlocher, H. P., Maraganore, J., and Lieberman, J. (2007) Interfering with disease: a progress report on siRNA-based therapeutics. *Nat. Rev. Drug Discov.* **6**, 443–453.
2. Elbashir, S. M., Harborth, J., Lendeckel, W., Yalcin, A., Weber, K., and Tuschl, T. (2001) Duplexes of 21-nucleotide RNAs mediate RNA interference in cultured mammalian cells. *Nature* **411**, 494–498.
3. Dykxhoorn, D. M., and Lieberman, J. (2005) The silent revolution: RNA interference as basic biology, research tool, and therapeutic. *Annu. Rev. Med.* **56**, 401–423.
4. Dykxhoorn, D. M., and Lieberman, J. (2006) Running interference: prospects and obstacles to using small interfering RNAs as small molecule drugs. *Annu. Rev. Biomed. Eng.* **8**, 377–402.
5. Peer, D., and Shimaoka, M. (2009) Systemic siRNA delivery to leukocyte-implicated diseases. *Cell Cycle* **8**, 853–859.
6. Peer, D., Zhu, P., Carman, C. V., Lieberman, J., and Shimaoka, M. (2007) Selective gene

silencing in activated leukocytes by targeting siRNAs to the integrin lymphocyte function-associated antigen-1. *Proc. Natl. Acad. Sci. USA.* **104**, 4095–4100.

7. Peer, D., Park, E. J., Morishita, Y., Carman, C. V., and Shimaoka, M. (2008) Systemic leukocyte-directed siRNA delivery revealing cyclin D1 as an anti-inflammatory target. *Science* **319**, 627–630.
8. Goffinet, C., and Keppler, O. T. (2006) Efficient nonviral gene delivery into primary lymphocytes from rats and mice. *FASEB J.* **20**, 500–502.
9. Bartlett, G. R. (1959) Colorimetric assay methods for free and phosphorylated glyceric acids. *J. Biol. Chem.* **234**, 466–468.
10. Peer, D., and Margalit, R. (2000) Physicochemical evaluation of a stability-driven approach to drug entrapment in regular and in surface-modified liposomes. *Arch. Biochem. Biophys.* **383**, 185–190.
11. Olson, F., Hunt, C. A., Szoka, F. C., Vail, W. J., and Papahadjopoulos, D. (1979) Preparation of liposomes of defined size distribution by extrusion through polycarbonate membranes. *Biochim. Biophys. Acta.* **557**, 9–23.

Chapter 30

Humanized Mice for Studying Human Leukocyte Integrins In Vivo

Sang-Soo Kim, Priti Kumar, Chunting Ye, and Premlata Shankar

Abstract

Humanized mice have recently emerged as powerful translational animal models for studying human hematopoiesis, immune interactions, and diseases of the human immune system. Several important advances in the humanized mouse technology have been reported over the last few years, thereby resulting in improved engraftment, high levels of human chimerism, and sustained human hematopoiesis. This chapter describes the detailed procedures for generating various humanized mouse models including hu-PBL, hu-HSC, and BLT models and discusses considerations for choosing the appropriate model system.

Key words: Humanized mice, Immunodeficient mice, Integrin, Hematopoietic stem cells, Xenotransplantation, Stem cell transplantation, Human fetal tissue, hu-PBL, hu-HSC, BLT

1. Introduction

The lack of an appropriate small animal model is a major impediment for investigation of the human immune system in vivo. Over the years there has been a considerable effort to develop "humanized" mouse models harboring a functional human immune system. Although technically, it is possible to enforce human transgene expression in immunocompetent mice, only immunodeficient mice are receptive to engraftment with human hematopoietic or lymphoid cells or tissues (1). Such humanized mice provide valuable new tools to translate basic research findings into clinical applications for various human diseases including autoimmunity, transplantation, infectious diseases, and cancer for which appropriate small animal model are unavailable (2). HIV research, in

Motomu Shimaoka (ed.), *Integrin and Cell Adhesion Molecules: Methods and Protocols*,
Methods in Molecular Biology, vol. 757, DOI 10.1007/978-1-61779-166-6_30, © Springer Science+Business Media, LLC 2011

particular, has gained a tremendous boost with the development of humanized mouse models that are permissive to the uniquely human-tropic HIV-1 and display symptoms that closely mirror human HIV infection (3, 4).

Successful transplantation of human cells or fetal tissues with hematopoietic potential was first achieved in severe combined immunodeficient (SCID) mice, which lack functional T and B lymphocytes. However, these were not ideal animal models since they still possessed significant levels of active innate immune responses that hindered complete engraftment of human hematopoietic cells. Therefore, several other mouse strains were developed in the quest for a model that can support the stable engraftment and growth of human hematopoietic cells (extensively reviewed in (1)). NOD/SCID mice, which were created by back-crossing the SCID gene onto the non-obese diabetic (NOD) background, proved a better engrafting strain as they have impaired natural killer (NK) and antigen-presenting cell function in addition to lacking functional T and B lymphocytes. However, engraftment and long-term maintenance of human cells was still inadequate due to residual innate immune activity. A further advance that represents a major leap forward is the development of immunodeficient mouse strains lacking a functional IL-2 cytokine receptor gamma chain (IL2rγ^{null}), including NOD/Shi-Scid IL2rγ^{null} (5, 6), NOD/SCID IL2rγ^{null} (7), and BALB/c-Rag2^{null} IL2rγ^{null} mice (8). Since the IL2rγ functions as a common component of receptors for the interleukins 2, 4, 7, 9, 15, and 21, the IL2rγ^{null} mice completely lack adaptive immune function and display multiple defects in innate immunity, features that allow heightened levels of human hematopoietic/lymphoid cell engraftment. In addition to better engraftment, many of the recently described IL2rγ^{null} mouse strains also have a longer life span, so they can serve as better models for human diseases compared to the previously available strains (1).

IL2rγ^{null} strains can be used for short-term reconstitution with human peripheral blood mononuclear cells (PBMCs) or long-term de novo generation of immune progeny cells from engrafted human hematopoietic stem cells (HSCs). The PBMC model (hu-PBL) utilizes leukocytes isolated from peripheral whole blood for rapid reconstitution with functionally mature lymphocytes. The advantages of this model are the simple procedure of generation and quick availability for experimentation as animals are reconstituted within 3–4 days of PBMC transfer. This model is best suited for short-term experiments (upto 4 weeks) because of the short duration of human reconstitution and the development of graft-versus-host disease (GVHD). In this model, adoptively transferred human T cells undergo expansion in response to xenogenic stimulation so they display a predominantly activated phenotype. Therefore, the hu-PBL model is suitable for the study of activated human lymphocytes but not resting T cells. For example, the availability of antibodies that selectively target the activated form of the integrin leukocyte function-associated antigen-1 (LFA-1), would make it possible to evaluate targeted delivery approaches to activated leukocytes. The hu-PBL model would serve well for validating RNA interference (RNAi)-based interventions for autoimmune diseases, GVHD, and septic shock where selective delivery of small interference RNA (siRNA) to activated leukocytes can prove useful for silencing genes involved in the proinflammatory cascade and/or T cell activation (9).

The HSC model (hu-HSC) utilizes $CD34^+$ HSCs isolated from human cord blood or fetal liver tissue to reconstitute immunodeficient mice. The advantage of the hu-HSC model is that the human T and B lymphocytes that develop from human stem cells engrafted in these mice are tolerant of the mouse host, possibly because of negative selection during lymphocyte differentiation in the mouse thymus. In addition, in contrast to T cells from hu-PBL mice, which display a predominantly activated phenotype, the T cells in this hu-HSC model are predominantly naive and unactivated. This model allows for the investigations of hematopoietic lineage development, and the generation of primary immune responses by a naive immune system. Accordingly, novel drug delivery systems driven by antibodies targeting the LFA-1 integrin receptors in a nonactivated conformation can be tested for in vivo applicability in this model (10).

Although now widely popular, this model still suffers from drawbacks of inefficient human thymopoiesis in the xenogenic recipient thymus (11). In an earlier NOD/SCID hu/HSC model, human fetal liver derived HSCs were transferred into surgically implanted human fetal thymic tissue which resulted in successful development of human T cells (human thymopoiesis) within the autologous human thymus graft. However, the usefulness of the human fetal tissue grafted mice was limited because the distribution of human cells was confined to the implant only (12, 13). To mitigate this problem, the approach has been modified recently, with the additional step of intravenous transfer of autologous $CD34^+$ HSCs obtained from fetal liver along with the surgical implantation of human fetal thymus/liver tissues (11, 14). In this humanized mouse model, referred to as BLT (for bone marrow, liver, and thymus) mice, HSCs take up residence in the mouse bone marrow and give rise to immune progeny consisting of all major human hematopoietic lineages including T and B cells, monocytes, macrophages, and dendritic cells (11). Furthermore, BLT mice also exhibit extensive infiltration of organs including liver, lung, gastrointestinal tract, and female reproductive tissues with human immune cells (4). Education in the autologous human thymic milieu provided by the surgically implanted tissue is also likely to be lead to the proper development of the T cell repertoire in these mice. All these reasons uphold the BLT mice as a useful in vivo model for an authentic study of the human immune system. In fact, this mouse model has been demonstrated to successfully support long-lasting HIV infection with both macrophage and T cell tropic strains of HIV (4, 15, 16). A disadvantage of the hu-HSC and BLT models, in comparison to the hu-PBL model, is the longer reconstitution time (10–12 weeks) for the complete development of the human immune system from engrafted HSCs. These models also involve more complicated generation procedures.

While any of the IL2rγ^{null} mouse strains (5–8) are good hosts for generating humanized mouse models, this chapter will detail the generation of the hu-PBL, hu-HSC, and BLT models by engraftment of immunodeficient NOD/SCID IL2rγ^{null} mice with human PBMCs, HSCs, or fetal liver/thymus tissues and CD34^{+} fetal liver cells, respectively.

2. Materials

2.1. hu-PBL Model (Fig. 1a)

1. Healthy donor peripheral blood.
2. Dulbecco's phosphate-buffered saline (PBS) 1× sterile, without calcium and magnesium.
3. Complete RPMI 1640 medium: RPMI 1640 supplemented with 10% heat-inactivated fetal bovine serum (FBS), 50 U/ml penicillin, 50 mg/ml streptomycin, 2 mM glutamine, 1 mM sodium pyruvate.
4. Ficoll (GE Healthcare Bio-sciences, Piscataway, NJ).
5. Nonobese diabetic–severe combined immunodeficient NOD.cg-*Prkdc*^{scid}IL2rg$^{tm/Wjl}$/Sz (NOD/SCID IL2rγ^{null}) mice (male or female, 5–12 weeks old; Jackson Laboratory, Bar Harbor, ME) (see Note 1).
6. 25-G needles.
7. 1-cc syringes.
8. Alcohol swab.
9. Sterile gauze pad.

2.2. The hu-HSC Model (Fig. 1b)

Although successful hematopoiesis for most lineages is achieved by HSC engraftment of adult mice, T cell reconstitution is variable and not always efficient. HSC engraftment of newborn mice requires technical proficiency but results in more robust human HSC engraftment and consistent reconstitution of all hematopoietic lineages, including T cells.

1. Human umbilical cord blood (see Note 3).
2. Dulbecco's PBS 1× sterile, without calcium and magnesium.

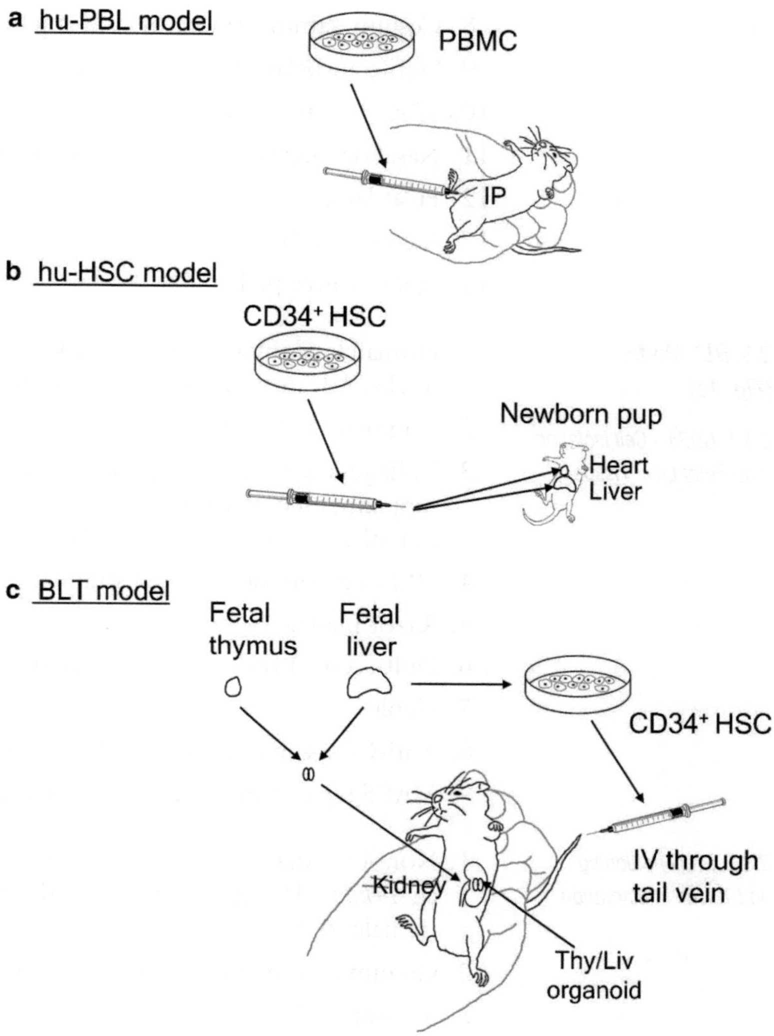

Fig. 1. Schematic representation of transplantation techniques for generation of humanized mice. (**a**) hu-PBL model. (**b**) hu-HSC model. (**c**) BLT model.

3. Ficoll.
4. Complete RPMI 1640 medium: RPMI 1640 supplemented with 10% heat-inactivated FBS, 50 U/ml penicillin, 50 mg/ml streptomycin, 2 mM glutamine, and 1 mM sodium pyruvate.
5. Anti-CD34 microbeads (Miltenyi Biotech, Auburn, CA): Used for magnetic-activated cell sorter (MACS) separation.
6. MACS separation apparatus and MACS LS separation columns (Miltenyi Biotech).
7. Nonobese diabetic–severe combined immunodeficient NOD.cg-*Prkdc*^{scid}IL2rg$^{tm/Wjl}$/Sz (NOD/SCID IL2rγ^{null}) breeder pairs.

8. Cesium gamma irradiator or X-ray biological irradiator.
9. Sterile autoclaved mouse housing device for irradiation.
10. 27-G 1/2-in. winged infusion set.
11. Nasal decongestant like Vicks Vaporub.
12. Heat lamp.
13. Alcohol swab.
14. Sterile gauze pad.

2.3. BLT Model (Fig. 1c)

2.3.1. CD34+ Cell Isolation from Fetal Liver Tissue

1. Human fetal thymus and liver tissues (gestational age of 17–20 weeks; Advanced Bioscience Resource, Alameda, CA).
2. Complete RPMI 1640 medium.
3. Collagenase/dispase solution: Complete RPMI 1640 medium supplemented with 1 mg/ml collagenase/dispase (Roche, Mannheim) and 0.5 U/ml DNase I (Roche).
4. 70-μm nylon mesh or cell strainer.
5. Razor blade.
6. Dulbecco's PBS 1× sterile, without calcium and magnesium.
7. Ficoll.
8. Anti-CD34 microbeads for MACS separation.
9. MACS separation apparatus and MACS LS separation columns.

2.3.2. Preconditioning and Tissue Implantation

1. Nonobese diabetic–severe combined immunodeficient NOD. cg-*Prkdc*^{scid}IL2rg$^{tm/Wjl}$/Sz (NOD/SCID IL2rγ^{null}) mice (female, 6–8 weeks old).
2. Cesium gamma irradiator or X-ray biological irradiator.
3. Ketamine/Xylazine.
4. Buprenorphine: Add 200 μl of 0.3 mg/ml buprenorphine to 5.8 ml sterile saline.
5. Alcohol swab.
6. Ophthalmic ointment (Vetropolycin; Pharmaderm Animal Health, Melville, NY).
7. Electric hair clipper.
8. Temperature-controlled heating pad.
9. Betadine (PDI Inc., Orengeburg, NY).
10. Surgical scalpel.
11. Sterile gauze pad.
12. Sterile saline.
13. Autoclaved tip with a blunt ball-shaped front.
14. 18-G blunt head spinal needles.
15. 4–0 absorbable vicryl and non-absorbable sutures.

16. 29-G needle/syringe from ultra-fine II insulin syringes.
17. 1-cc syringes.
18. 27-G needles.
19. Heat lamp.
20. Mouse restrainer.

2.4. Analysis of Reconstitution

1. Isoflurane.
2. Vacutainer heparinized blood collection tube.
3. Heparinized micro-hematocrit capillary tubes.
4. Red blood cell lysing buffer.
5. Anti-human monoclonal antibodies to CD3, CD4, CD8, CD11c, CD19, CD34, CD45, and CD45RA; anti-mouse monoclonal antibody to CD45; and isotype control monoclonal antibodies (BD PharMingen, San Diego, CA).
6. FACSCanto II and FACSDiva software (Becton Dickinson, Mountain View, CA).

3. Methods

3.1. hu-PBL Model

1. Isolate PBMC from whole blood by Ficoll density gradient centrifugation.
2. Resuspend cells at $5–10 \times 10^6$ per 0.5 ml of sterile RPMI media.
3. Inject 0.5 ml of cell suspension ($5–10 \times 10^6$ PBMCs) into peritoneal cavity using 1-cc syringe with 25-G needle.
4. Monitor mice daily for signs of distress (see Note 2).
5. Test cell engraftment 3–4 days after PBMC-transplantation.

3.2. The hu-HSC Model

1. Monitor breeder pairs for the birth of new litters. Engraftment procedures are performed on 24–48 h postnatal newborn pups.
2. Isolate mononuclear cells from human umbilical cord blood using Ficoll density gradient.
3. Isolate $CD34^+$ cells from mononuclear cells using the MACS separation system with anti-CD34 microbeads.
4. Resuspend isolated $CD34^+$cells in a complete RPMI media (6×10^5 cells per 0.1 ml RPMI media) and keep on ice until transplanted.
5. Stain a small proportion of the isolated $CD34^+$ cells with the mouse anti-human monoclonal antibodies to CD34 and CD3, and analyze by flow cytometry for CD34 expression

and residual CD3+ T cells to assess the efficacy of magnetic bead isolation (see Note 4).

6. Place pups in autoclaved sterile irradiation container.
7. Irradiate pups with 100 cGy whole-body irradiation.
8. Place irradiated pups on a gauze pad on ice for 5–10 min, until gross movement ceases (ensure that there is no direct contact with ice).
9. Load the CD34+ HSC cell suspension into a 1-ml syringe fitted with a 27-G 1/2-in. winged infusion set.
10. CD34+ cells can be transplanted through several routes; however, the two described below are the most reproducible and yield the best efficiencies
    (a) Intracardiac injection: Upon sufficient anesthesia, visualize the apex of the heart and insert the needle slightly above the apex in an upward direction. A little flicker of blood will appear at the base of the needle. Slowly inject 3×10^5 $CD34^+$ cells in a 50-μl volume.
    (b) Intrahepatic injection: 3×10^5 CD34+ cells in a 50-μl volume directly into the liver taking care not to pierce the entire liver.
11. Place the pups on a warming pad or under an incandescent lamp for 1–2 min.
12. Apply a small amount of nasal decongestant to the snout of both parents before returning pups to prevent scents that have transferred to the pups during handling.
13. Wean pups between 3–4 weeks of age and verify engraftment at 12 weeks of age.

3.3. BLT Model

3.3.1. CD34+ Cell Isolation from Fetal Liver Tissue

1. Cut fetal liver tissue into small pieces (~1 mm^3) using an autoclaved surgical scalpel (see Notes 5 and 6).
2. Digest small tissue pieces using sterile filtered complete RPMI supplemented with 1 mg/ml collagenase/dispase and 0.5 U/ml DNase I at 37°C for 1 h.
3. Gently disrupt the tissues by mixing every 15 min.
4. Filter the cell suspension through a 70-μm mesh.
5. Apply filtered cells on Ficoll to separate mononuclear cells.
6. Isolate $CD34^+$ cells from mononuclear cells using the magnetic-activated cell sorter (MACS) separation system with anti-CD34 microbeads.
7. Resuspend cells in a complete RPMI media ($1–5 \times 10^5$ cells per 0.2 ml RPMI media) and keep on ice until transplanted.
8. Stain a small proportion of the isolated $CD34^+$ cells with the mouse anti-human monoclonal antibodies to CD34 and CD3, and analyze by flow cytometry for CD34 expression

and residual CD3+ T cells to assess the efficacy of magnetic bead isolation.

3.3.2. Preparation of Immunodeficient Mice

1. Place mice in an autoclaved irradiation container and condition with sublethal whole-body irradiation using X-ray Biological Irradiator or Cesium gamma irradiator.
2. The dose of irradiation varies from 240 to 375 cGy depending on the mouse strain and irradiation source (see Note 7).
3. Anesthetize mice with ketamine (80 mg/kg) and xylazine (12 mg/kg) by intraperitoneal injection in PBS (10 ml/kg) based on individual animal body weight (see Notes 8 and 9).

3.3.3. Human Fetal Tissue Implantation

1. After inducing anesthesia, shave anterior abdominal skin using an electric hair clipper.
2. Place the body on a sterile support and fix the limbs with scotch tape.
3. Disinfect the skin of abdomen with betadine.
4. Make a 2–3 cm long midline incision of the abdominal skin and peritoneal membrane using a surgical scalpel.
5. Gently pull the intestines and spleen to the opposite side of the kidney and cover them with pre-warmed moisturized gauze pad.
6. After exposing the surface of the left kidney completely, gently puncture the renal capsule and make a tiny hole (1 mm) with a pair of fine forceps.
7. Insert an autoclaved micropipette tip with a blunt ball-shaped front (made by melting a normal pipette tip) into the hole to gently separate the renal capsule from the kidney parenchyma in order to make space to contain the fetal tissue.
8. Suck both the human fetal thymus and liver fragments measuring about 1 mm^3 together into an 18-G blunt head spinal needle attached to a 1 ml syringe.
9. Insert the needle through the hole into the space between the renal capsule and kidney parenchyma.
10. Push the fetal liver and thymus tissues deep inside the renal capsule and slowly remove the needle from the kidney (see Note 10).
11. Allow the bleeding in the surgical area to coagulate and carefully put back replace all the intestines into the abdominal cavity.
12. Clean the muscle and skin and separately suture with absorbable sutures.
13. Administer 40 mg/kg dose of gentamycin (intramuscular) and 0.05 mg/kg dose of buprenorphine (subcutaneous) after the surgery (see Note 11).

14. Keep the animals warm with a temperature-controlled heating pad until the mice regain consciousness.

3.3.4. Intravenous Transfer of CD34+ Cells

1. Mice subjected to tissue implantation receive cell preparations that were purified from the same donor on the day of human tissue transplantation.
2. Put mice under a heat lamp to increase blood flow to the tail vein and then transfer to a holding device, which restrains the mouse while allowing access to the tail vein.
3. Inject 0.2 ml cell suspension ($1–5 \times 10^5$ $CD34^+$ cells) into the lateral tail vein of recipient mouse (maximum volume 10% body weight) using 1-cc syringe with 27-G needle (see Note 12).
4. Monitor mice daily for signs of distress.
5. Allow human HSCs to engraft in mice for 10–12 weeks.
6. Test human cell engraftment.

3.4. Analysis of Reconstitution

1. Mice should be monitored for engraftment levels of human cells over a 5–15 day period for Hu-PBL mice and 10- to 12-week

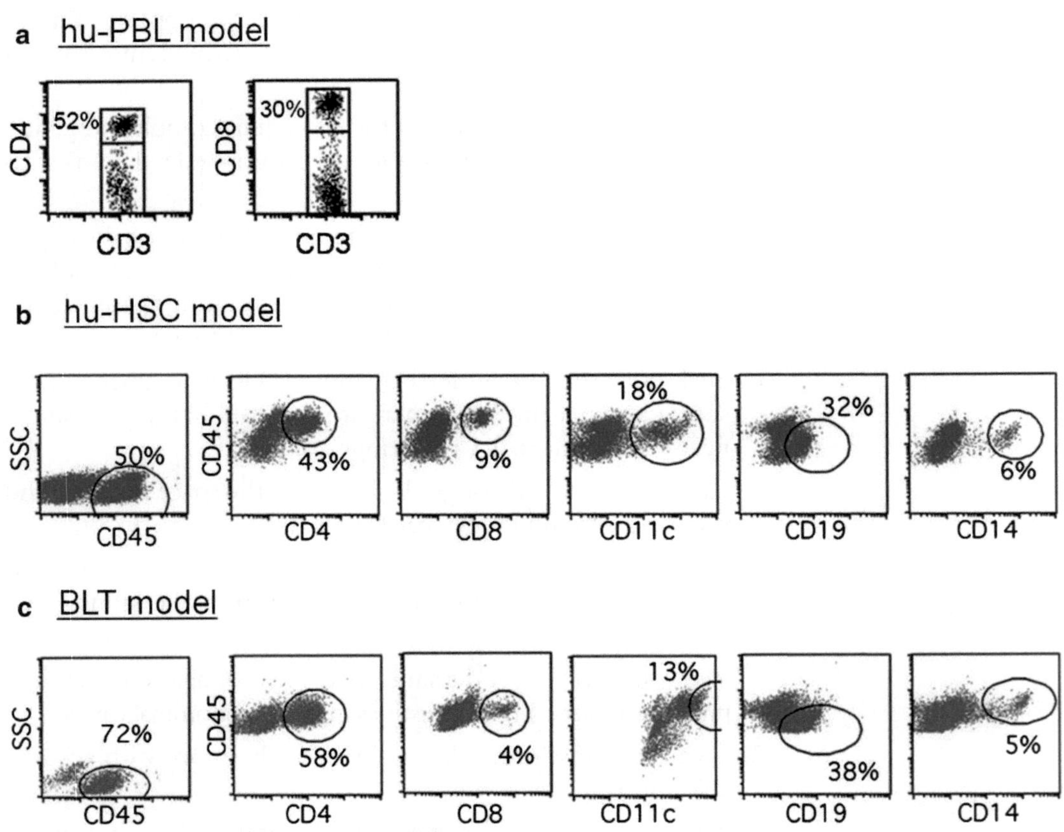

Fig. 2. Flow cytometric analysis showing reconstitution efficiencies in (**a**) Hu-PBL mice and (**b**) Hu-HSC and (**c**) BLT mice. Data were obtained 5 days (**a**) and 12 weeks (**b** and **c**) after the transplantation.

period for Hu-HSC by multicolor flow cytometric analysis for detection of lineage markers on human hemato-poietic cells in the mouse peripheral blood.

2. Induce anesthesia using isoflurane chamber.
3. Collect whole blood (~100 μl) from all mice to be tested by retro-orbital bleeding using a heparinized capillary tube.
4. Isolate mononuclear cells by ficoll centrifugation of peripheral blood after lysing red blood cells using red blood cell lysing buffer and resuspend in PBS containing 2% FBS.
5. Add appropriate antibody mix to the cell suspension and incubate for 30 min at 4°C to label the antibody.
6. Wash cells in PBS containing 2% FBS.
7. Resuspend cells in 200 μl of PBS containing 2% FBS for flow cytometric analysis.
8. Collect and analyze flow cytometric data using FACSCanto II and FACSDiva software (Fig. 2).

4. Notes

1. Immunodeficient mice should be housed in a specific pathogen free (SPF) environment, using sterile techniques and in micro-isolator cages. All the animals are fed with sterile food and water. General health monitoring. Mice should be monitored daily and must be cared for if they show signs of ill health such as hunched posture, inactivity or loss in appetite/dehydration.
2. The use of human hematopoietic stem cells requires appropriate Institutional Review Board (IRB) approval.
3. Most experiments should be done within 4 weeks post-PBMC injection before xenogenic GVHD symptoms develop. If there are signs of GVHD development (weight loss of >15%), animals should be euthanized.
4. This protocol yields greater than 90–95% pure CD34+ cells and less than 1.0% CD3+ T cells. If necessary, depletion of CD3 T cells can be performed using anti-CD3 MACS beads.
5. Human tissues and mice engrafted with human tissues should be considered potential biological hazards and handled with proper personal protective equipment at animal biosafety level 2 (ABSL2), in accordance with governmental and institutional biosafety guidelines.
6. All cell isolation and surgical procedure must use sterile techniques.
7. The sublethal irradiation dose should be adapted based on the irradiator and age of the mouse used. The transfer of mice

from the microisolator cages to the sterile autoclaved container should be done in a laminar flow biocontainment hood to minimize exposure of the immunodeficient mice to environmental infectious agents.

8. General anesthesia suppresses the heat-regulating mechanisms of the body. Thus, it is important to maintain body temperature in appropriate thermostatically controlled incubators or by other heat sources during and after operative procedures.
9. Smear sterile ophthalmic ointment onto the eyes to prevent drying during anesthesia.
10. The fetal tissues can be easily seen through the kidney capsule after injection and should remain compact after injection.
11. Buprenorphine is given as an analgesic and additional doses should be given every 6–8 h during the first 48 h following the surgery.
12. Surgical implantation of fetal tissue for the BLT protocol may be performed on the same day as, or up to 72 h following irradiation.

References

1. Shultz, L. D., Ishikawa, F., and Greiner, D. L. (2007) Humanized mice in translational biomedical research. *Nat. Rev. Immunol.* **7**, 118–130.
2. Legrand, N., Ploss, A., Balling, R., Becker, P. D., Borsotti, C., Brezillon, N., Debarry, J., de Jong, Y., Deng, H., Di Santo, J. P., Eisenbarth, S., Eynon, E., Flavell, R. A., Guzman, C. A., Huntington, N. D., Kremsdorf, D., Manns, M. P., Manz, M. G., Mention, J. J., Ott, M., Rathinam, C., Rice, C. M., Rongvaux, A., Stevens, S., Spits, H., Strick-Marchand, H., Takizawa, H., van Lent, A. U., Wang, C., Weijer, K., Willinger, T., and Ziegler, P. (2009) Humanized mice for modeling human infectious disease: challenges, progress, and outlook. *Cell Host Microbe.* **6**, 5–9.
3. Kumar, P., Ban, H. S., Kim, S. S., Wu, H., Pearson, T., Greiner, D. L., Laouar, A., Yao, J., Haridas, V., Habiro, K., Yang, Y. G., Jeong, J. H., Lee, K. Y., Kim, Y. H., Kim, S. W., Peipp, M., Fey, G. H., Manjunath, N., Shultz, L. D., Lee, S. K., and Shankar, P. (2008) T cell-specific siRNA delivery suppresses HIV-1 infection in humanized mice. *Cell* **134**, 577–586.
4. Denton, P. W., Estes, J. D., Sun, Z., Othieno, F. A., Wei, B. L., Wege, A. K., Powell, D. A., Payne, D., Haase, A. T., and Garcia, J. V. (2008) Antiretroviral pre-exposure prophylaxis prevents vaginal transmission of HIV-1 in humanized BLT mice. *PLoS Med.* **5**, e16.
5. Hiramatsu, H., Nishikomori, R., Heike, T., Ito, M., Kobayashi, K., Katamura, K., and Nakahata, T. (2003) Complete reconstitution of human lymphocytes from cord blood CD34+ cells using the NOD/SCID/gammacnull mice model. *Blood* **102**, 873–880.
6. Yahata, T., Ando, K., Nakamura, Y., Ueyama, Y., Shimamura, K., Tamaoki, N., Kato, S., and Hotta, T. (2002) Functional human T lymphocyte development from cord blood CD34+ cells in nonobese diabetic/Shi-scid, IL-2 receptor gamma null mice. *J. Immunol.* **169**, 204–209.
7. Ishikawa, F., Yasukawa, M., Lyons, B., Yoshida, S., Miyamoto, T., Yoshimoto, G., Watanabe, T., Akashi, K., Shultz, L. D., and Harada, M. (2005) Development of functional human blood and immune systems in NOD/SCID/IL2 receptor {gamma} chain(null) mice. *Blood* **106**, 1565–1573.
8. Traggiai, E., Chicha, L., Mazzucchelli, L., Bronz, L., Piffaretti, J. C., Lanzavecchia, A., and Manz, M. G. (2004) Development of a human adaptive immune system in cord blood cell-transplanted mice. *Science* **304**, 104–107.
9. Peer, D., Park, E. J., Morishita, Y., Carman, C. V., and Shimaoka, M. (2008) Systemic leukocyte-directed siRNA delivery revealing cyclin D1 as an anti-inflammatory target. *Science* **319**, 627–630.

10. Peer, D., Zhu, P., Carman, C. V., Lieberman, J., and Shimaoka, M. (2007) Selective gene silencing in activated leukocytes by targeting siRNAs to the integrin lymphocyte function-associated antigen-1. *Proc. Natl. Acad. Sci. USA.* **104**, 4095–4100.
11. Lan, P., Tonomura, N., Shimizu, A., Wang, S., and Yang, Y. G. (2006) Reconstitution of a functional human immune system in immuno-deficient mice through combined human fetal thymus/liver and CD34+ cell transplantation. *Blood* **108**, 487–492.
12. McCune, J. M., Namikawa, R., Kaneshima, H., Shultz, L. D., Lieberman, M., and Weissman, I. L. (1988) The SCID-hu mouse: murine model for the analysis of human hematolymphoid differentiation and function. *Science* **241**, 1632–1639.
13. McCune, J. M. (1997) Animal models of HIV-1 disease. *Science* **278**, 2141–2142.
14. Brainard, D. M., Seung, E., Frahm, N., Cariappa, A., Bailey, C. C., Hart, W. K., Shin, H. S., Brooks, S. F., Knight, H. L., Eichbaum, Q., Yang, Y. G., Sykes, M., Walker, B. D., Freeman, G. J., Pillai, S., Westmoreland, S. V., Brander, C., Luster, A. D., and Tager, A. M. (2009) Induction of robust cellular and humoral virus-specific adaptive immune responses in human immunodeficiency virus-infected humanized BLT mice. *J. Virol.* **83**, 7305–7321.
15. Berges, B. K., Akkina, S. R., Folkvord, J. M., Connick, E., and Akkina, R. (2008) Mucosal transmission of R5 and X4 tropic HIV-1 via vaginal and rectal routes in humanized Rag2−/− gammac −/− (RAG-hu) mice. *Virology* **373**, 342–351.
16. Melkus, M. W., Estes, J. D., Padgett-Thomas, A., Gatlin, J., Denton, P. W., Othieno, F. A., Wege, A. K., Haase, A. T., and Garcia, J. V. (2006) Humanized mice mount specific adaptive and innate immune responses to EBV and TSST-1. *Nat. Med.* **12**, 1316–1322.

Index

A

B

C

Motomu Shimaoka (ed.), *Integrin and Cell Adhesion Molecules: Methods and Protocols*,
Methods in Molecular Biology, vol. 757, DOI 10.1007/978-1-61779-166-6,

D

E

J

K

L

M

N

O

P

Q

R

S

T

U

V

W

X

Y

Z

MIX
Papier aus verantwortungsvollen Quellen
Paper from responsible sources
FSC® C105338

If you have any concerns about our products,
you can contact us on
ProductSafety@springernature.com

In case Publisher is established outside the EU,
the EU authorized representative is:
Springer Nature Customer Service Center GmbH
Europaplatz 3, 69115 Heidelberg, Germany

Printed by Libri Plureos GmbH
in Hamburg, Germany